BARRON'S

AP

STATISTICS

4TH EDITION

Martin Sternstein, Ph.D.
Professor of Mathematics
Ithaca College
Ithaca, New York

BARRON'S

AUTHOR'S NOTE

The number of students who take a statistics course in college will soon surpass the number who take a calculus course. High schools across the country have recognized this trend and are developing and expanding their statistics offerings. The Advanced Placement (AP) Program of the College Entrance Examination Board has lent their thoughtful and powerful support by developing a suggested course outline and by giving an AP examination in statistics for the first time in May 1997. Prompted by the AP exam, by several excellent textbooks, and by conscientious high school teachers across the country, a statistics course continues to evolve that stresses how to pose incisive questions, how to gather cogent information, how to think about and analyze this information, and how to test conjectures based on well-designed studies.

This Barron's book is intended both as a brush-up text and also to provide additional insights for students enrolled in a high school statistics course equivalent to a one-semester introductory non-calculus-based college course in statistics. Ultimately, the focus of the book is the sample practice questions and step-by-step solutions with detailed explanations.

Special thanks are due to David Bock, Rhonda Weissman, Lee Kucera, Jane Viau, Diann Resnick, and Peg Becker for their many useful suggestions regarding the text and for pretesting and commenting on the practice questions. Thanks to my brother, Allan, and my friends, Steve and Jan Hanson, for their always cheerful and heartfelt support. And I thank Linda Turner, senior editor at Barron's, for her guidance.

Thanks are also due to my sons, Jonathan and Jeremy, for their love and for the lessons they gave me, teaching me new computer techniques and teaching me humility and the value of keeping in good shape by crushing me on the basketball court. Most thanks of all are due to my wife, Faith, whose warm encouragement and always calm and optimistic perspective on life provide a home environment in which deadlines can be met and goals easily achieved.

Ithaca College
Spring 2007

Martin Sternstein

© Copyright 2007 by Barron's Educational Series, Inc.
© Copyright 2004, 2000, 1998 by Barron's Educational Series, Inc., under the title *How to Prepare for the AP Advanced Placement Exam in Statistics.*

All inquiries should be addressed to:
Barron's Educational Series, Inc.
250 Wireless Boulevard
Hauppauge, New York 11788
http://www.barronseduc.com

Library of Congress Control Number: 2007009074

ISBN-13 (book only): 978-0-7641-3683-2
ISBN-10 (book only): 0-7641-3683-6
ISBN-13 (with CD-ROM): 978-0-7641-9333-0
ISBN-10 (with CD-ROM): 0-7641-9333-3

Library of Congress Cataloging-in-Publication Data
Sternstein, Martin.
    Barron's AP Statistics / Martin Sternstein, Ph.D.—4th ed.
        p. cm.
    Includes index.
    ISBN-13: 978-0-7641-3683-2
    ISBN-10: 0-7641-3683-6
    1. Mathematical statistics—Examinations, questions, etc.   2. College entrance achievement tests—Study guides.   I. Title.   II. Title: AP Statistics.

    QA276.2.S74 2007
    519.5076—dc22

2007009074

PRINTED IN THE UNITED STATES OF AMERICA
9 8 7 6 5 4 3

Contents

Topic Fifteen: Tests of Significance—Chi-square and Slope of Least Squares Line   367

Chi-square test for goodness of fit; chi-square test for independence; chi-square test for homogeneity of proportions; hypothesis test for slope of least squares line.

Practice Examinations   405

Appendix   581

Introduction

The contents of this book cover the topics recommended by the AP Statistics Development Committee. A review of each of the 15 topics is followed by multiple-choice and free-response questions on that topic. Detailed explanations are provided for all answers. It should be noted that some of the topic questions are not typical AP exam questions but rather are intended to help review the topic. Finally, there are six full-length practice exams, made up of 276 questions, all with instructive, complete answers. An optional disk contains two new, full-length exams with 92 more questions.

Several points with regard to particular answers should be noted. First, step-by-step calculations using the given tables sometimes give minor differences from calculator answers due to round-off error. Second, there are examples where the case may be made for using t-scores or z-scores, again with minor differences in resulting answers. Third, calculator packages easily handle degrees of freedom that are not whole numbers, also resulting in minor answer differences. In all of the above cases, multiple-choice answers in this book have only one reasonable correct answer, and written explanations are necessary when answering free-response questions.

Students taking the AP Statistics Examination will be furnished with a list of formulas (from descriptive statistics, probability, and inferential statistics) and tables (including standard normal probabilities, t-distribution critical values, χ^2 critical values, and random digits). While students will be expected to bring a graphing calculator with statistics capabilities to the examination, answers should not be in terms of calculator syntax. Furthermore, many students have commented that calculator usage was less than they had anticipated. However, even though the calculator is simply a tool, to be used sparingly, as needed, students should be proficient with this technology.

The examination will consist of two parts: a 90-minute section with 40 multiple-choice problems and a 90-minute free-response section with five open-ended questions and an investigative task to complete. In grading, the two sections of the exam will be given equal weight. Students have remarked that the first section involves "lots of reading," while the second section involves "lots of writing." The percentage of questions from each content area is approximately 25% data analysis, 15% experimental design, 25% probability, and 35% inference. Questions in both sections may involve reading generic computer output.

A correction factor compensates for random guessing in the multiple-choice section; however, students should guess if they can eliminate even one incorrect choice. In this section the questions are much more conceptual than computational, and thus use of the calculator is minimal.

In the free-response section, students must show all their work, and communication skills go hand in hand with statistical knowledge. Methods must be clearly indicated, as the problems will be graded on the correctness of the methods as well as on the accuracy of the results and explanation. That is, the free-response answers should address *why* a particular test was chosen, not just *how* the test is performed. Even if a statistical test is performed on a calculator such as the TI-84, formulas should still be stated. Choice of test, in inference, must include confirmation of underlying assumptions, and answers must be stated in context, not just as numbers. Work is graded *holistically*, that is, a student's complete response is considered as a whole.

Multiple-choice questions are scored as the number of correct answers minus one-quarter the number of incorrect answers. Blank answers are ignored. Free-response questions are scored on a 0 to 4 scale, with each open-ended question counting 15% of the total free-response score and the investigative task counting 25% of the free-response score. The first open-ended question is typically the most straightforward, and after doing this one to build confidence, students might consider looking at the investigative task since it counts more. Each completed AP examination paper will receive a grade based on a 5-point scale, with 5 the highest score and 1 the lowest score. Most colleges and universities accept a grade of 3 or better for credit or advanced placement or both.

While a review book such as this can be extremely useful in helping prepare students for the AP exam (practice problems, practice more problems, and practice even more problems are the three strongest pieces of advice), nothing can substitute for a good high school teacher and a good textbook. This author personally recommends the following texts from among the many excellent books on the market: *Workshop Statistics* by Rossman, *Activity-Based Statistics* by Scheaffer, *Introduction to Statistics and Data Analysis* by Devore, Olsen, and Peck, *Stats, Modeling the World* by Bock, Velleman, and DeVeaux, *Statistics: The Art and Science of Learning from Data* by Agretsi and Franklin, *Introduction to the Practice of Statistics* by Moore and McCabe, *Statistics in Action: Understanding a World of Data* by Cobb, Scheaffer, and Watkins and *The Practice of Statistics: TI-83/89 Graphing Calculator Enhanced* by Yates, Moore, and Starnes.

Other wonderful sources of information are the College Board's Web sites: *www.collegeboard.com* for students and parents, and *www.apcentral.collegeboard.com* for teachers. After registering at the AP Central site, teachers should especially note the sections (1) AP Statistics Course Description, (2) The Statistics Exam, and (3) Exam Tips: Statistics.

A good piece of advice is for the student from day one to develop critical practices (like checking assumptions and conditions), to acquire strong technical skills, and to always write clear and thorough, yet to the point, interpretations in context. Final answers to most problems should not be numbers, but rather sentences explaining and analyzing numerical results. To help develop skills and insights to tackle AP free response questions (which often choose contexts students haven't seen before), pick up newspapers and magazines and figure out how to apply what you are learning to better understand articles in print that reference numbers, graphs, and statistical studies.

The student who uses this Barron's review book should study the text and illustrative examples carefully and try to complete the practice problems before referring to the solution keys. Simply reading the detailed explanations to the answers without first striving to work through the problems on one's own is not the best approach. There is an old adage: *Mathematics is not a spectator sport!* Teachers clearly may use this book with a class in many profitable ways. Ideally, each individual topic review, together with practice problems, should be assigned after the topic has been covered in class. The full-length practice exams should be reserved for final review shortly before the AP examination.

THEME ONE: EXPLORITORY ANALYSIS

Graphical Displays

> - Dotplots
> - Bar Charts
> - Histograms
> - Cumulative Frequency Plots
> - Stemplots
>
> - Center and Spread
> - Clusters and Gaps
> - Outliers
> - Modes
> - Shape

There are a variety of ways to organize and arrange data. Much information can be put into tables, but these arrays of bare figures tend to be spiritless and sometimes even forbidding. Some form of graphical display is often best for seeing patterns and shapes and for presenting an immediate impression of everything about the data. Among the most common visual representations of data are dotplots, bar charts, histograms, and stemplots. It is important to remember that all graphical displays should be clearly labeled, leaving no doubt what the picture represents—AP Statistics scoring guides harshly penalize the lack of titles and labels!

DOTPLOTS

Dotplots and bar charts are particularly useful with regard to *categorical* (or *qualitative*) *variables*, that is, variables that note the category to which each individual belongs. This is in contrast to *quantitative variables*, which take on numerical values.

EXAMPLE 1.1

Suppose that in a class of 35 students, 10 choose basketball as their favorite sport while 7 pick baseball, 6 pick football, 5 pick tennis, 5 pick soccer, and 2 pick hockey. These data can be displayed in the following *dotplot*:

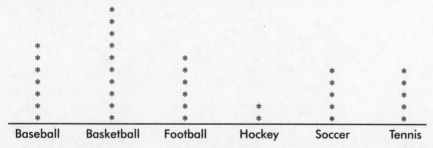

The *frequency* of each result is indicated by the *number of dots* representing that result.

The dotplot can also be drawn with a vertical axis and horizontal rows of dots. In fact, in almost all displays, vertical and horizontal can be switched depending upon which picture seems easier to read or simply which better fits the page.

BAR CHARTS

A common visual display to compare the sizes of categories or groups is the *bar chart*. Sizes can be measured as frequencies or as percents.

EXAMPLE 1.2

In a survey taken during the first week of January 1999, 55% of those surveyed wanted the Clinton impeachment trial to end immediately, 15% wanted it to continue with witnesses, 25% wanted the trial to continue without calling witnesses, and 5% expressed no opinion. These data can be displayed in the following bar chart (or *bar graph*):

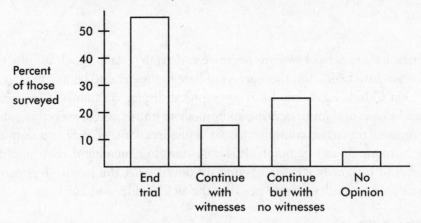

The relative frequencies of different results are indicated by the heights of the bars representing these results.

HISTOGRAMS

Histograms, useful for large data sets involving quantitative variables, show counts or percents falling either at certain values or between certain values. While the AP Statistics Exam does not stress construction of histograms, there are often questions on interpreting given histograms.

EXAMPLE 1.3

Suppose there are 2000 families in a small town and the distribution of children among them is as follows: 300 families are childless, 400 have one child, 700 have two children, 300 have three, 100 have four, 100 have five, and 100 have six. These data can be displayed in the following histogram:

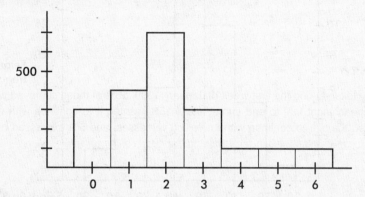

Sometimes, instead of labeling the vertical axis with frequencies, it is more convenient or more meaningful to use *relative frequencies*, that is, frequencies divided by the total number in the population.

Number of Children	Frequency	Relative Frequency
0	300	300/2000 = .150
1	400	400/2000 = .200
2	700	700/2000 = .350
3	300	300/2000 = .150
4	100	100/2000 = .050
5	100	100/2000 = .050
6	100	100/2000 = .050

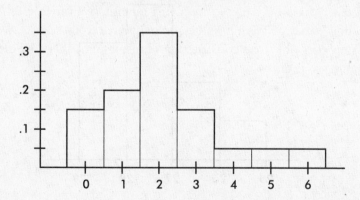

Note that the shape of the histogram is the same whether the vertical axis is labeled with frequencies or with relative frequencies. Sometimes we show both frequencies and relative frequencies on the same graph.

EXAMPLE 1.4

Consider the following histogram displaying 40 salaries paid to the top-level executives of a large company.

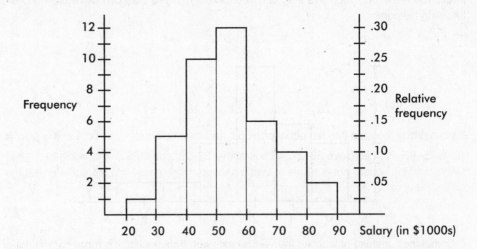

What can we learn from this histogram? For example, none of the executives earned more than $90,000 or less than $20,000. Twelve earned between $50,000 and $60,000. Twenty-five percent earned between $40,000 and $50,000. Note how this histogram shows the number of items (salaries) falling *between* certain values, whereas the preceding histogram showed the number of items (families) falling *at* each value. For example, in Example 1.4 we see that ten salaries fell somewhere between $40,000 and $50,000, while in Example 1.3 we see that 700 families had exactly two children.

EXAMPLE 1.5

Consider the following histogram where the vertical axis has not been labeled. What can we learn from this histogram?

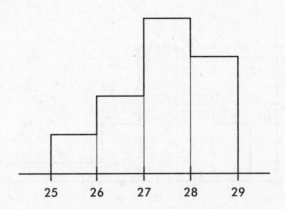

(continued)

Answer: It is impossible to determine the actual frequencies; however, we can determine the relative frequencies by noting the fraction of the total *area* that is over any interval:

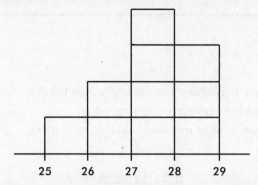

We can divide the area into ten equal portions, and then note that $\frac{1}{10}$ or 10% of the area is above 25–26, 20% is above 26–27, 40% is above 27–28, and 30% is above 28–29.

Although it is usually not possible to divide histograms so nicely into ten equal areas, the principle of relative frequencies corresponding to relative areas still applies.

Relative frequencies are the usual choice when comparing distributions of different size populations.

CUMULATIVE FREQUENCY PLOTS

Sometimes we sum frequencies and show the result visually in a *cumulative frequency plot* (also known as an *ogive*).

EXAMPLE 1.6

The following graph shows 1996 school enrollment in the United States by age.

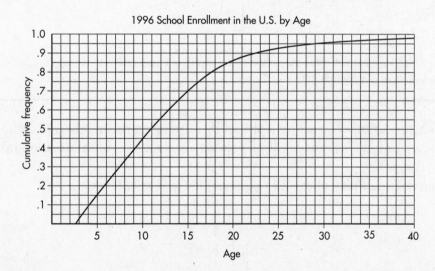

What can we learn from this cumulative frequency plot? For example, going up to the graph from age 5, we see that .15 or 15% of school enrollment is below age 5. Going over to the graph from .5 on the vertical axis, we see that 50% of the school enrollment is below and 50% is above a middle age of 11. Going up from age 30, we see that .95 or 95% of the enrollment is below age 30, and thus 5% is above age 30. Going over from .25 and .75 on the ver-

(continued)

tical axis, we see that the middle 50% of school enrollment is between ages 6 and 7 at the lower end and age 16 at the upper end.

STEMPLOTS

Although a histogram may show how many scores fall into each grouping or interval, the exact values of individual scores are often lost. An alternative pictorial display, called a stemplot, retains this individual information.

EXAMPLE 1.7

Consider the set {17, 17, 18, 13, 28, 38, 31, 27, 35, 50, 43, 37, 24} of percentages of three-point shots made by Michael Jordan during his 13 years with the Bulls. Let 1, 2, 3, 4, and 5 be placeholders for 10, 20, 30, 40, and 50. List the last digit of each value from the original set after the appropriate placeholder.

The result is a *stemplot* (also called a *stem and leaf display*) of these data:

Stems	Leaves
1	7 7 8 3
2	8 7 4
3	8 1 5 7
4	3
5	0

Drawing a continuous line around the leaves would result in a horizontal histogram:

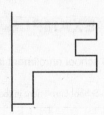

Note that the stemplot gives the shape of the histogram and, unlike the histogram, indicates the values of the original data.

Sometimes further structure is shown by rearranging the numbers in each row in ascending order. This ordered display shows a second level of information from the original stemplot.

The revised display of the data in Example 1.7 is as follows:

1	3 7 7 8
2	4 7 8
3	1 5 7 8
4	3
5	0

EXAMPLE 1.8

Suppose the distribution of 25 advertised home prices (in thousands of dollars) in a certain community is given by the set {56, 89, 165, 73, 83, 145, 90, 189, 127, 77, 110, 112, 132, 120, 94, 130, 84, 65, 99, 154, 86, 120, 122, 103, 130}. One possible stemplot of these data is

50–74	56	65	73					
75–99	77	83	84	86	89	90	94	99
100–124	03	10	12	20	20	22		
125–149	27	30	30	32	45			
150–174	54	65						
175–200	89							

Note that the stems were chosen to be intervals of length 25 and that the 1 in 100 is left out of the leaves so that they align vertically.

CENTER AND SPREAD

Looking at a graphical display, we see that two important aspects of the overall pattern are

1. the *center*, which separates the values (or area under the curve in the case of a histogram) roughly in half, and
2. the *spread*, that is, the scope of the values from smallest to largest.

In the histogram of Example 1.3, the center is 2 children while the spread is from 0 to 6 children.

In the histogram of Example 1.4 the center is between $50,000 and $60,000, and the spread is from $20,000 to $90,000; in the histogram of Example 1.5, the center is between 27 and 28, and the spread is from 25 to 29.

In the cumulative frequency plot of Example 1.6, the center is between 10 and 11, and the spread is from 3 to 40.

In the stemplot of Example 1.7, the center is 28%, and the spread is from 13% to 50%; in the stemplot of Example 1.8, the center is $110,000, and the spread is from $56,000 to $189,000.

CLUSTERS AND GAPS

Other important aspects of the overall pattern are

1. *clusters*, which show natural subgroups into which the values fall (for example, the salaries of teachers in Ithaca, NY, fall into three overlapping clusters, one for public school teachers, a higher one for Ithaca College professors, and an even higher one for Cornell University professors), and
2. *gaps*, which show holes where no values fall (for example, the Office of the Dean sends letters to students being put on the honor roll and to those being put on academic warning for low grades; thus the GPA distribution of students receiving letters from the Dean has a huge middle gap).

EXAMPLE 1.9

Consider the following histogram:

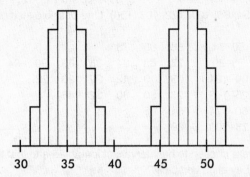

Simply saying that the center of the distribution is around 42 and the spread is from 31 to 52 clearly misses something. The values fall into two distinct clusters with a gap between.

OUTLIERS

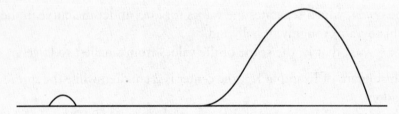

Extreme values, called *outliers*, are found in many distributions. Sometimes they are the result of errors in measurements and deserve scrutiny; however, outliers can also be the result of natural chance variation. Outliers may occur on one side or both sides of a distribution.

MODES

Some distributions have one or more major peaks, called *modes*. With exactly one or two such peaks, the distribution is said to be *unimodal* or *bimodal*, respectively.

EXAMPLE 1.10

Consider the histogram of the yearly average assists per game figures for Jeff Hornacek (Phoenix and Houston):

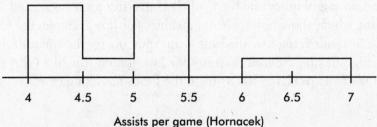

Assists per game (Hornacek)

(continued)

Note that this is a bimodal distribution. The four assist averages between 4 and 4.5 actually correspond to the four seasons from 1994–95 through 1997–98, while the four between 5 and 5.5 correspond to middle career seasons.

EXAMPLE 1.11

Consider the histogram of the yearly average points per game figures for Karl Malone (Utah) from 1985–86 through 1997–98.

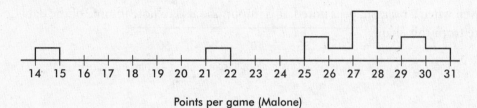

Points per game (Malone)

Note that this is a unimodal distribution with two outliers. The outliers, as might be guessed, correspond to Malone's first two seasons as a pro during which he averaged only 14.9 and 21.7 points per game. Disregarding these two outliers, we say that the distribution is almost *symmetric*.

Note that, as illustrated above, it is usually instructive to look for reasons behind outliers and modes.

SHAPE

Distributions come in an endless variety of shapes; however, certain common patterns are worth special mention:

1. A *symmetric* distribution is one in which the two halves are mirror images of each other. For example, the weights of all people in some organizations fall into symmetric distributions with two mirror-image bumps, one for men's weights and one for women's weights.
2. A distribution is *skewed to the right* if it spreads far and thinly toward the higher values. For example, ages of nonagenarians (people in their 90s) is a distribution with sharply decreasing numbers as one moves from 90-year-olds to 99-year-olds.
3. A distribution is *skewed to the left* if it spreads far and thinly toward the lower values. For example, scores on an easy exam show a distribution bunched at the higher end with few low values.
4. A *bell-shaped* distribution is symmetric with a center mound and two sloping tails. For example, the distribution of IQ scores across the general population is roughly symmetric with a center mound at 100 and two sloping tails.
5. A distribution is *uniform* if its histogram is a horizontal line. For example, tossing a fair die and noting how many spots (pips) appear on top yields a uniform distribution with 1 through 6 all equally likely.

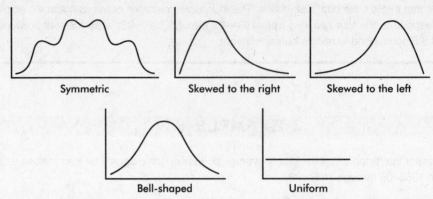

Even when a basic shape is noted, it is important also to note if some of the data deviate from this shape.

Questions on Topic 1: Graphical Displays

Multiple-Choice Questions

Directions: The questions or incomplete statements that follow are each followed by five suggested answers or completions. Choose the response that best answers the question or completes the statement.

1. Following are the SAT math scores for an AP Statistics class of 20 students: 664, 658, 610, 670, 640, 643, 675, 650, 676, 575, 660, 661, 520, 667, 668, 635, 671, 673, 645, and 650. The distribution of scores is

   (A) symmetric.
   (B) skewed to the left.
   (C) skewed to the right.
   (D) uniform.
   (E) bell-shaped.

2. Which of the following statements are true?

   I. Stemplots are useful both for quantitative and categorical data sets.
   II. Stemplots are equally useful for small and very large data sets.
   III. Stemplots can show symmetry, gaps, clusters, and outliers.

   (A) I only
   (B) II only
   (C) III only
   (D) I and II
   (E) I and III

3. Consider the following picture:

   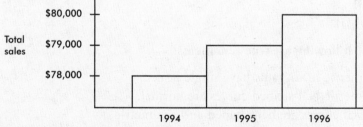

   Which of the following statements are true?

   I. Total sales in 1995 were two times total sales in 1994, while total sales in 1996 were three times the 1994 total.
   II. The choice of labeling for the vertical axis results in a misleading sales picture.
   III. A histogram showing the same information, but this time with a vertical axis starting at $78,000, would be less misleading.

   (A) I only
   (B) II only
   (C) III only
   (D) II and III
   (E) None of the above gives the complete set of true responses.

4. Consider the following histogram:

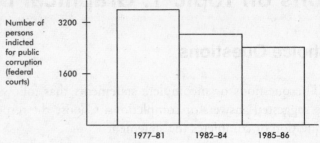

Which of the following statements are true?

 I. Each year from 1977 to 1986 the number of indictments has steadily decreased.
 II. While the number of indictments has decreased each year, the amount of decrease has lessened.
 III. The labeling of the horizontal axis has resulted in a misleading picture.

(A) I only
(B) II only
(C) III only
(D) I and II
(E) None of the above gives the complete set of true responses.

5. Which of the following are true statements?

 I. Both dotplots and stemplots can show symmetry, gaps, clusters, and outliers.
 II. In histograms, relative areas correspond to relative frequencies.
 III. In histograms, frequencies can be determined from relative heights.

(A) II only
(B) I and II
(C) I and III
(D) II and III
(E) I, II, and III

6. Which of the following are true statements?

 I. All symmetric histograms have single peaks.
 II. All symmetric bell-shaped curves are normal.
 III. All normal curves are bell-shaped and symmetric.

(A) I only
(B) II only
(C) III only
(D) I and II
(E) None of the above gives the complete set of true responses.

7. Which of the following distributions are more likely to be skewed to the right than skewed to the left?

   I. Household incomes
   II. Home prices
   III. Ages of teenage drivers

   (A) II only
   (B) I and II
   (C) I and III
   (D) II and III
   (E) I, II, and III

8. Which of the following are true statements?

   I. Two students working with the same set of data may come up with histograms that look different.
   II. Displaying outliers is less problematic when using histograms than when using stemplots.
   III. Histograms are more widely used than stemplots or dotplots because histograms display the values of individual observations.

   (A) I only
   (B) II only
   (C) III only
   (D) I and II
   (E) II and III

9. Following is a histogram of test scores.

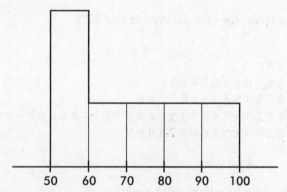

   Which of the following statements are true?

   I.   The middle (median) score was 75.
   II. If the passing score was 60, most students failed.
   III. More students scored between 50 and 60 than between 90 and 100.

   (A) I only
   (B) II only
   (C) III only
   (D) II and III
   (E) I, II, and III

10. Which is a histogram for the ages of all family members in households where the parents have been married less than 3 years and a few of the families have small children?

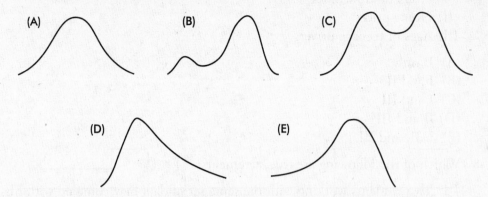

11. A histogram of the educational level (in number of years of schooling) of the adult population of the United States would have which of the following characteristics?

    I. Symmetry
    II. Clusters
    III. Skewness to the left

    (A) II only
    (B) I and II
    (C) I and III
    (D) II and III
    (E) I, II, and III

12. Following is a stemplot of a data set of size 125.

```
1 | 8
2 | 0 0 1 3 5 6 7 7 8 9
3 | 0 1 1 2 3 3 3 4 4 5 6 6 6 6 7 8 8 9 9
4 | 0 0 0 1 1 1 2 2 2 3 3 3 4 4 5 6 6 7 7 8 8
5 | 0 0 0 0 2 2 2 3 3 3 3 4 4 4 4 5 5 5 6 6 6 7 7 7 7 7 8 8 8 8 8 9 9 9 9 9 9 9 9 9 9
6 | 0 0 0 1 1 1 2 2 2 3 4 4 5 5 6 6 7 7 8 8 9 9
7 | 0 1 2 5 6 6 7 9
8 | 1 2
```

What is the middle (median) of this set?

    (A) 53
    (B) 54
    (C) 62
    (D) 63
    (E) 64

13. The following histogram indicates the relative frequencies of ages of U.S. scientists in 1967.

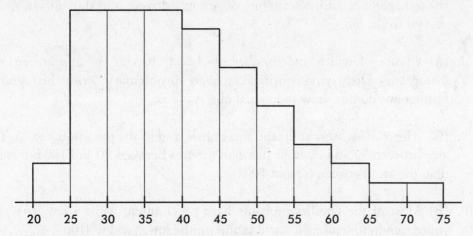

Approximately what percentage of the scientists are between 25 and 40 years of age?

(A) 10%
(B) 15%
(C) 25%
(D) 40%
(E) 50%

Answer Key

1. **B**	4. **C**	7. **B**	10. **B**	13. **E**
2. **C**	5. **B**	8. **A**	11. **D**	
3. **B**	6. **C**	9. **C**	12. **B**	

Answers Explained

1. **(B)** The low scores of 520, 575, and 610, far from the remaining scores which are all between 635 and 680, make this distribution skewed to the left.

2. **(C)** Stemplots are not used for categorical data sets and are too unwieldy to be used for very large data sets.

3. **(B)** The vertical axis, starting at $77,000, results in a misleading sales picture. It would be better to start at $0, not $78,000.

4. **(C)** Labeling the horizontal axis with different year spans results in a misleading picture. The number of indictments per year is actually increasing.

5. **(B)** In general, histograms give information about relative frequencies, not actual frequencies.

6. **(C)** Symmetric histograms can have any number of peaks. All normal curves are bell-shaped and symmetric, but not all symmetric bell-shaped curves are normal.

7. **(B)** Incomes and home prices tend to have a few very high scores that make the distributions skewed to the right. Teenage drivers mostly have ages in the last teenage years with a scattering of younger drivers, and thus a distribution skewed to the left.

8. **(A)** Choice of width and number of classes changes the appearance of a histogram. Displaying outliers is *more* problematic with histograms. Histograms do not show individual observations.

9. **(C)** The median score splits the area in half, and so the median is not 75. The area between 50 and 60 is greater than the area between 90 and 100 but is less than the area between 60 and 100.

10. **(B)** Most of the families probably have no children, while a few have very young children, hence the small bump on the lower end of (B).

11. **(D)** There will be clusters around 8, 12, and 16 years of schooling. Small segments of the population will have very few years of schooling, and so the distribution will be skewed to the left.

12. **(B)** The 125 scores are already arranged in ascending order. Counting up or down to the sixty-third score shows the median to be 54.

13. **(E)**

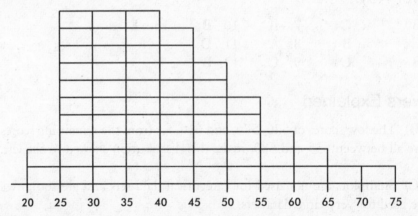

If we compare areas (or count small rectangles!), we can conclude that 50% of the scientists are between 25 and 40 years of age.

Free-Response Questions

Directions: You must show all work and indicate the methods you use. You will be graded on the correctness of your methods and on the accuracy of your final answers.

Two Open-Ended Questions

1. According to the *1992 NAEP Trial State Assessment* the average mathematics proficiency scores in eighth grade for 41 states were as follows:

AL	251	IN	269	MO	270	PA	271
AZ	265	IA	283	NE	277	RI	265
AR	255	KY	261	NH	278	SC	260
CA	260	LA	249	NJ	271	TN	258
CO	272	ME	278	NM	259	TX	264
CT	273	MD	264	NY	266	UT	274
DE	262	MA	272	NC	258	VA	267
FL	259	MI	267	ND	283	WV	258
GA	259	MN	282	OH	267	WI	277
HI	257	MS	246	OK	267	WY	274
ID	274						

What does a stemplot show about the shape of this distribution?

2. Consider the following murder rates (per 100,000 people).

AL	13.3	IN	6.2	NE	2.9	RI	4.0
AK	12.9	IA	2.6	NV	15.5	SC	11.5
AZ	9.4	KS	5.7	NH	1.4	SD	1.9
AR	9.1	SY	8.9	NJ	5.4	TN	9.4
CA	11.7	LA	15.8	NM	10.2	TX	14.2
CO	7.3	ME	2.7	NY	10.3	UT	3.7
CT	4.2	MD	8.2	NC	10.8	VT	3.3
DE	6.7	MA	3.7	ND	1.2	VA	8.8
FL	11.0	MI	10.6	OH	6.9	WA	4.6
GA	14.4	MN	2.0	OK	8.5	WV	6.8
HI	6.7	MS	12.6	OR	5.0	WI	2.5
ID	5.4	MO	10.4	PA	6.2	WY	7.1
IL	9.9	MT	4.8				

What does a histogram show about the shape of this distribution?

Answers Explained

1.

```
28 | 2 3 3
27 | 0 1 1 2 2 3 4 4 4 7 7 8 8
26 | 0 0 1 2 4 4 5 5 6 7 7 7 7 9
25 | 1 5 7 8 8 8 9 9 9
24 | 6 9
```

The data are roughly symmetric and bell-shaped; they are centered in the 260s, with a spread from 246 to 283 and no outliers, clusters, or gaps.

2. The data range from 1.9 to 15.8. Picking six class intervals, each of width 3, between 0 and 18 seems reasonable, although many other choices are possible. On the TI-83, for example, putting the data into L1, turning on Plot1 in STAT PLOT, picking the histogram from among the six type choices, designating L1 for Xlist and 1 for Freq, then either using ZoomStat or in WINDOW, setting Xmin = 0, Xmax = 18, Xscl = 3, Ymin = 0, and Ymax = 15 results in the following GRAPH:

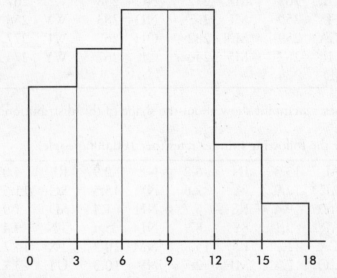

The distribution is not symmetric but rather has a slight skew to the right, indicating that there are a few states with much higher murder rates than found in the remaining bulk of the states.

Summarizing Distributions

• Measuring the Center	• Histograms
• Measuring Spread	• Cumulative Frequency
• Measuring Position	• Boxplots
• Empirical Rule	• Changing Units

Given a raw set of data, often we can detect no overall pattern. Perhaps some values occur more frequently, a few extreme values may stand out, and the range of values is usually apparent. The presentation of data, including summarizations and descriptions, and involving such concepts as representative or average values, measures of dispersion, positions of various values, and the shape of a distribution, falls under the broad topic of *descriptive statistics*. This aspect of statistics is in contrast to *statistical analysis*, the process of drawing inferences from limited data, a subject discussed in later topics.

MEASURING THE CENTER: MEDIAN AND MEAN

The word *average* is used in phrases common to everyday conversation. People speak of bowling and batting averages or the average life expectancy of a battery or a human being. Actually the word *average* is derived from the French *avarie*, which refers to the money that shippers contributed to help compensate for losses suffered by other shippers whose cargo did not arrive safely (i.e., the losses were shared, with everyone contributing an average amount). In common usage *average* has come to mean a representative score or a typical value or the center of a distribution. Mathematically, there are a variety of ways to define the average of a set of data. In practice, we use whichever method is most appropriate for the particular case under consideration. However, beware of a headline with the word *average*; the writer has probably chosen the method that emphasizes the point he wishes to make.

In the following paragraphs we consider the two primary ways of denoting an average:

1. The *median*, which is the middle number of a set of numbers arranged in numerical order.
2. The *mean*, which is found by summing items in a set and dividing by the number of items.

EXAMPLE 2.1

Consider the following set of home run distances (in feet) to center field in 13 ballparks: {387, 400, 400, 410, 410, 410, 414, 415, 420, 420, 421, 457, 461}. What is the average?

Answer: The median is 414 (there are six values below 414 and six values above), while the mean is

$$\frac{387 + 400 + 400 + 410 + 410 + \cdots + 457 + 461}{13} = 417.3 \text{ feet}$$

Median

The word *median* is derived from the Latin *medius* which means "middle." The values under consideration are arranged in ascending or descending order. If there is an odd number of values, the median is the middle one. If there is an even number, the median is found by adding the two middle values and dividing by 2. Thus the median of a set has the same number of elements above it as below it.

The median is not affected by exactly how large the larger values are or by exactly how small the smaller values are. Thus it is a particularly useful measurement when the extreme values, called *outliers*, are in some way suspicious or when we want do diminish their effect. For example, if ten mice try to solve a maze, and nine succeed in less than 15 minutes while one is still trying after 24 hours, the most representative value is the median (not the mean, which is over 2 hours). Similarly, if the salaries of four executives are each between $240,000 and $245,000 while a fifth is paid less than $20,000, again the most representative value is the median (the mean is under $200,000). It is often said that the median is "resistant" to extreme values.

In certain situations the median offers the most economical and quickest way to calculate an average. For example, suppose 10,000 lightbulbs of a particular brand are installed in a factory. An average life expectancy for the bulbs can most easily be found by noting how much time passes before exactly one-half of them have to be replaced. The median is also useful in certain kinds of medical research. For example, to compare the relative strengths of different poisons, a scientist notes what dosage of each poison will result in the death of exactly one-half the test animals. If one of the animals proves especially susceptible to a particular poison, the median lethal dose is not affected.

Mean

While the median is often useful in descriptive statistics, the *mean,* or more accurately, the *arithmetic mean,* is most important for statistical inference and analysis. Also, for the layperson, the average is usually understood to be the mean.

The mean of a *whole population* (the complete set of items of interest) is often denoted by the Greek letter μ (mu), while the mean of a *sample* (a part of a population) is often denoted by $\bar{x}$. For example, the mean value of the set of all houses in the United States might be μ = $56,400, while the mean value of 100 randomly chosen houses might be $\bar{x}$ = $52,100 or perhaps $\bar{x}$ = $63,800 or even $\bar{x}$ = $124,000.

In statistics we learn how to estimate a population mean from a sample mean. Throughout this book, the word *sample* often implies a *simple random sample* (SRS), that is, a sample selected in such a way that every possible sample of the desired size has an equal chance of being included. (It is also true that each element of the population will have an equal chance of being included.) In the real world, this process of random selection is often very difficult to achieve, and so we proceed, with caution, as long as we have good reason to believe that our sample is representative of the population.

Mathematically, the mean $= \frac{\Sigma x}{n}$, where Σx represents the sum of all the elements of the set under consideration and n is the actual number of elements. Σ is the uppercase Greek letter sigma.

EXAMPLE 2.2

Suppose that the numbers of unnecessary procedures recommended by five doctors in a 1-month period are given by the set {2, 2, 8, 20, 33}. Note that the median is 8 and the mean is $\frac{2+2+8+20+33}{5} = 13$. If it is discovered that the fifth doctor also recommended an additional 25 unnecessary procedures, how will the median and mean be affected?

Answer: The set is now {2, 2, 8, 20, 58}. The median is still 8; however, the mean changes to $\frac{2+2+8+20+58}{5} = 18$.

The above example illustrates how the mean, unlike the median, is sensitive to a change in any value.

EXAMPLE 2.3

Suppose the salaries of six employees are $3000, $7000, $15,000, $22,000, $23,000, and $38,000, respectively.

a. What is the mean salary?

Answer:

$$\frac{3000 + 7000 + 15,000 + 22,000 + 23,000 + 38,000}{6} = \$18,000$$

b. What will the new mean salary be if everyone receives a $3000 increase?

Answer:

$$\frac{6000 + 10,000 + 18,000 + 25,000 + 26,000 + 41,000}{6} = \$21,000$$

Note that $18,000 + $3000 = $21,000.

c. What if everyone receives a 10% raise?

Answer:

$$\frac{3300 + 7700 + 16,500 + 24,200 + 25,300 + 41,800}{6} = \$19,800$$

Note that 110% of $18,000 is $19,800.

The above example illustrates how adding the same constant to each value increases the mean by a like amount. Similarly, multiplying each value by the same constant multiplies the mean by a like amount.

MEASURING SPREAD: RANGE, INTERQUARTILE RANGE, VARIANCE, AND STANDARD DEVIATION

In describing a set of numbers, not only is it useful to designate an average value but it is also important to be able to indicate the *variability* or the *dispersion* of the measurements. A producer of time bombs aims for small variability—it would not be good for his 30-minute fuses actually to have a range of 10–50 minutes before detonation. On the other hand, a teacher interested in distinguishing better students from poorer students aims to design exams with large variability in results—it would not be helpful if all her students scored exactly the same. The players on two basketball teams may have the same average height, but this observation doesn't tell the whole story. If the dispersions are quite different, one team may have a 7-foot player, whereas the other has no one over 6 feet tall. Two Mediterranean holiday cruises may advertise the same average age for their passengers. One, however, may have only passengers between 20 and 25 years old, while the other has only middle-aged parents in their forties together with their children under age 10.

There are four primary ways of describing variability or dispersion:

1. The *range*, which is the difference between the largest and smallest values
2. The *interquartile range*, IQR, which is the difference between the largest and smallest values after removing the lower and upper quarters (i.e., IQR is the range of the middle 50%); that is, IQR = $Q_3 - Q_1$ = 75th percentile minus 25th percentile
3. The *variance*, which is determined by averaging the squared differences of all the values from the mean
4. The *standard deviation*, which is the square root of the variance.

EXAMPLE 2.4

The monthly rainfall in Monrovia, Liberia, where May through October is the rainy season and November through April the dry season, is as follows:

Month:	Jan	Feb	Mar	Apr	May	June	July	Aug	Sept	Oct	Nov	Dec
Rain (in.):	1	2	4	6	18	37	31	16	28	24	9	4

The mean is

$$\frac{1+2+4+6+18+37+31+16+28+24+9+4}{12} = 15 \text{ inches}$$

What are the measures of variability?

Answer: Range: The maximum is 37 inches (June), and the minimum is 1 inch (January). Thus the range is 37 − 1 = 36 inches of rain.

Interquartile range: Removing the lower and upper quarters leaves 4, 6, 9, 16, 18, and 24. Thus the interquartile range is 24 − 4 = 20. [The interquartile range is sometimes calculated as follows: The median of the lower half is $Q_1 = \frac{4+4}{2} = 4$, the median of the upper half is

(continued)

$Q_3 = \frac{24+28}{2} = 26$, and the interquartile range is $Q_3 - Q_1 = 22$. When there is a large number of values in the set, the two methods give the same answer.]

Variance:

$$\frac{14^2 + 13^2 + 11^2 + 9^2 + 3^2 + 22^2 + 16^2 + 1^2 + 13^2 + 9^2 + 6^2 + 11^2}{12} = 143.7$$

Standard deviation: $\sqrt{143.7} = 12.0$ inches

Range

The simplest, most easily calculated measure of variability is the *range*. The difference between the largest and smallest values can be noted quickly, and the range gives some impression of the dispersion. However, it is entirely dependent on the two extreme values and is insensitive to the ones in the middle.

One use of the range is to evaluate samples with very few items. For example, some quality control techniques involve taking periodic small samples and basing further action on the range found in several such samples.

Interquartile Range

Finding the *interquartile range* is one method of removing the influence of extreme values on the range. It is calculated by arranging the data in numerical order, removing the upper and lower quarters of the values, and noting the range of the remaining values. That is, it is the range of the middle 50% of the values.

A numerical rule sometimes used for designating outliers is to calculate 1.5 times the interquartile range (IQR) and then call a value an outlier if it is more than $1.5 \times$ IQR below the first quartile or $1.5 \times$ IQR above the third quartile.

EXAMPLE 2.5

Suppose that farm sizes in a small community have the following characteristics: the smallest value is 16.6 acres, 10% of the values are below 23.5 acres, 25% are below 41.1 acres, the median is 57.6 acres, 60% are below 87.2 acres, 75% are below 101.9 acres, 90% are below 124.0 acres, and the top value is 201.7 acres.

a. What is the range?
   Answer: The range is 201.7 − 16.6 = 185.1 acres.

b. What is the interquartile range?
   Answer: The interquartile range, with the highest and lowest quarters of the values removed, is 101.9 − 41.1 = 60.8 acres. Thus, while the largest farm is 185 acres more than the smallest, the middle 50% of the farm sizes range over a 61-acre interval.

(continued)

c. When the numerical rule is used for outliers, should either the smallest or largest value be called an outlier?

Answer: $1.5 \times IQR = 1.5 \times 60.8 = 91.2$. If a number is more than 91.2 below the first quartile, 41.1, or more than 91.2 above the third quartile, 101.9, then it will be called an outlier. Since the largest value, 201.7, is greater than $101.9 + 91.2 = 193.1$, it is considered an outlier by the numerical rule.

Variance

Dispersion is often the result of various chance happenings. For example, consider the motion of microscopic particles suspended in a liquid. The unpredictable motion of any particle is the result of many small movements in various directions caused by random bumps from other particles. If we average the total displacements of all the particles from their starting points, the result will not increase in direct proportion to time. If, however, we average the *squares* of the total displacements of all the particles, this result will increase in direct proportion to time.

The same holds true for the movement of paramecia. Their seemingly random motions as seen under a microscope can be described by the observation that the average of the squares of the displacements from their starting points is directly proportional to time. Also, consider ping-pong balls dropped straight down from a high tower and subjected to chance buffeting in the air. We can measure the deviations from a center spot on the ground to the spots where the balls actually strike. As the height of the tower is increased, the average of the squared deviations increases proportionately.

In a wide variety of cases we are in effect trying to measure dispersion from the mean due to a multitude of chance effects. The proper tool in these cases is the average of the squared deviations from the mean; it is called the *variance* and is denoted by σ^2 (σ is the lowercase Greek letter sigma):

$$\sigma^2 = \frac{\Sigma(x - \mu)^2}{n}$$

For circumstances specified later, the variance of a sample, denoted by s^2, is calculated as

$$s^2 = \frac{\Sigma(x - \bar{x})^2}{n - 1}$$

EXAMPLE 2.6

During the years 1929–39 of the Great Depression, the weekly average hours worked in manufacturing jobs were 45, 43, 41, 39, 39, 35, 37, 40, 39, 36, and 37, respectively. What is the variance?

(continued)

Answer: The variance can be quickly found on any calculator with a simple statistical package, or it can be found as follows:

$$\mu = \frac{45 + 43 + 41 + 39 + 39 + 35 + 37 + 40 + 39 + 36 + 37}{11} = 39.2 \text{ hours}$$

$$\sigma^2 = \frac{(45 - 39.2)^2 + (43 - 39.2)^2 + \cdots + (36 - 39.2)^2 + (37 - 39.2)^2}{11} = 8.1$$

EXAMPLE 2.7

Let $X = \{3, 7, 15, 23\}$. What is the variance?

Answer: Again, use the statistical package on your calculator, or

$$\Sigma x = 3 + 7 + 15 + 23 = 48$$
$$\mu = \frac{\Sigma x}{n} = \frac{48}{4} = 12$$

The variance can be calculated from its definition:

$$\sigma^2 = \frac{\Sigma (x - \mu)^2}{n} = \frac{(3 - 12)^2 + (7 - 12)^2 + (15 - 12)^2 + (23 - 12)^2}{4} = 59$$

Standard Deviation

Suppose we wish to pick a representative value for the variability of a certain population. The preceding discussions indicate that a natural choice is the value whose square is the average of the squared deviations from the mean. Thus we are led to consider the square root of the variance. This value is called the *standard deviation*, is denoted by σ, and is calculated on your calculator or as follows:

$$\sigma = \sqrt{\frac{\Sigma (x - \mu)^2}{n}}$$

Similarly, the standard deviation of a sample is denoted by s and is calculated on your calculator or as follows:

$$s = \sqrt{\frac{\Sigma (x - \bar{x})^2}{n - 1}}$$

While variance is measured in square units, standard deviation is measured in the same units as are the data.

For the various x-values, the deviations $x - \bar{x}$ are called *residuals*, and s is a "typical value" for the residuals. While s is not the average of the residuals (the average of the residuals is always 0), s does give a measure of the spread of the x-values around the sample mean.

EXAMPLE 2.8

Putting the data {1, 6, 3, 8} into a calculator such as the TI-84 gives what value for the standard deviation?

Answer: The TI-84 gives

```
1-Var Stats
X̄ =4.5
∑x=18
∑x²=110
Sx=3.109126351
σx=2.692582404
```

Thus, if the data are a population, the standard deviation is $\sigma = 2.693$, while if the data are a sample, the standard deviation is $s = 3.109$.

MEASURING POSITION: SIMPLE RANKING, PERCENTILE RANKING, AND *Z*-SCORE

We have seen several ways of choosing a value to represent the center of a distribution. We also need to be able to talk about the *position* of any other values. In some situations, such as wine tasting, simple rankings are of interest. Other cases, for example, evaluating college applications, may involve positioning according to percentile rankings. There are also situations in which position can be specified by making use of measurements of both central tendency and variability.

There are three important, recognized procedures for designating position:

1. *Simple ranking*, which involves arranging the elements in some order and noting where in that order a particular value falls
2. *Percentile ranking*, which indicates what percentage of all values fall below the value under consideration
3. The *z-score*, which states very specifically by how many standard deviations a particular value varies from the mean.

EXAMPLE 2.9

The water capacities (in gallons) of the 57 major solid-fuel boilers sold in the United States are 6.3, 7.4, 8.6, 10, 12.1, 50, 8.2, 9.8, 11.4, 12.9, 14.5, 16.1, 26, 21, 27, 40, 55, 30, 35, 55, 65, 18.8, 23.3, 28.3, 33.8, 26.4, 33, 50, 35, 21, 21, 18.5, 26.4, 37, 12, 12, 50, 65, 65, 56, 66, 60, 70, 27.7, 34.3, 42, 46.2, 33, 25, 29, 40, 19, 9, 12.5, 15.8, 24.5, and 16.5, respectively (John W. Bartok, *Solid-Fuel Furnaces and Boilers,* Garden Way, Charlotte, Vermont). What is the position of the Passat HO-45, which has a capacity of 46.2 gallons?

Answer: Since there are 12 boilers with higher capacities on the list, the Passat has a simple ranking of thirteenth (out of 57). Forty-four boilers have lower capacities, and so the Passat has a percentile ranking of $\frac{44}{57} = 77.2\%$. The above list has a mean of 30.2 and a standard deviation of 17.8, so the Passat has a z-score of $\frac{46.2-30.3}{17.8} = 0.89$.

Simple Ranking

Simple ranking is easily calculated and easily understood. We know what it means for someone to graduate second in a class of 435, or for a player from a team of size 30 to have the seventh-best batting average. Simple ranking is useful even when no numerical values are associated with the elements. For example, detergents can be ranked according to relative cleansing ability without any numerical measurements of strength.

Percentile Ranking

Percentile ranking, another readily understood measurement of position, is helpful in comparing positions with different bases. We can more easily compare a rank of 176 out of 704 with a rank of 187 out of 935 by noting that the first has a rank of 75%, and the second, a rank of 80%. Percentile rank is also useful when the exact population size is not known or is irrelevant. For example, it is more meaningful to say that Jennifer scored in the 90th percentile on a national exam rather than trying to determine her exact ranking among some large number of test takers.

The *quartiles*, Q_1 and Q_3, lie one-quarter and three-quarters of the way up a list, respectively. Their percentile ranks are 25% and 75%, respectively. The interquartile range defined earlier can also be defined to be $Q_3 - Q_1$. The *deciles* lie one-tenth and nine-tenths of the way up a list, respectively, and have percentile ranks of 10% and 90%.

z-Score

The *z-score* is a measure of position that takes into account both the center and the dispersion of the distribution. More specifically, the *z*-score of a value tells how many standard deviations the value is from the mean. Mathematically, $x - \mu$ gives the raw distance from μ to x; dividing by σ converts this to number of standard deviations. Thus $z = \frac{x-\mu}{\sigma}$, where x is the raw score, μ is the mean, and σ is the standard deviation. If the score x is greater than the mean μ, then z is positive; if x is less than μ, then z is negative.

Given a *z*-score, we can reverse the procedure and find the corresponding raw score. Solving for x gives $x = \mu + z\sigma$.

EXAMPLE 2.10

Suppose the average (mean) price of gasoline in a large city is $1.80 per gallon with a standard deviation of $0.05. Then $1.90 has a *z*-score of $\frac{1.90-1.80}{0.05} = +2$, while $1.65 has a *z*-score of $\frac{1.65-1.80}{0.05} = -3$. Alternatively, a *z*-score of +2.2 corresponds to a raw score of 1.80 + 2.2(0.05) = 1.80 + 0.11 = 1.91, while a *z*-score of −1.6 corresponds to 1.80 − 1.6(0.05) = 1.72.

It is often useful to portray integer *z*-scores and the corresponding raw scores as follows:

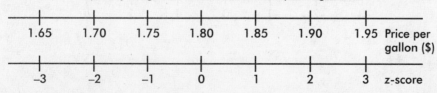

EXAMPLE 2.11

Suppose the attendance at a movie theater averages 780 with a standard deviation of 40. Adding multiples of 40 to and subtracting multiples of 40 from the mean 780 gives

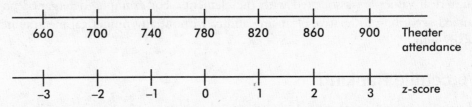

A theater attendance of 835 is converted to a z-score as follows: $\frac{835-780}{40} = \frac{55}{40} = 1.375$. A z-score of −2.15 is converted to a theater attendance as follows: $780 - 2.15(40) = 694$.

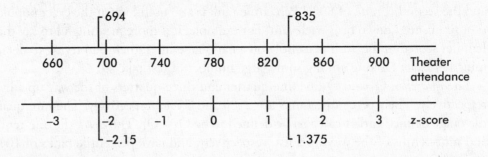

EMPIRICAL RULE

The *empirical rule* (also called the *68-95-99.7 rule*) applies specifically to symmetric bell-shaped data. In this case, about 68% of the values lie within 1 standard deviation of the mean, about 95% of the values lie within 2 standard deviations of the mean, and more than 99% of the values lie within 3 standard deviations of the mean.

In the following figure the horizontal axis shows z-scores:

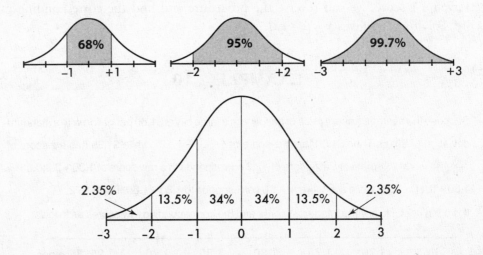

EXAMPLE 2.12

Suppose that taxicabs in New York City are driven an average of 75,000 miles per year with a standard deviation of 12,000 miles. What information does the empirical rule give us?

Answer: Assuming that the distribution is bell-shaped, we can conclude that approximately 68% of the taxis are driven between 63,000 and 87,000 miles per year, approximately 95% are driven between 51,000 and 99,000 miles, and virtually all are driven between 39,000 and 111,000 miles.

The empirical rule also gives a useful quick estimate of the standard deviation in terms of the range. We can see in the figure above that 95% of the data fall within a span of 4 standard deviations (from −2 to +2 on the z-score line) and 99.7% of the data fall within 6 standard deviations (from −3 to +3 on the z-score line). It is therefore reasonable to conclude that for these data the standard deviation is roughly between one-fourth and one-sixth of the range. Since we can find the range of a set almost immediately, the empirical rule technique for estimating the standard deviation is often helpful in pointing out gross arithmetic errors.

EXAMPLE 2.13

If the range of a data set is 60, what is an estimate for the standard deviation?

Answer: By the empirical rule, the standard deviation is expected to be between $\left(\frac{1}{6}\right)60 = 10$ and $\left(\frac{1}{4}\right)60 = 15$. If the standard deviation is calculated to be 0.32 or 87, there is probably an arithmetic error; a calculation of 12, however, is reasonable.

However, it must be stressed that the above use of the range is not intended to provide an accurate value for the standard deviation. It is simply a tool for pointing out unreasonable answers rather than, for example, blindly accepting computer outputs.

HISTOGRAMS AND MEASURES OF CENTRAL TENDENCY

Suppose we have a detailed histogram such as

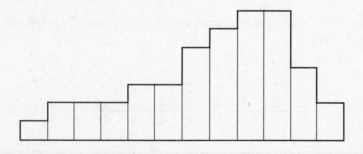

Our measures of central tendency fit naturally into such a diagram.

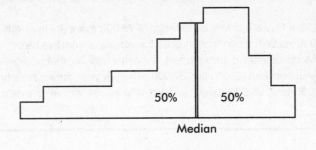

The *median* divides a distribution in half, so it is represented by a line that divides the area of the histogram in half.

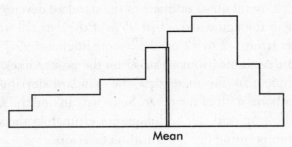

The *mean* is affected by the spacing of all the values. Therefore, if the histogram is considered to be a solid region, the mean corresponds to a line passing through the center of gravity, or balance point.

The above distribution, spread thinly far to the low side, is said to be *skewed to the left*. Note that in this case the mean is usually less than the median. Similarly, a distribution spread far to the high side is *skewed to the right*, and its mean is usually greater than its median.

EXAMPLE 2.14

Suppose that the faculty salaries at a college have a median of $32,500 and a mean of $38,700. What does this indicate about the shape of the distribution of the salaries?

Answer: The median is less than the mean, and so the salaries are probably skewed to the right. There are a few highly paid professors, with the bulk of the faculty at the lower end of the pay scale.

It should be noted that the above principle is a useful, but not hard-and-fast, rule.

EXAMPLE 2.15

The set given by the dotplot below is skewed to the right; however, its median (3) is greater than its mean (2.97).

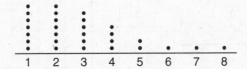

HISTOGRAMS, *Z*-SCORES, AND PERCENTILE RANKINGS

We have seen that relative frequencies are represented by relative areas, and so labeling the vertical axis is not crucial. If we know the standard deviation, the horizontal axis can be labeled in terms of *z*-scores. In fact, if we are given the percentile rankings of various *z*-scores, we can construct a histogram.

EXAMPLE 2.16

Suppose we are asked to construct a histogram from these data:

z-score:	−2	−1	0	1	2
Percentile ranking:	0	20	60	70	100

We note that the entire area is less than *z*-score +2 and greater than *z*-score −2. Also, 20% of the area is between *z*-scores −2 and −1, 40% is between −1 and 0, 10% is between 0 and 1, and 30% is between 1 and 2. Thus the histogram is as follows:

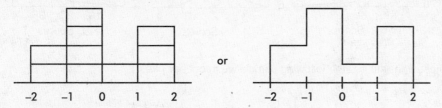

or

Now suppose we are given four in-between *z*-scores as well:

z-score	Percentile Ranking
2.0	100
1.5	80
1.0	70
0.5	65
0.0	60
−0.5	30
−1.0	20
−1.5	5
−2.0	0

Then we have:

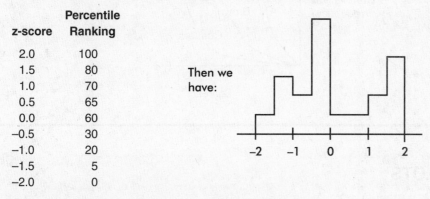

With 1000 *z*-scores perhaps the histogram would look like

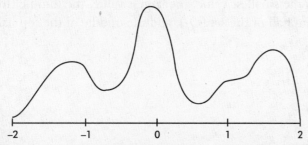

The height at any point is meaningless; what is important is relative areas. For example, in the final diagram above, what percentage of the area is between *z*-scores of +1 and +2?
Answer: Still 30%.

What percent is to the left of 0?
Answer: Still 60%.

CUMULATIVE FREQUENCY AND SKEWNESS

A distribution skewed to the left has a cumulative frequency plot that rises slowly at first and then steeply later, while a distribution skewed to the right has a cumulative frequency plot that rises steeply at first and then slowly later.

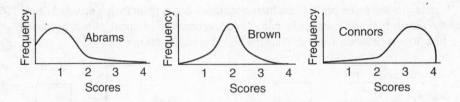

EXAMPLE 2.17

Consider the essay grading policies of three teachers, Abrams, who gives very high scores, Brown, who gives equal numbers of low and high scores, and Connors, who gives very low scores. Histograms of the grades (with 1 the highest score and 4 the lowest score) are as follows:

These translate into the following cumulative frequency plots:

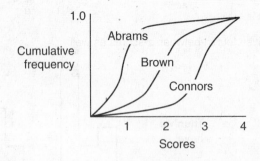

BOXPLOTS

A *boxplot* (also called a *box and whisker display*) is a visual representation of dispersion that shows the smallest value, the largest value, the middle (median), the middle of the bottom half of the set (Q_1), and the middle of the top half of the set (Q_3).

EXAMPLE 2.18

The total farm product indexes for the years 1919–45 (with 1910–14 as 100) are 215, 210, 130, 140, 150, 150, 160, 150, 140, 150, 150, 125, 85, 70, 75, 90, 115, 120, 125, 100, 95, 100, 130, 160, 200, 200, 210, respectively. (Note the instability of prices received by farmers!) The largest value is 215, the smallest is 70, the middle is 140, the middle of the top half is 160, and the middle of the bottom half is 100. A boxplot of these five numbers is

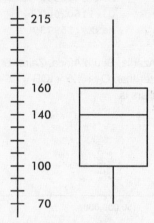

Note that the display consists of two "boxes" together with two "whiskers"—hence the alternative name. The boxes show the spread of the two middle quarters; the whiskers show the spread of the two outer quarters. This relatively simple display conveys information not immediately available from histograms or stem and leaf displays.

Putting the above data into a list, for example, L1, on the TI-84, not only gives the five-number summary

```
1-Var Stats
minX=70
Q1=100
Med=140
Q3=160
MaxX=215
```

but also gives the boxplot itself using STAT PLOT, choosing the boxplot from among the six type choices, and then using ZoomStat or in WINDOW letting Xmin=0 and Xmax=225.

When a distribution is strongly skewed, or when it has pronounced outliers, drawing a boxplot with its five-number summary including median, quartiles, and extremes, gives a more useful description than calculating a mean and a standard deviation.

Sometimes values more than $1.5 \times$ IQR (1.5 times the interquartile range) outside the two boxes are plotted separately as possible outliers. (The TI-84 has a modified boxplot option. Note the two options in the second row of Type in StatPlot.)

EXAMPLE 2.19

Inputting the 1990 populations of the 58 countries of Africa results in a calculator output of

```
1-Var Stats
minX=67000
Q1=1119000
Med=5412500
Q3=11602000
MaxX=115973000
```

There were five countries (Tanzania, South Africa, Zaire, Ethiopia, and the largest country, Nigeria) with populations greater than $Q_3 + 1.5 \times IQR = Q_3 + 1.5(Q_3 - Q_1) = 27{,}326{,}500$. A boxplot indicating possible outliers is

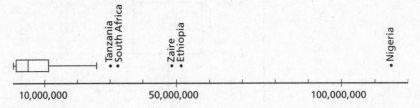

Note: some computer output shows *two* levels of outliers—mild (between 1.5 IQR and 3 IQR) and extreme (more than 3 IQR).

It should be noted that two sets can have the same five-number summary and thus the same boxplots but have dramatically different distributions.

EXAMPLE 2.20

Let $A = \{0, 5, 10, 15, 25, 30, 35, 40, 45, 50, 71, 72, 73, 74, 75, 76, 77, 78, 100\}$ and $B = \{0, 22, 23, 24, 25, 26, 27, 28, 29, 50, 55, 60, 65, 70, 75, 85, 90, 95, 100\}$. Simple inspection indicates very different distributions, however the TI-84 gives identical boxplots with Min = 0, Q_1 = 25, Med = 50, Q_3 = 75, and Max = 100 for each.

EFFECT OF CHANGING UNITS

Changing units, for example, from dollars to rubles or from miles to kilometers, is common in a world that seems to become smaller all the time. It is instructive to note how measures of center and spread are affected by such changes.

Adding the same constant to every value increases the mean and median by that same constant; however, the distances between the increased values stay the same, and so the range and standard deviation are unchanged.

EXAMPLE 2.21

A set of experimental measurements of the freezing point of an unknown liquid yield a mean of 25.32 degrees Celsius with a standard deviation of 1.47 degrees Celsius. If all the measurements are converted to the Kelvin scale, what are the new mean and standard deviation?

Answer: Kelvins are equivalent to degrees Celsius plus 273.16. The new mean is thus 25.32 + 273.16 = 298.48 kelvins. However, the standard deviation remains numerically the same, 1.47 kelvins. Graphically, you should picture the whole distribution moving over by the constant 273.16; the mean moves, but the standard deviation (which measures spread) doesn't change.

Multiplying every value by the same constant multiplies the mean, median, range, and standard deviation all by that constant.

EXAMPLE 2.22

Measurements of the sizes of farms in an upstate New York county yield a mean of 59.2 hectares with a standard deviation of 11.2 hectares. If all the measurements are converted from hectares (metric system) to acres (one acre was originally the area a yoke of oxen could plow in one day), what are the new mean and standard deviation?

Answer: One hectare is equivalent to 2.471 acres. The new mean is thus 2.471 × 59.2 = 146.3 acres with a standard deviation of 2.471 × 11.2 = 27.7 acres. Graphically, multiplying each value by the constant 2.471 both moves and spreads out the distribution.

Questions on Topic Two:
Summarizing Distributions

Multiple-Choice Questions

Directions: The questions or incomplete statements that follow are each followed by five suggested answers or completions. Choose the response that best answers the question or completes the statement.

1. In the second game of the 1989 World Series between Oakland and San Francisco (played 2 days before the northern California earthquake postponed the series), ten players went hitless, eight players had one hit apiece, and one player (Rickey Henderson) had three hits. What were the mean and median number of hits?

   (A) 11/19, 0
   (B) 11/19, 1
   (C) 1, 1
   (D) 0, 1
   (E) 1, 0

2. Which of the following are true statements?

   I. The range of the sample data set is never greater than the range of the population.
   II. The interquartile range is half the distance between the first quartile and the third quartile.
   III. While the range is affected by outliers, the interquartile range is not.

   (A) I only
   (B) II only
   (C) III only
   (D) I and II
   (E) I and III

   Problems 3–5 refer to the following five boxplots.

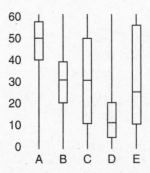

3. To which of the above boxplots does the following histogram correspond?

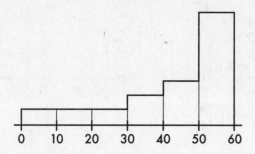

(A) A
(B) B
(C) C
(D) D
(E) E

4. To which of the above boxplots does the following histogram correspond?

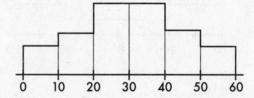

(A) A
(B) B
(C) C
(D) D
(E) E

5. To which of the above boxplots does the following histogram correspond?

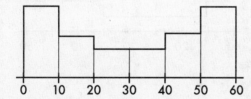

(A) A
(B) B
(C) C
(D) D
(E) E

Problems 6–8 refer to the following five histograms:

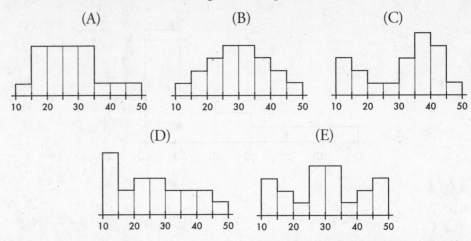

6. To which of the above histograms does the following boxplot correspond?

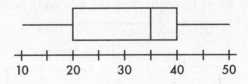

(A) A
(B) B
(C) C
(D) D
(E) E

7. To which of the above histograms does the following boxplot correspond?

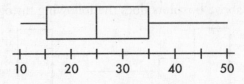

(A) A
(B) B
(C) C
(D) D
(E) E

8. To which of the above histograms does the following boxplot correspond?

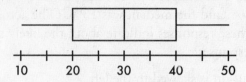

   (A) A
   (B) B
   (C) C
   (D) D
   (E) E

9. Suppose the average score on a national test is 500 with a standard deviation of 100. If each score is increased by 25, what are the new mean and standard deviation?

   (A) 500, 100
   (B) 500, 125
   (C) 525, 100
   (D) 525, 105
   (E) 525, 125

10. Suppose the average score on a national test is 500 with a standard deviation of 100. If each score is increased by 25%, what are the new mean and standard deviation?

    (A) 500, 100
    (B) 525, 100
    (C) 625, 100
    (D) 625, 105
    (E) 625, 125

11. If quartiles $Q_1 = 20$ and $Q_3 = 30$, which of the following must be true?

    I. The median is 25.
    II. The mean is between 20 and 30.
    III. The standard deviation is at most 10.

    (A) I only
    (B) II only
    (C) III only
    (D) All are true.
    (E) None are true.

12. A 1995 poll by the Program for International Policy asked respondents what percentage of the U.S. budget they thought went to foreign aid. The mean response was 18%, and the median was 15%. (The actual amount is less than 1%.) What do these responses indicate about the likely shape of the distribution of all the responses?

    (A) The distribution is skewed to the left.
    (B) The distribution is skewed to the right.
    (C) The distribution is symmetric around 16.5%.
    (D) The distribution is bell-shaped with a standard deviation of 3%.
    (E) The distribution is uniform between 15% and 18%.

13. Assuming that batting averages have a bell-shaped distribution, arrange in ascending order:

    I. An average with a z-score of -1.
    II. An average with a percentile rank of 20%.
    III. An average at the first quartile, Q_1.

    (A) I, II, III
    (B) III, I, II
    (C) II, I, III
    (D) II, III, I
    (E) III, II, I

14. Which of the following are true statements?

    I. If the sample has variance zero, the variance of the population is also zero.
    II. If the population has variance zero, the variance of the sample is also zero.
    III. If the sample has variance zero, the sample mean and the sample median are equal.

    (A) I and II
    (B) I and III
    (C) II and III
    (D) I, II, and III
    (E) None of the above gives the complete set of true responses.

15. When there are multiple gaps and clusters, which of the following is the best choice to give an overall picture of a distribution?

    (A) Mean and standard deviation
    (B) Median and interquartile range
    (C) Boxplot with its five-number summary
    (D) Stemplot or histogram
    (E) None of the above are really helpful in showing gaps and clusters.

16. Suppose the starting salaries of a graduating class are as follows:

Number of Students	Starting Salary ($)
10	15,000
17	20,000
25	25,000
38	30,000
27	35,000
21	40,000
12	45,000

What is the mean starting salary?

(A) $30,000
(B) $30,533
(C) $32,500
(D) $32,533
(E) $35,000

17. When a set of data has suspect outliers, which of the following are preferred measures of central tendency and of variability?

(A) mean and standard deviation
(B) mean and variance
(C) mean and range
(D) median and range
(E) median and interquartile range

18. If the standard deviation of a set of observations is 0, you can conclude

(A) that there is no relationship between the observations.
(B) that the average value is 0.
(C) that all observations are the same value.
(D) that a mistake in arithmetic has been made.
(E) none of the above.

19. A teacher is teaching two AP Statistics classes. On the final exam, the 20 students in the first class averaged 92 while the 25 students in the second class averaged only 83. If the teacher combines the classes, what will the average final exam score be?

(A) 87
(B) 87.5
(C) 88
(D) None of the above
(E) More information is needed to make this calculation.

20. The 60 longest rivers in the world have lengths distributed as follows:

Length (mi):	1000–1499	1500–1999	2000–2499	2500–2999	3000–3499	3500–3999	4000–4499
Number of rivers:	21	22	4	8	2	2	1

(The Nile is the longest with a length of 4145 miles, and the Amazon is the second longest at 3900 miles.)

Which of the following best describes these data?

(A) Skewed distribution, mean greater than median
(B) Skewed distribution, median greater than mean
(C) Symmetric distribution, mean greater than median
(D) Symmetric distribution, median greater than mean
(E) Symmetric distribution with outliers on high end

21. In 1993 the seven states with the fewest business bankruptcies were Vermont (900), Alaska (1000), North Dakota (1100), Wyoming (1300), South Dakota (1400), Hawaii (1500), and Delaware (1600).

Which of the following are reasonable conclusions about the distribution of bankruptcies throughout the 50 states in 1993?

  I. The total number of bankruptcies in the United States in 1993 was approximately $50(1300) = 65,000$.
  II. Because of these low values, the distribution was skewed to the left.
  III. The range of the distribution was approximately $50(1600 - 1000) = 30,000$.

(A) I only
(B) II only
(C) III only
(D) All are reasonable.
(E) None are reasonable.

22. Suppose 10% of a data set lie between 40 and 60. If 5 is first added to each value in the set and then each result is doubled, which of the following is true?

(A) 10% of the resulting data will lie between 85 and 125.
(B) 10% of the resulting data will lie between 90 and 130.
(C) 15% of the resulting data will lie between 80 and 120.
(D) 20% of the resulting data will lie between 45 and 65.
(E) 30% of the resulting data will lie between 85 and 125.

23. A stemplot for the 1988 per capita personal income (in hundreds of dollars)
    for the 50 states is

11	0 7
12	6 2 7 8 2 7 5 7 8 5 0
13	3 7 7
14	9 7 8 1 6
15	0 9 0 5 2 5 0 4 4
16	4 5 9 4 8 2 8 6
17	7 6 4 6
18	9
19	5 3 0 3
20	7
21	9
22	8

    Which of the following best describes these data?

    (A) Skewed distribution, mean greater than median
    (B) Skewed distribution, median greater than mean
    (C) Symmetric distribution, mean greater than median
    (D) Symmetric distribution, median greater than mean
    (E) Symmetric distribution with outliers on high end

24. Which of the following statements are true?

    I. If the right and left sides of a histogram are mirror images of each other,
    the distribution is symmetric.
    II. A distribution spread far to the right side is said to be skewed to the right.
    III. If a distribution is skewed to the right, its mean is often greater than its
    median.

    (A) I only
    (B) I and II
    (C) I and III
    (D) II and III
    (E) None of the above gives the complete set of true responses.

25. Which of the following statements are true?

    I. In a stemplot the number of leaves equals the size of the set of data.
    II. Both the dotplot and the stemplot are useful in identifying outliers.
    III. Histograms do not retain the identity of individual scores; however, dot-
    plots, stemplots, and boxplots all do.

    (A) I and II
    (B) I and III
    (C) II and III
    (D) I, II, and III
    (E) None of the above gives the complete set of true responses.

26. The 70 highest dams in the world have an average height of 206 meters with a standard deviation of 35 meters. The Hoover and Grand Coulee dams have heights of 221 and 168 meters, respectively. The Russian dams, the Nurek and Charvak, have heights with z-scores of +2.69 and −1.13, respectively. List the dams in order of ascending size.

    (A)   Charvak, Grand Coulee, Hoover, Nurek
    (B)   Charvak, Grand Coulee, Nurek, Hoover
    (C)   Grand Coulee, Charvak, Hoover, Nurek
    (D)   Grand Coulee, Charvak, Nurek, Hoover
    (E)   Grand Coulee, Hoover, Charvak, Nurek

27. According to *Consumer Reports* magazine, the costs per pound of protein for 20 major-brand beef hot dogs are $14.23, $21.70, $14.49, $20.49, $14.47, $15.45, $25.25, $24.02, $18.86, $18.86, $30.65, $25.62, $8.12, $12.74, $14.21, $13.39, $22.31, $19.95, $22.90, and $19.78, respectively. If the least expensive is considered the top-ranked, what is the position in terms of a z-score of Thorn Apple Valley beef hot dogs at $14.23 per pound of protein?

    (A) −2.67
    (B) −0.85
    (C) 0.85
    (D) 1.42
    (E) 2.67

28. The first 115 Kentucky Derby winners by color of horse were as follows: roan, 1; gray, 4; chestnut, 36; bay, 53; dark bay, 17; and black, 4. (You should "bet on the bay!") Which of the following visual displays is most appropriate?

    (A) Bar chart
    (B) Histogram
    (C) Stemplot
    (D) Boxplot
    (E) Time plot

For Questions 29 and 30 consider the following: The graph below shows cumulative proportions plotted against grade point averages for a large public high school.

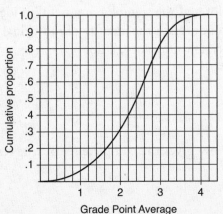

29. What is the median grade point average?

    (A) 0.8
    (B) 2.0
    (C) 2.4
    (D) 2.5
    (E) 2.6

30. What is the interquartile range?

    (A) 1.0
    (B) 1.8
    (C) 2.4
    (D) 2.8
    (E) 4.0

31. The following graph shows cumulative proportions with regard to outstanding balances on credit cards.

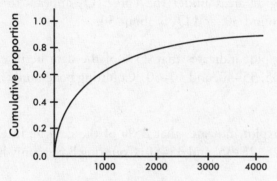

    What is the approximate interquartile range?

    (A) $200
    (B) $600
    (C) $1550
    (D) $1750
    (E) $2000

Answer Key

1. **A**	8. **E**	14. **C**	20. **A**	26. **A**	
2. **E**	9. **C**	15. **D**	21. **E**	27. **B**	
3. **A**	10. **E**	16. **B**	22. **B**	28. **A**	
4. **B**	11. **E**	17. **E**	23. **A**	29. **C**	
5. **C**	12. **B**	18. **C**	24. **E**	30. **A**	
6. **C**	13. **A**	19. **A**	25. **A**	31. **C**	
7. **D**					

Answers Explained

1. **(A)** The set is {0, 0, 0, 0, 0, 0, 0, 0, 0, 0, 1, 1, 1, 1, 1, 1, 1, 1, 3}, so the mean is $\frac{11}{19}$ and the median is 0.

2. **(E)** All elements of the sample are taken from the population, and so the smallest value in the sample cannot be less than the smallest value in the population; similarly, the largest value in the sample cannot be greater than the largest value in the population. The interquartile range is the full distance between the first quartile and the third quartile. Outliers are extreme values, and while they may affect the range, they do not affect the interquartile range when the lower and upper quarters have been removed before calculation.

3. **(A)** The value 50 seems to split the area under the histogram in two, so the median is about 50. Furthermore, the histogram is skewed to the left with a tail from 0 to 30.

4. **(B)** Looking at areas under the curve, Q_1 appears to be around 20, the median is around 30, and Q_3 is about 40.

5. **(C)** Looking at areas under the curve, Q_1 appears to be around 10, the median is around 30, and Q_3 is about 50.

6. **(C)** The boxplot indicates that 25% of the data lie in each of the intervals 10–20, 20–35, 35–40, and 40–50. Counting boxes, only histogram C has this distribution.

7. **(D)** The boxplot indicates that 25% of the data lie in each of the intervals 10–15, 15–25, 25–35, and 35–50. Counting boxes, only histogram D has this distribution.

8. **(E)** The boxplot indicates that 25% of the data lie in each of the intervals 10–20, 20–30, 30–40, and 40–50. Counting boxes, only histogram E has this distribution.

9. **(C)** Adding the same constant to every value increases the mean by that same constant; however, the distances between the increased values and the increased mean stay the same, and so the standard deviation is unchanged. Graphically, you should picture the whole distribution as moving over by a constant; the mean moves, but the standard deviation (which measures spread) doesn't change.

10. **(E)** Multiplying every value by the same constant multiplies both the mean and the standard deviation by that constant. Graphically, increasing each value by 25% (multiplying by 1.25) both moves and spreads out the distribution.

11. **(E)** The median is somewhere between 20 and 30, but not necessarily at 25. Even a single very large score can result in a mean over 30 and a standard deviation over 10.

12. **(B)** The median is less than the mean, and so the responses are probably skewed to the right; there are a few high guesses, with most of the responses on the lower end of the scale.

13. **(A)** Given that the empirical rule applies, a z-score of -1 has a percentile rank of about 16%. The first quartile Q_1 has a percentile rank of 25%.

14. **(C)** If the variance of a set is zero, all the values in the set are equal. If all the values of the population are equal, the same holds true for any subset; however, if all the values of a subset are the same, this may not be true of the whole population. If all the values in a set are equal, the mean and the median both equal this common value and so equal each other.

15. **(D)** Stemplots and histograms can show gaps and clusters that are hidden when one simply looks at calculations such as mean, median, standard deviation, quartiles, and extremes.

16. **(B)** There are a total of $10 + 17 + 25 + 38 + 27 + 21 + 12 = 150$ students. Their total salary is $10(15,000) + 17(20,000) + 25(25,000) + 38(30,000) + 27(35,000) + 21(40,000) + 12(45,000) = \$4,580,000$. The mean is $\frac{4,580,000}{150} = \$30,533$.

17. **(E)** The mean, standard deviation, variance, and range are all affected by outliers; the median and interquartile range are not.

18. **(C)** Because of the squaring operation in the definition, the standard deviation (and also the variance) can be zero only if all the values in the set are equal.

19. **(A)** The sum of the scores in one class is $20 \times 92 = 1840$, while the sum in the other is $25 \times 83 = 2075$. The total sum is $1840 + 2075 = 3915$. There are $20 + 25 = 45$ students, and so the average score is $\frac{3915}{45} = 87$.

20. **(A)** A distribution spread thinly on the high end is a skewed distribution with the mean greater than the median.

21. **(E)** None are reasonable because we are not looking at a random sample of states but rather only at the seven lowest values, that is, at the very low end of the tail of the whole distribution. (The distribution was actually skewed to the right, with California having 159,700 bankruptcies, followed by New York with 51,300, and all other states below 50,000.)

22. **(B)** Increasing every value by 5 gives 10% between 45 and 65, and then doubling gives 10% between 90 and 130.

23. **(A)** A distribution spread thinly on the high end is a skewed distribution with the mean greater than the median.

24. **(E)** All three statements are true.

25. **(A)** Dotplots and stemplots retain the identity of individual scores; however, histograms and boxplots do not.

26. **(A)** $206 + 2.69(35) = 300$; $206 - 1.13(35) = 166$.

27. **(B)** A calculator gives a mean of 18.875 with a standard deviation of 5.472, and so the z-score is $\frac{14.23 - 18.875}{5.472} = -0.85$. Note that in this case "top-ranked" brands have more negative z-scores.

28. **(A)** Bar charts are used for categorical variables.

29. **(C)** The median corresponds to the 0.5 cumulative proportion.

30. **(A)** The 0.25 and 0.75 cumulative proportions correspond to $Q_1 = 1.8$ and $Q_3 = 2.8$, respectively, and so the interquartile range is $2.8 - 1.8 = 1.0$.

31. **(C)** The cumulative proportions of 0.25 and 0.75 correspond to $Q_1 = 200$ and $Q_3 = 1750$, respectively, and so the interquartile range is 1550.

Free-Reponse Questions

Directions: You must show all work and indicate the methods you use. You will be graded on the correctness of your methods and on the accuracy of your final answers.

Four Open-Ended Questions

1. According to a 1988 *New York Times* article, the ten car models with the highest theft rates were as follows:

Vehicle Model	Thefts per 1000 Cars
Pontiac Firebird	30.14
Chevrolet Camaro	26.02
Chevrolet Monte Carlo	20.28
Toyota MR2	19.25
Buick Regal	14.70
Mitsubishi Starion	14.70
Ferrari Mondial	13.60
Mitsubishi Mirage	12.80
Pontiac Fiero	12.68
Oldsmobile Cutlass	11.73

   What are the mean, range, and standard deviation of these theft rates? Explain how each of these values changes if each theft rate increases by 1.5 and if each increases by 15%.

2. In 1977 the 88 quarterbacks on National Football League (NFL) rosters earned a mean average salary of either $58,750 or $89,354. The number that is not the mean is the median. Which number is the mean salary? Justify your answer.

3. A survey is conducted to determine the amount a high school student spends monthly on entertainment. The survey is stratified by class and is summarized as follows: for freshmen and sophomores the distributions are symmetric; however, the mean spent by sophomores is greater than the mean spent by freshmen. For juniors the mean spent is greater than the median, while for seniors the mean spent is less than the median. Draw histograms to illustrate these facts. Describe/explain each.

4. Suppose a distribution has mean 180 and standard deviation 20. If the z-score of Q_1 is -0.7 and the z-score of Q_3 is 0.8, what values would be considered outliers?

Answers Explained

1. Use your calculator to obtain a mean of 17.59 and a standard deviation of 5.94. The range is $30.14 - 11.73 = 18.41$. Increasing each value by 1.5 increases the mean by 1.5 to 19.09 and leaves the range and standard deviation unchanged. Increasing each value by 15% increases the mean, range, and standard deviation each by 15% to 20.23, 21.17, and 6.83, respectively.

2. If the median were greater than the mean, the salaries would be skewed to the left with a few very low salaries dragging the mean down while the bulk of the salaries would be above the mean. However, it would be difficult to construct a data set where reasonable, but low, salaries dragged the mean down to $58,750 while half the salaries were above $89,354. It is much more likely that the median was less than the mean and the salary distribution was skewed to the right with a few high salaries dragging the mean up while the bulk of the salaries were below the mean. We conclude that the mean was $89,354 (and that the median was $58,750).

3. There are many different possible answers. One is as follows:

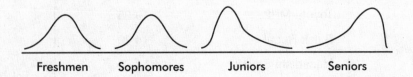

Freshmen   Sophomores   Juniors   Seniors

Note that the freshman and sophomore distributions do not have to be bell-shaped, but each does have to be symmetric, with the freshman distribution appearing to the left of the sophomore distribution. The mean greater than the median indicates a probable right skew, while the mean less than the median indicates a probable left skew.

4. Z-scores give the number of standard deviations from the mean, so

   $Q_1 = 180 - 0.7(20) = 166$ and $Q_3 = 180 + 0.7(20) = 194$.

   The interquartile range is $IQR = 194 - 166 = 28$, and $1.5(IQR) = 1.5(28) = 42$.

   The standard definition of outliers encompasses all values less than $Q_1 - 42 = 124$ and all values greater than $Q_3 + 42 = 236$.

Comparing Distributions

- Dotplots
- Double Bar Charts
- Back-to-back Stemplots
- Parallel Boxplots
- Cumulative Frequency Plots

Many real-life applications of statistics involve comparisons of *two* populations. Such comparisons can involve modifications of graphical displays such as dotplots, bar charts, stemplots, boxplots, and cumulative frequency plots to portray both sets simultaneously.

DOTPLOTS

EXAMPLE 3.1

The caloric intakes of 25 people on each of two weight loss programs are recorded as follows:

Program A: 1000, 1000, 1100, 1100, 1100, 1200, 1200, 1200, 1200, 1300, 1300, 1300, 1300, 1300, 1400, 1400, 1400, 1400, 1400, 1500, 1500, 1600, 1600, 1700, 1900

Program B: 1000, 1100, 1100, 1200, 1200, 1200, 1300, 1300, 1300, 1400, 1400, 1400, 1400, 1500, 1500, 1500, 1500, 1500, 1600, 1600, 1600, 1700, 1700, 1800, 1800

These data can be compared with dotplots, one above the other, using the same horizontal scale.

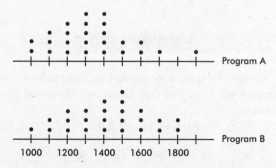

Program A appears to be associated with a lower average caloric intake than Program B. Comparing shape, center, and spread, we have:

Shape: We see that both sets of data are roughly bell-shaped (the empirical rule applies), and Program A has an outlier at 1900 calories (while Program B has no outliers).

(continued)

Center: Visually, or by counting dots, the centers of the two distributions are 1300 and 1400 calories, for Programs A and B respectively. (A calculator gives means of $\overline{x}_A = 1340$ and $\overline{x}_B = 1424$.) By any method, the center for Program B is higher.

Spread: The spreads are approximately the same, 1000 to 1900 calories for Program A and 1000 to 1800 calories for Program B. (A calculator gives standard deviations of $s_A = 229$ and $s_B = 218$.)

DOUBLE BAR CHARTS

EXAMPLE 3.2

A study tabulated the percentages of young adults who recognized various photographs as follows: Joe Stalin (10%), Joe Camel (95%), Senator Simpson of Wyoming (5%), Bart Simpson (80%), Al Gore (30%), Al Bundy (60%), Mickey Mantle (20%), Mickey Mouse (100%), Charlie Chaplin (25%), Charlie the Tuna (90%). These data can be illustrated with a bar chart appropriately displayed in pairs of bars.

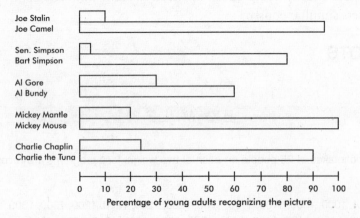

The pairs of bar graphs visually indicate the reason for concern felt by some educators.

BACK-TO-BACK STEMPLOTS

EXAMPLE 3.3

In a 40-year study, survival years were measured for cancer patients undergoing one of two different chemotherapy treatments. The data for 25 patients on the first drug and 30 on the second were as follows:

Drug A: 5, 10, 17, 39, 29, 25, 20, 4, 8, 31, 21, 3, 12, 11, 19, 10, 4, 22, 17, 18, 13, 28, 11, 14, 21

Drug B: 19, 12, 20, 28, 22, 35, 1, 21, 21, 26, 18, 28, 29, 20, 15, 32, 31, 24, 22, 26, 18, 20, 22, 35, 30, 18, 25, 24, 19, 21

(continued)

In drawing a back-to-back stemplot of the above data, we place a vertical line on each side of the column of stems and then arrange one set of leaves extending out to the right while the other extends out to the left.

	Drug B			Drug A	
	1		0		3 4 4 5 8
9 9 8 8 8 5 2		1		0 0 1 1 2 3 4 7 7 8 9
9 8 8 6 6 5 4 4 2 2 2 1 1 1 0 0 0		2		0 1 1 2 5 8 9
5 5 2 1 0			3		1 9

Note that even though drug A showed the longest-surviving patient (39 years) and drug B showed the shortest-surviving patient (1 year), the back-to-back stemplot indicates that the bulk of patients on drug B survived longer than the bulk of patients on drug A.

Comparing shape, center, and spread, we have:

Shape: Both distributions are roughly bell-shaped (the empirical rule applies). The drug A distribution appears to have a high outlier at 39, while the drug B distribution appears to have a low outlier at 1.

Center: Visually, or by counting values, the centers of the two distributions are 17 and 22, respectively. (A calculator gives means of $\bar{x}_A = 16.48$ and $\bar{x}_B = 22.73$.) By either method, the drug B distribution has a greater center.

Spread: The spreads are 3 to 39 survival years for drug A and 1 to 35 survival years for drug B. (A calculator gives standard deviations of $s_A = 9.25$ and $s_B = 6.97$.) By either method, the drug A distribution has a greater spread than the drug B distribution.

PARALLEL BOXPLOTS

EXAMPLE 3.4

Mail-order labs and 1-hour minilabs were compared with regard to price for developing and printing one 24-exposure roll of 35-millimeter color-print film. Prices included shipping and handling charges where applicable. Following is a computer output describing the results:

For mail-order labs:

```
Mean = 5.37    Standard deviation = 1.92    Min = 3.51
Max = 8.00   N = 18   Median = 4.77
Quartiles = 3.92, 6.45
```

For 1-hour minilabs:

```
Mean = 10.11    Standard deviation = 1.32    Min = 8.58
Max = 11.95   N = 15   Median = 10.08
Quartiles = 8.97, 11.51
```

(continued)

In drawing parallel boxplots (also called side-by-side boxplots) of the above data, we place both on the same diagram:

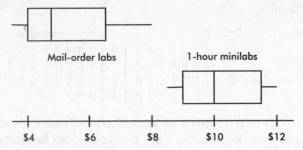

Boxplots show the minimum, maximum, median, and quartile values. The distribution of mail-order lab prices is lower and more spread out than that of prices of 1-hour minilabs. Both are slightly skewed toward the upper end (the skewness can also be noted from the computer output showing the mean to be greater than the median in both cases.)

Parallel boxplots are useful in presenting a picture of the comparison of several distributions.

EXAMPLE 3.5

Following are parallel boxplots showing the daily price fluctuations of a certain common stock over the course of 5 years. What trends do the boxplots show?

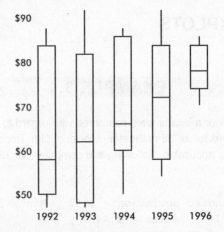

The parallel boxplots show that from year to year the median daily stock price has steadily risen 20 points from about $58 to about $78, the third quartile value has been roughly stable at about $84, the yearly low has never decreased from that of the previous year, and the interquartile range has never increased from one year to the next.

CUMULATIVE FREQUENCY PLOTS

EXAMPLE 3.6

The graph below compares cumulative frequency plotted against age for the U.S. population in 1860 and in 1980.

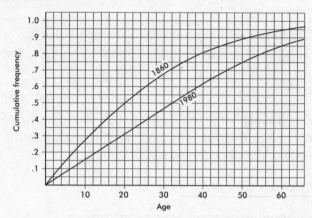

How do the medians and interquartile ranges compare?

Answer: Looking across from .5 on the vertical axis, we see that in 1860 half the population was under the age of 20, while in 1980 all the way up to age 32 must be included to encompass half the population. Looking across from .25 and .75 on the vertical axis, we see that for 1860, $Q_1 = 9$ and $Q_3 = 35$ and so the interquartile range is $35 - 9 = 26$ years, while for 1980, $Q_1 = 16$ and $Q_3 = 50$ and so the interquartile range is $50 - 16 = 34$ years.

Questions on Topic Three: Comparing Distributions

Multiple-Choice Questions

Directions: The questions or incomplete statements that follow are each followed by five suggested answers or completions. Choose the response that best answers the question or completes the statement.

1. The following double bar graph shows the choice of heating fuel for new homes completed during the indicated years. In each pair the top bar indicates electricity while the lower bar indicates gas.

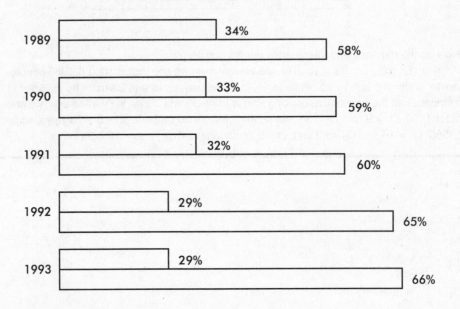

Which of the following are true statements?

  I. The percentage of new homes using electricity has never risen from one year to another.
  II. The percentage of new homes using gas has increased every year.
  III. The percentage of new homes using fuels other than electricity and gas has never risen from one year to another.

(A) I and II
(B) II and III
(C) I and III
(D) I, II, and III
(E) None of the above gives the complete set of true responses.

For Questions 2 and 3 consider the following two histograms:

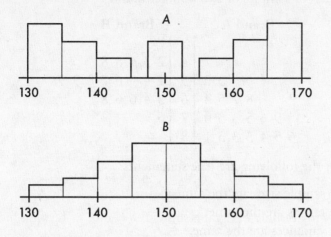

2. Which of the following statements are true?

   I. Both sets have the same mean.
   II. Both sets have the same range.
   III. Both sets have the same variance.

  (A) I only
  (B) I and II
  (C) I and III
  (D) I, II, and III
  (E) None of the above gives the complete set of true responses.

3. Which of the following statements are true?

   I. The empirical rule applies only to set A.
   II. You can be sure that the standard deviation of set A is greater than 5.
   III. You can be sure that the standard deviation of set B is greater than 5.

  (A) I only
  (B) II only
  (C) I and II
  (D) I and III
  (E) II and III

4. Consider the following back-to-back stemplots comparing car battery lives (in months) of samples of two popular brands.

Brand A		Brand B
	3	7
7	4	2 3 4 8 8
3 2	5	1 4 5 6 7 8 9 9
8 7 5 4	6	3 4 6 6 8
9 6 5 3 3 0	7	6
6 5 4 3 3 3 1	8	

Which of the following are true statements?

   I. The sample sizes are the same.
   II. The ranges are the same.
   III. The variances are the same.
   IV. The means are the same.
   V. The medians are the same.

(A) I and II
(B) I and IV
(C) II and V
(D) III and V
(E) I, II, and III

5. Consider the following parallel boxplots illustrating the daily temperatures (in degrees Fahrenheit) of an upstate New York city during January and July.

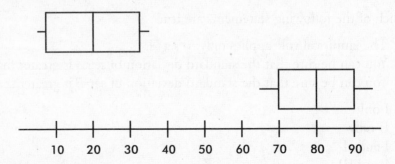

Which of the following are true statements?

   I. The ranges are the same.
   II. The interquartile ranges are the same.
   III. Because of symmetry, the medians are the same.

(A) I only
(B) II only
(C) I and II
(D) I and III
(E) II and III

For Questions 6 and 7 consider the following parallel boxplots of gasoline mileage for three car makes:

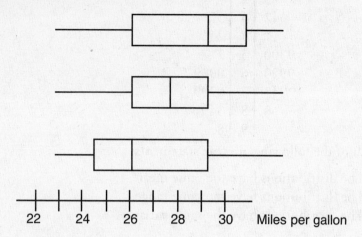

6. Which of the following are true statements?

   I. All three have the same range.
   II. All three have the same interquartile range.
   III. The difference in the medians between the first and third distributions is equal to the interquartile range of the second distribution.

   (A) I and II
   (B) I and III
   (C) II and III
   (D) I, II, and III
   (E) None of the above gives the complete set of true responses.

7. Which of the following are true statements?

   I. All three are symmetric.
   II. The first is skewed to the left while the third is skewed to the right.
   III. The second is skewed on both sides.

   (A) I only
   (B) II only
   (C) III only
   (D) II and III
   (E) None of the above gives the complete set of true responses.

8. Consider the following back-to-back stemplot:

```
      73 | 2 |
     642 | 3 | 37
       7 | 4 | 246
    9300 | 5 | 7
    9920 | 6 | 0039
     943 | 7 | 0299
       8 | 8 | 349
         | 9 | 8
```

Which of the following are true statements?

I. The distributions have the same mean.
II. The distributions have the same range.
III. The distributions have the same variance.

(A) II only
(B) I and II
(C) I and III
(D) II and III
(E) I, II, and III

Answer Key

1. **D**	3. **E**	5. **A**	7. **B**
2. **B**	4. **A**	6. **B**	8. **D**

Answers Explained

1. **(D)** The percentage of homeowners choosing electricity has gone down from 34% to 33% to 32% to 29%, where it has settled for another year. The percentage of homeowners using gas has risen steadily from 58% to 66%. Together, electricity and gas have captured from 92% to 95% of the market, leaving 8%, 8%, 8%, 6%, and 5% of the market for other heating fuels such as oil.

2. **(B)** Both sets are symmetric about 150 and so have the same mean. Both sets have the range 170–130 = 40. Set *A* is much more spread out than set *B*, and so set *A* has the greater variance.

3. **(E)** The empirical rule applies to bell-shaped data like those found in set *B*, not in set *A*. For bell-shaped data, 95% of the values fall within two standard deviations of the mean, and 99.7% within three. However, in the histogram for set *B* one sees that 95% of the data are not between 140 and 160 and 99.7% are not between 135 and 165. Thus the standard deviation for set *B* must be greater than 5. The standard deviation for set *A* is even larger, and so it too must be greater than 5.

4. **(A)** Both sets have 20 elements. The ranges, 76–37 = 39 and 86–47 = 39, are equal. Brand A clearly has the larger mean and median, and with its skewness it also has the larger variance.

5. **(A)** Both ranges are 30. January has an interquartile range of 25, while July has an interquartile range of 20 (but to compare the interquartile ranges, you need only visually compare the lengths of the boxes). The median temperatures for January and July are 20 and 80, respectively.

6. **(B)** The range is the distance between the tips of the two whiskers, and so all three ranges are equal. The interquartile range is the length of the box, and so these three values are not all equal. The median of the first distribution is equal to Q_3 of the second, while the median of the third distribution is equal to Q_1 of the second, and so the difference of the medians is $Q_3 - Q_1$, which is the interquartile range of the second.

7. **(B)** In the first distribution the median is far to the right, and so the scores are concentrated there, while both the lower whisker and between the median and first quartile are spread out; thus it is skewed to the left. Similarly the third distribution is skewed to the right. Only the middle distribution looks symmetric around its median. There is no such thing as being skewed to both sides.

8. **(D)** The two sets of data have different means, but they have identical shapes and thus the same variability, including both range and variance (and standard deviation). Note that adding 10 to each score in the left set results in the right set. Thus the means differ by 10, but the measures of variability remain the same.

Free-Response Questions

> ***Directions:*** You must show all work and indicate the methods you use. You will be graded on the correctness of your methods and on the accuracy of your final answers.

Four Open-Ended Questions

1. Census data was used to compare the marital status versus age of residents of a small town in rural upstate New York. Resulting histograms are given below.

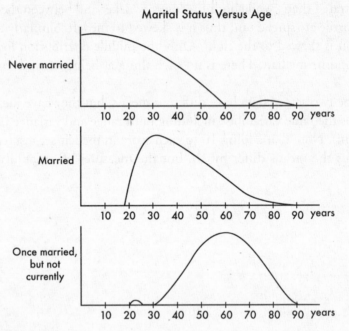

Write a few sentences comparing the distributions of ages of people in this town who have never been married, who are currently married, and who are no longer married.

2. During one season, 20 top NBA rebounding leaders averaged the following numbers of rebounds per game: 9.4, 9.6, 9.7, 10.3, 10.4, 10.6, 10.6, 10.6, 10.8, 10.8, 10.9, 10.9, 10.9, 11.0, 11.1, 11.4, 12.6, 13.4, 14.1, and 16.8, while 20 top assist leaders averaged the following number of assists per game: 6.1, 6.2, 6.4, 6.9, 7.1, 7.2, 7.4, 7.5, 7.6, 7.7, 7.7, 7.9, 8.2, 8.3, 8.7, 8.8, 9.3, 9.4, 10.2, and 12.3. Is it easier for the best NBA players to get a rebound or an assist? Answer the question and compare these data using back-to-back stemplots.

3. In 1980 automobile registrations by states ranged from a low of 262,000 (Alaska) to a high of 16,873,000 (California) with a median of 2,329,000 (Iowa), a 25th percentile of 834,000, and a 75th percentile of 3,749,000. In 1990 the registrations ranged from 462,000 (Vermont) to 21,926,000 (California) with a median of 2,649,000 (Oklahoma), a 25th percentile of

1,054,000, and a 75th percentile of 4,444,000. Draw parallel boxplots and compare the distributions. Comment on whether or not any of the six named states are outliers in their respective distributions.

4. Cumulative frequency graphs of the ages of people on three different Caribbean cruises (A, B, and C) are given below:

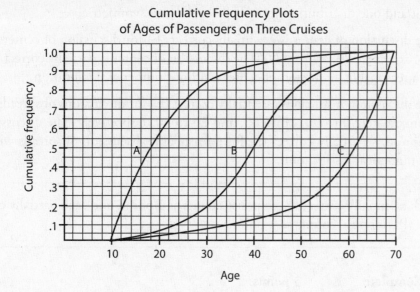

Cumulative Frequency Plots
of Ages of Passengers on Three Cruises

Write a few sentences comparing the distributions of ages of people on the three cruises.

Answers Explained (with illustrations of AP scoring)

1. A complete answer compares shape, center, and spread.

   Shape: The "never married" and "married" distributions are skewed right (toward the higher ages), while the "once married" distribution appears more bell-shaped and symmetric. The "never married" has a gap between 60 and 70 with a cluster between 70 and 90, while the "once married" has a small gap between 25 and 30 with perhaps outliers between 20 and 25.

   Center: Considering the center to be a value separating the area under the histogram roughly in half, the centers of the "never married," "married," and "once married" distributions are approximately 20, 40, and 60, respectively.

   Spread: The spreads of the "never married," "married," and "once married" distributions are approximately 0–90, 18–85, and 20–90, respectively.

Scoring Guide

The discussion of shape is essentially correct (E) for mention of two distributions skewed right and one distribution more bell-shaped and symmetric, and some mention of gaps, clusters, or possible outliers. The discussion of shape is partially correct (P) for mention just of two distributions skewed right and one distribution more bell-shaped and symmetric.

The discussion of center is essentially correct (E) for discussion of centers of approximately 20, 40, and 60. The discussion of center is partially correct (P) for a reasonable discussion of centers without numerical justification.

The discussion of spread is essentially correct (E) for discussions of spreads of approximately 0 to 90, 18 to 85, and 20 to 90, respectively. The discussion of spread is partially correct (P) for a reasonable discussion of spreads without numerical justification.

Each essentially correct (E) response counts as 1 point, and each partially correct (P) response counts as ½ point.

---

4 Complete   3 points.

3 Substantial   2½ points.

2 Developing   2 points.

1 Minimal   1 points.

0   0 points.

If the total is ½ point or 1½ points, look at the strength of the overall communication in deciding whether or not to round up or down.

2.

Assists per game		Rebounds per game

```
          9 4 2 1 | 6  |
  9 7 7 6 5 3 2 1 | 7  |
          8 7 3 2 | 8  |
              4 3 | 9  | 4 6 7
                2 | 10 | 3 4 6 6 6 8 8 9 9 9
                  | 11 | 0 1 4
                3 | 12 | 6
                  | 13 | 4
                  | 14 | 1
                  | 15 |
                  | 16 | 8
```

Clearly the number of rebounds that the best rebounders get per game is greater than the number of assists that the best assist leaders get per game. A complete answer compares shape, center, and spread.

Shape: Both distributions are skewed right (toward higher values). The lower 16 values of each distribution look symmetric and bell-shaped. Each distribution has a clear outlier on the upper end (summary statistics were not asked for, but would actually have shown four upper-end outliers in the "rebound" distribution).

Center: The center of the "assists" distribution is a little under 8, while the center of the "rebounds" distribution is a little under 11.

Spread: The spreads of the "assists" and "rebounds" distributions are 6.1 to 12.3 and 9.4 to 16.8, respectively.

Scoring Guide

The back-to-back stemplot is essentially correct (E) for a correct plot and labeling, but is only partially correct (P) if the labeling is missing.

The discussion of shape is essentially correct (E) for stating that both are skewed right and both appear to have outliers on the high end. The discussion of shape is partially correct (P) for correctly discussing skewness but not mentioning possible outliers.

The discussion of center is essentially correct (E) for stating tht the center of the "assists" distribution is less than the center of the "rebounds" distribution and giving some numerical justification. The discussion of center is partially correct (P) for a correct comparison of centers without a numerical justification.

The discussion of spread is essentially correct (E) for stating that the spreads are approximately the same or that the spread of the "assists" distribution is less than the spread of the "rebounds" distribution and giving some numerical justification. The discussion of spread is partially correct (P) for a correct comparison of spread without a numerical justification.

Each essentially correct (E) response counts as 1 point, and each partially correct (P) response counts as ½ point.

4 Complete	4 points
3 Substantial	3 points
2 Developing	2 points
1 Minimal	1 point
0	0 points

If a total is between two scores (for example, 3½ points), look at strength of the overall communication in deciding whether or not to round up or down.

3.

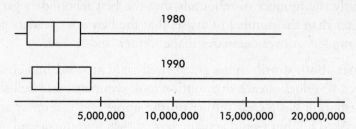

A complete answer compares shape, center, and spread, and notes outliers.

Shape: The most striking aspect is how skewed to the right (towards higher values) both sets of data appear to be.

Center: The median value changed very little from 1980 to 1990 (rising slightly from 2,329,000 to 2,649,000.

Spread: The difference between the lowest value and the median changed very little between 1980 and 1990, while the upper-half spread increased substantially.

Outliers: For 1980, IQR = $Q_3 - Q_1$ = 3,749,000 − 834,000 = 2,915,000, and outliers are any values less than $Q_1 - 1.5(IQR)$ = −3,538,500 or greater than $Q_3 + 1.5(IQR)$ = 8,121,500, so California does represent an outlier. For 1990, IQR = $Q_3 - Q_1$ = 4,444,000 − 1,054,000 = 3,390,000, and outliers are any values less than $Q_1 - 1.5(IQR)$ = −4,031,000 or greater than $Q_3 + 1.5(IQR)$ = 9,529,000, so again California represents an outlier.

Scoring Guide

Correctly drawing both boxplots (unnecessary to show outliers) gives an essentially correct (E), while correctly drawing only one gives a partially correct (P).

Stating that both distributions are skewed right gives an essentially correct (E).

The discussion of center is essentially correct (E) for noting that the median value changed very little from 1980 to 1990 (rising slightly from 2,329,000 to 2,649,000). The discussion of center is partially correct (P) for a correct conclusion about median with no numerical justification.

The discussion of spread is essentially correct (E) for noting that the difference between the lowest value and the median changed very little between 1980 and 1990, while the upper-half spread increased substantially. The discussion of spread is partially correct (P) for simply noting that the spread has increased.

The discussion of outliers is essentially correct (E) for noting that CA is an outlier in both distributions and giving a numerical justification. The discussion of outliers is partially correct (P) for noting that CA is an outlier in both distributions but not giving a numerical justification.

Each essentially correct (E) response counts as 1 point, and each partially correct (P) response counts as ½ point.

---

4 **Complete**	4½ or 5 points	
3 **Substantial**	3½ or 4 points	
2 **Developing**	2½ or 3 points	
1 **Minimal**	1½ or 2 points	
0	0 or ½ point	

If the total is 1 point, look at strength of the overall communication in deciding whether or not to round up or down.

4. A complete answer compares shape, center, and spread.

Shape: Cruise A, for which the cumulative frequency plot rises steeply at first, has more younger passengers, and thus a distribution skewed to the right (towards the higher ages). Cruise C, for which the cumulative frequency plot rises slowly at first and then steeply towards the end, has more older passengers, and thus a distribution skewed to the left (towards the younger ages). Cruise B, for which the cumulative frequency plot rises slowly at each end and steeply in the middle, has a more bell-shaped distribution.

Center: Considering the center to be a value separating the area under the histogram roughly in half, the centers will correspond to a cumulative frequency of 0.5. Reading across from 0.5 to the intersection of each graph, and then down to the *x*-axis, shows centers of approximately 18, 40, and 61 years, respectively.

Spread: The spreads of the age distributions of all three cruises are the same: from 10 to 70 years.

Scoring Guide

The discussion of shape is essentially correct (E) for correctly identifying which distribution is skewed left, skewed right, and more bell-shaped, and gives a correct justification based on the cumulative frequency plots. The discussion of shape is partially correct (P) for correctly identifying which distribution is skewed left, skewed right, and more bell-shaped without giving a good justification.

The discussion of center is essentially correct (E) for correctly noting that Cruise A has the lowest center and Cruise C the greatest center, and giving some numerical justification. The discussion of center is partially correct (P) for correctly noting that Cruise A has the lowest center and Cruise C the greatest center but without giving a good justification.

The discussion of spread is essentially correct (E) for correctly noting that all three cruises have the same spread and giving some numerical justification. The discussion of spread is partially correct (P) for correctly noting that all three cruises have the same spread but without a good justification.

Each essentially correct (E) response counts as 1 point, and each partially correct (P) response counts as ½ point.

4 **Complete**   3 points

3 **Substantial**   2½ points

2 **Developing**   2 points

1 **Minimal**   1 point

0   0 points

If the total is ½ point or 1½, look at strength of the overall communication in deciding whether or not to round up or down.

An Investigative Task

1. Draw a back-to-back stemplot of two sets of data whose parallel boxplots are the following:

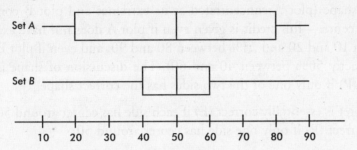

Solution

Both sets appear to be roughly symmetric with median 50, low score 10, and high score 90. Set *A* is concentrated at the extremes, with 25% of its scores between 10 and 20 and 25% between 80 and 90. Set *B* is concentrated in the middle, with 50% of its scores between 40 and 60. Some care has to be taken in the choice of scores. One of many possible stemplots is

Set *A*		Set *B*
9 8 7 6 5 4 3 2 1 0	1	0 1 2
3 2 1	2	3 4 5 6
7 6 5 4	3	7 8 9
9 9 8	4	1 2 3 4 5 6 7 8 9 9
2 1 1	5	1 1 2 3 4 5 6 7 8 9
6 5 4 3	6	1 2 3
9 8 7	7	4 5 6
9 8 7 6 5 4 3 2 1	8	7 8 9
0	9	0

Scoring Guide

The shape is essentially correct (E) for a back-to-back stemplot correctly illustrating shape (plot A concentrated at its extremes and plot B concentrated near its center—full credit is given even if plot A does not have exactly 25% between 10 and 20 and 25% between 80 and 90, and even if plot B does not have exactly 50% between 40 and 60). The discussion of shape is partially correct (P) if only one of the two sides has the correct shape.

The center is essentially correct (E) if each side has center around 50 and partially correct (P) if only one side has center around 50.

The spread is essentially correct if each side has a minimum value of 10 and a maximum value of 90 and partially correct (P) if only one side has these values.

Each essentially correct (E) response counts as 1 point, and each partially correct (P) response counts as ½ point.

4	**Complete**	3 points
3	**Substantial**	2½ points
2	**Developing**	2 points
1	**Minimal**	1 point
0		0 points

If the total is ½ point or 1½ points, look at the strength of the overall communication in deciding whether or not to round up or down.

Exploring Bivariate Data

- Scatterplots
- Correlation and Linearity
- Least Squares Regression Line
- Residual Plots
- Outliers and Influential Points
- Transformations to Achieve Linearity

Our studies so far have been concerned with measurements of a single variable. However, many important applications of statistics involve examining whether two or more variables are related to one another. For example, is there a relationship between the smoking histories of pregnant women and the birth weights of their children? Between SAT scores and success in college? Between amount of fertilizer used and amount of crop harvested?

Two questions immediately arise. First, how can the strength of an apparent relationship be measured? Second, how can an observed relationship be put into functional terms? For example, a real estate broker might not only wish to determine whether a relationship exists between the prime rate and the number of new homes sold in a month but might also find useful an expression with which to predict the number of home sales given a particular value of the prime rate.

A graphical display, called a *scatterplot*, gives an immediate visual impression of a possible relationship between two variables, while a numerical measurement, called a *correlation coefficient*, is often used as a quantitative value of the strength of a linear relationship. In either case, evidence of a relationship is not evidence of causation.

SCATTERPLOTS

Suppose a relationship is perceived between two quantitative variables X and Y, and we graph the pairs (x, y). We are interested in the strength of this relationship, the scatterplot arising from the relationship, and any deviation from the basic pattern of this relationship. In this topic we examine whether the relationship can be reasonably explained in terms of a linear function, that is, one whose graph is a straight line.

For example, we might be looking at a plot such as

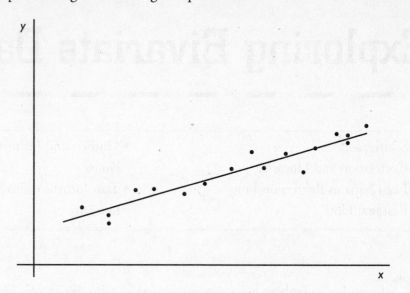

We need to know what the term *best-fitting straight line* means and how we can find this line. Furthermore, we want to be able to gauge whether the relationship between the variables is strong enough so that finding and making use of this straight line is meaningful.

Patterns in Scatterplots

When larger values of one variable are associated with larger values of a second variable, the variables are called *positively associated*. When larger values of one are associated with smaller values of the other, the variables are called *negatively associated*.

EXAMPLE 4.1

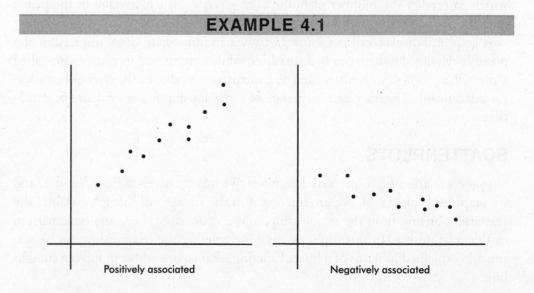

The strength of the association is gauged by how close the plotted points are to a straight line.

EXAMPLE 4.2

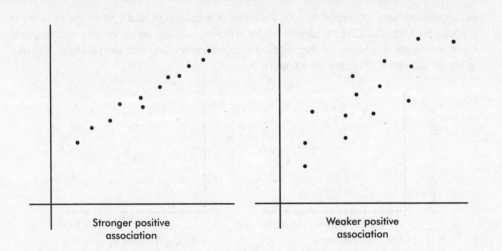

Stronger positive
association

Weaker positive
association

Sometimes different dots in a scatterplot are labeled with different symbols or different colors to show a categorical variable. The resulting *labeled scatterplot* might distinguish between men and women, between stocks and bonds, and so on.

EXAMPLE 4.3

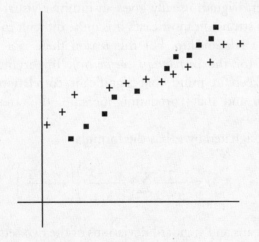

The above diagram is a labeled scatterplot distinguishing men with plus signs and women with square dots to show a categorical variable.

When analyzing the overall pattern in a scatterplot, it is also important to note *clusters* and *outliers*.

EXAMPLE 4.4

An experiment was conducted to note the effect of temperature and light on the potency of a particular antibiotic. One set of vials of the antibiotic was stored under different temperatures, but under the same lighting, while a second set of vials was stored under different lightings, but under the same temperature.

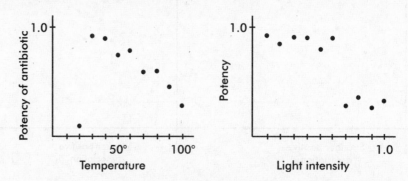

In the first scatterplot note the linear pattern with one outlier far outside this pattern. A possible explanation is that the antibiotic is more potent at lower temperatures, but only down to a certain temperature at which it drastically loses potency.

In the second histogram note the two clusters. It appears that below a certain light intensity the potency is one value, while above that intensity it is another value. In each cluster there seems to be no association between intensity and potency.

CORRELATION AND LINEARITY

Although a scatter diagram usually gives an intuitive visual indication when a linear relationship is strong, in most cases it is quite difficult to visually judge the specific strength of a relationship. For this reason there is a mathematical measure called *correlation* (or the *correlation coefficient*). Important as correlation is, we always need to keep in mind that significant correlation does not necessarily indicate *causation* and that correlation measures the strength only of a *linear* relationship.

Correlation, designated by r, has the formula

$$r = \frac{1}{n-1} \sum \left(\frac{x_i - \overline{x}}{s_x} \right) \left(\frac{y_i - \overline{y}}{s_y} \right)$$

in terms of the means and standard deviations of the two sets. We note that the formula is actually the sum of the products of the corresponding z-scores divided by 1 less than the sample size. However, you should be able to quickly calculate correlation using the statistical package on your calculator.

Note from the formula that correlation does not distinguish between which variable is called x and which is called y. The formula is also based on standardized scores (z-scores), and so changing units does not change the correlation. Finally, since means and standard deviations can be strongly influenced by outliers, correlation is also strongly affected by extreme values.

The value of r always falls between −1 and +1, with −1 indicating perfect negative correlation and +1 indicating perfect positive correlation. It should be stressed that a correlation at or near zero doesn't mean there isn't a relationship between the variables; there may still be a strong *nonlinear* relationship.

EXAMPLE 4.5

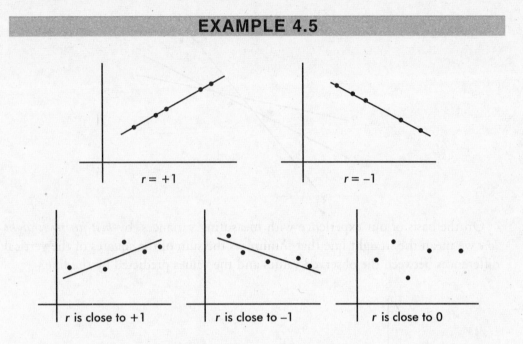

It can be shown that r^2, called the *coefficient of determination*, is the ratio of the variance of the predicted values $\hat{y}$ to the variance of the observed values y. That is, there is a partition of the y-variance, and r^2 is the proportion of this variance that is predictable from a knowledge of x. Alternatively, we can say that r^2 gives the percentage of variation in y that is explained by the variation in x. In either case, always interpret r^2 in context of the problem. Remember when calculating r from r^2 that r may be positive or negative.

While the correlation r is given as a decimal between −1.0 and 1.0, the coefficient of determination r^2 is usually given as a percentage. An r^2 of 100% is a perfect fit, with all the variation in y explained by variation in x. How large a value of r^2 is desirable depends on the application under consideration. While scientific experiments often aim for an r^2 in the 90% or above range, observational studies with r^2 of 10% to 20% might be considered informative. Note that while a correlation of .6 is twice a correlation of .3, the corresponding r^2 of 36% is *four* times the corresponding r^2 of 9%.

LEAST SQUARES REGRESSION LINE

What is the best-fitting straight line that can be drawn through a set of points?

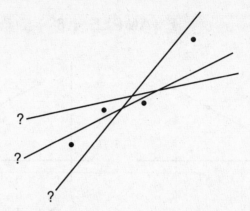

On the basis of our experience with measuring variances, by *best-fitting straight line* we mean the straight line that minimizes the sum of the squares of the vertical differences between the observed values and the values predicted by the line.

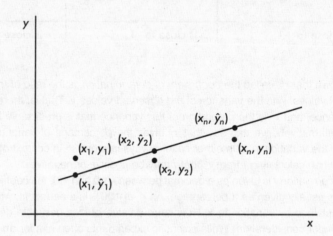

That is, in the above figure, we wish to minimize

$$(y_1 - \hat{y}_1)^2 + (y_2 - \hat{y}_2)^2 + \cdots + (y_n - \hat{y}_n)^2$$

It is reasonable, intuitive, and correct that the best-fitting line will pass through $(\bar{x}, \bar{y})$ where $\bar{x}$ and $\bar{y}$ are the means of the variables X and Y. Then, from the basic expression for a line with a given slope through a given point, we have

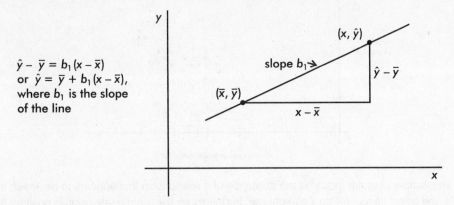

$\hat{y} - \bar{y} = b_1(x - \bar{x})$
or $\hat{y} = \bar{y} + b_1(x - \bar{x})$,
where b_1 is the slope
of the line

The slope b_1 can be determined from the formula

$$b_1 = r\frac{s_y}{s_x}$$

where r is the correlation and s_x and s_y are the standard deviations of the two sets. That is, each standard deviation change in x results in a change of r standard deviations in y. If you graph z-scores for the y-variable against z-scores for the x-variable, the slope of the regression line is precisely r, and, in fact, the linear equation becomes $z_y = rz_x$.

This best-fitting straight line, that is, the line that minimizes the sum of the squares of the differences between the observed values and the values predicted by the line, is called the *least squares regression line* or simply the *regression line*. It can be calculated directly by entering the two data sets and using the statistics package on your calculator.

EXAMPLE 4.6

An insurance company conducts a survey of 15 of its life insurance agents. The average number of minutes spent with each potential customer and the number of policies sold in a week are noted for each agent. Letting X and Y represent the average number of minutes and the number of sales, respectively, we have

X: 25 23 30 25 20 33 18 21 22 30 26 26 27 29 20
Y: 10 11 14 12   8 18   9 10 10 15 11 15 12 14 11

Find the equation of the best-fitting straight line for the data.

(continued)

Answer: Plotting the 15 points (25, 10), (23, 11), . . . , (20, 11) gives an intuitive visual impression of the relationship:

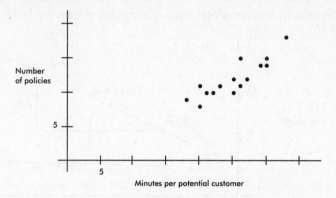

This scatter diagram indicates the existence of a relationship that appears to be *linear*; that is, the points lie roughly on a straight line. Furthermore, the linear relationship is *positive*; that is, as one variable increases, so does the other (the straight line slopes upward).

Using a calculator, we find the correlation to be $r = .8836$, the coefficient of determination to be $r^2 = .78$ (indicating that 78% of the variation in the number of policies is explained by the variation in the number of minutes spent), and the regression line to be

$$\hat{y} = \bar{y} + b_1(x - \bar{x})$$
$$= 12 + 0.5492(x - 25)$$
$$= -1.73 + 0.5492x$$

We also write: Policies = $-1.73 + 0.5492$ Minutes

Adding this to our scatterplot yields

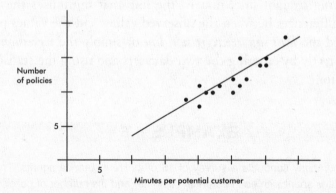

Thus, for example, we might predict that agents who average 24 minutes per customer will average $0.5492(24) - 1.73 = 11.45$ sales per week. We also note that each additional minute spent seems to produce an average 0.5492 extra sale.

EXAMPLE 4.7

Following are advertising expenditures and total sales for six detergent products:

| Advertising ($1000) (x): | 2.3 | 5.7 | 4.8 | 7.3 | 5.9 | 6.2 |
| Total sales ($1000) (y): | 77 | 105 | 96 | 118 | 102 | 95 |

Predict the total sales if $5000 is spent on advertising and interpret the slope of the regression line. What if $100,000 is spent on advertising?

Answer: With your calculator, the equation of the regression line is found to be

$$\hat{y} = 98.833 + 7.293(x - 5.367) = 59.691 + 7.293x$$

(A calculator like the TI-84, with less round-off error, directly gives $\hat{y} = 59.683 + 7.295x$.)

It is also worthwhile to replace the x and y with more appropriately named variables, resulting, for example, in

$$Sales = 59.691 + 7.293 \text{ Advcost}$$

The regression line predicts that if $5000 is spent on advertising, the resulting total sales will be 7.293(5) + 59.691 = 96.156 thousands of dollars ($96,156).

The slope of the regression line indicates that every extra $1000 spent on advertising will result in $7293 of added sales.

If $100,000 is spent on advertising, we calculate 7.293(100) + 59.691 = 788.991 thousands of dollars (≈$789,000). How much confidence should we have in this answer? Not much! We are trying to use the regression line to predict a value far outside the range of the data values. This procedure is called *extrapolation* and must be used with great care.

It should be noted that when we use the regression line to predict a y-value for a given x-value, we are actually predicting the *mean* y-value for that given x-value. For any given x-value, there are many possible y-values, and we are predicting their mean. So if $5000 is spent many times on advertising, various resulting total sales figures may result, but their predicted average is $96,156.

EXAMPLE 4.8

A random sample of 30 U.S. farm regions surveyed during the summer of 2003 produced the following statistics:

Average temperature (°F) during growing season: $\bar{x} = 81$, $s_x = 3$
Average corn yield per acre (bu.): $\bar{y} = 131$, $s_y = 5$
Correlation $r = .32$

Based on this study, what is the mean predicted corn yield for a region where the average growing season temperature is 76.5°F?

Answer: 76.5 is $\frac{76.5 - 81}{3} = -1.5$ standard deviations below the average temperature reported in the study, so with a correlation of $r = .32$, the predicted corn yield is .32(–1.5) = –0.48 standard deviations below the average corn yield, or 131 – 0.48(5) = 128.6 bushels per acre.

Alternatively, we could have found the linear regression equation relating these variables: $slope = r\frac{s_y}{s_x} = .32\left(\frac{5}{3}\right) \approx 0.533$, intercept ≈ 131 – 81(0.533) ≈ 87.8, and thus Yield = 0.533 Temp + 87.8. Then 0.533(76.5) + 87.8 ≈ 128.6.

RESIDUAL PLOTS

The difference between an observed and predicted value is called the *residual*. When the regression line is graphed on the scatterplot, the residual of a point is the vertical distance the point is from the regression line.

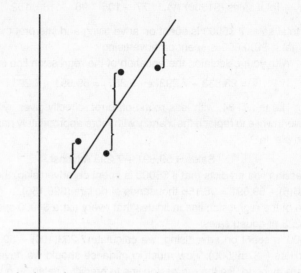

The regression line is the line that minimizes the sum of the squares of the residuals.

EXAMPLE 4.9

We calculate the predicted values from the regression line in Example 4.7 and subtract from the observed values to obtain the residuals:

x	2.3	5.7	4.8	7.3	5.9	6.2
y	77	105	96	118	102	95
$\hat{y}$	76.5	101.3	94.7	112.9	102.7	104.9
$y - \hat{y}$	0.5	3.7	1.3	5.1	−0.7	−9.9

Note that the sum of the residuals is

$$0.5 + 3.7 + 1.3 + 5.1 - 0.7 - 9.9 = 0.0$$

The above equation is true in general; that is, *the sum and thus the mean of the residuals is always zero.*

The notation for residuals is $\hat{e}_i = y_i - \hat{y}_i$ and so $\sum_{i=1}^{n} \hat{e}_i = 0$. The standard deviation of the residuals is calculated as follows:

$$s_e = \sqrt{\frac{\sum e_i^2}{n-2}} = \sqrt{\frac{\sum (y_i - \hat{y}_i)^2}{n-2}}$$

s_e gives a measure of how the points are spread around the regression line.

Plotting the residuals gives further information. In particular, a residual plot with a definite pattern is an indication that a nonlinear model will show a better fit to the data than the straight regression line. In addition to whether or not the residuals are randomly distributed, one should look at the balance between positive and

(continued)

negative residuals and also the size of the residuals in comparison to the associated *y*-values.

The residuals can be plotted against either the *x*-values or the $\hat{y}$-values (since $\hat{y}$ is a linear transformation of x, the plots are identical except for scale and a left-right reversal when the slope is negative).

It is also important to understand that a linear model may be appropriate, but weak, with a low correlation. And, alternatively, a linear model may not be the best model (as evidenced by the residual plot), but it still might be a very good model with high r^2.

EXAMPLE 4.10

Suppose the drying time of a paint product varies depending on the amount of a certain additive it contains.

Additive (oz), *x*:	1	2	3	4	5	6	7	8	9	10
Drying time (hr), *y*:	4	2.1	1.5	1	1.2	1.7	2.5	3.6	4.9	6.1

Using a calculator, we find the regression line $y = 0.327x + 1.062$, and we find the residuals:

x:	1	2	3	4	5	6	7	8	9	10
$y - \hat{y}$:	2.61	0.38	−0.54	−1.37	−1.5	−1.32	−0.85	−0.08	0.89	1.77

The resulting residual plot shows a strong pattern:

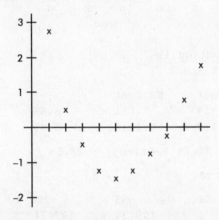

This pattern indicates that a nonlinear model will be a better fit than a straight-line model. A scatterplot of the original data shows clearly what is happening:

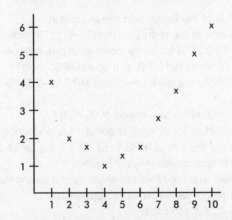

The ability to interpret computer output is important not only to do well on the AP Statistics exam, but also to understand statistical reports in the business and scientific world.

EXAMPLE 4.11

Miles per gallon versus speed for a new model automobile is fitted with a least squares regression line. The graph of the residuals and some computer output for the regression are as follows:

Regression Analysis: MPG Versus Speed

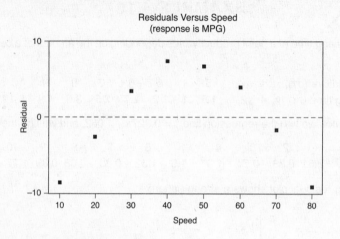

Residuals Versus Speed
(response is MPG)

```
The regression equation is
MPG = 38.9 - 0.218 Speed

Predictor        Coef      SE Coef           T          P
Constant       38.929        5.651        6.89      0.000
Speed         -0.2179        0.1119       -1.95      0.099

S = 7.252   R-Sq = 38.7%  R-Sq(adj) = 28.5%

Analysis of Variance

Source           DF          SS          MS         F    P
Regression        1      199.34      199.34      3.79  0.099
Residual Error    6      315.54       52.59
Total             7      514.88
```

a. Interpret the slope of the regression line in context.
   Answer: The slope of the regression line is −0.2179, indicating that, on average, the MPG drops by 0.2179 for every increase of one mile per hour in speed.
b. What is the mean predicted MPG at a speed of 30 mph?
   Answer: At 30 mph, the mean predicted MPG is −0.2179(30) + 38.929, or about 32.4 MPG.
c. What was the actual MPG at a speed of 30 mph?
   Answer: The residual for 30 mph is about +3.5, and since residual = actual − predicted, we estimate the actual MPG to be 32.4 + 3.5, or about 36 MPG.
d. Is a line the most appropriate model? Explain.
   Answer: The fact that the residuals show such a strong curved pattern indicates that a nonlinear model would be more appropriate.
e. What does "S = 7.252" refer to?
   Answer: The standard deviation of the residuals, $s_e = 7.252$, is a "typical value" of the residuals and gives a measure of how the points are spread around the regression line.

EXAMPLE 4.12

The number of youngsters playing Little League baseball in Ithaca, New York, during the years 1995–2003 is fitted with a least squares regression line. The graph of the residuals and some computer output for their regression are as follows:

Regression Analysis: Number of Players Versus Years Since 1995

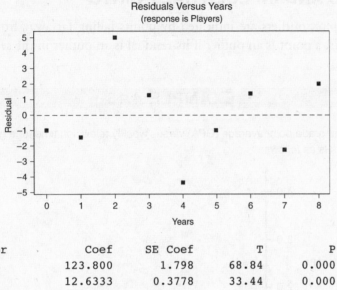

Predictor	Coef	SE Coef	T	P
Constant	123.800	1.798	68.84	0.000
Years	12.6333	0.3778	33.44	0.000

S = 2.926   R–Sq = 99.4%   R–Sq(adj) = 99.3%

Analysis of Variance

Source	DF	SS	MS	F	P
Regression	1	9576.1	9576.1	1118.45	0.000
Residual Error	7	59.9	8.6		
Total	8	9636.0			

a. Does it appear that a line is an appropriate model for the data? Explain.
    Answer: Yes. R–Sq = 99.4% is large, and the residual plot shows no pattern. Thus a linear model is appropriate.
b. What is the equation of the regression line (in context)?
    Answer: # of players = 123.8 + 12.6 (years since 1995)
c. Interpret the slope of the regression line in the context of the problem.
    Answer: The slope of the regression line is 12.6, indicating that, on average, the predicted number of children playing Little League baseball in Ithaca increased by 12 or 13 players per year during the 1995–2003 time period.
d. Interpret the *y*-intercept of the regression line in the context of the problem.
    Answer: The *y*-intercept, 123.8, refers to the year 1995. Thus the number of players in Little League in Ithaca in 1995 was predicted to be around 124.
e. What is the predicted number of players in 1997?
    Answer: For 1997, $x = 2$, so the predicted number of players is 12.6(2) + 123.8 = 149.
f. What was the actual number of players in 1997?
    Answer: The residual for 1997 ($x = 2$) from the residual plot is +5, so actual – predicted = 5, and thus the actual number of players in 1997 must have been 5 + 149 = 154.

(continued)

g. What years, if any, did the number of players decrease from the previous year? Explain.

Answer: The number would decrease if one residual were more than 12.6 greater than the next residual. This never happens, so the number of players never decreased.

OUTLIERS AND INFLUENTIAL POINTS

In a scatterplot, outliers are indicated by points falling far away from the overall pattern. That is, a point is an outlier if its residual is an outlier in the set of residuals.

EXAMPLE 4.13

A scatterplot of grade point average (GPA) versus weekly television time for a group of high school seniors is as follows:

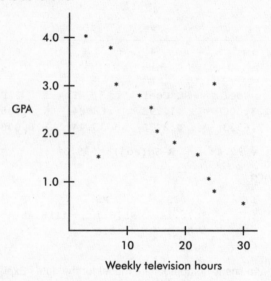

By direct observation of the scatterplot, we note that there are two outliers: one person who watches 5 hours of television weekly yet has only a 1.5 GPA, and another person who watches 25 hours weekly yet has a 3.0 GPA. Note also that while the value of 30 weekly hours of television may be considered an outlier for the television hours variable and the 0.5 GPA may be considered an outlier for the GPA variable, the point (30, 0.5) is *not* an outlier in the regression context because it does not fall off the straight-line pattern.

Scores whose removal would sharply change the regression line are called *influential scores.* Sometimes this description is restricted to points with extreme *x*-values. An influential score may have a small residual but still have a greater effect on the regression line than scores with possibly larger residuals but average *x*-values.

EXAMPLE 4.14

Consider the following scatterplot and regression line for starting salaries versus years of education:

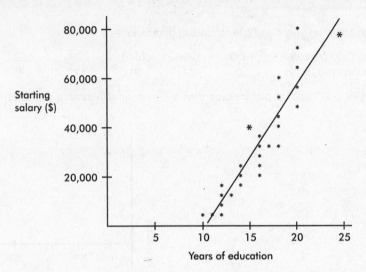

The dotted lines below show what happens if the points (15, 40,000) and (24, 76,000) are removed, respectively.

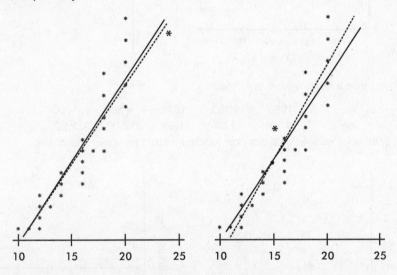

Note that the regression line is greatly affected by the removal of (24, 76,000) but not by the removal of (15, 40,000). Thus (24, 76,000) is an influential score, while (15, 40,000) is not. This is true in spite of the fact that (24, 76,000) is closer to the original regression line than (15, 40,000).

TRANSFORMATIONS TO ACHIEVE LINEARITY

Often a straight-line pattern is not the best model for depicting a relationship between two variables. A clear indication of this problem is when the scatterplot shows a distinctive curved pattern. Another indication is when the residuals show a distinctive pattern rather than a random scattering. In such a case, the nonlinear model can sometimes be revealed by transforming one or both of the variables and then noting a linear relationship. Useful transformations often result from using the *log* or *ln* buttons on your calculator to create new variables.

EXAMPLE 4.15

Consider the following years and corresponding populations:

Year, *x*:	1950	1960	1970	1980	1990
Population (1000s), *y*:	50	67	91	122	165

The scatterplot and residual plot indicate that a nonlinear relationship would be an even stronger model.

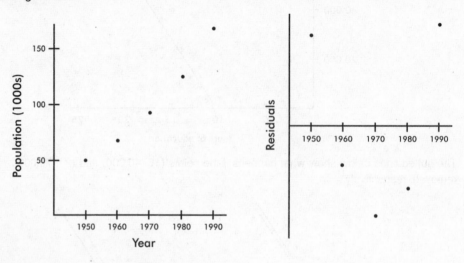

Letting log *y* be a new variable, we obtain

x:	1950	1960	1970	1980	1990
log *y*:	1.70	1.83	1.96	2.09	2.22

The scatterplot and residual plot now indicate a stronger linear relationship.

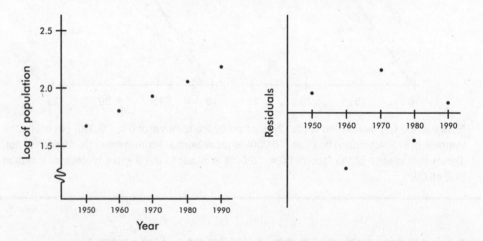

A regression analysis yields log $y = 0.013x - 23.65$. In context we have: log(Pop) = $-23.65 + 0.013$(Year). So, for example, the population predicted for the year 2000 would be calculated log(Pop) = $0.013(2000) - 23.65 = 2.35$, and so Pop = $10^{2.35} \approx 224$ thousand, or 224,000. The "linear" equation log $y = 0.013x - 23.65$ can be re-expressed as $y = 10^{0.013x-23.65}$ [= $10^{-23.65} \times (10^{0.013})^x = 2.2387E-24 \times 1.0304^x$].

There are many useful transformations. For example:

Log y as a linear function of x, log $y = ax + b$, re-expresses as an *exponential*:

$$y = 10^{ax+b} \text{ or } y = b_1 10^{ax} \text{ where } b_1 = 10^b$$

Log y as a linear function of log x, log $y = a \log x + b$, re-expresses as a *power*:

$$y = 10^{a \log x+b} \text{ or } y = b_1 x^a \text{ where } b_1 = 10^b$$

$\sqrt{y}$ as a linear function of x, $\sqrt{y} = ax + b$, re-expresses as a *quadratic*:

$$y = (ax + b)^2$$

$\frac{1}{y}$ as a linear function of x, $\frac{1}{y} = ax + b$, re-expresses as a *reciprocal*:

$$y = \frac{1}{ax + b}$$

y as a linear function of log x, $y = a \log x + b$, is a *logarithmic* function.

EXAMPLE 4.16

What are possible models for the following data?

x	1	2	3	4	5
y	20	60	120	190	280

Answer: A linear fit to x and y gives $\hat{y} = 65x - 61$ with $r = .99$.

A linear fit to x and log y gives log $y = 0.279x + 1.139$ with $r = .98$. This results in an *exponential* relationship:

$$\hat{y} = 10^{0.279x+1.139} = 13.77(10^{0.279x}) = 13.77(1.901^x)$$

A linear fit to log x and log y gives log $y = 1.639 \log x + 1.295$, also with $r = .99$. This results in a *power* relationship:

$$\hat{y} = 10^{1.639 \log x+1.295} = 19.72(x^{1.639})$$

All three models give high correlation and are reasonable fits. Further analysis can be done by examining the residual plots:

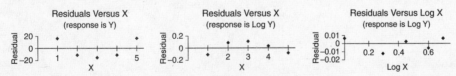

The first two residual plots have distinct curved patterns. The third residual plot illustrates both a more random pattern and smaller residuals. (One can also create residual plots using the derived curved models, but we choose to restrict our attention to residual plots of the linear models.) Among the above three models, the power model $\hat{y} = 19.72(x^{1.639})$, appears to be best.

Questions on Topic Four: Exploring Bivariate Data

Multiple-Choice Questions

Directions: The questions or incomplete statements that follow are each followed by five suggested answers or completions. Choose the response that best answers the question or completes the statement.

1. As reported in *The New York Times* (September 21, 1994, page C10), a study at the University of Toronto determined that, for every 10 grams of saturated fat consumed per day, a woman's risk of developing ovarian cancer rises 20%. What is the meaning of the slope of the appropriate regression line?

   (A) Taking in 10 grams of fat results in a 20% increased risk of developing ovarian cancer.
   (B) Consuming 0 grams of fat per day results in a zero increase in the risk of developing ovarian cancer.
   (C) Consuming 50 grams of fat doubles the risk of developing ovarian cancer.
   (D) Increased intake of fat causes higher rates of developing ovarian cancer.
   (E) A woman's risk of developing ovarian cancer rises 2% for every gram of fat consumed per day.

2. A simple random sample of 35 world-ranked chess players provides the following statistics:

   Number of hours of study per day: $\bar{x} = 6.2$, $s_x = 1.3$
   Yearly winnings: $\bar{y} = \$208{,}000$, $s_y = \$42{,}000$
   Correlation $r = .15$

   Based on this data, what is the resulting linear regression equation?

   (A) Winnings = 178,000 + 4850 Hours
   (B) Winnings = 169,000 + 6300 Hours
   (C) Winnings = 14,550 + 31,200 Hours
   (D) Winnings = 7750 + 32,300 Hours
   (E) Winnings = −52,400 + 42,000 Hours

3. A scatterplot of a company's revenues versus time indicates a possible exponential relationship. A linear regression on $Y = \log(\text{revenue in \$1000})$ against $X =$ years since 2000 gives $y = 0.67 + 0.82x$ with $r = .73$. Which of the following are valid conclusions?

   I. On the average, revenue goes up 0.82 thousand dollars (or \$820) per year.

  II. The predicted revenue in year 2005 is approximately 59 million dollars.

 III. 53% of the variation in revenue can be explained by variation in time.

(A) I only

(B) II only

(C) III only

(D) I and III

(E) None of the above are valid conclusions.

4. Consider the following three scatterplots:

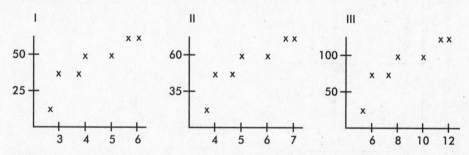

Which has the greatest correlation coefficient?

(A) I

(B) II

(C) III

(D) They all have the same correlation coefficient.

(E) This question cannot be answered without additional information.

5. Suppose the correlation is negative. Given two points from the scatterplot, which of the following is possible?

   I. The first point has a larger x-value and a smaller y-value than the second point.

  II. The first point has a larger x-value and a larger y-value than the second point.

 III. The first point has a smaller x-value and a larger y-value than the second point.

    (A)   I only

    (B)   II only

    (C)   III only

    (D)   I and III

    (E)   I, II, and III

6. Consider the following residual plot:

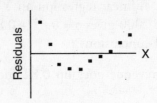

Which of the following scatterplots could have resulted in the above residual plot? (The *y*-axis scales are not the same in the scatterplots as in the residual plot.)

(A)

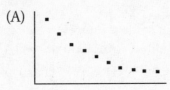

(B)

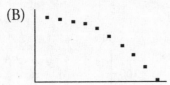

(C)

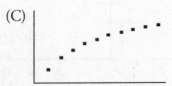

(D)

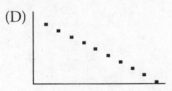

(E) None of these could result in the given residual plot.

7. Suppose the regression line for a set of data, $y = 3x + b$, passes through the point (2, 5). If $\bar{x}$ and $\bar{y}$ are the sample means of the *x*- and *y*-values, respectively, then $\bar{y} =$

(A) $\bar{x}$.

(B) $\bar{x} - 2$.

(C) $\bar{x} + 5$.

(D) $3\bar{x}$.

(E) $3\bar{x} - 1$.

8. Suppose a study finds that the correlation coefficient relating family income to SAT scores is $r = +1$. Which of the following are proper conclusions?

   I. Poverty causes low SAT scores.
   II. Wealth causes high SAT scores.
   III. There is a very strong association between family income and SAT scores.

   (A) I only
   (B) II only
   (C) III only
   (D) I and II
   (E) I, II, and III

9. A study of department chairperson ratings and student ratings of the performance of high school statistics teachers reports a correlation of $r = 1.15$ between the two ratings. From this information we can conclude that

   (A) chairpersons and students tend to agree on who is a good teacher.
   (B) chairpersons and students tend to disagree on who is a good teacher.
   (C) there is little relationship between chairperson and student ratings of teachers.
   (D) there is strong association between chairperson and student ratings of teachers, but it would be incorrect to infer causation.
   (E) a mistake in arithmetic has been made.

10. Which of the following statements about the correlation r are true?

    I. A correlation of .2 means that 20% of the points are highly correlated.
    II. The square of the correlation measures the proportion of the y-variance that is predictable from a knowledge of x.
    III. Perfect correlation, that is, when the points lie exactly on a straight line, results in $r = 0$.

    (A) I only
    (B) II only
    (C) III only
    (D) None of these statements is true.
    (E) None of the above gives the complete set of true responses.

11. Which of the following statements about the correlation r are true?

    I. It is not affected by changes in the measurement units of the variables.
    II. It is not affected by which variable is called x and which is called y.
    III. It is not affected by extreme values.

    (A) I and II
    (B) I and III
    (C) II and III
    (D) I, II, and III
    (E) None of the above gives the complete set of true responses.

12. With regard to regression, which of the following statements about outliers are true?

    I. Outliers in the *y*-direction have large residuals.
    II. A point may not be an outlier even though its *x*-value is an outlier in the *x*-variable and its *y*-value is an outlier in the *y*-variable.
    III. Removal of an outlier sharply affects the regression line.

    (A) I and II
    (B) I and III
    (C) II and III
    (D) I, II, and III
    (E) None of the above gives the complete set of true responses.

13. Which of the following statements about influential scores are true?

    I. Influential scores have large residuals.
    II. Removal of an influential score sharply affects the regression line.
    III. An *x*-value that is an outlier in the *x*-variable is more indicative that a point is influential than a *y*-value that is an outlier in the *y*-variable.

    (A) I and II
    (B) I and III
    (C) II and III
    (D) I, II, and III
    (F) None of the above gives the complete set of true responses.

14. Which of the following statements about residuals are true?

    I. The mean of the residuals is always zero.
    II. The regression line for a residual plot is a horizontal line.
    III. A definite pattern in the residual plot is an indication that a nonlinear model will show a better fit to the data than the straight regression line.

    (A) I and II
    (B) I and III
    (C) II and III
    (D) I, II, and III
    (E) None of the above gives the complete set of true responses.

15. Data are obtained for a group of college freshmen examining their SAT scores (math plus verbal) from their senior year of high school and their GPAs during their first year of college. The resulting regression equation is

    $$\text{GPA} = 1.35 + 0.00161 \, (\text{SAT total}) \quad \text{with} \quad r = .632$$

    What percentage of the variation in GPAs can be explained by looking at SAT scores?

    (A) 0.161%
    (B) 16.1%
    (C) 39.9%
    (D) 63.2%
    (E) This value cannot be computed from the information given.

Questions 16 and 17 are based on the following: The heart disease death rates per 100,000 people in the United States for certain years, as reported by the National Center for Health Statistics, were

Year:	1950	1960	1970	1975	1980
Death rate:	307.6	286.2	253.6	217.8	202.0

16. Which one of the following is a correct interpretation of the slope of the best-fitting straight line for the above data?

   (A) The heart disease rate per 100,000 people has been dropping about 3.627 per year.
   (B) The baseline heart disease rate is 7386.87.
   (C) The regression line explains 96.28% of the variation in heart disease death rates over the years.
   (D) The regression line explains 98.12% of the variation in heart disease death rates over the years.
   (E) Heart disease will be cured in the year 2036.

17. Based on the regression line, what is the predicted death rate for the year 1983?

   (A) 145.8 per 100,000 people
   (B) 192.5 per 100,000 people
   (C) 196.8 per 100,000 people
   (D) 198.5 per 100,000 people
   (E) None of the above

18. Consider the following scatterplot of midterm and final exam scores for a class of 15 students.

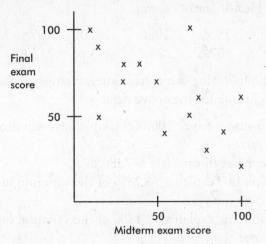

Which of the following are true statements?

I. The same number of students scored 100 on the midterm exam as scored 100 on the final exam.

II. Students who scored higher on the midterm exam tended to score higher on the final exam.

III. The scatterplot shows a moderate negative correlation between midterm and final exam scores.

(A) I and II
(B) I and III
(C) II and III
(D) I, II, and III
(E) None of the above gives the complete set of true responses.

19. If every woman married a man who was exactly 2 inches taller than she, what would the correlation between the heights of married men and women be?

(A) Somewhat negative
(B) 0
(C) Somewhat positive
(D) Nearly 1
(E) 1

20. Which of the following statements about the correlation r are true?

I. The correlation and the slope of the regression line have the same sign.

II. A correlation of $-.35$ and a correlation of $+.35$ show the same degree of clustering around the regression line.

III. A correlation of .75 indicates a relationship that is 3 times as linear as one for which the correlation is only .25.

(A) I and II
(B) I and III
(C) II and III
(D) I, II, and III
(E) None of the above gives the complete set of true responses.

21. Suppose the correlation between two variables is $r = .23$. What will the new correlation be if .14 is added to all values of the *x*-variable, every value of the *y*-variable is doubled, and the two variables are interchanged?

    (A) .23
    (B) .37
    (C) .74
    (D) −.23
    (E) −.74

22. As reported in the *Journal of the American Medical Association* (June 13, 1990, page 3031), for a study of ten nonagenarians (subjects were age 90 ± 1), the following tabulation shows a measure of strength (heaviest weight subject could lift using knee extensors) versus a measure of functional mobility (time taken to walk 6 meters). Note that the functional mobility is greater with lower walk times.

Strength (kg):	7.5	6	11.5	10.5	9.5	18	4	12	9	3
Walk time (s):	18	46	8	25	25	7	22	12	10	48

    What is the sign of the slope of the regression line and what does it signify?

    (A) The sign is positive, signifying a direct cause-and-effect relationship between strength and functional mobility.
    (B) The sign is positive, signifying that the greater the strength, the greater the functional mobility.
    (C) The sign is negative, signifying that the relationship between strength and functional mobility is weak.
    (D) The sign is negative, signifying that the greater the strength, the greater the functional mobility.
    (E) The slope is close to zero, signifying that the relationship between strength and functional mobility is weak.

23. Suppose the correlation between two variables is −.57. If each of the *y*-scores is multiplied by −1, which of the following is true about the new scatterplot?

    (A) It slopes up to the right, and the correlation is −.57.
    (B) It slopes up to the right, and the correlation is +.57.
    (C) It slopes down to the right, and the correlation is −.57.
    (D) It slopes down to the right, and the correlation is +.57.
    (E) None of the above is true.

24. A study of 100 elementary school children showed a strong positive correlation between weight and reading speed. Which of the following are proper conclusions?

    I. Heavier elementary school children tend to have higher reading speeds.
    II. Among elementary school children, faster readers tend to be heavier.
    III. If you want to improve the reading speed of elementary school children, you should feed them more.

    (A) I only
    (B) I and II
    (C) I, II, and III
    (D) None of the three statements is a proper conclusion.
    (E) None of the above gives the complete set of proper conclusions.

25. Consider the set of points $\{(2, 5), (3, 7), (4, 9), (5, 12), (10, n)\}$. What should n be so that the correlation between the x- and y-values is 1?

    (A) 21
    (B) 24
    (C) 25
    (D) A value different from any of the above.
    (E) No value for n can make $r = 1$.

26. A study is conducted relating GPA to number of study hours per week, and the correlation is found to be .5. Which of the following are true statements?

    I. On the average, a 30% increase in study time per week results in a 15% increase in GPA.
    II. Fifty percent of a student's GPA can be explained by the number of study hours per week.
    III. Higher GPAs tend to be associated with higher numbers of study hours.

    (A) I and II
    (B) I and III
    (C) II and III
    (D) I, II, and III
    (E) None of the above gives the complete set of true responses.

27. Consider the following three scatterplots:

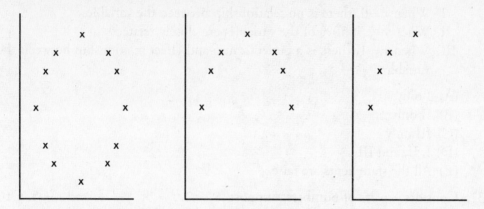

Which of the following is a true statement about the correlations for the three scatterplots?

(A) None are 0.
(B) One is 0, one is negative, and one is positive.
(C) One is 0, and both of the others are positive.
(D) Two are 0, and the other is 1.
(E) Two are 0, and the other is close to 1.

28. Consider the three points (2, 11), (3, 17), and (4, 29). Given any straight line, we can calculate the sum of the squares of the three vertical distances from these points to the line. What is the smallest possible value this sum can be?

(A) 6
(B) 9
(C) 29
(D) 57
(E) None of these values

29. Suppose that the scatterplot of log X and log Y shows a strong positive correlation close to 1. Which of the following is true?

   I. The variables X and Y also have a correlation close to 1.
   II. A scatterplot of the variables X and Y shows a strong nonlinear pattern.
   III. The residual plot of the variables X and Y shows a random pattern.

(A) I only
(B) II only
(C) III only
(D) I and II
(E) I, II, and III

30. Which of the following statements about the correlation r are true?

    I. When $r = 0$, there is no relationship between the variables.
    II. When $r = .5$, 50% of the variables are closely related.
    III. When $r = 1$, there is a perfect cause-and-effect relationship between the variables.

    (A) I only
    (B) II only
    (C) III only
    (D) I, II, and III
    (E) All the statements are false.

31. Consider n pairs of numbers. Suppose $\bar{x} = 2$, $s_x = 3$, $\bar{y} = 4$, and $s_y = 5$. Of the following, which could be the least squares line?

    (A) $\hat{y} = -2 + x$
    (B) $\hat{y} = 2x$
    (C) $\hat{y} = -2 + 3x$
    (D) $\hat{y} = \frac{5}{3} - x$
    (E) $\hat{y} = 6 - x$

Answer Key

1. **E**	8. **C**	14. **D**	20. **A**	26. **E**
2. **A**	9. **E**	15. **C**	21. **A**	27. **E**
3. **B**	10. **B**	16. **A**	22. **D**	28. **A**
4. **D**	11. **A**	17. **E**	23. **B**	29. **B**
5. **E**	12. **A**	18. **B**	24. **B**	30. **E**
6. **A**	13. **C**	19. **E**	25. **E**	31. **E**
7. **E**				

Answers Explained

1. **(E)** The slope is 20/10 = 2; that is, a woman's risk of developing ovarian cancer rises 2% for every gram of fat consumed per day. The other statements may be true, but they do not answer the question.

2. **(A)** Slope $= .15(\frac{42,000}{1.3}) \approx 4850$ and intercept $= 208,000 - 4850(6.2) \approx 178,000$.

3. **(B)** log(revenue in $1000), not revenue, goes up 0.82 per year.

    log(revenue in $1000) = 0.82(5) + 0.67 = 4.77 gives revenue = $10^{4.77}$ thousand dollars $\approx$ 59 million dollars.

    $r^2 = (.73)^2 = 53\%$ of the variation in log(revenue in $1000), not revenue, can be explained by variation in time.

4. **(D)** The correlation coefficient is not changed by adding the same number to each value of one of the variables or by multiplying each value of one of the variables by the same positive number.

5. **(E)** A negative correlation shows a tendency for higher values of one variable to be associated with lower values of the other; however, given any two points, anything is possible.

6. **(A)** This is the only scatterplot in which the residuals go from positive to negative and back to positive.

7. **(E)** Since (2, 5) is on the line $y = 3x + b$, we have $5 = 6 + b$ and $b = -1$. Thus the regression line is $y = 3x - 1$. The point $(\bar{x}, \bar{y})$ is always on the regression line, and so we have $\bar{y} = 3\bar{x} - 1$.

8. **(C)** The correlation r measures association, not causation.

9. **(E)** The correlation r cannot take a value greater than 1.

10. **(B)** It can be shown that r^2, called the coefficient of determination, is the ratio of the variance of the predicted values $\hat{y}$ to the variance of the observed values y. Alternatively, we can say that there is a partition of the y-variance and that r^2 is the proportion of this variance that is predictable from a knowledge of x. In the case of perfect correlation, $r = \pm 1$.

11. **(A)** Correlation has the formula

$$r = \frac{1}{n-1} \sum \left(\frac{x_i - \bar{x}}{s_x} \right) \left(\frac{y_i - \bar{y}}{s_y} \right)$$

We see that x and y are interchangeable, and so the correlation does not distinguish between which variable is called x and which is called y. The formula is also based on standardized scores (z-scores), and so changing units does not change the correlation. Finally, since means and standard deviations can be strongly influenced by outliers, the correlation is also strongly affected by extreme values.

12. **(A)** Removal of scores with large residuals but average x-values may not have a great effect on the regression line.

13. **(C)** An influential score may have a small residual but still have a greater effect on the regression line than scores with possibly larger residuals but average x-values.

14. **(D)** The sum and thus the mean of the residuals are always zero. In a good straight-line fit, the residuals show a random pattern.

15. **(C)** The coefficient of determination r^2 gives the proportion of the y-variance that is predictable from a knowledge of x. In this case $r^2 = (.632)^2 = .399$ or 39.9%.

16. **(A)** The regression equation is $\hat{y} = 7386.87 - 3.627x$, and so its slope is -3.627 and the y-value drops 3.627 for every unit increase in the x-value. Answer (C) is a true statement but doesn't pertain to the question.

17. **(E)** $-3.627(1983) + 7386.87 = 194.5$

18. **(B)** On each exam, two students had scores of 100. There is a general negative slope to the data showing a moderate negative correlation.

19. **(E)** On the scatterplot all the points lie perfectly on a line sloping up to the right, and so $r = 1$.

20. **(A)** The slope and the correlation are related by the formula

$$b_1 = r \frac{s_y}{s_x}$$

The standard deviations are always positive, and so b_1 and r have the same sign. Positive and negative correlations with the same absolute value indicate data having the same degree of clustering around their respective regression lines, one of which slopes up to the right and the other of which slopes down to the right. While $r = .75$ indicates a better fit with a linear model than $r = .25$ does, we cannot say that the linearity is threefold.

21. **(A)** The correlation is not changed by adding the same number to every value of one of the variables, by multiplying every value of one of the variables by the same positive number, or by interchanging the x- and y-variables.

22. **(D)** The slope is negative (-2.4348); that is, the regression line slopes down to the right, indicating that nonagenarians with greater strength have lower walk times and thus greater functional mobility.

23. **(B)** The slope and the correlation coefficient have the same sign. Multiplying every y-value by -1 changes this sign.

24. **(B)** Correlation shows association, not causation. In this example, older school children both weigh more and have faster reading speeds.

25. **(E)** A scatterplot readily shows that while the first three points lie on a straight line, the fourth point does not lie on this line. Thus no matter what the fifth point is, all the points cannot lie on a straight line, and so r cannot be 1.

26. **(E)** Only III is true. A positive correlation indicates that higher values of x tend to be associated with higher values of y.

27. **(E)** All three scatterplots show very strong nonlinear patterns; however, the correlation r measures the strength of only a linear association. Thus $r = 0$ in the first two scatterplots and is close to 1 in the third.

28. **(A)** Using your calculator, find the regression line to be $y = 9x - 8$. The regression line, also called the least squares regression line, minimizes the sum of the squares of the vertical distances between the points and the line. In this case $(2, 10)$, $(3, 19)$, and $(4, 28)$ are on the line, and so the minimum sum is $(10 - 11)^2 + (19 - 17)^2 + (28 - 29)^2 = 6$.

29. **(B)** When transforming the variables leads to a linear relationship, the original variables have a nonlinear relationship, their correlation (which measures linearity) is not close to 1, and the residuals do not show a random pattern.

30. **(E)** These are all misconceptions about correlation. Correlation measures only linearity, and so when $r = 0$, there still may be a nonlinear relationship. Correlation shows association, not causation.

31. **(E)** The least squares line passes through $(\overline{x}, \overline{y}) = (2, 4)$, and the slope b satisfies $b = r\frac{s_y}{s_x} = \frac{5r}{3}$. Since $-1 \le r \le 1$, we have $-\frac{5}{3} \le b \le \frac{5}{3}$.

Free-Response Questions

Directions: You must show all work and indicate the methods you use. You will be graded on the correctness of your methods and on the accuracy of your final answers.

Ten Open-Ended Questions

1. According to *The New York Times* (April 2, 1993, page A1), the average monthly rate for basic television cable service has increased as follows:

Year:	1986	1987	1988	1989	1990	1991	1992
Rate ($):	11.00	13.20	13.90	15.20	16.80	18.00	20.00

   a. Find the equation of the best-fitting straight line.
   b. Interpret the slope.
   c. Predict the average monthly rate in 1993 (actually, it was $21.00).
   d. Predict in what year the rate will reach $50.00.
   e. Comment on the accuracy of the predictions in parts *c* and *d.*

2. The shoe sizes and the number of ties owned by ten corporate vice presidents are as follows.

Shoe size, x:	8	9.5	9	11	9	9.5	8.5	9	9	9.5
Number of ties, y:	10	10	8	15	12	13	16	7	12	4

   a. Draw a scatterplot for these data.
   b. Find the correlation r.
   c. Can we find the best-fitting straight-line approximation to the above data? Does it make sense to use this equation to predict the number of ties owned by a corporate executive who wears size 10 shoes? Explain.

3. Following is a scatterplot of the average life expectancies and per capita incomes (in thousands of dollars) for people in a sample of 50 countries.

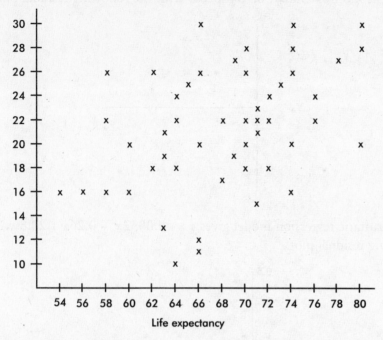

Life expectancy

   a. Estimate the mean for the set of 50 life expectancies and for the set of 50 per capita incomes.

   b. Estimate the standard deviation for the set of life expectancies and for the set of per capita incomes. Explain your reasoning.

   c. Does the scatterplot show a correlation between per capita income and life expectancy? Is it positive or negative? Is it weak or strong?

4. *a.* Find the correlation *r* for each of the three sets:

$$A = \{(5, 5), (5, 10), (10, 5), (10, 10)\}$$
$$B = \{(50, 50), (50, 55), (55, 50), (55, 55)\}$$
$$C = \{(90, 90), (90, 95), (95, 90), (95, 95)\}$$

   b. Find the correlation for the set consisting of the 12 scores from A, B, and C.

   c. Comment on the above results.

5. An outlier can have a striking effect on the correlation *r*. For example, comment on the following three scatterplots:

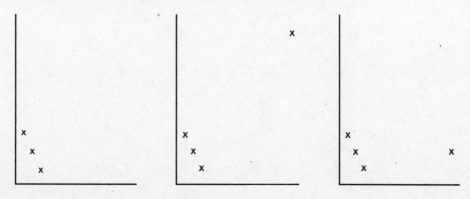

6. Fuel economy y (in miles per gallon) is tabulated for various speeds x (in miles per hour) for a certain car model. A linear regression model gives Fuel economy = 34.8 − 0.16 (Speed) with the following residual plot:

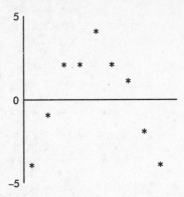

A quadratic regression model gives $y = -0.0032x^2 + 0.26x + 23.8$ with the following residual plot:

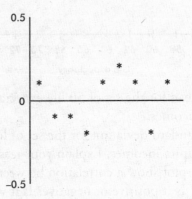

(a) What does each model predict for fuel economy at 50 miles per hour?
(b) Which model is a better fit? Explain.

7. The following scatterplot shows the grades for research papers for a sociology
   professor's class plotted against the lengths of the papers (in pages).

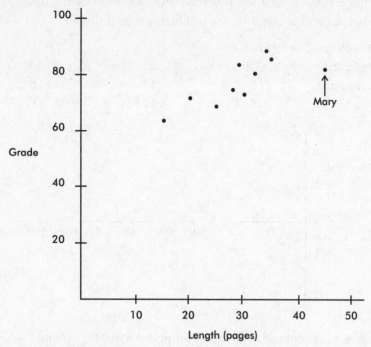

Mary turned in her paper late and was told by the professor that her grade
would have been higher if she had turned it in on time. A computer printout
fitting a straight line to the data (not including Mary's score) by the method of
least squares gives

```
Grade = 46.51 + 1.106 Length
R-sq = 74.6%
```

(a) Find the correlation coefficient for the relationship between grade and
    length of paper based on these data (excluding Mary's paper).
(b) What is the slope of the regression line and what does it signify?
(c) How will the correlation coefficient change if Mary's paper is included?
    Explain your answer.
(d) How will the slope of the regression line change if Mary's paper is
    included? Explain your answer.
(e) What grade did Mary receive? Predict what she would have received if
    her paper had been on time.

8. Data show a trend in winning long jump distances for an international competition over the years 1972–92. With jumps recorded in inches and dates in years since 1900, a least squares regression line is fit to the data. The computer output and a graph of the residuals are as follows:

```
R squared = 92.1%
Variable    Coefficient    SE of Coeff    t-ratio    Prob
Constant     256.576         11.59          22.1      0.0001
Year         0.95893          0.141          6.81      0.0024
```

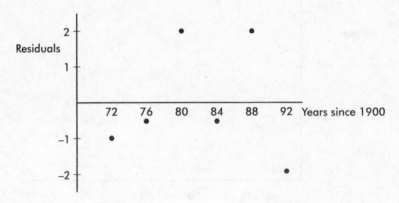

(a) Does a line appear to be an appropriate model? Explain.
(b) What is the slope of the least squares line? Give an interpretation of the slope.
(c) What is the correlation?
(d) What is the predicted winning distance for the 1980 competition?
(e) What was the actual winning distance in 1980?

9. A scatterplot of the number of accidents per day on a particular interstate highway during a 30-day month is as follows:

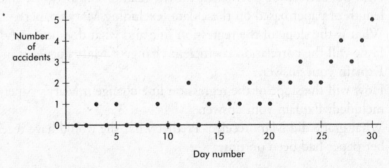

(a) Draw a histogram of the frequencies of the number of accidents.
(b) Draw a boxplot of the number of accidents.
(c) Name a feature apparent in the scatterplot but not in the histogram or boxplot.
(d) Name a feature clearly shown by the histogram and boxplot but not as obvious in the scatterplot.

10. The following scatterplot shows the pulse rate drop (in beats per minute) plotted against the amount of medication (in grams) of an experimental drug being field-tested in several hospitals.

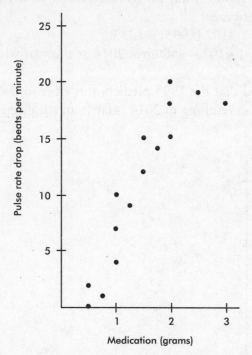

A computer printout showing the results of fitting a straight line to the data by the method of least squares gives

```
PulseRateDrop = -1.68 + 8.5 Grams
R-sq = 81.9%
```

(a) Find the correlation coefficient for the relationship between pulse rate drop and grams of medication.
(b) What is the slope of the regression line and what does it signify?
(c) Predict the pulse rate drop for a patient given 2.25 grams of medication.
(d) A patient given 5 grams of medication has his pulse rate drop to zero. Does this invalidate the regression equation? Explain.
(e) How will the size of the correlation coefficient change if the 3-gram result is removed from the data set? Explain.
(f) How will the size of the slope of the least squares regression line change if the 3-gram result is removed from the data set? Explain.

Answers Explained

1. *a.* A calculator readily gives $\hat{y} = -2790.47 + 1.4107x$.

   b. The average monthly rate for basic television cable service has increased by about $1.41 per year.

   c. $-2790.47 + 1.4107\,(1993) = \$21.06$

   d. $-2790.47 + 1.4107x = 50$ gives 2014 as the year when the rate will reach $50.00.

   e. It is expected that the 1993 prediction is close to the actual rate; however, the extrapolation resulting in 2014 is far from the given values and cannot be relied on.

2. *a.*

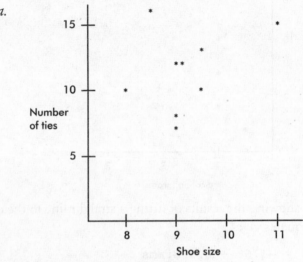

   b. A calculator gives $r = .1568$.

   c. The correlation r is low for this number of data scores, and the scatterplot shows no linear pattern whatsoever. Although theoretically we could use our techniques to find the best-fitting straight-line approximation, the result would be meaningless and should not be used for predictions.

3. *a.* By visual inspection $\bar{x} \approx 68$ and $\bar{y} \approx 21$.

   b. The range of the life expectancies is $80 - 54 = 26$, and so the standard deviation is roughly $\frac{26}{4} = 6.5$. Similarly the standard deviation of the per capita incomes is roughly $\frac{30-10}{4} = 5$.

   c. While the points generally fall from the lower left to the upper right, they are still widely scattered. Thus the scatterplot shows a weak positive correlation between per capita income and life expectancy.

4. *a.* The correlation for each of the three sets is 0.

   b. The correlation for the set consisting of all 12 scores is .9948.

   c. The data from each set taken separately show no linear pattern. However, together they show a strong linear fit. Note the positions of the data from the separate sets in the complete scatterplot.

5. In the first scatterplot, the points fall exactly on a downward sloping straight line, so $r = -1$. In the second scatterplot, the outlier is an influential point, and r is close to $+1$. In the third scatterplot, the outlier is also influential, and r is close to 0.

6. (a) $y = -0.16(50) + 34.8 = 26.8$ miles per gallon, and $y = -0.0032(50)^2 + 0.258(50) + 23.8 = 28.7$ miles per gallon.
   (b) Model 2 is the better fit. First, the residuals are much smaller for model 2, indicating that this model gives values much closer to the observed values. Second, a curved residual pattern like that in model 1 indicates that a nonlinear model would be better. A more uniform residual scatter as in model 2 indicates a better fit.

7. (a) The correlation coefficient is $r = \sqrt{.746} = .864$. It is positive because the slope of the regression line is positive.
   (b) The slope is 1.106, signifying that each additional page raises a grade by 1.106.
   (c) Including Mary's paper will lower the correlation coefficient because her result seems far off the regression line through the other points.
   (d) Including Mary's paper will swing the regression line down and lower the value of the slope.
   (e) From the graph, Mary received an 82. From the regression line, Mary would have received $y = 46.51 + 1.106(45) = 96.3$ if she had turned in her paper on time.

8. (a) Yes. The residual graph is not curved, does not show fanning, and appears to be random or scattered.
   (b) The slope is 0.95893, indicating that the winning jump improves 0.95893 inches per year or about 3.8 inches every four years.
   (c) With $r^2 = .921$, the correlation r is .96.
   (d) $0.95893(80) + 256.576 \approx 333.3$ inches
   (e) The residual for 1980 is $+2$, and so the actual winning distance must have been $333.3 + 2 = 335.3$ inches.

9. (a)

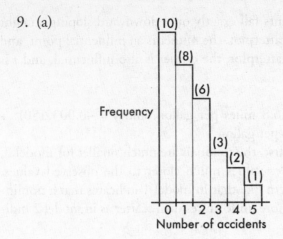

(b)

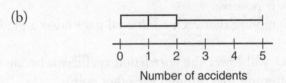

(c) There is a roughly linear trend with daily accidents increasing during the month.

(d) The daily number of accidents is strongly skewed to the right.

10. (a) The correlation coefficient $r = \sqrt{.819} = .905$. It is positive because the slope of the regression line is positive.

(b) The slope is 8.5, signifying that each gram of medication lowers the pulse rate by 8.5 beats per minute.

(c) $y = -1.68 + 8.5(2.25) = 17.4$ beats per minute.

(d) There is always danger in using a regression line to extrapolate beyond the values of x contained in the data. In this case, the 5 grams was an overdose, the patient died, and the regression line cannot be used for such values beyond the data set.

(e) Removing the 3-gram result from the data set will increase the correlation coefficient because the 3-gram result appears to be far off a regression line through the remaining points.

(f) Removing the 3-gram result from the data set will swing the regression line upward so that the slope will increase.

Three Investigative Tasks

1. The following table (*Statistical Abstract of the United States*) gives the U.S. population (in millions) for various years of the twentieth century.

1900	76.1	1950	151.9
1905	83.8	1955	165.1
1910	92.4	1960	180.0
1915	100.5	1965	193.5
1920	106.5	1970	204.0
1925	115.8	1975	215.5
1930	123.1	1980	227.2
1935	127.3	1985	237.9
1940	132.5	1990	249.4
1945	133.4		

   a. Find the equation of the regression line of population in terms of year, and calculate the correlation.

   b. To find an alternative nonlinear model, make a table of years *t* since 1900 versus the log of the above population values. Find the equation of the regression line of log *y* in terms of *t* and calculate the correlation.

   c. Use each model to predict what the population will be in the year 2100 and to predict when the population will reach 300 million.

   d. Comment on the accuracy of your answers in part *c*.

2. Following are the average distances from the sun and the times of revolution for the nine planets in our solar system:

Planet	Average Distance from Sun (million miles)	Time for Revolution (days)
Mercury	36	88
Venus	67	275
Earth	93	365
Mars	141	687
Jupiter	484	4,332
Saturn	888	10,826
Uranus	1,764	30,676
Neptune	2,790	59,911
Pluto	3,654	90,824

a. Letting *x* be the average distance and *y* the revolution time, find the equation of the regression line of log *y* in terms of log *x* and calculate the correlation.

b. Suppose a new planet is discovered at a distance of 5000 million miles from the sun. Using your regression line, extrapolate and predict the time of revolution.

c. Suppose a new planet is discovered with a revolution time of 45,000 days. Using your regression line, predict its average distance from the sun.

3. The 23 shuttle launches before *Challenger* each experienced from zero to three O-ring failures. There was some speculation that the number of O-ring failures was related to the temperature at liftoff. Find the correlation *r* and the regression equation given the following data:

Temperature at Liftoff	Number of O-Ring Failures	Temperature at Liftoff	Number of O-Ring Failures
66	0	67	0
70	1	53	3
69	0	67	0
68	0	75	0
67	0	70	0
72	0	81	0
73	0	76	0
70	0	79	0
57	1	75	2
63	1	76	0
70	1	58	1
78	0		

a. *Challenger* was launched at 31 degrees Fahrenheit. What would have been predicted for the number of O-ring failures using a linear model? What is the correlation *r*?

b. A scatterplot indicates a nonlinear model. Replace the *y*-values with log(*y* + 1) and predict the number of O-ring failures. Why can't you simply use log *y*? What is the correlation *r*?

c. Unfortunately, National Aeronautical and Space Administration (NASA) scientists made their calculations using only data from the seven prior launches that had had at least one O-ring failure. What correlation *r* did they calculate? What do you think was their conclusion?

Answers Explained

1. *a.* Population = −3593 + 1.926 Year; $r = .989$.
   b. On your calculator you can replace the y-list with log y and the x-list with $x − 1900$. Then the regression calculation gives log $y = 0.005639t + 1.905$, $r = .996$.
   c. Using the linear model, we calculate $1.926(2100) − 3593 = 452$ (million); and $1.926x − 3593 = 300$ gives $x = 2021$. Using the nonlinear model, we calculate log $y = 0.005639(200) + 1.905 = 3.0328$, and so $y = 1078$ (million); and log $300 = 0.005639t + 1.905$, and so $t = 101$ and the year is $1900 + 101 = 2001$.
   d. The nonlinear model showed a higher correlation and so should be more accurate. Predicting the year when the population will reach 300 million gives an answer not too far from the range of the given data. However, predicting the population in the year 2100 is a substantial extrapolation and must be viewed with some suspicion.

2. *a.* Replacing list X with log X and list Y with log Y and then running a linear regression program on a calculator yields log $y = −0.342 + 1.486$ log x, with $r = .9997$.
   b. log $y = −0.342 + 1.486$ log $5000 = 5.155$, and so $y = 142,780$ days.
   c. log $45,000 = −0.342 + 1.486$ log x, which gives log $x = 3.362$ and $x = 2300$ million miles.

3. *a.* Using the linear regression feature of your calculator, find $y = −0.063t + 4.8$, with $r = −.561$. At 31 degrees Fahrenheit this gives $−0.063(31) + 4.8 = 2.8$.
   b. We need to use $\log(y + 1)$ because $\log 0$ is not defined. Replacing the y-values with $\log(y + 1)$ and performing the linear regression calculation results in $\log(y + 1) = −0.015t + 1.15$, with $r = −.570$. At 31 degrees Fahrenheit, $\log(y + 1) = −0.015(31) + 1.15 = 0.685$, so $y + 1 = 10^{0.685} = 4.8$ and $y = 3.8$.
   c. Using only data from the seven prior launches that had had O-ring failures results in $r = −.263$. Looking at a correlation of only $−.263$, the scientists allowed *Challenger* to be launched. (Looking at a correlation of $−.570$, they might have decided differently!)

Exploring Categorical Data: Frequency Tables

> • Marginal Frequencies for Two-Way Tables
> • Conditional Relative Frequencies and Association

While many variables such as age, income, and years of education are quantitative or numerical in nature, others such as gender, race, brand preference, mode of transportation, and type of occupation are qualitative or categorical. Quantitative variables, too, are sometimes grouped into categorical classes.

MARGINAL FREQUENCIES FOR TWO-WAY TABLES

Qualitative data often encompass two categorical variables that may or may not have a dependent relationship. These data can be displayed in a *two-way contingency table*.

EXAMPLE 5.1

A 4-year study, reported in *The New York Times*, on men more than 70 years old analyzed blood cholesterol and noted how many men with different cholesterol levels suffered nonfatal or fatal heart attacks.

	Low cholesterol	Medium cholesterol	High cholesterol
Nonfatal heart attacks	29	17	18
Fatal heart attacks	19	20	9

Severity of heart attacks is the *row variable*, while cholesterol level is the *column variable*.

One method of analyzing these data involves first calculating the totals for each row and each column:

	Low cholesterol	Medium cholesterol	High cholesterol	Total
Nonfatal heart attacks	29	17	18	64
Fatal heart attacks	19	20	9	48
Total	48	37	27	112

These totals are placed in the right and bottom margins of the table and thus are called *marginal frequencies.*

These marginal frequencies are often put in the form of proportions or percentages. The *marginal distribution* of the cholesterol level is

Low: $\frac{48}{112} = .429 = 42.9\%$

Medium: $\frac{37}{112} = .330 = 33.0\%$

High: $\frac{27}{112} = .241 = 24.1\%$.

This distribution can also be displayed in a bar graph as follows:

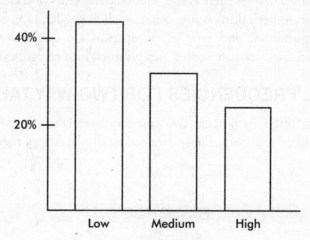

Cholesterol levels
(of elderly men who suffered heart attacks)

Similarly we can determine the marginal distribution for the severity of heart attacks:

Nonfatal: $\frac{64}{112} = .571 = 57.1\%$

Fatal: $\frac{48}{112} = .429 = 42.9\%$.

(continued)

The representative bar graph is

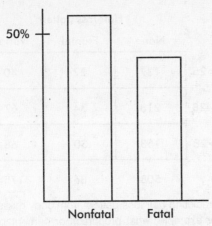

Severity of heart attacks
(of elderly men who suffered heart attacks)

CONDITIONAL RELATIVE FREQUENCIES AND ASSOCIATION

The marginal distributions described and calculated above do not describe or measure the relationship between the two categorical variables. For this we must consider the information in the body of the table, not just the sums in the margins.

EXAMPLE 5.2

Is hair loss pattern related to body mass index? One study (*Journal of the American Medical Association*, February 24, 1993, page 1000) of 769 men showed the following numbers:

		Hair loss pattern		
		None	Frontal	Vertex
	<25	137	22	40
Body mass index	25–28	218	34	67
	>28	153	30	68

(continued)

The analysis first involves finding the row and column totals as we did before.

Hair loss pattern

		None	Frontal	Vertex	
Body mass index	<25	137	22	40	199
	25–28	218	34	67	319
	>28	153	30	68	251
		508	86	175	769

We are interested in predicting hair loss pattern from body mass index, and so we look at each row separately. For example, what proportion or percentage of the 199 men with a body mass index less than 25 have each of the hair loss patterns?

None: $\frac{137}{199} = .688 = 68.8\%$

Frontal: $\frac{22}{199} = .111 = 11.1\%$

Vertex: $\frac{40}{199} = .201 = 20.1\%$.

These *conditional relative frequencies* can be displayed either with groupings of bars or by a segmented bar chart where each segment has a length corresponding to its relative frequency:

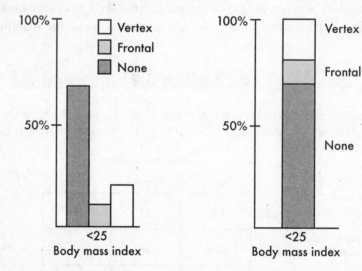

Similarly, the conditional relative frequencies for the 319 men with a body mass index between 25 and 28 are

None: $\frac{218}{319} = .683 = 68.3\%$

Frontal: $\frac{34}{319} = .107 = 10.7\%$

Vertex: $\frac{67}{319} = .210 = 21.0\%$.

(continued)

For the 251 men with a body mass index of more than 28 we have

None: $\frac{153}{251} = .610 = 61.0\%$

Frontal: $\frac{30}{251} = .120 = 12.0\%$

Vertex: $\frac{68}{251} = .271 = 27.1\%.$

Both of the following bar charts give good visual pictures:

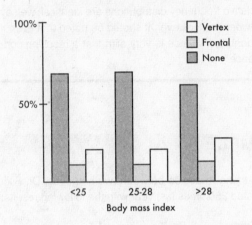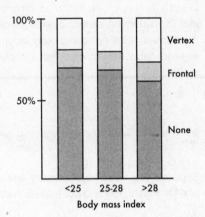

Segmented bar charts indicate a slight relationship between higher vertex pattern baldness and a body mass index of more than 28.

EXAMPLE 5.3

A study was made to compare year in high school with preference for vanilla or chocolate ice cream with the following results:

	Vanilla	Chocolate
Freshman	20	10
Sophomore	24	12
Junior	18	9
Senior	22	11

(continued)

What are the conditional relative frequencies for each class?

Freshmen: $\frac{20}{30} = .667$ prefer vanilla and $\frac{10}{30} = .333$ prefer chocolate.

Sophomores: $\frac{24}{36} = .667$ prefer vanilla and $\frac{12}{36} = .333$ prefer chocolate.

Juniors: $\frac{18}{27} = .667$ prefer vanilla and $\frac{9}{27} = .333$ prefer chocolate.

Seniors: $\frac{22}{33} = .667$ prefer vanilla and $\frac{11}{33} = .333$ prefer chocolate.

In such a case, where all the conditional relative frequency distributions are identical, we say that the two variables show *perfect independence.* (However, it should be noted that even if the two variables are completely independent, the chance is very slim that a resulting contingency table will show perfect independence.)

EXAMPLE 5.4

Suppose you need heart surgery and are trying to decide between two surgeons, Dr. Fixit and Dr. Patch. You find out that each operated 250 times last year with the following results:

	Dr. F	Dr. P
Died	60	50
Survived	190	200

Whom should you go to? Among Dr. Fixit's 250 patients 190 survived, for a survival rate of $\frac{190}{250} = .76$ or 76%, while among Dr. Patch's 250 patients 200 survived, for a survival rate of $\frac{200}{250} = .80$ or 80%. Your choice seems clear.

However, everything may not be so clear-cut. Suppose that on further investigation you determine that the surgeons operated on patients who were in either good or poor condition with the following results:

Good condition	Dr. F	Dr. P
Died	8	17
Survived	60	120

Poor condition	Dr. F	Dr. P
Died	52	33
Survived	130	80

Note that adding corresponding boxes from these two tables gives the original table above.

How do the surgeons compare when operating on patients in good health? Dr. Fixit's 68 patients in good condition have a survival rate of $\frac{60}{68} = .882$ or 88.2%, while Dr. Patch's 137 patients in good condition have a survival rate of $\frac{120}{137} = .876$ or 87.6%. Similarly, we note that Dr. Fixit's 182 patients in poor condition have a survival rate of $\frac{130}{182} = .714$ or 71.4%, while Dr. Patch's 113 patients in poor condition have a survival rate of $\frac{80}{113} = .708$ or 70.8%.

Thus Dr. Fixit does better with patients in good condition (88.2% versus Dr. Patch's 87.6%) and also does better with patients in poor condition (71.4% versus Dr. P's 70.8%). However, Dr. Fixit has a lower overall patient survival rate (76% versus Dr. Patch's 80%)! How can this be?

This problem is an example *of Simpson's paradox*, where a comparison can be reversed when more than one group is combined to form a single group. The effect of another variable, sometimes called a *lurking variable*, is masked when the groups are combined. In this particular example, closer scrutiny reveals that Dr. Fixit operates on many more patients in poor condition than Dr. Patch, and these patients in poor condition are precisely the ones with lower survival rates. Thus even though Dr. Fixit does better with all patients, his overall rating is lower. Our original table hid the effect of the lurking variable related to the condition of the patients.

Questions on Topic Five: Exploring Categorical Data

Multiple-Choice Questions

Directions: The questions or incomplete statements that follow are each followed by five suggested answers or completions. Choose the response that best answers the question or completes the statement.

Questions 1–5 are based on the following: To study the relationship between party affiliation and support for a balanced budget amendment, 500 registered voters were surveyed with the following results:

	For	Against	No opinion
Democrat	50	150	50
Republican	125	50	25
Independent	15	10	25

1. What percentage of those surveyed were Democrats?

   (A) 10%
   (B) 20%
   (C) 30%
   (D) 40%
   (E) 50%

2. What percentage of those surveyed were for the amendment and were Republicans?

   (A) 25%
   (B) 38%
   (C) 40%
   (D) 62.5%
   (E) 65.8%

3. What percentage of Independents had no opinion?

   (A) 5%
   (B) 10%
   (C) 20%
   (D) 25%
   (E) 50%

4. What percentage of those against the amendment were Democrats?

   (A) 30%
   (B) 42%
   (C) 50%
   (D) 60%
   (E) 71.4%

5. Voters of which affiliation were most likely to have no opinion about the amendment?

   (A) Democrat
   (B) Republican
   (C) Independent
   (D) Republican and Independent, equally
   (E) Democrat, Republican, and Independent, equally

Questions 6–10 are based on the following: A study of music preferences in three geographic locations resulted in the following segmented bar chart:

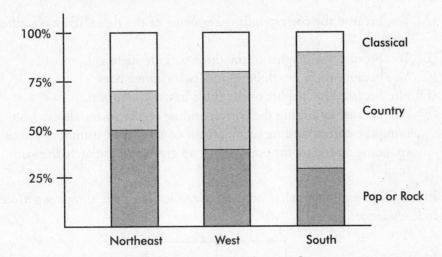

6. What percentage of those surveyed from the Northeast prefer country music?

   (A) 20%
   (B) 30%
   (C) 40%
   (D) 50%
   (E) 70%

7. Which of the following is greatest?

   (A) The percentage of those from the Northeast who prefer classical.
   (B) The percentage of those from the West who prefer country.
   (C) The percentage of those from the South who prefer pop or rock.
   (D) The above are all equal.
   (E) It is impossible to determine the answer without knowing the actual numbers of people involved.

8. Which of the following is greatest?

   (A) The number of people in the Northeast who prefer pop or rock.
   (B) The number of people in the West who prefer classical.
   (C) The number of people in the South who prefer country.
   (D) The above are all equal.
   (E) It is impossible to determine the answer without knowing the actual numbers of people involved.

9. All three bars have a height of 100%.

   (A) This is a coincidence.
   (B) This happened because each bar shows a complete distribution.
   (C) This happened because there are three bars each divided into three segments.
   (D) This happened because of the nature of musical patterns.
   (E) None of the above is true.

10. Based on the given segmented bar chart, does there seem to be a relationship between geographic location and music preference?

    (A) Yes, because the corresponding segments of the three bars have different lengths.
    (B) Yes, because the heights of the three bars are identical.
    (C) Yes, because there are three segments and three bars.
    (D) No, because the heights of the three bars are identical.
    (E) No, because summing the corresponding segments for classical, summing the corresponding segments for country, and summing the corresponding segments for pop or rock all give approximately the same total.

11. In the following table, what value for n results in a table showing perfect independence?

20	50
30	n

    (A) 10
    (B) 40
    (C) 60
    (D) 75
    (E) 100

12. A company employs both men and women in its secretarial and executive positions. In reports filed with the government, the company shows that the percentage of female employees who receive raises is higher than the percentage of male employees who receive raises. A government investigator claims that the percentage of male secretaries who receive raises is higher than the percentage of female secretaries who receive raises, and that the percentage of male executives who receive raises is higher than the percentage of female executives who receive raises. Is this possible?

(A) No, either the company report is wrong or the investigator's claim is wrong.

(B) No, if the company report is correct, then either a greater percentage of female secretaries than of male secretaries receive raises or a greater percentage of female executives than of male executives receive raises.

(C) No, if the investigator is correct, then by summation of the corresponding numbers, the total percentage of male employees who receive raises would have to be greater than the total percentage of female employees who receive raises.

(D) All of the above are true.

(E) It is possible for both the company report to be true and the investigator's claim to be correct.

Answer Key

1. **E**	4. **E**	7. **B**	10. **A**
2. **A**	5. **C**	8. **E**	11. **D**
3. **E**	6. **A**	9. **B**	12. **E**

Answers Explained

1. **(E)** Of the 500 people surveyed, $50 + 150 + 50 = 250$ were Democrats, and $\frac{250}{500} = .5$ or 50%.

2. **(A)** Of the 500 people surveyed, 125 were both for the amendment and were Republicans, and $\frac{125}{500} = .25$ or 25%.

3. **(E)** There were $15 + 10 + 25 = 50$ Independents; 25 of them had no opinion, and $\frac{25}{50} = .5$ or 50%.

4. **(E)** There were $150 + 50 + 10 = 210$ people against the amendment; 150 of them were Democrats, and $\frac{150}{210} = .714$ or 71.4%.

5. **(C)** The percentages of Democrats, Republicans, and Independents with no opinion are 20%, 12.5%, and 50%, respectively.

6. **(A)** In the bar corresponding to the Northeast, the segment corresponding to country music stretches from the 50% level to the 70% level, indicating a length of 20%.

7. **(B)** Based on lengths of indicated segments, the percentage from the West who prefer country is the greatest.

8. **(E)** The given bar chart shows percentages, not actual numbers.

9. **(B)** In a complete distribution, the probabilities sum to 1, and the relative frequencies total 100%.

10. **(A)** The different lengths of corresponding segments show that in different geographic regions different percentages of people prefer each of the music categories.

11. **(D)** Relative frequencies must be equal. Either looking at rows gives $\frac{20}{70} = \frac{30}{30+n}$ or looking at columns gives $\frac{20}{50} = \frac{50}{50+n}$. We could also set up a proportion $\frac{n}{30} = \frac{50}{20}$ or $\frac{n}{50} = \frac{30}{20}$. Solving any of these equations gives $n = 75$.

12. **(E)** It is possible for both to be correct, for example, if there were 11 secretaries (10 women, 3 of whom receive raises, and 1 man who receives a raise) and 11 executives (10 men, 1 of whom receives a raise, and 1 woman who does not receive a raise). Then 100% of the male secretaries receive raises while only 30% of the female secretaries do; and 10% of the male executives receive raises while 0% of the female executives do. However, overall 3 out of 11 women receive raises, while only 2 out of 11 men receive raises. This is an example of Simpson's paradox.

Free-Response Questions

Directions: You must show all work and indicate the methods you use. You will be graded on the correctness of your methods and on the accuracy of your final answers.

Two Open-Ended Questions

1. Suppose that in a telephone survey of 800 registered voters, the data are cross-classified both by gender of respondent and by respondent's opinion on an environmental bond issue.

| | Bond issue | |
	For	Against
Men	450	150
Women	160	40

Analyze this table looking both at marginal distributions and conditional distributions.

2. Following are the results of a study (*New England Journal of Medicine*, Vol. 319, No. 17, p. 1108, 1988) to determine whether taking an aspirin every other day reduces the risk of a heart attack in men:

	Aspirin	Placebo
Fatal attack	5	18
Nonfatal attack	99	171
No heart attack	10,933	10,845

Analyze this table by looking both at marginal distributions and conditional distributions, and give a conclusion in context.

Answers Explained

1. First we find the row and column totals:

	Bond issue		
	For	Against	
Men	450	150	600
Women	160	40	200
	610	190	800

We calculate the following marginal distributions:

Of the 800 people interviewed, $\frac{600}{800}$ or 75% are men and $\frac{200}{800}$ or 25% are women.

Of the 800 people interviewed, $\frac{610}{800}$ or 76.25% are for the bond issue and $\frac{190}{800}$ or 23.75% are against the bond issue.

Looking at the body of the table, we calculate the following conditional distributions:

Of the 600 men interviewed, $\frac{450}{600}$ or 75% are for the bond issue and $\frac{150}{600}$ or 25% are against it.

Of the 200 women interviewed, $\frac{160}{200}$ or 80% are for the bond issue and $\frac{40}{200}$ or 20% are against it.

Of the 610 people for the bond issue, $\frac{450}{610}$ or 73.77% are men and $\frac{160}{610}$ or 26.23% are women.

Of the 190 people against the bond issue, $\frac{150}{190}$ or 78.95% are men and $\frac{40}{190}$ or 21.05% are women.

There doesn't seem to be much of a relationship between the gender of a respondent and the respondent's opinion on the environmental bond issue.

2. First we find the row and column totals:

	Aspirin	Placebo	
Fatal attack	5	18	23
Nonfatal attack	99	171	270
No heart attack	10,933	10,845	21,778
	11,037	11,034	22,071

We calculate the following marginal distributions:

Of the 22,071 men in the study, $\frac{11,037}{22,071}$ or 50% took aspirin, and $\frac{11,034}{22,071}$ or 50% took the placebo.

Of the 22,071 men in the study, $\frac{23}{22,071}$ or .10% had a fatal heart attack, $\frac{270}{22,071}$ or 1.22% had a nonfatal attack, and $\frac{21,778}{22,071}$ or 98.67% had no heart attack.

Looking at the body of the table, we calculate the following conditional distributions:

Of the 11,037 men taking aspirin, $\frac{5}{11,037}$ or .045% had a fatal heart attack, $\frac{99}{11,037}$ or .897% had a nonfatal attack and $\frac{10,933}{11,037}$ or 99.058% had no heart attack.

Of the 11,034 men taking the placebo, $\frac{18}{11,034}$ or .163% had a fatal heart attack, $\frac{171}{11,034}$ or 1.550% had a nonfatal attack, and $\frac{10,845}{11,034}$ or 98.287% had no heart attack.

Of the 23 men who had a fatal heart attack, $\frac{5}{23}$ or 21.74% had taken aspirin, while $\frac{18}{23}$ or 78.26% had taken the placebo.

Of the 270 men who had a nonfatal heart attack, $\frac{99}{270}$ or 36.67% had taken aspirin, while $\frac{171}{270}$ or 63.33% had taken the placebo.

Of the 21,778 men who had no heart attack, $\frac{10,933}{21,778}$ or 50.2% had taken aspirin, while $\frac{10,845}{21,778}$ or 49.8% had taken the placebo.

Men not taking aspirin have almost a fourfold chance of a fatal heart attack (.163% versus .045%) and almost a twofold chance of a nonfatal heart attack (1.550% versus .897%) when compared to men who do take aspirin.

It seems clear that in men there is a relationship between taking an aspirin every other day and the risk of having a heart attack.

Two Investigative Tasks

1. The graduate school at the University of California at Berkeley reported that in 1973 they accepted 44% of 8442 male applicants and 35% of 4321 female applicants. Concerned that one of their programs was guilty of gender bias, the graduate school analyzed admissions to the six largest graduate programs and obtained the following results:

Program	Men Accepted	Men Rejected	Women Accepted	Women Rejected
A	511	314	89	19
B	352	208	17	8
C	120	205	202	391
D	137	270	132	243
E	53	138	95	298
F	22	351	24	317

  a. Find the percentage of men and the percentage of women accepted by each program. Comment on any pattern or bias you see.

  b. Find the percentage of men and the percentage of women accepted overall by these six programs. Does this appear to contradict the results from part *a*?

  c. If you worked in the Graduate Admissions Office, what would you say to an inquiring reporter who is investigating gender bias in graduate admissions?

2. Suppose you are a baseball scout and are interested in comparing two relief pitchers, Curver and Knuckler, both of whom play college ball. In particular you are interested in their strikeout ability. A quick glance at their records shows you that each came in to relieve in a similar number of games and that each faced 150 batters during his last season. However, Curver struck out 95 batters, while Knuckler struck out only 85 batters. When you tell their coaches that you're probably going to pick Curver, Knuckler's coach asks you to please look at their separate records against right- and left-handed batters. You do the required calculations and are surprised to find out that Knuckler has a higher strikeout rate against right-handed batters than does Curver, and that Knuckler also has a higher strikeout rate against left-handed batters than does Curver!

Construct a set of hypothetical data to show how this is possible and explain the apparent paradox. *Hint:* One way of doing this involves having one pitcher face many more right-handed batters than the second, while the second faces many more left-handed batters than the first.

Answers Explained

1. *a.*

Program	Percentage of Men Accepted (%)	Percentage of Women Accepted (%)
A	62	82
B	63	68
C	37	34
D	33	35
E	28	24
F	6	7

There doesn't appear to be any real pattern; however, women seem to be favored in four of the programs, while men seem to be slightly favored in the other two programs.

b. Overall, 1195 out of 2681 male applicants were accepted, for a 45% acceptance rate, while 559 out of 1835 female applicants were accepted, for a 30% acceptance rate. This appears to contradict the results from part *a.*

c. You should tell the reporter that while it is true that the overall acceptance rate for women is 30% compared to the 44% acceptance rate for men, program by program women have either higher acceptance rates or only slightly lower acceptance rates than men. The reason behind this apparent paradox is that most men applied to programs A and B, which are easy to get into and have high acceptance rates. However, most women applied to programs C, D, E, and F, which are much harder to get into and have low acceptance rates.

2. One possible set of data is

Against right-handed batters

	Curver	Knuckler
Strikeout	80	45
Non-strikeout	20	5

Against left-handed batters

	Curver	Knuckler
Strikeout	15	40
Non-strikeout	35	60

Note that Knuckler struck out $\frac{45}{50}$ or 90% of the right-handed batters he faced, while Curver struck out only $\frac{80}{100}$ or 80% of the right-handed batters he faced. Similarly, Knuckler struck out $\frac{40}{100}$ or 40% of the left-handed batters he faced, while Curver struck out only $\frac{15}{50}$ or 30% of the left-handed batters he faced. So Knuckler did better against right-handed and against left-handed batters than Curver did, even though Curver seemed to do better overall. The reason for the apparent paradox is that both pitchers seem to have more trouble against left-handed batters, and Knuckler faced many more of these than Curver did. Curver was able to fatten up his overall strikeout percentage by pitching more often against the easier right-handed batters!

Overview of Methods of Data Collection

- Census
- Sample Survey
- Experiment
- Observational Study

I n the real world, time and cost considerations usually make it impossible to analyze an entire population. Does the government question you and your parents before announcing the monthly unemployment rates? Does a television producer check every household's viewing preferences before deciding whether a pilot program will be continued? In studying statistics we learn how to estimate *population* characteristics by considering a *sample*. For example, later in this book we will see how to estimate population means and proportions by looking at sample means and proportions.

To derive conclusions about the larger population, we need to be confident that the sample we have chosen represents that population fairly. Analyzing the data with computers is often easier than gathering the data, but the frequently quoted "Garbage in, garbage out" applies here. Nothing can help if the data are badly collected. Unfortunately, many of the statistics with which we are bombarded by newspapers, radio, and television are based on poorly designed data collection procedures.

CENSUS

A *census* is a complete enumeration of an entire population. In common use, it is often thought of as an official attempt to contact every member of the population, usually with details regarding age, marital status, race, gender, occupation, income, years of school completed, and so on. Every 10 years the U.S. Bureau of the Census divides the nation into nine regions and attempts to gather information about everyone in the country. A massive amount of data is obtained, but even with the resources of the U.S. government, the census is not complete. For example, many homeless people are always missed, or counted at two temporary residences, and there are always households that do not respond even after repeated requests for information. It was estimated that the 1990 census missed about 1.6% of the population as a whole and may have missed up to 5.0% of Hispanics and 4.6% of African Americans.

In most studies, both in the private and public sectors, a complete census is unreasonable because of time and cost involved. Furthermore, attempts to gather complete data have been known to lead to carelessness. Finally, and most important, a well-designed, well-conducted sample survey is far superior to a poorly designed study involving a complete census. For example, a poorly worded question might give meaningless data even if everyone in the population answers.

SAMPLE SURVEY

The census tries to count everyone; it is not a sample. A sample survey aims to obtain information about a whole population by studying a part of it, that is, a sample. The goal is to gather information without disturbing or changing the population. Numerous procedures are used to collect data through sampling, and much of the statistical information distributed to us comes from sample surveys. Often, controlled experiments are later undertaken to demonstrate relationships suggested by sample surveys.

However, the one thing that most quickly invalidates a sample and makes useful information impossible to obtain is *bias*. A sample is biased if in some critical way it does not represent the population. The main technique to avoid bias is to incorporate *randomness* into the selection process. Randomization protects us from effects and influences, both known and unknown. Finally, the larger the sample, the better the results, but what is critical is the sample size, not the percentage or fraction of the population. That is, a random sample of size 500 from a population of size 100,000 is just as representative as a random sample of size 500 from a population of size 1,000,000.

EXPERIMENT

In a controlled study, called an *experiment*, the researcher should randomly divide subjects into appropriate groups. Some action is taken on one or more of the groups, and the response is observed. For example, patients may be randomly given unmarked capsules of either aspirin or acetaminophen and the effects of the medication measured. Experiments often have a *treatment group* and a *control group*; in the ideal situation, neither the subjects nor the researcher knows which group is which. The Salk vaccine experiment of the 1950s, in which half the children received the vaccine and half were given a placebo, with not even their doctors knowing who received what, is a classic example of this *double-blind* approach. Controlled experiments can indicate cause-and-effect relationships.

The critical principles behind good experimental design include *control* (outside of who receives what treatments, conditions should be as similar as possible for all involved groups), *blocking* (the subjects can be divided into representative groups to bring certain differences directly into the picture), *randomization* (unknown and uncontrollable differences are handled by randomizing who receives what treatments), *replication* (treatments need to be repeated on a sufficient number of subjects), and *generalizability* (ability to repeat an experiment in a variety of settings).

OBSERVATIONAL STUDY

Sample surveys are one example of what are called *observational studies*. In observational studies there is no choice in regard to who goes into the treatment and control groups. For example, a researcher cannot ethically tell 100 people to smoke three packs of cigarettes a day and 100 others to smoke only one pack per day; he can only observe people who habitually smoke these amounts. In observational studies the researcher strives to determine which variables affect the noted response. While results may suggest relationships, it is difficult to conclude cause and effect.

Questions on Topic Six: Overview of Methods of Data Collection

Multiple-Choice Questions

Directions: The questions or incomplete statements that follow are each followed by five suggested answers or completions. Choose the response that best answers the question or completes the statement.

1. When travelers change airlines during connecting flights, each airline receives a portion of the fare. Several years ago, the major airlines used a sample trial period to determine what percentage of certain fares each should collect. Using these statistical results to determine fare splits, the airlines now claim huge savings over previous clerical costs. Which of the following is true?

   I. The airlines ran an experiment using a trial period for the control group.
   II. The airlines ran an observational study using the calculations from a trial period as a sample.
   III. The airlines feel that any monetary error in fare splitting resulting from using a statistical sample is smaller than the previous clerical costs necessary to calculate exact fare splits.

   (A) I only
   (B) II only
   (C) III only
   (D) I and III
   (E) II and III

2. Which of the following are true statements?

   I. In an experiment some treatment is intentionally forced on one group to note the response.
   II. In an observational study information is gathered on an already existing situation.
   III. Sample surveys are observational studies, not experiments.

   (A) I and II
   (B) I and III
   (C) II and III
   (D) I, II, and III
   (E) None of the above gives the complete set of true responses.

3. Which of the following are true statements?

 I. In an experiment researchers decide how people are placed in different groups.
 II. In an observational study, the participants select which group they are in.
 III. A control group is most often a self-selected grouping in an experiment.

 (A) I and II
 (B) I and III
 (C) II and III
 (D) I, II, and III
 (E) None of the above gives the complete set of true responses.

4. In one study on the effect of niacin on cholesterol level, 100 subjects who acknowledged being long-time niacin takers had their cholesterol levels compared with those of 100 people who had never taken niacin. In a second study, 50 subjects were randomly chosen to receive niacin and 50 were chosen to receive a placebo.

 (A) The first study was a controlled experiment, while the second was an observational study.
 (B) The first study was an observational study, while the second was a controlled experiment.
 (C) Both studies were controlled experiments.
 (D) Both studies were observational studies.
 (E) Each study was part controlled experiment and part observational study.

5. In one study subjects were randomly given either 500 or 1000 milligrams of vitamin C daily, and the number of colds they came down with during a winter season was noted. In a second study people responded to a questionnaire asking about the average number of hours they sleep per night and the number of colds they came down with during a winter season.

 (A) The first study was an experiment without a control group, while the second was an observational study.
 (B) The first study was an observational study, while the second was a controlled experiment.
 (C) Both studies were controlled experiments.
 (D) Both studies were observational studies.
 (E) None of the above is a correct statement.

6. In a 1992 London study, 12 out of 20 migraine sufferers were given chocolate whose flavor was masked by peppermint, while the remaining eight sufferers received a similar-looking, similar-tasting tablet that had no chocolate. Within 1 day, five of those receiving chocolate complained of migraines, while no complaints were made by any of those who did not receive chocolate. Which of the following is a true statement?

   (A) This study was an observational study of 20 migraine sufferers in which it was noted how many came down with migraines after eating chocolate.
   (B) This study was a sample survey in which 12 out of 20 migraine sufferers were picked to receive peppermint-flavored chocolate.
   (C) A census of 20 migraine sufferers was taken, noting how many were given chocolate and how many developed migraines.
   (D) A study was performed using chocolate as a placebo to study one cause of migraines.
   (E) An experiment was performed comparing a treatment group that was given chocolate to a control group that was not.

7. Suppose you wish to compare the average class size of mathematics classes to the average class size of English classes in your high school. Which is the most appropriate technique for gathering the needed data?

   (A) Census
   (B) Sample survey
   (C) Experiment
   (D) Observational study
   (E) None of these methods is appropriate.

8. Which of the following are true statements?

   I. Based on careful use of control groups, experiments can often indicate cause-and-effect relationships.
   II. While observational studies may suggest relationships, great care must be taken in concluding that there is cause and effect because of the lack of control over lurking variables.
   III. A complete census is the only way to establish a cause-and-effect relationship absolutely.

   (A) I and II
   (B) I and III
   (C) II and III
   (D) I, II, and III
   (E) None of the above gives the complete set of true responses.

9. Two studies are run to compare the experiences of families living in high-rise public housing to those of families living in townhouse subsidized rentals. The first study interviews 25 families who have been in each government program for at least 1 year, while the second randomly assigns 25 families to each program and interviews them after 1 year. Which of the following is a true statement?

   (A) Both studies are observational studies because of the time period involved.
   (B) Both studies are observational studies because there are no control groups.
   (C) The first study is an observational study, while the second is an experiment.
   (D) The first study is an experiment, while the second is an observational study.
   (E) Both studies are experiments.

10. Two studies are run to determine the effect of low levels of wine consumption on cholesterol level. The first study measures the cholesterol levels of 100 volunteers who have not consumed alcohol in the past year and compares these values with their cholesterol levels after 1 year, during which time each volunteer drinks one glass of wine daily. The second study measures the cholesterol levels of 100 volunteers who have not consumed alcohol in the past year, randomly picks half the group to drink one glass of wine daily for a year while the others drink no alcohol for the year, and finally measures their levels again. Which of the following is a true statement?

    (A) The first study is an observational study, while the second is an experiment.
    (B) The first study is an experiment, while the second is an observational study.
    (C) Both studies are observational studies, but only one uses both randomization and a control group.
    (D) The first study is a census of 100 volunteers, while the second study is an experiment.
    (E) Both studies are experiments.

Answer Key

1. **E**	3. **A**	5. **A**	7. **A**	9. **C**
2. **D**	4. **B**	6. **E**	8. **A**	10. **E**

Answers Explained

1. **(E)** This study is not an experiment in which responses are being compared. It is an observational study in which the airlines use split fare calculations from a trial period as a sample to indicate the pattern of all split fare transactions. They claim that this leads to "huge savings."

2. **(D)** I and II can be considered part of the definitions of *experiment* and *observational study*. A sample survey does not impose any treatment; it simply counts a certain outcome, and so it is an observational study, not an experiment.

3. **(A)** In an experiment a control group is the untreated group picked by the researchers. It is usually best when the selection process involves chance.

4. **(B)** The first study was observational because the subjects were not chosen for treatment.

5. **(A)** The first study was an experiment with two treatment groups and no control group. The second study was observational; the researcher did not randomly divide the subjects into groups and have each group sleep a designated number of hours per night.

6. **(E)** This study was an experiment in which the researchers divided the subjects into treatment and control groups. A census would involve a study of all migraine sufferers, not a sample of 20. The response of the treatment group receiving chocolate was compared to the response of the control group receiving a placebo. The peppermint tablet with no chocolate was the placebo.

7. **(A)** The main office at your school should be able to give you the class sizes of every math and English class. If need be, you can check with every math and English teacher.

8. **(A)** A complete census can provide much information about a population, but it doesn't necessarily establish a cause-and-effect relationship among seemingly related population parameters.

9. **(C)** In the first study the families were already in the housing units, while in the second study one of two treatments was applied to each family.

10. **(E)** Both studies apply treatments and measure responses, and so both are experiments.

Planning and Conducting Surveys

- Simple Random Sampling
- Characteristics of a Well-Designed, Well-Conducted Survey
- Sampling Error
- Sources of Bias
- Other Sampling Methods

Most data collection involves observational studies, not controlled experiments. Furthermore, while most data collection has some purpose, many studies come to mind after the data have been assembled and examined. For data collection to be useful, the resulting sample must be representative of the population under consideration.

SIMPLE RANDOM SAMPLING

How can a good, that is, a representative, sample be chosen? The most accurate technique would be to write the name of each member of the population on a card, mix the cards thoroughly in a large box, and pull out a specified number of cards. This method would give everyone in the population an equal chance of being selected as part of the sample. Unfortunately, this method is usually too time-consuming and too costly, and bias might still creep in if the mixing is not thorough. A *simple random sample*, that is, **one in which every possible sample of the desired size has an equal chance of being selected**, can more easily be obtained by assigning a number to everyone in the population and using a random number table or having a computer generate random numbers to indicate choices.

EXAMPLE 7.1

Suppose 80 students are taking an AP Statistics course and the teacher wants to randomly pick out a sample of 10 students to try out a practice exam. She first assigns the students numbers 01, 02, 03, . . . , 80. Reading off two digits at a time from a random number table, she ignores any over 80 and ignores repeats, stopping when she has a set of ten. If the table began 75425 56573 90420 48642 27537 61036 15074 84675, she would choose the students numbered 75, 42, 55, 65, 73, 04, 27, 53, 76, and 10. Note that 90 and 86 are ignored because they are over 80, and the second and third occurrences of 42 are ignored because they are repeats.

CHARACTERISTICS OF A WELL-DESIGNED, WELL-CONDUCTED SURVEY

A well-designed survey always incorporates chance, such as using random numbers from a table or a computer. However, the use of probability techniques is not enough to ensure a representative sample. Often we don't have a complete listing of the population, and so we have to be careful about exactly how we are applying "chance." Even when subjects are picked by chance, they may choose not to respond to the survey or they may not be available to respond, thus calling into question how representative the final sample really is. The wording of the questions must be neutral—subjects give different answers depending on the phrasing.

EXAMPLE 7.2

Suppose we are interested in determining the percentage of adults in a small town who eat a nutritious breakfast. How about randomly selecting 100 numbers out of the telephone book, calling each one, and asking whether the respondent is intelligent enough to eat a nutritious breakfast every morning?

Answer: Random selection is good, but a number of questions should be addressed. For example, are there many people in the town without telephones or with unlisted numbers? How will the time of day the calls are made affect whether the selected people are reachable? If people are unreachable, will replacements be randomly chosen in the same way or will this lead to a certain class of people being underrepresented? Finally, even if these issues are satisfactorily addressed, the wording of the question is clearly not neutral—unless the phrase *intelligent enough* is dropped, answers will be almost meaningless.

SAMPLING ERROR: THE VARIATION INHERENT IN A SURVEY

No matter how well-designed and well-conducted a survey is, it still gives a sample *statistic* as an estimate for a population *parameter*. Different samples give different sample statistics, all of which are estimates for the same population parameter, and so error, called *sampling error*, is naturally present. This error can be described using probability; that is, we can say how likely we are to have a certain size error. Generally, the chance of this error occurring is smaller when the sample size is larger. However, the way the data are obtained is crucial—a large sample size cannot make up for a poor survey design or faulty collection techniques.

EXAMPLE 7.3

Each of four major news organizations surveys likely voters and separately reports that the percentage favoring the incumbent candidate is 53.4%, 54.1%, 52.0%, and 54.2%, respectively. What is the correct percentage? Did three or more of the news organizations make a mistake?

Answer: There is no way of knowing the correct population percentage from the information given. The four surveys led to four statistics, each an estimate of the population parameter. No one made a mistake unless there was a bad survey, for example, one without the use of chance, or not representative of the population, or with poor wording of the question. Sampling differences are natural.

SOURCES OF BIAS IN SURVEYS

Poorly designed sampling techniques result in *bias,* that is, in a tendency to favor the selection of certain members of a population. An often-cited example of *selection bias* is the *Literary Digest* opinion poll that predicted a landslide victory for Alfred Landon over Franklin D. Roosevelt in the 1936 presidential election. The *Digest* surveyed people with cars and telephones, but in 1936 only the wealthy minority, who mainly voted Republican, had cars and telephones. Examples of *nonresponse bias* are present in most mailed questionnaires. They tend to have very low response percentages, and it is often unclear which part of the population is responding. *Unintentional bias* often creeps in when the surveyor tries to systematically pick people representative of the whole population. For example, in the 1948 presidential election, the *Chicago Tribune* incorrectly called Thomas E. Dewey the winner over Harry S. Truman. Here the mistake was partly due to misleading polls based on *quota sampling* that left the interviewers too much free choice in picking people to fill their quotas.

The wording of questions or the very questions themselves can lead to *response bias.* People often don't want to be perceived as having unpopular, unsavory, or illegal views and so may respond untruthfully when face to face with an interviewer or when filling out a questionnaire that is not anonymous. Sometimes people chosen for a survey simply refuse to respond or are unreachable or too difficult to contact. These situations lead to *nonresponse bias.* At other times certain people are left out of consideration, which results in *undercoverage bias.* For example, telephone surveys simply ignore all those possible subjects who don't have telephones, and door-to-door surveys ignore the homeless.

Two common sampling techniques that often result in flawed conclusions are voluntary response samples and convenience samples. *Voluntary response samples,* based on individuals who offer to participate, typically give too much emphasis to persons with strong opinions. For example, radio call-in programs about controversial topics such as gun control, abortion, and school segregation lead to *voluntary response bias,* and do not produce meaningful data on what proportion of the population favor or oppose related issues. *Convenience samples,* based on choosing individuals who are easy to reach, are also suspect. For example, interviews at shopping malls tend to produce data highly unrepresentative of the entire population, often leading to *undercoverage bias.*

EXAMPLE 7.4

In response to student protests over the firing of Coach Niceguy, the university public relations spokesperson released the following statement: "A questionnaire mailed to 120,000 alumni produced over 5000 responses, 60% of which were unfavorable to Coach Niceguy. Furthermore, these respondents contributed an average gift of over $200 last year, including Mr. Mon E. Bags' contribution of a million dollars for a new luxury box on the fifty yard line. Finally, 30% of the faculty have expressed unhappiness with Niceguy, so that a total of 90% of the alumni and faculty want him to go. We had no choice but to fire him." What do you think of the university's explanation of the firing?

Answer: There are so many examples of improper reasoning above that it is hard to know where to begin. Only 5000 of 120,000 alumni responded (nonresponse bias). The survey was unrepresentative, as most respondents seem to have been contributors (voluntary response survey). The average contribution of $200 is meaningless if one of the contributions was $1 million. You can't add 60% of one group and 30% of another group and come up with 90% of anything.

OTHER SAMPLING METHODS

Time- and cost-saving modifications are often used to implement sampling procedures other than simple random samples.

Systematic sampling involves listing the population in some order (for example, alphabetically), choosing a random point to start, and then picking every tenth (or hundredth, or thousandth, or *k*th) person from the list. This gives a reasonable sample as long as the original order of the list is not in any way related to the variables under consideration.

In *stratified sampling* the population is divided into *homogeneous* groups called *strata*, and random samples of persons from all strata are chosen. For example, we can stratify by age or gender or income level or race and pick a sample of people from each stratum. Note that all individuals in a given stratum have a characteristic in common. We could further do *proportional sampling,* where the sizes of the random samples from each stratum depend on the proportion of the total population represented by the stratum.

In *cluster sampling* the population is divided into *heterogeneous* groups called *clusters,* and we then take a random sample of clusters from among all the clusters. For example, to survey high school seniors we could randomly pick several senior class homerooms in which to conduct our study. Note that each cluster should resemble the entire population.

Multistage sampling refers to a procedure involving two or more steps, each of which could involve any of the various sampling techniques.

EXAMPLE 7.5

Suppose a sample of 100 high school students from a school of size 5000 is to be chosen to determine their views on the death penalty. One method would be to have each student write his or her name on a slip of paper, put the papers in a box, and have the principal reach in and pull out 100 of the papers. However, questions could arise regarding how well the papers are mixed up in the box. For example, how might the outcome be affected if all students in one homeroom toss in their names at the same time so that their papers are clumped together? Another method would be to assign each student a number from 1 to 5000 and then use a random number table, picking out four digits at a time and tossing out repeats and numbers over 5000 (simple random sampling). What are alternative procedures?

Answer: From a list of the students, the surveyor could simply note every fiftieth name (systematic sampling). Since students in each class have certain characteristics in common, the surveyor could use a random selection method to pick 25 students from each of the separate lists of freshmen, sophomores, juniors, and seniors (stratified sampling). The researcher could separate the homerooms by classes; then randomly pick five freshmen homerooms, five sophomore homerooms, five junior homerooms, and five senior homerooms (cluster sampling); and then randomly pick five students from each of the homerooms (multistage sampling). The surveyor could separately pick random samples of males and females (stratified sampling), the size of each of the two samples chosen according to the proportion of male and female students attending the school (proportional sampling).

It should be noted that none of the alternative procedures in the above example result in a *simple random sample* because every possible sample of size 100 does not have an equal chance of being selected.

Questions on Topic Seven: Planning and Conducting Surveys

Multiple-Choice Questions

Directions: The questions or incomplete statements that follow are each followed by five suggested answers or completions. Choose the response that best answers the question or completes the statement.

1. Ann Landers, who wrote a daily advice column appearing in newspapers across the country, once asked her readers, "If you had it to do over again, would you have children?" Of the more than 10,000 readers who responded, 70% said no. What does this show?

   (A) The survey is meaningless because of voluntary response bias.
   (B) No meaningful conclusion is possible without knowing something more about the characteristics of her readers.
   (C) The survey would have been more meaningful if she had picked a random sample of the 10,000 readers who responded.
   (D) The survey would have been more meaningful if she had used a control group.
   (E) This was a legitimate sample, randomly drawn from her readers and of sufficient size to allow the conclusion that most of her readers who are parents would have second thoughts about having children.

2. Which of the following are true statements?

   I. If bias is present in a sampling procedure, it can be overcome by dramatically increasing the sample size.
   II. There is no such thing as a "bad sample."
   III. Sampling techniques that use probability techniques effectively eliminate bias.

   (A) I only
   (B) II only
   (C) III only
   (D) None of the statements are true.
   (E) None of the above gives the complete set of true responses.

3. Two possible wordings for a questionnaire on gun control are as follows:

   I. The United States has the highest rate of murder by handguns among all countries. Most of these murders are known to be crimes of passion or crimes provoked by anger between acquaintances. Are you in favor of a 7-day cooling-off period between the filing of an application to purchase a handgun and the resulting sale?

   II. The United States has one of the highest violent crime rates among all countries. Many people want to keep handguns in their homes for self-protection. Fortunately, U.S. citizens are guaranteed the right to bear arms by the Constitution. Are you in favor of a 7-day waiting period between the filing of an application to purchase a needed handgun and the resulting sale?

   One of these questions showed that 25% of the population favored a 7-day waiting period between application for purchase of a handgun and the resulting sale, while the other question showed that 70% of the population favored the waiting period. Which produced which result and why?

   (A) The first question probably showed 70% and the second question 25% because of the lack of randomization in the choice of pro-gun and anti-gun subjects as evidenced by the wording of the questions.

   (B) The first question probably showed 25% and the second question 70% because of a placebo effect due to the wording of the questions.

   (C) The first question probably showed 70% and the second question 25% because of the lack of a control group.

   (D) The first question probably showed 25% and the second question 70% because of response bias due to the wording of the questions.

   (E) The first question probably showed 70% and the second question 25% because of response bias due to the wording of the questions.

4. Which of the following are true statements?

   I. Voluntary response samples often underrepresent people with strong opinions.

   II. Convenience samples often lead to undercoverage bias.

   III. Questionnaires with nonneutral wording are likely to have response bias.

   (A) I and II
   (B) I and III
   (C) II and III
   (D) I, II, and III
   (E) None of the above gives the complete set of true responses.

5. Each of the 29 NBA teams has 12 players. A sample of 58 players is to be chosen as follows. Each team will be asked to place 12 cards with their players names into a hat and randomly draw out two names. The two names from each team will be combined to make up the sample. Will this method result in a simple random sample of the 348 basketball players?

   (A) Yes, because each player has the same chance of being selected.
   (B) Yes, because each team is equally represented.
   (C) Yes, because this is an example of stratified sampling, which is a special case of simple random sampling.
   (D) No, because the teams are not chosen randomly.
   (E) No, because not each group of 58 players has the same chance of being selected.

6. To survey the opinions of bleacher fans at Wrigley Field, a surveyor plans to select every one-hundredth fan entering the bleachers one afternoon. Will this result in a simple random sample of Cub fans who sit in the bleachers?

   (A) Yes, because each bleacher fan has the same chance of being selected.
   (B) Yes, but only if there is a single entrance to the bleachers.
   (C) Yes, because the 99 out of 100 bleacher fans who are not selected will form a control group.
   (D) Yes, because this is an example of systematic sampling, which is a special case of simple random sampling.
   (E) No, because not every sample of the intended size has an equal chance of being selected.

7. Which of the following are true statements about sampling error?

   I. Sampling error can be eliminated only if a survey is both extremely well designed and extremely well conducted.
   II. Sampling error concerns natural variation between samples, is always present, and can be described using probability.
   III. Sampling error is generally smaller when the sample size is larger.

   (A) I and II
   (B) I and III
   (C) II and III
   (D) I, II, and III
   (E) None of the above gives the complete set of true responses.

8. What fault do all these sampling designs have in common?

   I. The *Wall Street Journal* plans to make a prediction for a presidential election based on a survey of its readers.
   II. A radio talk show asks people to phone in their views on whether the United States should pay off its huge debt to the United Nations.
   III. A police detective, interested in determining the extent of drug use by teenagers, randomly picks a sample of high school students and interviews each one about any illegal drug use by the student during the past year.

   (A) All the designs make improper use of stratification.
   (B) All the designs have errors that can lead to strong bias.
   (C) All the designs confuse *association* with *cause and effect*.
   (D) None of the designs satisfactorily controls for sampling error.
   (E) None of the designs makes use of chance in selecting a sample.

9. A state auditor is given an assignment to choose and audit 26 companies. She lists all companies whose name begins with *A*, assigns each a number, and uses a random number table to pick one of these numbers and thus one company. She proceeds to use the same procedure for each letter of the alphabet and then combines the 26 results into a group for auditing. Which of the following are true statements?

   I. Her procedure makes use of chance.
   II. Her procedure results in a simple random sample.
   III. Each company has an equal probability of being audited.

   (A) I and II
   (B) I and III
   (C) II and III
   (D) I, II, and III
   (E) None of the above gives the complete set of true responses.

10. A researcher planning a survey of heads of households in a particular state has census lists for each of the 23 counties in that state. The procedure will be to obtain a random sample of heads of households from each of the counties rather than grouping all the census lists together and obtaining a sample from the entire group. Which of the following is a true statement about the resulting stratified sample.

    I. It is not a simple random sample.
    II. It is easier and less costly to obtain than a simple random sample.
    III. It gives comparative information that a simple random sample wouldn't give.

    (A) I only
    (B) I and II
    (C) I and III
    (D) I, II, and III
    (E) None of the above gives the complete set of true responses.

11. To find out the average occupancy size of student-rented apartments, a researcher picks a simple random sample of 100 such apartments. Even after one follow-up visit, the interviewer is unable to make contact with anyone in 27 of these apartments. Concerned about nonresponse bias, the researcher chooses another simple random sample and instructs the interviewer to continue this procedure until contact is made with someone in a total of 100 apartments. The average occupancy size in the final 100-apartment sample is 2.78. Is this estimate probably too low or too high?

    (A) Too low, because of undercoverage bias.
    (B) Too low, because convenience samples overestimate average results.
    (C) Too high, because of undercoverage bias.
    (D) Too high, because convenience samples overestimate average results.
    (E) Too high, because voluntary response samples overestimate average results.

12. To conduct a survey of long-distance calling patterns, a researcher opens a telephone book to a random page, closes his eyes, puts his finger down on the page, and then reads off the next 50 names. Which of the following are true statements?

    I. The survey design incorporates chance.
    II. The procedure results in a simple random sample.
    III. The procedure could easily result in selection bias.

    (A) I and II
    (B) I and III
    (C) II and III
    (D) I, II, and III
    (E) None of the above gives the complete set of true responses.

13. Which of the following are true statements about sampling?

    I. Careful analysis of a given sample will indicate whether or not it is random.
    II. Sampling error implies an error, possibly very small but still an error, on the part of the surveyor.
    III. Data obtained while conducting a census are always more accurate than data obtained from a sample, no matter how careful the design of the sample.

    (A) I only
    (B) II only
    (C) III only
    (D) None of these statements are true.
    (E) None of the above gives the complete set of true responses.

14. Consider the following three events:

    I. Although 18% of the student body are minorities, in a random sample of 20 students, 5 are minorities.

    II. In a survey about sexual habits, an embarrassed student deliberately gives the wrong answers.

    III. A surveyor mistakenly records answers to one question in the wrong space.

    Which of the following correctly characterizes the above?

    (A) I, sampling error; II, response bias; III, human mistake
    (B) I, sampling error; II, nonresponse bias; III, hidden error
    (C) I, hidden bias; II, voluntary sample bias; III, sampling error
    (D) I, undercoverage error; II, voluntary error; III, unintentional error
    (E) I, small sample error; II, deliberate error; III, mistaken error

15. A researcher plans a study to examine the depth of belief in God among the adult population. He obtains a simple random sample of 100 adults as they leave church one Sunday morning. All but one of them agree to participate in the survey. Which of the following are true statements?

    I. Proper use of chance as evidenced by the simple random sample makes this a well-designed survey.

    II. The high response rate makes this a well-designed survey.

    III. Selection bias makes this a poorly designed survey.

    (A) I only
    (B) II only
    (C) III only
    (D) I and II
    (E) None of these statements is true.

Answer Key

1. **A**	4. **C**	7. **C**	10. **D**	13. **D**
2. **D**	5. **E**	8. **B**	11. **C**	14. **A**
3. **E**	6. **E**	9. **E**	12. **B**	15. **C**

Answers Explained

1. **(A)** This survey provides a good example of voluntary response bias, which often overrepresents negative opinions. The people who chose to respond were most likely parents who were very unhappy, and so there is very little chance that the 10,000 respondents were representative of the population. Knowing more about her readers, or taking a sample of the sample would not have helped.

2. **(D)** If there is bias, taking a larger sample just magnifies the bias on a larger scale. If there is enough bias, the sample can be worthless. Even when the subjects are chosen randomly, there can be bias due, for example, to non-response or to the wording of the questions.

3. **(E)** The wording of the questions can lead to response bias. The neutral way of asking this question would simply have been: Are you in favor of a 7-day waiting period between the filing of an application to purchase a handgun and the resulting sale?

4. **(C)** Voluntary response samples, like radio call-in surveys, are based on individuals who offer to participate, and they typically overrepresent persons with strong opinions. Convenience samples, like shopping mall surveys, are based on choosing individuals who are easy to reach, and they typically miss a large segment of the population. Nonneutral wording can readily lead to response bias, and if surveyors want a particular result, they deliberately build bias into the wording.

5. **(E)** In a simple random sample, every possible group of the given size has to be equally likely to be selected, and this is not true here. For example, with this procedure it will be impossible for all the Bulls to be together in the final sample. This procedure is an example of stratified sampling, but stratified sampling does not result in simple random samples.

6. **(E)** In a simple random sample, every possible group of the given size has to be equally likely to be selected, and this is not true here. For example, with this procedure it will be impossible for all the early arrivals to be together in the final sample. This procedure is an example of systematic sampling, but systematic sampling does not result in simple random samples.

7. **(C)** Different samples give different sample statistics, all of which are estimates of a population parameter. Sampling error relates to natural variation between samples, can never be eliminated, can be described using probability, and is generally smaller if the sample size is larger.

8. **(B)** The *Wall Street Journal* survey has strong selection bias; that is, people who read the *Journal* are not very representative of the general population. The talk show survey results in a *voluntary response sample*, which typically gives too much emphasis to persons with strong opinions. The police detective's survey has strong response bias in that students may not give truthful responses to a police detective about their illegal drug use.

9. **(E)** Only I is a true statement. While the auditor does use chance, each company will have the same chance of being audited only if the same number of companies have names starting with each letter of the alphabet. This will not result in a simple random sample because each possible set of 26 companies does not have the same chance of being picked as the sample. For example, a group of companies whose names all start with *A* will not be chosen.

10. **(D)** This is not a simple random sample because all possible sets of the required size do not have the same chance of being picked. For example, a set of households all from just half the counties has no chance of being picked to be the sample. Stratified samples are often easier and less costly to obtain and

also make comparative data available. In this case responses can be compared among various counties.

11. **(C)** It is most likely that the apartments at which the interviewer had difficulty finding someone home were apartments with fewer students living in them. Replacing these with other randomly picked apartments most likely replaces smaller-occupancy apartments with larger-occupancy ones.

12. **(B)** While the procedure does use some element of chance, all possible groups of size 50 do not have the same chance of being picked, and so the result is not a simple random sample. There is a very real chance of selection bias. For example, a number of relatives with the same name and similar long-distance calling patterns might be selected.

13. **(D)** To determine if a sample is random, one must analyze the procedure by which it was obtained. Sampling error is natural variation among samples; it is not an error committed by any person. If a census is poorly run, it will provide less information and be less accurate than a well-designed survey. For example, having the principal ask every single student in the school whether or not he or she regularly cheats on exams produces less useful data than a carefully worded anonymous questionnaire filled out by a randomly selected sample of the student body.

14. **(A)** The natural variation in samples is called sampling error. Embarrassing questions and resulting untruthful answers are an example of response bias. Inaccuracies and mistakes due to human error are one of the real concerns of researchers.

15. **(C)** Surveying people coming out of church results in a very unrepresentative sample of the adult population, especially given the question under consideration. Using chance and obtaining a high response rate will not change the selection bias and make this into a well-designed survey.

Free-Response Questions

Directions: You must show all work and indicate the methods you use. You will be graded on the correctness of your methods and on the accuracy of your final answers.

Five Open-Ended Questions

1. A questionnaire is being designed to determine whether most people are or are not in favor of legislation protecting the habitat of the spotted owl. Give two examples of poorly worded questions, one biased toward each response.

2. To obtain a sample of 25 students from among the 500 students present in school one day, a surveyor decides to pick every twentieth student waiting on line to attend a required assembly in the gym.

   a. Explain why this procedure will not result in a simple random sample of the students present that day.

   b. Describe a procedure that will result in a simple random sample of the students present that day.

3. A hot topic in government these days is welfare reform. Suppose a congress-woman wishes to survey her constituents concerning their opinions on whether the federal government should turn welfare over to the states. Discuss possible sources of bias with regard to the following four options: (1) conducting a survey via random telephone dialing into her district, (2) sending out a mailing using a registered voter list, (3) having a pollster interview everyone who walks past her downtown office, and (4) broadcasting a radio appeal urging interested citizens in her district to call in their opinions to her office.

4. You and nine friends go to a restaurant and check your coats. You all forget to pick up the ticket stubs, and so when you are ready to leave, Hilda, the hat-check girl, randomly gives each of you one of the ten coats. You are surprised that one person actually receives the correct coat. You would like to explore this further and decide to use a random number table to simulate the situation. Describe how the random number table can be used to simulate one trial of the coat episode. Explain what each of the digits 0 through 9 will represent.

5.

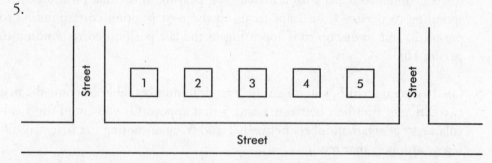

   You are supposed to interview the residents of two of the above five houses.

   (a) How would you choose which houses to interview?

   (b) You plan to visit the homes at 9 a.m. If someone isn't home, explain the reasons for and against substituting another house.

   (c) Are there any differences you might expect to find among the residents based on the above sketch?

Answers Explained

1. There are many possible examples, such as Are you in favor of protecting the habitat of the spotted owl, which is almost extinct and desperately in need of help from an environmentally conscious government? and Are you in favor of protecting the habitat of the spotted owl no matter how much unemployment and resulting poverty this causes among hard-working loggers?

2. *a.* To be a simple random sample, every possible group of size 25 has to be equally likely to be selected, and this is not true here. For example, if there are 40 students who always rush to be first in line, this procedure will allow for only 2 of them to be in the sample. Or if each homeroom of size 20 arrives as a unit, this procedure will allow for only 1 person from each homeroom to be in the sample.

   b. A simple random sample of the students can be obtained by numbering them from 001 to 500 and then picking three digits at a time from a random number table, ignoring numbers over 500 and ignoring repeats, until a group of 25 numbers is obtained. The students corresponding to these 25 numbers will be a simple random sample.

3. The direct telephone and mailing options will both suffer from undercoverage bias. For example, especially affected by the legislation under discussion are the homeless, and they do not have telephones or mailing addresses. The pollster interviews will result in a convenience sample, which can be highly unrepresentative of the population. In this case, there might be a real question concerning which members of her constituency spend any time in the downtown area where her office is located. The radio appeal will lead to a voluntary response sample, which typically gives too much emphasis to persons with strong opinions.

4. In numbering the people 0 through 9, each digit stands for whose coat someone receives. Pick the digits, omitting repeats, until a group of ten different digits is obtained. Check for a match (1 appearing in the first position corresponding to person 1, or 2 appearing in the next position corresponding to person 2, and so on, up to 0 appearing in the last position corresponding to person 10).

5. (a) To obtain an SRS, you might use a random number table and note the first two different numbers between 1 and 5 that appear. Or you could use a calculator to generate numbers between 1 and 5, again noting the first two different numbers that result.

   (b) Time and cost considerations would be the benefit of substitution. However, substitution rather than returning to the same home later could lead to selection bias because certain types of people are not and will not be home at 9 a.m. With substitution the sample would no longer be a simple random sample.

   (c) Corner lot homes like homes 1 and 5 might have different residents (perhaps with higher income levels) than other homes.

An Investigative Task

1. Suppose that in your statistics class of 30 students, 2 students discover they have the same birthday (month and date). You decide to test whether this is an unusual occurrence by using a random number table to simulate picking 30 randomly selected birth dates.

   a. Clearly explain how to perform this test and then do so, showing your procedure step by step and your result.

   b. Do you feel justified in making some conclusion based on the result of your test? Explain.

   c. Can you make any definite conclusion about the probability of finding two students with the same birth month? Explain.

   84177 06757 17613 15582 51506 81435 41050 92031 06449 05059
   59884 31180 53115 84469 94868 57967 05811 84514 75011 13006
   63395 55041 15866 06589 13119 71020 85940 91932 06488 74987
   54355 52704 90359 02649 47496 71567 94268 08844 26294 64759
   08989 57024 97284 00637 89283 03514 59195 07635 03309 72605
   29357 23737 67881 03668 33876 35841 52869 23114 15864 38942

Answers Explained

1. *a.* Let the birth dates correspond to the numbers 001 (January 1) through 365 (December 31). Go through the random number table, picking three digits at a time, and throwing away 000 and any numbers over 365. After listing a group of 30 such numbers, check for duplicates.

   Looking at groups of three digits, and throwing away numbers over 365 gives {l55, 150, 203, 106, 050, 180, 158, 184, 111, 300, 131, 197, 102, 085, 320, 270, 359, 026, 294, 024, 063, 283, 035, 145, 024, 097, 260, 357, 237, 103}. Note that there is a duplicate, namely, 024.

   b. There happens to be a number (birthday) that is duplicated among the sample of 30 numbers from the random number table. Clearly there won't always be duplicates. Before we can draw any conclusion, we should repeat this procedure, picking groups of 30 random numbers between 001 and 365 many more times. Only then might we be able to estimate the probability of finding a duplicate birthday in a room of 30 people.

   c. The probability is 1 that 2 of the 30 students will have the same birth month because there is no way of spreading 30 students among 12 months without having duplicates.

Planning and Conducting Experiments

- Experiments versus Observational Studies
- Confounding, Control Groups, Placebo Effects, Blinding
- Treatments, Experimental Units, Randomization
- Replication, Blocking, Generalizability of Results

There are several primary principles dealing with the proper planning and conducting of experiments. First, possible confounding variables must be controlled for, usually through the use of comparison. Second, chance should be used in assigning which subjects are to be placed in which groups for which treatment. Third, natural variation in outcomes can be lessened by using more subjects.

EXPERIMENTS VERSUS OBSERVATIONAL STUDIES VERSUS SURVEYS

In an experiment we impose some change or treatment and measure the result or response. In an observational study we simply observe and measure something that has taken place or is taking place, while trying not to cause any changes by our presence. A sample survey is an observational study in which we draw conclusions about an entire population by considering an appropriately chosen sample to look at. An experiment often suggests a causal relationship, while an observational study may show only the existence of associations.

EXAMPLE 8.1

A study is to be designed to determine whether daily calcium supplements benefit women by increasing bone mass. How can an observational study be performed? An experiment? Which is more appropriate here?

Answer: An observational study might interview and run tests on women seen purchasing calcium supplements in a pharmacy. Or perhaps all patients hospitalized during a particular time period could be interviewed with regard to taking calcium and then their bone mass measured. The bone mass measurements of those taking calcium supplements could then be compared to that of those not taking supplements.

(continued)

An experiment could be performed by selecting some number of subjects, using chance to pick half to receive calcium supplements while the other half receives similar-looking placebos, and noting the difference in bone mass before and after treatment for each group.

The experimental approach is more appropriate here. With the observational study there could be many explanations for any bone mass difference noted between patients who take calcium and those who don't. For example, women who have voluntarily been taking calcium supplements might be precisely those who take better care of themselves in general and thus have higher bone mass for other reasons. The experiment tries to control for lurking variables by randomly giving half the subjects calcium.

EXAMPLE 8.2

A study is to be designed to examine the life expectancies of tall people versus those of short people. Which is more appropriate, an observational study or an experiment?

Answer: An observational study, examining medical records of heights and ages at time of death, seems straightforward. An experiment where subjects are randomly chosen to be made short or tall, followed by recording age at death, would be groundbreaking (and, of course, nonsensical).

EXAMPLE 8.3

A study is to be designed to examine the GPAs of students who take marijuana regularly and those who don't. Which is more appropriate, an observational study or an experiment?

Answer: As much as some researchers might want to randomly require half the subjects to take an illegal drug, this would be unethical. The proper procedure here is an observational study, having students anonymously fill out questionnaires asking about marijuana usage and GPA.

Experiments involve *explanatory variables*, called *factors*, that are believed to have an effect on *response variables*. A group is treated with some level of the explanatory variable, and the outcome on the response variable is measured.

EXAMPLE 8.4

To test the value of help sessions outside the classroom, students could be divided into three groups, with one group receiving 4 hours of help sessions per week outside the classroom, a second group receiving 2 hours of help sessions outside the classroom, and a third group receiving no help outside the classroom. What are the explanatory and response variables and what are the levels?

Answer: The explanatory variable, help sessions outside the classroom, is being given at three levels: 4 hours weekly, 2 hours weekly, and 0 hours weekly. The response variable is not specified but might be a final exam score or performance on a particular test.

The different factor-level combinations are called *treatments*. In Example 8.4, there are three treatments (corresponding to the three levels of the one factor). Suppose the students were further randomly divided into a morning class and an afternoon class. There would then be two factors, one with three levels and one with two levels, and a total of six treatments (AM class with 4 hrs help, AM class with 2 hrs help, AM class with 0 hrs help, PM class with 4 hrs help, PM class with 2 hrs help, and PM class with 0 hrs help).

CONFOUNDING, CONTROL GROUPS, PLACEBO EFFECTS, AND BLINDING

When there is uncertainty with regard to which variable is causing an effect, we say the variables are *confounded.* For example, suppose two fertilizers require different amounts of watering. In an experiment it might be difficult to determine if the difference in fertilizers or the difference in watering is the real cause of observed differences in plant growth. Sometimes we can control for confounding. For example, we can have many test plots using one or the other of the fertilizers, with equal numbers of sunny and shady plots for each fertilizer, so that fertilizer and sun are not confounded.

A *lurking variable* is a variable that drives two other variables, creating the mistaken impression that the two other variables are related by cause and effect. For example, elementary school students with larger shoe sizes appear to have higher reading levels. However, there is a lurking variable, age, which drives both the other variables. That is, older students tend to wear larger shoes than younger students, and older students also tend to have higher reading levels. Wearing larger shoes will not improve reading skills!

In an experiment there is a group that receives the treatment, and there is a *control group* that doesn't. The experiment compares the responses in the treatment group to the responses in the control group. Randomly putting subjects into treatment and control groups can help reduce the problems posed by confounding and lurking variables. Thus these problems are easier to control for when doing experiments than when doing observational studies.

It is a fact that many people respond to any kind of perceived treatment. This is called the *placebo effect.* For example, when given a sugar pill after surgery but told that it is a strong pain reliever, many patients feel immediate relief from their pain. In many studies, subjects appear to consciously or subconsciously want to help the researcher prove a point. Thus when responses are noticed in any experiment, there is concern whether real physical responses are being caused by the psychological placebo effect. *Blinding* occurs when the subjects or the response evaluators don't know which subjects are receiving different treatments such as placebos.

EXAMPLE 8.5

A study is intended to test the effects of vitamin E and beta carotene on heart attack rates. How should it be set up?

Answer: Using randomization, the subjects should be split into four groups: those who will be given just vitamin E, just beta carotene, both vitamin E and beta carotene, and neither vitamin E nor beta carotene. For example, as each subject joins the test, the next digit in a random number table can be read off, ignoring 0 and 5–9, and with 1, 2, 3, and 4 designating which group the subject is placed in. Or if the total number of subjects is known and available, for example, 800, then each can be assigned a number and three digits at a time be read off the random number table. With repeats and numbers over 800 thrown away, the first 200 numbers picked represent one group, the next 200 another, and so on. More meaningful results will be obtained if the study is double-blind, that is, if not only are the subjects unaware of what kind of tablets they are taking but neither are the doctors evaluating whether or not they have heart problems. Many diagnoses are not clear-cut, and doctors can be influenced if they know exactly which potential preventative their patients are taking.

TREATMENTS, EXPERIMENTAL UNITS, AND RANDOMIZATION

An experiment is performed on objects called *experimental units*, and if the units are people, they are called *subjects*. The experimental units or subjects are typically divided into two groups. One group receives a *treatment* and is called the *treatment group*. A comparison is made between the response noted in the treatment group and the response noted in the *control group*, the group that receives no treatment.

To help minimize the effect of lurking variables, and of confounding, it is important to use *randomization*, that is, to use chance in deciding which subjects go into which group. It is not sufficient to try to systematically match characteristics between the two groups. It seems reasonable, for example, to hand-sort subjects so that both the treatment group and the control group have the same number of women, the same number of Catholics, the same number of Hispanics, the same number of short people, and so on, but this method does not work well. There are always other variables that one might not think of considering until after the results of the experiment start coming in. The best method to use is randomization employing a computer, a hat with names in it, or a random number table.

Note that *randomization* usually refers to how given subjects are assigned to treatments, not to how a group of subjects are chosen from an entire population. The object of an experiment is to see if different treatments lead to different responses, and so we randomly assign subjects to treatments to balance unknown sources of variability. Random assignment to treatments is critical, especially if the subjects are not randomly selected, as is the case in medical/drug experiments. Generalizing the findings of the study is a separate question, one that depends on how the initial group of subjects was assembled.

COMPLETELY RANDOMIZED DESIGN FOR TWO TREATMENTS

Comparing two treatments using randomization is often the design of choice. To help minimize hidden bias, it is best if subjects do not know which treatment they are receiving. This is called *single-blinding*. Another precaution is the use of *double-blinding*, in which neither the subjects nor those evaluating their responses know who is receiving which treatment.

EXAMPLE 8.6

There is a pressure point on the wrist that some doctors believe can be used to help control the nausea experienced following certain medical procedures. The idea is to place a band containing a small marble firmly on a patient's wrist so that the marble is located directly over the pressure point. Describe how an experiment might be run on 50 postoperative patients.

Answer: Assign each patient a number from 01 to 50. From a random number table read off two digits at a time, throwing away repeats, 00, and numbers over 50, until 25 numbers have been selected. Put wristbands with marbles over the pressure point on the patients with these assigned numbers. Put wristbands with marbles on the remaining patients also, but *not* over the pressure point. Have a researcher check by telephone with all 50 patients at designated time intervals to determine the degree of nausea being experienced. Neither the patients nor the researcher on the telephone should know which patients have the marbles over the correct pressure point.

EXAMPLE 8.7

A chemical fertilizer company wishes to test whether using their product results in superior vegetables. After dividing a large field into small plots, how might the experiment proceed?

Answer: If the company has one recommended fertilizer application level, half the plots can be randomly selected (assigning the plots numbers and using a random number table) to receive the prescribed dosage of fertilizer. This random selection of plots is to ensure that neither fertilized plants nor unfertilized plants are inadvertently given land with better rainfall, sunshine, soil type, and so on. To avoid possible bias on the part of employees who will weed and water the plants, they should not know which plots have received the fertilizer. It might be necessary to have containers, one for each plot, of a similar-looking, similar-smelling sub-stance, half of which contain the fertilizer while the rest contain a chemically inactive mate-rial. Finally, if the vegetables are to be judged by quantity and size, the measurements will be less subject to bias. However, if they are to be judged qualitatively, for example, by taste, the judges should not know which vegetables were treated with the fertilizer and which were not.

If the researchers also wish to consider level, that is, the amount of fertilizer, randomiza-tion should be used for more groupings. For example, if there are 60 plots on which to test four levels of fertilizer, the first 12 different two-digit numbers in the range 01–60 appearing on a random number table might receive one level, the next 12 new two-digit numbers a sec-ond level, and so on, with the last 12 plots receiving the "placebo" treatment.

RANDOMIZED PAIRED COMPARISON DESIGN

Two treatments can be compared based on the responses of paired subjects, one of whom receives one treatment while the other receives the second treatment. Often the paired subjects are really single subjects who are given both treatments, one at a time.

EXAMPLE 8.8

The famous Pepsi-Coke tests had subjects compare the taste of samples of each drink. How could such a paired comparison test be set up?

Answer: It is crucial that such a test be blind, that is, that the subjects not know which cup contains which drink. Furthermore, to help avoid hidden bias, which drink the subjects taste first should be decided by chance. For example, as each subject arrives, the researcher could read off the next digit from a random number table, with the subject receiv-ing Pepsi or Coke first depending on whether the digit is odd or even.

EXAMPLE 8.9

Does seeing pictures of accidents caused by drunk drivers influence one's opinion on penal-ties for drunk drivers? How could a comparison test be designed?

Answer: The subjects could be asked questions about drunk driving penalties before and then again after seeing the pictures, and any change in answers noted. This would be a poor design because there is no control group, there is no use of randomization, and sub-jects might well change their answers because they realize that that is what is expected of them after seeing the pictures.

(continued)

A better design is to use randomization to split the subjects into two groups, half of whom simply answer the questions while the other half first see the pictures and then answer the questions.

Another possibility is to use a group of twins as subjects. One of each set of twins is randomly picked (e.g., based on choosing an odd or even digit from a random number table) to answer the questions without seeing the pictures, while the others first see the pictures and then answer the questions. The answers could be compared from each set of twins. This is a paired comparison test that might help minimize lurking variables due to family environment, heredity, and so on.

REPLICATION, BLOCKING, AND GENERALIZABILITY OF RESULTS

When differences are observed in a comparison test, the researcher must decide whether these differences are *statistically significant* or whether they can be explained by natural variation. One important consideration is the size of the sample—the larger the sample, the more significant the observation. This is the principle of *replication*; that is, the treatment should be repeated on a sufficient number of subjects so that real response differences are more apparent.

Just as stratification in sampling design first divides the population into representative groups called strata, *blocking* in experiment design first divides the subjects into representative groups called blocks. One can think of blocking as running a separate experiment on each block. This technique helps control certain lurking variables by bringing them directly into the picture and helps make conclusions more specific. The paired comparison design is a special case of blocking in which each pair (or each subject if the subjects serve as their own controls) can be considered a block.

EXAMPLE 8.10

There is a rising trend for star college athletes to turn professional without finishing their degrees. A study is performed to assess whether reading an article about professional salaries has an impact on such decisions. Randomization can be used to split the subjects into two groups, and those in one group given the article before answering questions. How can a block design be incorporated into the design of this experiment?

Answer: The subjects can be split into two blocks, underclass and upperclass, before using randomization to assign some to read the article before questioning. With this design, the impact of the salary article on freshmen and sophomores can be distinguished from the impact on juniors and seniors.

Similarly, blocking can be used to separately analyze men and women, those with high GPAs and those with low GPAs, those in different sports, those with different majors, and so on.

A major goal of experiments is to be able to *generalize* the results to broader populations. Often an experiment must be repeated in a variety of settings. For example, it is hard to generalize from the effect a television commercial has on students at a private midwestern high school to the effect the same commercial has on retired senior citizens in Florida. Generally, comparison and randomization are important, blinding is sometimes critical, and taking care to avoid hidden bias as much as possible is always indicative of a well-designed experiment. However, knowledge of the subject so that realistic situations can be created in testing should also be emphasized. Testing and experimenting on people does not put them in natural states, and this situation can lead to artificial responses.

Questions on Topic Eight: Planning and Conducting Experiments

Multiple-Choice Questions

Directions: The questions or incomplete statements that follow are each followed by five suggested answers or completions. Choose the response that best answers the question or completes the statement.

1. A study is made to determine whether studying Latin helps students achieve higher scores on the verbal section of the SAT exam. In comparing records of 200 students, half of whom have taken at least 1 year of Latin, it is noted that the average SAT verbal score is higher for those 100 students who have taken Latin than for those who have not. Based on this study, guidance counselors begin to recommend Latin for students who want to do well on the SAT exam. Which of the following are true statements?

    I. While this study indicates a relation, it does not prove causation.
   II. There could well be a confounding variable responsible for the seeming relationship.
  III. Self-selection here makes drawing the counselors' conclusion difficult.

    (A) I and II
    (B) I and III
    (C) II and III
    (D) I, II, and III
    (E) None of the above gives the complete set of true responses.

2. In a 1927–32 Western Electric Company study on the effect of lighting on worker productivity, productivity increased with each increase in lighting but then also increased with every decrease in lighting. If it is assumed that the workers knew a study was in progress, this is an example of

    (A) the effect of a treatment unit.
    (B) the placebo effect.
    (C) the control group effect.
    (D) sampling error.
    (E) voluntary response bias.

3. When the estrogen-blocking drug tamoxifen was first introduced to treat breast cancer, there was concern that it would cause osteoporosis as a side effect. To test this concern, cancer subjects were randomly selected and given tamoxifen, and their bone density was measured before and after treatment. Which of the following is a true statement?

    I. This study was an observational study.

    II. This study was a sample survey of randomly selected cancer patients.

    III. This study was an experiment in which the subjects were used as their own controls.

    (A) I only

    (B) II only

    (C) III only

    (D) I and II

    (E) None of the above gives the complete set of true responses.

4. In designing an experiment, blocking is used

    (A) to reduce bias.

    (B) to reduce variation.

    (C) as a substitute for a control group.

    (D) as a first step in randomization.

    (E) to control the level of the experiment.

5. Which of the following are true statements about blocking?

    I. Blocking is to experiment design as stratification is to sampling design.

    II. By controlling certain variables, blocking can make conclusions more specific.

    III. The paired comparison design is a special case of blocking.

    (A) I and II

    (B) I and III

    (C) II and III

    (D) I, II, and III

    (E) None of the above gives the complete set of true responses.

6. Consider the following studies being run by three different nursing home establishments.

    I. One nursing home has pets brought in for an hour every day to see if patient morale is improved.

    II. One nursing home allows hourly visits every day by kindergarten children to see if patient morale is improved.

    III. One nursing home administers antidepressants to all patients to see if patient morale is improved.

Which of the following is true?

    (A) None of these studies uses randomization.

    (B) None of these studies uses control groups.

    (C) None of these studies uses blinding.

    (D) Important information can be obtained from all these studies, but none will be able to establish causal relationships.

    (E) All of the above

7. A consumer product agency tests miles per gallon for a sample of automobiles using each of four different octanes of gasoline. Which of the following is true?

   (A) There are four explanatory variables and one response variable.
   (B) There is one explanatory variable with four levels of response.
   (C) Miles per gallon is the only explanatory variable, but there are four response variables corresponding to the different octanes.
   (D) There are four levels of a single explanatory variable.
   (E) Each explanatory level has an associated level of response.

8. Which of the following are true statements?

   I. In general, strong association implies causation.
   II. In well-designed, well-conducted experiments, strong association implies causation.
   III. Causation and association are unrelated concepts.

   (A) I only
   (B) II only
   (C) III only
   (D) I and II
   (E) I, II, and III

9. Which of the following are true statements?

   I. In well-designed observational studies, responses are systematically influenced during the collection of data.
   II. In well-designed experiments, the treatments result in responses that are as similar as possible.
   III. A well-designed experiment always has a single treatment but may test that treatment at different levels.

   (A) I only
   (B) II only
   (C) III only
   (D) II and III
   (E) None of these statements is true.

10. Which of the following are important in the design of experiments?

   I. Control of confounding variables
   II. Randomization in assigning subjects to different treatments
   III. Replication of the experiment using sufficient numbers of subjects

   (A) I and II
   (B) I and III
   (C) II and III
   (D) I, II, and III
   (E) None of the above gives the complete set of true responses.

11. Which of the following are true about the design of matched-pair experiments?

    I. Each subject might receive both treatments.
    II. Each pair of subjects receives the identical treatment, and differences in their responses are noted.
    III. Blocking is one form of matched-pair design.

    (A) I only
    (B) II only
    (C) III only
    (D) I and III
    (E) II and III

12. A nutritionist believes that having each player take a vitamin pill before a game enhances the performance of the football team. During the course of one season, each player takes a vitamin pill before each game, and the team achieves a winning season for the first time in several years. Is this an experiment or an observational study?

    (A) An experiment, but with no reasonable conclusion possible about cause and effect
    (B) An experiment, thus making cause and effect a reasonable conclusion
    (C) An observational study, because there was no use of a control group
    (D) An observational study, but a poorly designed one because randomization was not used
    (E) An observational study, thus allowing a reasonable conclusion of association but not of cause and effect

13. Some researchers believe that too much iron in the blood can raise the level of cholesterol. The iron level in the blood can be lowered by making periodic blood donations. A study is performed by randomly selecting half of a group of volunteers to give periodic blood donations while the rest do not. Is this an experiment or an observational study?

    (A) An experiment with control group and blinding
    (B) An experiment with blocking
    (C) An observational study with comparison and randomization
    (D) An observational study with little if any bias
    (E) None of the above

Answer Key

1. **D**	4. **B**	7. **D**	10. **D**	13. **E**
2. **B**	5. **D**	8. **B**	11. **A**	
3. **C**	6. **E**	9. **E**	12. **A**	

Answers Explained

1. **(D)** It may well be that very bright students are the same ones who both take Latin and do well on the SAT verbal exam. If students could be randomly assigned to take or not take Latin, the results would be more meaningful. Of course, ethical considerations might make it impossible to isolate the confounding variable in this way.

2. **(B)** The desire of the workers for the study to be successful led to a placebo effect.

3. **(C)** In experiments on people, the subjects can be used as their own controls, with responses noted before and after the treatment. However, with such designs there is always the danger of a placebo effect. Thus the design of choice would involve a separate control group to be used for comparison.

4. **(B)** Blocking divides the subjects into groups, such as men and women, or political affiliations, and thus reduces variation.

5. **(D)** Blocking in experiment design first divides the subjects into representative groups called blocks, just as stratification in sampling design first divides the population into representative groups called strata. This procedure can control certain variables by bringing them directly into the picture, and thus conclusions are more specific. The paired comparison design is a special case of blocking in which each pair can be considered a block.

6. **(E)** None of the studies has any controls, such as randomization, control groups, or blinding, and so while they may give valuable information, they cannot establish cause and effect.

7. **(D)** Octane is the only explanatory variable, and it is being tested at four levels. Miles per gallon is the single response variable.

8. **(B)** Well-designed experiments can show cause and effect.

9. **(E)** In good observational studies, the responses are not influenced during the collecting of data. In good experiments, treatments are compared as to differences in responses. In an experiment, there can be many treatments, each at a different level.

10. **(D)** Control, randomization, and replication are all important aspects of well-designed experiments.

11. **(A)** Each subject might receive both treatments, as, for example, in the Pepsi-Coke taste comparison study. The point is to give each subject in a matched pair a different treatment and note any difference in responses. Matched-pair experiments are a particular example of blocking, not vice versa.

12. **(A)** This study was an experiment because a treatment (vitamin pills) was imposed on the subjects. However, it was a poorly designed experiment with no use of randomization and no control over lurking variables, and so the results are meaningless.

13. **(E)** This study is an experiment because a treatment (periodic removal of a pint of blood) is imposed. There is no blinding because the subjects clearly know whether or not they are giving blood. There is no blocking because the subjects are not divided into blocks before random assignment to treatments. For example, blocking would have been used if the subjects had been separated by gender or age before random assignment to give or not give blood donations.

Free-Response Questions

Directions: You must show all work and indicate the methods you use. You will be graded on the correctness of your methods and on the accuracy of your final answers.

Eleven Open-Ended Questions

1. Some health care professionals recommend the use of melatonin to promote better sleep patterns. To test this idea the manufacturer has 100 of its employees fill out a questionnaire about their sleeping patterns, once at the beginning of the study and then again after taking a 3-milligram melatonin capsule every night at bedtime for a week. Comment on the design of this experiment.

2. Suppose a new drug is developed that appears in laboratory settings to completely prevent people who test positive for human immunodeficiency virus (HIV) from ever developing full-blown acquired immunodeficiency syndrome (AIDS). Putting all ethical considerations aside, design an experiment to test the drug. What ethical considerations might arise during the testing that would force an early end to the experiment?

3. A new weight-loss supplement is to be tested at three different levels (once, twice, and three times a day). Design an experiment, including a control group and including blocking for gender, for 80 overweight volunteers, half of whom are men. Explain carefully how you will use randomization.

4. Two studies are run to measure the health benefits of long-time use of daily high doses of vitamin C. Researchers in the first study send a questionnaire to all 50,000 subscribers to a health magazine, asking whether they have taken large doses of vitamin C for at least a 2-year period and what they perceive to be the health benefits, if any. The response rate is 80%. The 10,000 people who did not respond to the first mailing receive follow-up telephone calls, and eventually responses are registered from 98% of the magazine subscribers. Researchers in a second study take a group of 200 volunteers and randomly

select 100 to receive high doses of vitamin C while the others receive a similar-looking, similar-tasting placebo. The volunteers are not told whether they are receiving the vitamin, but their doctors know and are asked to note health changes during a 2-year period. Comment on the designs of the two studies, remarking on their good points and on possible sources of error.

5. Explain how you would design an experiment to evaluate whether subliminal advertising (flashing "BUY POPCORN" on the screen for a fraction of a second) results in more popcorn being sold in a movie theater. Show how you will incorporate comparison, randomization, and blinding.

6. Throughout history millions of people have used garlic to obtain a variety of perceived health benefits. A vitamin production company decides to run a scientific test to assess the value of garlic in promoting a general sense of well-being. They randomly pick 250 of their employees, and once a day for 2 months the employees fill out questionnaires about their sense of well-being that day. For the next 2 months the employees take garlic capsules daily and again fill out the same questionnaires. Finally, for 2 concluding months the employees stop taking the pills and continue to fill out the daily questionnaires. Comment on the design of this experiment.

7. A new pain control procedure has been developed in which the patient uses a small battery pack to vary the intensity and duration of electric signals to electrodes surgically embedded in the afflicted area. Putting all ethical considerations aside, design an experiment to test the procedure. What ethical considerations might arise during the testing that would force an early end to the experiment?

8. A new vegetable fertilizer is to be tested at two different levels (regular concentration and double concentration). Design an experiment, including a control, for 30 test plots, half of which are in shade. Explain carefully how you will use randomization.

9. Two studies are run to measure the extent to which taking zinc lozenges helps to shorten the duration of the common cold. Researchers in the first study send questionnaires to all 5000 employees of a major teaching hospital asking whether they have taken zinc lozenges to fight the common cold and what they perceive to be the benefits, if any. The response rate is 90%. The 500 people who did not respond to the first mailing receive follow-up telephone calls, and eventually responses are obtained from over 99% of the hospital employees. Researchers in the second study take a group of 100 volunteers and randomly select 50 to receive zinc lozenges while the others receive a similar-looking, similar-tasting placebo. The volunteers are not told whether they are taking the zinc lozenges, but their doctors know and are asked to accurately measure the duration of common cold symptoms experienced by the volunteers. Comment on the designs of the two studies, remarking on their good points and on possible sources of error.

10. Explain how you would design an experiment to evaluate whether praying for a hospitalized heart attack patient leads to a speedier recovery. Show how you would incorporate comparison, randomization, and blinding.

11. The computer science department plans to offer three introductory-level CS courses: one using Pascal, one using C++, and one using Java.
    (a) The department chairperson plans to give all students the same general programming exam at the end of the year and to compare the relative effectiveness of using each of the programming languages by comparing the mean grades of the students from each course. What is wrong, if anything, with the chairperson's plan?
    (b) The chairperson also wishes to determine whether math majors or science majors do better in the courses. Suppose he calculates that the average grade of science majors was higher than the average grade of math majors in each of the courses. Does it follow that the average grade of all the science majors taking the three courses must be higher than the average grade of all the math majors? Explain.
    (c) Suppose 300 students wish to take introductory programming. How would you randomly assign 100 students to each of the three courses?
    (d) How would you randomly assign students to the three courses if you wanted the assignment to be independent from student to student with each student in turn having a one-third probability of taking each of the three classes.
    (e) Name a lurking variable that all the above methods miss.

Answers Explained

1. Any conclusions would probably be meaningless. There is a substantial danger that the placebo effect will occur here; that is, real physical responses can be caused by the psychological effect of knowing the intent of the research. The experiment would be considerably strengthened by using a control group taking a look-alike capsule. Any conclusions would be further suspect because of the choice of subjects. Rather than making some random (or at least representative) selection from the intended population, the company is using its own employees, a sample almost guaranteed to have concerns, interests, and backgrounds that will confound the responses or limit the generalizability of the results.

2. Ask doctors, hospitals, or blood testing laboratories to make known that you are looking for HIV-positive volunteers. As the volunteers arrive, use a random number table to give each one the drug or a placebo (e.g., if the next digit in the table is odd, the volunteer gets the drug, while if the next digit is even, the volunteer gets a placebo). Use double-blinding; that is, both the volunteers and their doctors should not know if they are receiving the drug or the placebo. Ethical considerations will arise, for example, if the drug is very successful. If volunteers on the placebo are steadily developing full-blown AIDS while no one on the drug is, then ethically the test should be stopped and everyone put on the drug. Or if most of the volunteers on the drug are dying from an unexpected fatal side effect, the test should be stopped and everyone taken off the drug.

3. To achieve blocking by gender, first separate the men and women. Label the 40 men 01 through 40. Use a random number table to pick two digits at a

time, ignoring 00 and numbers greater than 40, and ignoring repeats, until a group of ten such numbers is obtained. These men will receive the supplement at the once-a-day level. Follow along in the table, continuing to ignore repeats, until another group of ten is selected. These men will receive the supplement at the twice-a-day level. Again ignore repeats until a third group of ten is selected to receive the supplement at the three-times-a-day level, while the remaining men will be a control group and not receive the supplement. Now repeat the entire procedure, starting by labeling the women 01 through 40. A decision should be made whether or not to use a placebo and have all participants take "something" three times a day. Weigh all 80 overweight volunteers before and after a predetermined length of time. Calculate the change in weight for each individual. Calculate the average change in weight among the ten people in each of the eight groups. Compare the four averages from each block (men and women) to determine the effect, if any, of different levels of the supplement for men and for women.

4. The first study, an observational study, does not suffer from nonresponse bias, as do most mailed questionnaires, because it involved follow-up telephone calls and achieved a high response rate. However, this study suffers terribly from selection bias because people who subscribe to a health magazine are not representative of the general population. One would expect most of them to strongly believe that vitamins improve their health. The second study, a controlled experiment, used comparison between a treatment group and a control group, used randomization in selecting who went into each group, and used blinding to control for a placebo effect on the part of the volunteers. However, it did not use double-blinding; that is, the doctors knew whether their patients were receiving the vitamin, and this could have introduced hidden bias when they made judgments regarding their patients' health.

5. Every day for some specified period of time, look at the next digit on a random number table. If it is odd, flash the subliminal message all day on the screen, while if it is even, don't flash the message that day (randomization). Don't let the customers know what is happening (blinding) and don't let the clerks selling the popcorn know what is happening (double-blinding). Compare the quantity of popcorn bought by the treatment group, that is, by the people who receive the subliminal message, to the quantity bought by the control group, the people who don't receive the message (comparison).

6. Any conclusions would probably be meaningless. There is a substantial danger of the placebo effect here; that is, real physical responses could be caused by the psychological effect of knowing the intent of the research. The experiment would be considerably strengthened by using a control group taking a look-alike capsule. Any conclusions are further suspect because of the choice of subjects. Rather than making a random selection from the intended population, the company is using a sample from its own employees, a sample almost guaranteed to have concerns, interests, and backgrounds that will confound the responses or limit their generalizability.

7. Ask doctors and hospitals to make known that you are looking for volunteers from among intractable pain sufferers. As the volunteers arrive, use a random number table to decide which will have the electrodes properly embedded in their pain centers and which will have the electrodes harmlessly embedded in wrong positions. For example, if the next digit in the table is odd, the volunteer receives the proper embedding, while if the next digit is even, the volunteer does not. Use double-blinding; that is, both the volunteers and their doctors should not know if the volunteers are receiving the proper embedding. Ethical considerations will arise, for example, if the procedure is very successful. If volunteers with the wrong embeddings are in constant pain, while everyone with proper embedding is pain-free, then ethically the test should be stopped and everyone given the proper embedding. Or if most of the volunteers with proper embedding develop an unexpected side effect of the pain spreading to several nearby sites, then the test should be stopped and the procedure discontinued for everyone.

8. To achieve blocking by sunlight, first separate the sunlit and shaded plots. Label the 15 sunlit plots 01 through 15. Using a random number table, pick two digits at a time, ignoring 00 and numbers above 15 and ignoring repeats, until a group of five such numbers is obtained. These sunlit plots will receive the fertilizer at regular concentration. Continue in the table, ignoring repeats, until another group of five is selected. These sunlit plots will receive the fertilizer at double concentration, while the remaining sunlit plots will be a control group receiving no fertilizer. Now repeat the procedure, this time labeling the shaded plots 01 through 15. Assuming size is the pertinent outcome, weigh all vegetables at the end of the season, compare the average weights among the three sunlit groups, and compare the average weights among the three shaded groups to determine the effect of the fertilizer, if any, at different levels on sunlit plots and separately on shaded plots.

9. The first study, an observational study, does not suffer from nonresponse bias, as do most studies involving mailed questionnaires, because the researchers made follow-up telephone calls and achieved a very high response rate. However, the first study suffers terribly from selection bias. People who work at a teaching hospital are not representative of the general population. One would expect many of them to have heard about how zinc coats the throat to hinder the propagation of viruses. The second study, a controlled experiment, used comparison between a treatment group and a control group, used randomization in selecting who went into each group, and used blinding to control for a placebo effect on the part of the volunteers. However, they did not use double-blinding; that is, the doctors knew whether their patients were receiving the zinc lozenges, and this could introduce hidden bias as the doctors make judgments about their patients' health.

10. For each new heart attack patient entering the hospital, look at the next digit from a random number table. If it is odd, give the name to a group of people who will pray for the patient throughout his or her hospitalization, while if it is even, don't ask the group to pray (randomization). Don't let the patients

know what is happening (blinding) and don't let the doctors know what is happening (double-blinding). Compare the lengths of hospitalization of patients who receive prayers with those of control group patients who don't receive prayers (comparison).

11. (a) Allowing the students to self-select which class to take leads to a lurking variable that could be significant. For example, perhaps the brighter students all want to learn a certain one of the three languages.

(b) It is possible for the average score of all science majors to be lower than the average for all math majors even though the science majors averaged higher in each class. For example, suppose that the students taking Java scored much higher than the students in the other two classes. Furthermore, only one science major took Java, and she scored tops in the class. Then the overall average of the math majors could well be higher. This is an example of Simpson's paradox, in which a comparison can be reversed when more than one group is combined to form a single group.

(c) Number the students 001 through 300. Read off three digits at a time from a random number table, noting all triplets between 001 and 300 and ignoring repeats, until 100 such numbers have been selected. Keep reading off three digits, ignoring repeats, until 100 new numbers between 001 and 300 are selected. These get C++, while the remaining 100 get Java. Even quicker would be to use a calculator to generate random digits between 001 and 300.

(d) Go through the list of students, flipping a die for each. If a 1 or a 2 shows, the student takes Pascal, if a 3 or a 4 shows, C++, and if a 5 or a 6 shows, Java.

(e) A possible lurking variable are the teachers. For example, perhaps the better teachers teach Java.

Two Investigative Tasks

1. A high school offers two precalculus courses, one that uses a traditional lecture and drill method, and a second that divides students into small groups to work on open-ended problems. To compare the effectiveness of the two methods, the administration proposes to compare average SAT math scores for the students in the two courses.

a. What is wrong with the administration's proposal?

b. Suppose a group of 50 students are willing to take either course. Explain how you would use a random number table to set up an experiment comparing the effectiveness of the two courses.

c. Apply your setup procedure to the given random number table:

```
84177 06757 17613 15582 51506 81435 41050 92031 06449
05059 59884 31180 53115 84469 94868 57967 05811 84514
75011 13006 63395 55041 15866 06589 13119 71020 85940
91932 06488 74987 54355 52704 90359 02649 47496 71567
94268 08844 26294 64759 08989 57024 97284 00637 89283
03514 59195 07635 03309 72605 29357 23737 67881 03668
33876 35841 52869 23114 15864 38942
```

d. Discuss any lurking variables that your setup doesn't consider.

2. Suppose you are head of admissions at a college. Two of your employees come up with ideas for distributing recruiting brochures, and you would like to see which idea results in more follow-up contacts. You print test quantities of version A and version B and must decide how to distribute them as requests come in.

   a. Design an experiment with instructions to a secretary explaining how to decide which brochure should be sent out as each individual request comes in.

   b. Design an experiment with instructions to a secretary explaining how to evenly distribute brochures to each set of 20 requests that come in.

   c. Suppose that 20 people (Jones, Smith, Black, Berry, Stein, Gold, West, Tomas, Johns, Little, Jagne, Hurd, Tastle, Novak, Bass, Yolt, Cross, Nash, Dole, and Uvay) request brochures. Using the random number table, list who will receive each version of the brochure under the experimental design in part *a*. Then do the same for the experimental design in part *b*.

   84177 06757 17613 15582 51506 81435 41050 92031 06449
   05059 59884 31180 53115 84469 94868 57967 05811 84514
   75011 13006 63395 55041 15866 06589 13119 71020 85940
   91932 06488 74987 54355 52704 90359 02649 47496 71567

Answers Explained

1. *a.* There is no reason to believe that there was anything random about which students took which course. Perhaps all the weaker students self-selected or were advised to choose the traditional course.

   b. The students could be labeled 01 through 50. Pairs of digits could then be read off a random number table, ignoring numbers over 50 and ignoring duplicates, until a set of 25 numbers is obtained. The students corresponding to these numbers could be enrolled in the traditional course, and the remaining students in the other.

   c. Applying the above procedure results in {17, 31, 14, 35, 41, 05, 09, 20, 06, 44, 50, 43, 11, 18, 45, 01, 13, 33, 04, 19, 02, 08, 40, 49, 03}. Enroll the students with these numbers in the traditional course.

   d. Which teachers teach which courses is not considered. Perhaps the more interesting, exciting teachers teach the new version. Even though a control group is selected, there is no blinding, and so students in the new version might work harder because they realize they are part of an experiment.

2. *a.* As each request comes in, the secretary looks at the next digit in the random number table. If the digit is odd, the person is sent version A, while if the digit is even, the person is sent version B.

   b. The secretary gives each person a number from 01 to 20. Two digits at a time are then read off the random number table, and he copies down all resulting pairs between 01 and 20, ignoring those larger than 20 and ignoring repeats, until ten such numbers have been selected. The people corresponding to these ten numbers are sent version A, while the remaining ten people are sent version B. For the next set of 20 people, the secretary resumes reading numbers where he left off the previous time.

c. With the first procedure, 8 is even and so Jones gets version B, 4 is even and so Smith gets version B, 1 is odd and so Black gets version A, and so on, resulting in

Version A: Black, Berry, Stein, Tomas, Johns, Little, Jagne, Hurd, Novak, Bass, Yolt, Cross, Nash

Version B: Jones, Smith, Gold, West, Tastle, Dole, Uvay

With the second procedure, each person is first assigned a number between 01 and 20: Jones (01), Smith (02), Black (03), Berry (04), Stein (05), Gold (06), West (07), Tomas (08), Johns (09), Little (10), Jagne (11), Hurd (12), Tastle (13), Novak (14), Bass (15), Yolt (16), Cross (17), Nash (18), Dole (19), and Uvay (20). Reading pairs of digits from the table, ignoring those larger than 20 and ignoring repeats, and underlining the selections give

84<u>17</u>7 06757 17613 15582 51506 8<u>14</u>35 41<u>050</u> <u>92</u>031 06449
05059 59884 <u>31</u>1<u>80</u> 53115 84469 94868 57967 0581<u>1</u> <u>84</u>514
75<u>011</u> 13006 63395 55041 <u>15</u>866 06589 13119 71020 85940
91932 06488 74987 54355 52704 90359 02649 47496 71567

The resulting set, {17, 14, 05, 09, 20, 06, 11, 18, 01, 15}, corresponds to those receiving version A, while the rest receive version B:

Version A: Cross, Novak, Stein, Johns, Uvay, Gold, Jagne, Nash, Jones, Bass
Version B: Smith, Black, Berry, West, Tomas, Little, Hurd, Tastle, Yolt, Dole

THEME THREE: PROBABILITY

Probability as Relative Frequency

- Law of Large Numbers
- Addition Rule
- Multiplication Rule
- Conditional Probabilities
- Independence
- Multistage Probability Calculations
- Discrete Random Variables and
  Probability Distributions

In the world around us, unlikely events sometimes take place. At other times, events that seem inevitable do not occur. Because of the myriad and minute origins of various happenings, it is often impracticable, or simply impossible, to predict exact outcomes. However, while we may not be able to foretell a specific result, we can sometimes assign what is called a *probability* to indicate the likelihood that a particular event will occur.

THE LAW OF LARGE NUMBERS

The *relative frequency* of an event is the proportion of times the event happened, that is, the number of times the event happened divided by the total number of trials.

EXAMPLE 9.1

If there were 12 cloudy days during a 30-day period, the relative frequency of cloudy days was $\frac{12}{30} = .4$.

Relative frequencies may change every time an experiment is performed. However, when an experiment is performed a large number of times, the relative frequency of an event tends to become closer to what is called the probability of the event. (In some cases, *probability* can be defined as long-term relative frequency.)

EXAMPLE 9.2

A nurse notes that on one day 23 out of 50 new patients had a fever, while on another day 28 out of 56 had a fever. The two relative frequencies $\frac{19}{50} = .38$ and $\frac{28}{56} = .50$, are different. However, summing up these numbers day after day for many days, she might be able to conclude that the probability of a patient arriving with a fever is, for example, about .412.

EXAMPLE 9.3

Suppose the numbers in the following random number table correspond to people arriving for work at a large factory. Let 0, 1, and 2 be smokers and 3–9 be nonsmokers. After each ten arrivals, calculate the total relative frequency of smokers and graph the results.

84177	06757	17613	15582	51506	81435	41050	92031	06449	05059
59884	31180	53115	84469	94868	57967	05811	84514	75011	13006
63395	55041	15866	06589	13119	71020	85940	91932	06488	74987
54355	52704	90359	02649	47496	71567	94268	08844	26294	64759
08989	57024	97284	00637	89283	03514	59195	07635	03309	72605
29357	23737	67881	03668	33876	35841	52869	23114	15864	38942

Answer: The cumulative relative frequencies are $\frac{2}{10} = .2$, $\frac{6}{20} = .3$, $\frac{9}{30} = .3$, $\frac{15}{40} = .375$, $\frac{18}{50} = .36$, $\frac{21}{60} = .35$, $\frac{23}{70} = .329$, $\frac{23}{80} = .288$, $\frac{27}{90} = .3$, $\frac{33}{100} = .33$, and so on. Graphing these and further values gives

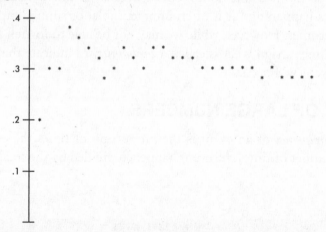

Note how the cumulative relative frequency at first fluctuates considerably but then seems to level off. The concept that the relative frequency tends closer and closer to a certain number (the probability) as an experiment is repeated more and more times is called the *law of large numbers.*

The *probability* of a particular outcome of an experiment is a mathematical statement about the likelihood of that event occurring. Probabilities are always between 0 and 1, with a probability close to 0 meaning that an event is unlikely to occur and a probability close to 1 meaning that the event is likely to occur. The sum of the probabilities of all the separate outcomes of an experiment is always 1.

ADDITION RULE, MULTIPLICATION RULE, CONDITIONAL PROBABILITIES, AND INDEPENDENCE

Complementary Events

The probability that an event will not occur, that is, the probability of its complement, is equal to 1 minus the probability that the event will occur:

$$P(A^C) = 1 - P(A)$$

EXAMPLE 9.4

If the probability that a company will win a contract is .3, what is the probability that it will not win the contract?

Answer: 1 − .3 = .7

Addition Principle

If two events are mutually exclusive, that is, they cannot occur simultaneously, the probability that at least one event will occur is equal to the sum of the respective probabilities of the two events:

$$\text{If } P(A \cap B) = 0 \quad \text{then} \quad P(A \cup B) = P(A) + P(B)$$

where $A \cap B$, (read "*A* intersect *B*") means that both *A* and *B* occur, while $A \cup B$ (read "*A* union *B*") means that either *A* or *B* or both occur.

EXAMPLE 9.5

If the probabilities that Jane, Tom, and Mary will be chosen chairperson of the board are .5, .3, and .2, respectively, the probability that the chairperson will be either Jane or Mary is .5 + .2 = .7.

When two events are not mutually exclusive, the sum of their probabilities counts their shared occurrence twice. This leads to the following rule.

General Addition Rule

For any pair of events *A* and *B*,

$$P(A \cup B) = P(A) + P(B) - P(A \cap B)$$

EXAMPLE 9.6

Suppose the probability that a construction company will be awarded a certain contract is .25, the probability that it will be awarded a second contract is .21, and the probability that it will get both contracts is .13. What is the probability that the company will win at least one of the two contracts?

Answer: .25 + .21 − .13 = .33

Multiplication Rule

If the chance that one event will happen is not influenced by whether or not a second event happens, the probability that *both* events will occur is the product of their separate probabilities.

EXAMPLE 9.7

The probability that a student will receive a state grant is $\frac{1}{3}$, while the probability that she will be awarded a federal grant is $\frac{1}{2}$. If whether or not she receives one grant is not influenced by whether or not she receives the other, what is the probability of her receiving both grants?

Answer: $\frac{1}{3} \times \frac{1}{2} = \frac{1}{6}$

The multiplication rule can be extended to more than two events; that is, given a sequence of *independent* events, the probability that *all* will happen is equal to the product of their individual probabilities.

EXAMPLE 9.8

Suppose a reputed psychic in an extrasensory perception (ESP) experiment has called heads or tails correctly on ten successive tosses of a coin, What is the probability that guessing would have yielded this perfect score?

Answer: $\left(\frac{1}{2}\right)\left(\frac{1}{2}\right) \cdots \left(\frac{1}{2}\right) = \left(\frac{1}{2}\right)^{10} = \frac{1}{1024}$

To best understand independence as mentioned above, we first examine the idea of *conditional probability*, that is, the probability of an event given that another event has occurred.

EXAMPLE 9.9

The table below gives the results of a survey of the drinking and smoking habits of 1200 college students. Rows and columns have also been summed.

	Drink beer	Don't drink	
Smoke	315	165	480
Don't smoke	585	135	720
	900	300	1200

What is the probability that someone in this group smokes?

Answer: $P(\text{smokes}) = \frac{480}{1200} = .4$

What is the probability a student smokes given that she is a beer drinker?

Answer: We now narrow our attention to the 900 beer drinkers, and thus we calculate $\frac{315}{900} = .35$. We use the following notation: P(smoke|drink) (read "the probability that people smoke given that they drink").

(continued)

What is the probability that someone drinks?

Answer: $P(\text{drink}) = \frac{900}{1200} = .75$

What is the probability that students drink given that they smoke?

Answer: $P(\text{drink}|\text{smoke}) = \frac{315}{480} = .65625$

A formula for conditional probability is given by

$$P(A|B) = \frac{P(A \cap B)}{P(B)}$$

EXAMPLE 9.10

If 90% of the households in a certain region have answering machines and 50% have both answering machines and call waiting, what is the probability that a household chosen at random and found to have an answering machine also has call waiting?

Answer:

$$P(\text{call waiting}|\text{answering machine})$$
$$= \frac{P(\text{call waiting} \cap \text{answering machine})}{P(\text{answering machine})} = \frac{.5}{.9} = \frac{5}{9}$$

EXAMPLE 9.11

A psychologist interested in right-handedness versus left-handedness and in IQ scores collected the following data from a random sample of 2000 high school students.

	Right-handed	Left-handed	
High IQ	190	10	200
Normal IQ	1710	90	1800
	1900	100	2000

What is the probability that a student from this group has a high IQ?

Answer: $P(\text{highIQ}) = \frac{200}{2000} = .1$

What is the probability that a student has a high IQ given that she is left-handed?

Answer: $P(\text{highIQ}|\text{left-handed}) = \frac{10}{100} = .1$

Note that the probability of having a high IQ is not affected by the fact that the student is left-handed. We say that the two events, high IQ and left-handed, are *independent*.

That is, A and B are independent if $P(A|B) = P(A)$, and earlier it was noted that in this case we also have

$$P(A \cap B) = P(A)P(B)$$

Whether events are mutually exclusive or are independent are two very different properties! One refers to events being disjoint, the other to an event having no effect on whether or not the other occurs. Note that mutually exclusive (disjoint) events are *not* independent (except in the special case that one of the events has probability 0). That is, mutually exclusive gives that $P(A \cap B) = 0$, while independence gives that $P(A \cap B) = P(A)P(B)$ (and the only way these are ever simultaneously true is in the very special case when $P(A) = 0$ or $P(B) = 0$).

MULTISTAGE PROBABILITY CALCULATIONS

EXAMPLE 9.12

A videocassette recorder (VCR) manufacturer receives 70% of his parts from factory F1 and the rest from factory F2. Suppose that 3% of the output from F1 are defective, while only 2% of the output from F2 are defective. What is the probability a received part is defective?

Answer: In such problems it is helpful to draw a tree diagram like the following.

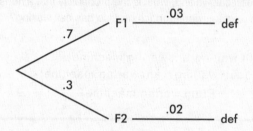

We then have

$$P(F1 \cap def) = P(F1)P(def|F1) = (.7)(.03) = .021$$
$$P(F2 \cap def) = P(F2)P(def|F2) = (.3)(.02) = .006$$

At this stage a Venn diagram is helpful in finishing the problem:

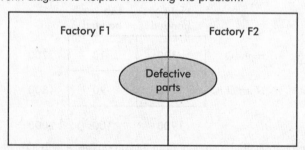

$$P(def) = P(F1 \cap def) + P(F2 \cap def) = .021 + .006 = .027$$

We can take the above analysis one stage further and answer such questions as: If a randomly chosen part is defective, what is the probability it came from factory F1? From factory F2?

$$P(F1|def) = \frac{P(F1 \cap def)}{P(def)} = \frac{.021}{.027} \approx .778$$

$$P(F2|def) = \frac{P(F2 \cap def)}{P(def)} = \frac{.006}{.027} \approx .222$$

EXAMPLE 9.13

Suppose that three branches of a local bank average, respectively, 120, 180, and 100 clients per day. Suppose further that the probabilities that a client will transact business involving more than $100 during a visit are, respectively, .5, .6, and .7. A client is chosen at random. What is the probability that the client will transact business involving over $100? What is the probability that the client went to the first branch given that she transacted business involving over $100?

Answer: $P(B1) = \frac{120}{400} = .3$, $P(B2) = \frac{180}{400} = .45$, and $P(B3) = \frac{100}{400} = .25$. Then we have

$$P(B1 \cap >\$100) = (.3)(.5) = .15$$

$$P(B2 \cap >\$100) = (.45)(.6) = .27$$

$$P(B3 \cap >\$100) = (.25)(.7) = .175$$

Now $P(>\$100) = .15 + .27 + .175 = .595$, and $P(B1 \mid >\$100) = \frac{.15}{.595} \approx .252$.

DISCRETE RANDOM VARIABLES AND THEIR PROBABILITY DISTRIBUTIONS

Often each outcome of an experiment has not only an associated probability but also an associated *real number*. For example, the probability may be $\frac{1}{2}$ that there are five defective batteries, the probability may be .01 that a company receives seven contracts, or the probability may be .95 that three people recover from a disease. If X represents the different numbers associated with the potential outcomes of some chance situation, we call X a *random variable*.

EXAMPLE 9.14

A prison official knows that $\frac{1}{2}$ of the inmates he admits stay only 1 day, $\frac{1}{4}$ stay 2 days, $\frac{1}{5}$ stay 3 days, and $\frac{1}{20}$ stay 4 days before they are either released or are sent on to the county jail. If X represents the number of days, then X is a random variable that takes the value 1 with probability $\frac{1}{2}$, the value 2 with probability $\frac{1}{4}$, the value 3 with probability $\frac{1}{5}$, and the value 4 with probability $\frac{1}{20}$.

The random variable in Example 9.14 is called *discrete* because it can assume only a countable number of values. The random variable in Example 9.15, however, is said to be *continuous* because it can assume values associated with a whole line interval.

EXAMPLE 9.15

Let *X* be a random variable whose values correspond to the speeds at which a jet plane can fly. The jet may be traveling at 623.478 . . . miles per hour or at any other speed in some whole interval. We might ask what the probability is that the plane is flying at between 300 and 400 miles per hour.

A *probability distribution* for a discrete variable is a listing or formula giving the probability for each value of the random variable.

In many applications, such as coin tossing, there are only *two* possible outcomes. For example, on each toss either the psychic guesses correctly or she guesses incorrectly, either a radio is defective or it is not defective, either the workers will go on strike or they will not walk out, either the manager's salary is above $50,000 or it does not exceed $50,000. In some applications such two-outcome situations are repeated many times. For example, we can ask about the chances that at most one out of four tires is defective, that at least three out of five unions will vote to go on strike, or that exactly two out of three executives have salaries above $50,000. For applications in which a two-outcome situation is repeated a certain number of times and the probability of each of the two outcomes remains the same for each repetition, the resulting calculations involve what are known as *binomial probabilities*.

EXAMPLE 9.16

Suppose the probability that a lightbulb is defective is .1. What is the probability that four lightbulbs are all defective?

Answer: Because of independence (i.e., whether one lightbulb is defective is not influenced by whether any other lightbulb is defective), we can multiply individual probabilities of being defective to find the probability that all the bulbs are defective:

$$(.l)(.1)(.l)(.l) = (.1)^4 = .0001$$

EXAMPLE 9.17

Again suppose the probability that a lightbulb is defective is .1. What is the probability that exactly two out of three lightbulbs are defective?

Answer: We subdivide the problem as follows: The probability that the first two bulbs are defective and the third is good is $(.1)(.1)(.9) = .009$. (Note that if the probability of being defective is .1, the probability of being good is .9.) The probability that the first bulb is good and the other two are defective is $(.9)(.1)(.1) = .009$. Finally, the probability that the second bulb is good and the other two are defective is $(.1)(.9)(.1) = .009$. Summing, we find that the probability that exactly two out of three bulbs are defective is $.009 + .009 + .009 = .027$.

EXAMPLE 9.18

If the probability that a lightbulb is defective is .1, what is the probability that exactly three out of eight lightbulbs are defective?

Answer: Again we can subdivide the problem. For example, the probability that the first, third, and seventh bulbs are defective and the rest are good is

$$(.1)(.9)(.1)(.9)(.9)(.9)(.1)(.9) = (.1)^3(.9)^5 = .00059049$$

The probability that the second, third, and fifth bulbs are defective and the rest are good is

$$(.9)(.1)(.1)(.9)(.1)(.9)(.9)(.9) = (.1)^3(.9)^5 = .00059049$$

As can be seen, the probability of any particular arrangement of three defective and five good bulbs is $(.1)^3(.9)^5 = .00059049$. How many such arrangements are there? In other words, in how many ways can we pick three of eight positions for the defective bulbs (the remaining five positions are for good bulbs)? The answer is given by *combinations*:

$$\binom{8}{3} = \frac{8!}{5!3!} = \frac{8 \times 7 \times 6}{3 \times 2} = 56$$

Each of these 56 arrangements has a probability of .00059049. Thus, the probability that exactly three out of eight lightbulbs are defective is $56 \times .00059049 = .03306744$:

$$\binom{8}{3}(.1)^3(.9)^5 = \frac{8!}{5!3!}(.1)^3(.9)^5 = .03306744$$

[On the TI-84 one can calculate binompdf(8, .1, 3) = .03306744.]

EXAMPLE 9.19

Suppose 30% of the employees in a large factory are smokers. What is the probability that there will be exactly two smokers in a randomly chosen five-person work group?

Answer: We reason as follows. The probability that a person smokes is 30% = .3, and so the probability that he or she does not smoke is $1 - .3 = .7$. The probability of a particular arrangement of two smokers and three nonsmokers is $(.3)^2(.7)^3 = .03087$. The number of such arrangements is

$$\binom{5}{2} = \frac{5!}{2!3!} = 10$$

Each such arrangement has probability .03087, and so the final answer is $10 \times .03087 = .3087$.

$$\binom{5}{2}(.3)^2(.7)^3 = \frac{5!}{2!3!}(.3)^2(.7)^3 = .3087$$

[Or binompdf(5, .3, 2) = .3087.]

We can state the general principle as follows.

Binomial Formula

Suppose an experiment has two possible outcomes, called *success* and *failure*, with the probability of success equal to p and the probability of failure equal to q (of course, $p + q = 1$). Suppose further that the experiment is repeated n times and that the outcome at any particular time has no influence over the outcome at any other time. Then the probability of exactly k successes (and thus $n–k$ failures) is

$$\binom{n}{k} p^k q^{n-k} = \frac{n!}{k!(n-k)!} p^k q^{n-k}$$

EXAMPLE 9.20

A manager notes that there is a .125 probability that any employee will arrive late for work. What is the probability that exactly one person in a six-person department will arrive late?

   Answer: If the probability of being late is .125, the probability of being on time is $1 - .125$ = .875. If one person out of six is late, $6 - 1 = 5$ will be on time. Thus the desired probability is

$$\binom{6}{1}(.125)^1(.875)^5 = 6(.125)(.875)^5 = .385$$

[Or going to 2nd DISTR on the TI-84, binompdf(6, .125, 1) $\approx$ .385.]

Many, perhaps most, applications of probability involve such phrases as *at least*, *at most*, *less than*, and *more than*. In these cases, solutions involve summing two or more cases.

EXAMPLE 9.21

A manufacturer has the following quality control check at the end of a production line: If at least eight of ten randomly picked articles meet all specifications, the whole shipment is approved. If, in reality, 85% of a particular shipment meet all specifications, what is the probability that the shipment will make it through the control check?

   Answer: The probability of an item meeting specifications is .85, and so the probability of it not meeting specifications must be .15. We want to determine the probability that at least eight out of ten articles will meet specifications, that is, the probability that exactly eight or exactly nine or exactly ten articles will meet specifications. We sum the three binomial probabilities:

Exactly 8 of 10 meet specifications	Exactly 9 of 10 meet specifications	Exactly 10 of 10 meet specifications

$$\binom{10}{8}(.85)^8(.15)^2 \quad + \binom{10}{9}(.85)^9(.15)^1 \quad + \binom{10}{10}(.85)^{10}(.15)^0$$

$$= \frac{10!}{8!2!}(.85)^8(.15)^2 + 10(.85)^9(.15) + (.85)^{10} \approx .820$$

[On the TI-84 one can calculate $1 -$ binomcdf(10, .85, 7) $\approx$ .820 or binomcdf(10, .15, 2) $\approx$ .820.]

EXAMPLE 9.22

For the problem in Example 9.21, what is the probability that a shipment in which only 70% of the articles meet specifications will make it through the control check?

Answer:

$$\binom{10}{8}(.7)^8(.3)^2 + \binom{10}{9}(.7)^9(.3)^1 + \binom{10}{10}(.7)^{10}(.3)^0$$
$$= 45(.7)^8(.3)^2 + 10(.7)^9(.3) + (.7)^{10}$$
$$\approx .383$$

[Or binomcdf(10, .3, 2) ≈ .383.]

In some situations it is easier to calculate the probability of the complementary event and subtract this value from 1.

EXAMPLE 9.23

Joe DiMaggio had a career batting average of .325. What was the probability that he would get at least one hit in five official times at bat?

Answer: We could sum the probabilities of exactly one hit, two hits, three hits, four hits, and five hits. However, the complement of "at least one hit" is "zero hits." The probability of no hit is

$$\binom{5}{0}(.325)^0(.675)^5 = (.675)^5 \approx .140$$

and thus the probability of at least one hit in five times at bat is 1 − .140 = .860.

[Or binomcdf(5, .675, 4) ≈ .860.]

EXAMPLE 9.24

A grocery store manager notes that 35% of customers who buy a particular product make use of a store coupon to receive a discount. If seven people purchase the product, what is the probability that fewer than four will use a coupon?

Answer: In this situation, "fewer than four" means zero, one, two, or three.

$$\binom{7}{0}(.35)^0(.65)^7 + \binom{7}{1}(.35)^1(.65)^6 + \binom{7}{2}(.35)^2(.65)^5 + \binom{7}{3}(.35)^3(.65)^4$$
$$= (.65)^7 + 7(.35)(.65)^6 + 21(.35)^2(.65)^5 + 35(.35)^3(.65)^4$$
$$\approx .800$$

[Or binomcdf(7, .35, 3) ≈ .800.]

Sometimes we are asked to calculate the probability of each of the possible outcomes (the results should sum to 1).

EXAMPLE 9.25

If the probability that a male birth will occur is .51, what is the probability that a five-child family will have all boys? Exactly four boys? Exactly three boys? Exactly two boys? Exactly one boy? All girls?

Answer:

$$P(5 \text{ boys}) = \binom{5}{5} (.51)^5(.49)^0 = \quad (.51)^5 \quad = .0345$$

$$P(4 \text{ boys}) = \binom{5}{4} (.51)^4(.49)^1 = 5(.51)^4(.49) = .1657$$

$$P(3 \text{ boys}) = \binom{5}{3} (.51)^3(.49)^2 = 10(.51)^3(.49)^2 = .3185$$

$$P(2 \text{ boys}) = \binom{5}{2} (.51)^2(.49)^3 = 10(.51)^2(.49)^3 = .3060$$

$$P(1 \text{ boys}) = \binom{5}{1} (.51)^1(.49)^4 = 5(.51)(.49)^4 = .1470$$

$$P(0 \text{ boys}) = \binom{5}{0} (.51)^0(.49)^5 = \quad (.49)^5 \quad = \frac{.0283}{1.0000}$$

[Or binompdf(5, .49) = {.0345 .1657 .3185 .3060 .1470 .0283}.]

A list such as the one in Example 9.25 shows the entire *probability distribution*, which in this case refers to a listing of all outcomes and their probabilities.

Geometric Probabilities

Suppose an experiment has two possible outcomes, called *success* and *failure*, with the probability of success equal to p and the probability of failure equal to $q = 1 - p$, and the trials are independent. Then the probability that the first success is on trial number $X = k$ is

$$q^{k-1}p$$

EXAMPLE 9.26

Suppose only 12% of men in ancient Greece were honest. What is the probability that the first honest man Diogenes encounters will be the third man he meets?

Answer: $(.88)^2(.12) = .092928$ [or geometpdf (.12, 3) = .092928]

What is the probability that the first honest man he encounters will be no later than the fourth man he meets?

Answer: $(.12) + (.88)(.12) + (.88)^2(.12) + (.88)^3(.12) = .40030464$

[or geometcdf (.12, 4) = .40030464]

SIMULATION OF PROBABILITY DISTRIBUTIONS, INCLUDING BINOMIAL AND GEOMETRIC

Instead of algebraic calculations, sometimes we can use simulation to answer probability questions.

EXAMPLE 9.27

A study reported by Liu and Schutz states that in hockey the better team has a 65% chance of scoring the first goal in overtime. If a team is considered the better team in all five of its overtime games, is it reasonable to believe that it will score the first goal in at least four of these games? Answer the question using simulation.

Answer: Since the team will score first 65% of the time, let the digits 01–65 represent this outcome. To simulate five games select ten digits from the random number table and look at them two at a time. Among the five pairs, note how many times 01–65 appear. (Repeats are OK.) Underlining these pairs gives

84<u>17</u>706<u>57</u>	<u>17</u>6<u>13155</u>82	<u>5150</u>68<u>1435</u>	<u>41</u>050920<u>31</u>	06<u>44</u>90<u>5059</u>
59<u>88431180</u>	<u>5311584469</u>	9486857967	0<u>581184514</u>	75<u>0111300</u>6
<u>6339555041</u>	<u>1586606589</u>	<u>1311</u>97<u>1020</u>	8594<u>091932</u>	06<u>488749</u>87
<u>5435552704</u>	90<u>3590264</u>9	<u>4749</u>67<u>1567</u>	94<u>26</u>808<u>844</u>	<u>2629464759</u>
0<u>898957024</u>	97<u>28400637</u>	89<u>28303514</u>	<u>5919507635</u>	03<u>30972605</u>
<u>2935723737</u>	6788<u>103668</u>	<u>3387635841</u>	<u>5286923114</u>	<u>1586438942</u>

For example, in the first group of five pairs, two (17 and 57) out of five are between 01 and 65. In the second group, four (17, 61, 31, and 55) out of five are in the range. Tabulating the frequencies of each result gives

Number of First Goals	Frequency
0	1
1	0
2	4
3	8
4	13
5	4

In this simulation we see that the better team will score the first goal in at least four of the five overtime games 17 out of 30 times. Thus the answer to the original question is that it is not unreasonable to believe that the better team will win at least four of the five games.

It is easy to confuse *number of samples* with *sample size*. Note that above we have simulated 30 samples of five games each.

In Example 9.27 simulation is used with regard to a binomial distribution, while in the example below it is used with regard to a geometric distribution. A *geometric distribution* shows the number of trials needed until a success is achieved.

EXAMPLE 9.28

Suppose the probability that a company will land a contract is .3. Use simulation to find the probability that the company will first land a contract on the first bid, on the second, on the third, and so on.

(continued)

Answer: Let 0, 1, and 2 represent landing a contract, while 3–9 represent not getting the contract. Read the random number table, one digit at a time, noting how many digits must be read until a 0, 1, or 2 is encountered. Using the table in Example 9.24 results in

841(first success in third try)
770(first success in third try)
67571(first success in fifth try)
761(first success in third try)
31(first success in second try).

Continuing in this fashion and tabulating the results, we have

Number of the Try at Which the First Success Occurred	Frequency
1	21
2	19
3	16
4	7
5	9
Over 5	12

Thus we estimate the probabilities as follows:

Number of the Try at Which the First Success Occurred	Estimated Probability
1	$\frac{21}{84} = .25$
2	$\frac{19}{84} = .23$
3	$\frac{16}{84} = .19$
4	$\frac{7}{84} = .08$
5	$\frac{9}{84} = .11$
Over 5	$\frac{12}{84} = .14$

The actual probabilities are $.3$, $(.7)(.3) = .21$, $(.7)^2(.3) = .147$, $(.7)^3(.3) = .1029$, $(.7)^4(.3) = .07203$, and $1 - (.3 + .21 + .147 + .1029 + .07203) = .16807$.

[On the TI-84 one can calculate the geometric probabilities geometpdf(.3, 1) = .3, . . . , geometpdf(.3, 5) = .07203, and 1 – geometcdf(.3, 5) = .16807.]

One example of simulating a nonbinomial, nongeometric distribution is the following.

EXAMPLE 9.29

If a two-person committee is chosen at random from a group consisting of three supervisors and six employees, what is the probability that the committee will have exactly one supervisor and one employee? Use simulation to estimate the answer.

Answer: Let the digits 1, 2, and 3 represent the supervisors, and the digits 4, 5, 6, 7, 8, and 9 represent the employees. Our procedure is to pick digits one at a time from a random number table, ignoring repeats and ignoring 0's, until two different digits have been chosen. Reading from the random number table in Exercise 9.26 results in the committees (8, 4) with two employees, (1, 7) with one of each, (7, 6) with two employees, (7, 5) with two employees, (7, 1) with one of each, (7, 6) with two employees, (1, 3) with two supervisors, and so on. Tabulating from the table gives

Two supervisors: 11
One of each: 53
Two employees: 66.

Thus an estimate of the probability that the committee will consist of one supervisor and one employee is $\frac{53}{130} = .41$.

[The actual probability is $\dfrac{\binom{3}{1}\binom{6}{1}}{\binom{9}{2}} = \dfrac{18}{36} = .5$.]

In performing a simulation, you must:

1. Set up a correspondence between outcomes and random numbers.
2. Give a procedure for choosing the random numbers (for example, pick three digits at a time from a designated row in a random number table).
3. Give a stopping rule.
4. Note what is to be counted (what is the purpose of the simulation), and give the count if requested.

MEAN (EXPECTED VALUE) AND STANDARD DEVIATION OF A RANDOM VARIABLE, INCLUDING BINOMIAL

A bettor placing a chip on a number in roulette has a small chance of winning a lot and a large chance of losing a little. To determine whether the bet is "fair," we must be able to calculate the *expected value* of the game to the bettor. An insurance company in any given year pays out a large sum of money to each of a relatively small number of people, while collecting a small sum of money from each of many other individuals. To determine policy premiums an actuary must calculate the expected values relating to deaths in various age groups of policyholders.

EXAMPLE 9.30

Concessionaires know that attendance at a football stadium will be 60,000 on a clear day, 45,000 if there is light snow, and 15,000 if there is heavy snow. Furthermore, the probability of clear skies, light snow, or heavy snow on any particular day is $\frac{1}{2}$, $\frac{1}{3}$, and $\frac{1}{6}$, respectively. (Here we have a random variable X that takes the values 60,000, 45,000, and 15,000.)

(continued)

Outcome:	Clear skies	Light snow	Heavy snow
Probability:	$\frac{1}{2}$,	$\frac{1}{3}$,	$\frac{1}{6}$,
Random variable:	60,000	45,000	15,000

What average attendance should be expected for the season?

Answer: We reason as follows. Suppose on, say, 12 game days we have clear skies $\frac{1}{2}$ of the time (6 days), a light snow $\frac{1}{3}$ of the time (4 days), and a heavy snow $\frac{1}{6}$ of the time (2 days). Then the average attendance will be

$$\frac{(6 \times 60,000) + (4 \times 45,000) + (2 \times 15,000)}{12} = 47,500$$

Note that we could have divided the 12 into each of the three terms in the numerator to obtain

$$\left(\frac{6}{12} \times 60,000\right) + \left(\frac{4}{12} \times 45,000\right) + \left(\frac{2}{12} \times 15,000\right)$$

or, equivalently,

$$\left(\frac{1}{2} \times 60,000\right) + \left(\frac{1}{3} \times 45,000\right) + \left(\frac{1}{6} \times 15,000\right) = 47,500$$

Actually, there was no need to consider 12 days. We could have simply multiplied probabilities by corresponding attendances and summed the resulting products.

The final calculation in Example 9.30 motivates our definition of expected value. The *expected value* (or *average* or *mean*) of a random variable X is the sum of the products obtained by multiplying each value x_i by the corresponding probability p_i:

$$E(X) = \mu_x = \sum x_i p_i$$

EXAMPLE 9.31

In a lottery, 10,000 tickets are sold at $1 each with a prize of $7500 for one winner. What is the average result for each bettor?

Answer: The actual winning payoff is $7499 because the winner paid $1 for a ticket, so we have

Outcome:	Win	Lose
Probability:	$\frac{1}{10,000}$	$\frac{9,999}{10,000}$
Random variable:	7499	−1

$$\text{Expected Value} = 7499\left(\frac{1}{10,000}\right) + (-1)\left(\frac{9,999}{10,000}\right) = -0.25$$

Thus the *average* result for each person betting on the lottery is a $0.25 loss. Alternatively, we can say that the expected payoff to the lottery system is $0.25 for each ticket sold.

EXAMPLE 9.32

A manager must choose among three options. Option A has a 10% chance of resulting in a $250,000 gain but otherwise will result in a $10,000 loss. Option B has a 50% chance of gaining $40,000 and a 50% chance of losing $2000. Finally, option C has a 5% chance of gaining $800,000 but otherwise will result in a loss of $20,000. Which option should the manager choose?

Answer: Calculate the expected values of the three options:

	Option A		Option B		Option C	
Outcome	Gain	Loss	Gain	Loss	Gain	Loss
Probability	.10	.90	.50	.50	.05	.95
Random variable	250,000	−10,000	40,000	−2,000	800,000	−20,000

$$E(A) = .10(250,000) + .90(-10,000) = \$16,000$$

$$E(B) = .50(40,000) + .50(-2000) = \$19,000$$

$$E(C) = .05(800,000) + .95(-20,000) = \$21,000$$

The manager should choose option C! However, although option C has the greatest mean (expected value), the manager may well wish to consider the relative riskiness of the various options. If, for example, a $5000 loss would be disastrous for the company, the manager might well decide to choose option B with its maximum possible loss of $2000. (It should be intuitively clear that some concept of variance would be helpful here in measuring the risk; later in this topic the variance of a random variable will be defined.)

Suppose we have a *binomial random variable,* that is, a random variable whose values are the numbers of "successes" in some binomial probability distribution.

EXAMPLE 9.33

Of the automobiles produced at a particular plant, 40% had a certain defect. Suppose a company purchases five of these cars. What is the expected value for the number of cars with defects?

Answer: We might guess that the average or mean or expected value is 40% of 5 = $.4 \times 5 = 2$, but let's calculate from the definition. Letting X represent the number of cars with the defect, we have

$$P(0) = \binom{5}{0}(.4)^0(.6)^5 = (.6)^5 = .07776$$

$$P(1) = \binom{5}{1}(.4)^1(.6)^4 = 5(.4)(.6)^4 = .25920$$

$$P(2) = \binom{5}{2}(.4)^2(.6)^3 = 10(.4)^2(.6)^3 = .34560$$

$$P(3) = \binom{5}{3}(.4)^3(.6)^2 = 10(.4)^3(.6)^2 = .23040$$

$$P(4) = \binom{5}{4}(.4)^4(.6)^1 = 5(.4)^4(.6) = .07680$$

$$P(5) = \binom{5}{5}(.4)^5(.6)^0 = (.4)^5 = .01024$$

(continued)

Outcome (no. of cars):	0	1	2	3	4	5
Probability:	.07776	.25920	.34560	.23040	.07680	.01024
Random variable:	0	1	2	3	4	5

$$E(X) = 0(.07776) + 1(.25920) + 2(.34560) + 3(.23040)$$
$$+ 4(.07680) + 5(.01024)$$
$$= 2$$

Thus the answer turns out to be the same as would be obtained by simply multiplying the probability of a "success" by the number of cases.

The following is true: If we have a binomial probability situation with the probability of a success equal to *p* and the number of trials equal to *n,* the expected value or mean number of successes for the *n* trials is *np.*

EXAMPLE 9.34

An insurance salesperson is able to sell policies to 15% of the people she contacts. Suppose she contacts 120 people during a 2-week period. What is the expected value for the number of policies she sells?

Answer: We have a binomial probability with the probability of a success .15 and the number of trials 120, and so the mean or expected value for the number of successes is $120 \times .15 = 18$.

We have seen that the mean of a random variable is $\Sigma x_i p_i$. However, not only is the mean important, but also we would like to measure the variability for the values taken on by a random variable. Since we are dealing with chance events, the proper tool is variance (along with standard deviation). *Variance* was defined in Topic Two to be the mean average of the squared deviations $(x - \mu)^2$. If we regard the $(x - \mu)^2$ terms as the values of some random variable (whose probability is the same as the probability of *x*), the mean of this new random variable is $\Sigma(x_i - \mu_x)^2 p_i$, which is precisely how we define the variance σ^2 of a random variable:

$$var(X) = \sigma_x^2 = \sum (x_i - \mu_x)^2 p_i$$

As before, the standard deviation σ is the square root of the variance.

EXAMPLE 9.35

A highway engineer knows that his crew can lay 5 miles of highway on a clear day, 2 miles on a rainy day, and only 1 mile on a snowy day. Suppose the probabilities are as follows:

Outcome:	Clear	Rain	Snow
Probability:	.6	.3	.1
Random variable (miles of highway):	5	2	1

(continued)

What are the mean (expected value) and the variance?
Answer:

$$\mu_x = \sum x_i p_i = 5(.6) + 2(.3) + 1(.1) = 3.7$$

$$\sigma_x^2 = \sum (x_i - \mu_x)^2 p_i = (5 - 3.7)^2 (.6) + (2 - 3.7)^2 (.3)$$
$$+ (1 - 3.7)^2 (.1)$$
$$= 2.61$$

EXAMPLE 9.36

Look again at Example 9.33. We calculated the mean to be 2. What is the variance?
Answer: $\sigma^2 = (0 - 2)^2(.07776) + (1 - 2)^2(.2592) + (2 - 2)^2(.3456) + (3 - 2)^2(.2304) + (4 \times 2)^2(.0768) + (5 - 2)^2(.01024) = 1.2$

Could we have calculated the above result more easily? In this case, we have a binomial random variable and the variance can be simply calculated as $npq = np(1 - p)$.

EXAMPLE 9.37

How can we use this method to calculate the variance in Example 9.35 more simply?
Answer: $np(1 - p) = 5(.4)(.6) = 1.2$

Thus, for a random variable X,

Mean or expected value: $\qquad \mu_x = \Sigma x_i p_i$

Variance: $\qquad \sigma_x^2 = \Sigma(x_i - \mu_x)^2 p_i$

Standard deviation: $\qquad \sigma_x = \sqrt{\sum (x_i - \mu_x)^2 p_i}.$

In the case of a binomial probability distribution with probability of a success equal to p and number of trials equal to n, if we let X be the number of successes in the n trials, the above equations become

Mean or expected value: $\qquad \mu_x = np$

Variance: $\qquad \sigma_x^2 = npq = np(1 - p)$

Standard deviation: $\qquad \sigma_x = \sqrt{npq} = \sqrt{np(1 - p)}.$

EXAMPLE 9.38

Sixty percent of all buyers of new cars choose automatic transmissions. For a group of five buyers of new cars, calculate the mean and standard deviation for the number of buyers choosing automatic transmissions.
Answer:

$$\mu_x = np = 5(.6) = 3.0$$
$$\sigma_x = \sqrt{np(1 - p)} = \sqrt{5(.6)(.4)} \approx 1.1$$

Questions on Topic Nine: Probability as Relative Frequency

Multiple-Choice Questions

1. Of the coral reef species on the Great Barrier Reef off Australia, 73% are poisonous. If a tourist boat taking divers to different points off the reef encounters an average of 25 coral reef species, what are the mean and standard deviation for the expected number of poisonous species seen?

   (A) $\mu_x = 6.75$, $\sigma_x = 4.93$
   (B) $\mu_x = 18.25$, $\sigma_x = 2.22$
   (C) $\mu_x = 18.25$, $\sigma_x = 4.93$
   (D) $\mu_x = 18.25$, $\sigma_x = 8.88$
   (E) None of the above gives a set of correct answers.

2. In the November 27, 1994, issue of *Parade* magazine, the "Ask Marilyn" section contained this question: "Suppose a person was having two surgeries performed at the same time. If the chances of success for surgery A are 85%, and the chances of success for surgery B are 90%, what are the chances that both would fail?" What do you think of Marilyn's solution: $(.15)(.10) = .015$ or 1.5%?

   (A) Her solution is mathematically correct but not explained very well.
   (B) Her solution is both mathematically correct and intuitively obvious.
   (C) Her use of complementary events is incorrect.
   (D) Her use of the general addition formula is incorrect.
   (E) She assumed independence of events, which is most likely wrong.

3. In November 1994, Intel announced that a "subtle flaw" in its Pentium chip would affect 1 in 9 billion division problems. Suppose a computer performs 20 million divisions (a not unreasonable number) in the course of a particular program. What is the probability of no error? Of at least one error?

   (A) .99778, .00222
   (B) .000000000111, .00222
   (C) .000000000111, .999999999889
   (D) .999999999889, .000000000111
   (E) .999999999889, .00222

4. According to a CBS/*New York Times* poll taken in 1992, 15% of the public have responded to a telephone call-in poll. In a random group of five people, what is the probability that exactly two have responded to a call-in poll?

   (A) .138
   (B) .165
   (C) .300
   (D) .835
   (F) .973

5. In a 1974 "Dear Abby" letter a woman lamented that she had just given birth to her eighth child, and all were girls! Her doctor had assured her that the chance of the eighth child being a girl was only 1 in 100.

   a. What was the real probability that the eighth child would be a girl?
   b. Before the birth of the first child, what was the probability that the woman would give birth to eight girls in a row?

   (A) .5, .0039
   (B) .0039, .0039
   (C) .5, .5
   (D) .0039, .4
   (E) .5, .01

6. The yearly mortality rate for American men from prostate cancer has been constant for decades at about 25 of every 100,000 men. (This rate has not changed in spite of new diagnostic techniques and new treatments.) In a group of 100 American men, what is the probability that at least 1 will die from prostate cancer in a given year?

   (A) .00025
   (B) .0247
   (C) .025
   (D) .9753
   (E) .99975

7. Alan Dershowitz, one of O. J. Simpson's lawyers, has stated that only 1 out of every 1000 abusive relationships ends in murder each year. If he is correct, and if there are approximately 1.5 million abusive relationships in the United States, what is the expected value for the number of people who are killed each year by an abusive partner?

   (A) 1
   (B) 500
   (C) 1000
   (D) 1500
   (E) None of the above

8. A television game show has three payoffs with the following probabilities:

Payoff ($):	0	1000	10,000
Probability:	.6	.3	.1

What are the mean and standard deviation for the payoff variable?

(A) $\mu_x = 1300$, $\sigma_x = 2934$
(B) $\mu_x = 1300$, $\sigma_x = 8802$
(C) $\mu_x = 3667$, $\sigma_x = 4497$
(D) $\mu_x = 3667$, $\sigma_x = 5508$
(E) None of the above gives a set of correct answers.

9. Suppose that among the 6000 students at a high school, 1500 are taking honors courses and 1800 prefer watching basketball to watching football. If taking honors courses and preferring basketball are independent, how many students are both taking honors courses and prefer basketball to football?

(A) 300
(B) 330
(C) 450
(D) 825
(E) There is insufficient information to answer this question.

10. As reported in *The New York Times* (February 19, 1995, p. 12), the Russian Health Ministry announced that one-quarter of the country's hospitals had no sewage system and one-seventh had no running water. What is the probability that a Russian hospital will have at least one of these problems

   a. if the two problems are independent?
   b. if hospitals with a running water problem are a subset of those with a sewage problem?

(A) 11/28, 1/4
(B) 11/28, 1/7
(C) 9/28, 1/4
(D) 9/28, 1/7
(E) 5/14, 1/4

11. An inspection procedure at a manufacturing plant involves picking three items at random and then accepting the whole lot if at least two of the three items are in perfect condition. If in reality 90% of the whole lot are perfect, what is the probability that the lot will be accepted?

(A) .600
(B) .667
(C) .729
(D) .810
(E) .972

12. Suppose that, in a certain part of the world, in any 50-year period the probability of a major plague is .39, the probability of a major famine is .52, and the probability of both a plague and a famine is .15. What is the probability of a famine given that there is a plague?

    (A) .240
    (B) .288
    (C) .370
    (D) .385
    (E) .760

13. Suppose that, for any given year, the probabilities that the stock market declines, that women's hemlines are lower, and that both events occur are, respectively, .4, .35, and .3. Are the two events independent?

    (A) Yes, because $(.4)(.35) \neq .3$.
    (B) No, because $(.4)(.35) \neq .3$.
    (C) Yes, because $.4 > .35 > .3$.
    (D) No, because $.5(.3 + .4) = .35$.
    (E) There is insufficient information to answer this question.

14. If $P(A) = .2$ and $P(B) = .1$, what is $P(A \cup B)$ if A and B are independent?

    (A) .02
    (B) .28
    (C) .30
    (D) .32
    (E) There is insufficient information to answer this question.

15. The following data are from *The Commissioner's Standard Ordinary Table of Mortality:*

Age	Number Surviving
0	10,000,000
20	9,664,994
40	9,241,359
70	5,592,012

    What is the probability that a 20-year-old will survive to be 70?

    (A) .407
    (B) .421
    (C) .559
    (D) .579
    (E) .966

16. At a warehouse sale 100 customers are invited to choose one of 100 identical boxes. Five boxes contain $700 color television sets, 25 boxes contain $540 camcorders, and the remaining boxes contain $260 cameras. What should a customer be willing to pay to participate in the sale?

    (A) $260
    (B) $352
    (C) $500
    (D) $540
    (E) $699

17. There are two games involving flipping a coin. In the first game you win a prize if you can throw between 40% and 60% heads. In the second game you win if you can throw more than 75% heads. For each game would you rather flip the coin 50 times or 500 times?

    (A) 50 times for each game
    (B) 500 times for each game
    (C) 50 times for the first game, and 500 for the second
    (D) 500 times for the first game, and 50 for the second
    (E) The outcomes of the games do not depend on the number of flips.

18. Which of the following are true statements?

    I. The probability of an event is always at least 0 and at most 1.
    II. The probability that an event will happen is always 1 minus the probability that it won't happen.
    III. If two events cannot occur simultaneously, the probability that at least one event will occur is the sum of the respective probabilities of the two events.

    (A) I and II
    (B) I and III
    (C) II and III
    (D) I, II, and III
    (E) None of the above gives the complete set of true responses.

19. Suppose there are five outcomes to an experiment and a computer calculates the respective probabilities of the outcomes to be .4, .5, .3, 0, and − .2. The proper conclusion is that

    (A) the sum of the individual probabilities is 1.
    (B) one of the outcomes will never occur.
    (C) one of the outcomes will occur 50% of the time.
    (D) all of the above are true.
    (E) there is an error in the computer program.

20. The average annual incomes of high school and college graduates in a midwestern town are $21,000 and $35,000, respectively. If a company hires only personnel with at least a high school diploma and 20% of its employees have been through college, what is the mean income of the company employees?

    (A) $23,800
    (B) $27,110
    (C) $28,000
    (D) $32,200
    (E) $56,000

21. An insurance company charges $800 annually for car insurance. The policy specifies that the company will pay $1000 for a minor accident and $5000 for a major accident. If the probability of a motorist having a minor accident during the year is .2, and of having a major accident, .05, how much can the insurance company expect to make on a policy?

    (A) $200
    (B) $250
    (C) $300
    (D) $350
    (E) $450

22. Which of the following are true statements?

    I. By the law of large numbers, the mean of a random variable will get closer and closer to a specific value.
    II. The standard deviation of a random variable is never negative.
    III. The standard deviation of a random variable is 0 only if the random variable takes a lone single value.

    (A) I and II
    (B) I and III
    (C) II and III
    (D) I, II, and III
    (E) None of the above gives the complete set of true responses.

23. You can choose one of three boxes. Box A has four $5 bills and a single $100 bill, box B has 400 $5 bills and 100 $100 bills, and box C has 24 $1 bills. You can have all of box C or blindly pick one bill out of either box A or box B. Which offers the greatest expected winning?

    (A) Box A
    (B) Box B
    (C) Box C
    (D) Either A or B, but not C
    (E) All offer the same expected winning.

24. Given that 52% of the U.S. population are female and 15% are older than age 65, can we conclude that $(.52)(.15) = 7.8\%$ are women older than age 65?

    (A) Yes, by the multiplication rule.
    (B) Yes, by conditional probabilities.
    (C) Yes, by the law of large numbers.
    (D) No, because the events are not independent.
    (E) No, because the events are not mutually exclusive.

25. Companies proved to have violated pollution laws are being fined various amounts with the following probabilities:

Fine ($):	1000	10,000	50,000	100,000
Probability:	.4	.3	.2	.1

    What are the mean and standard deviation for the fine variable?

    (A) $\mu_x = 40{,}250$, $\sigma_x = 39{,}118$
    (B) $\mu_x = 40{,}250$, $\sigma_x = 45{,}169$
    (C) $\mu_x = 23{,}400$, $\sigma_x = 31{,}350$
    (D) $\mu_x = 23{,}400$, $\sigma_x = 85{,}185$
    (E) None of the above gives a set of correct answers.

Questions 26–30 refer to the following study: One thousand students at a city high school were classified both according to GPA and whether or not they consistently skipped classes.

	GPA			
	<2.0	2.0–3.0	>3.0	
Many skipped classes	80	25	5	110
Few skipped classes	175	450	265	890
	255	475	270	

26. What is the probability that a student has a GPA between 2.0 and 3.0?

    (A) .025
    (B) .227
    (C) .450
    (D) .475
    (E) .506

27. What is the probability that a student has a GPA under 2.0 and has skipped many classes?

    (A) .080
    (B) .281
    (C) .285
    (D) .314
    (E) .727

28. What is the probability that a student has a GPA under 2.0 or has skipped many classes?

    (A) .080
    (B) .281
    (C) .285
    (D) .314
    (E) .727

29. What is the probability that a student has a GPA under 2.0 given that he has skipped many classes?

    (A) .080
    (B) .281
    (C) .285
    (D) .314
    (E) .727

30. Are "GPA between 2.0 and 3.0" and "skipped few classes" independent?

    (A) No, because .475 ≠ .506.
    (B) No, because .475 ≠ .890.
    (C) No, because .450 ≠ .475.
    (D) Yes, because of conditional probabilities.
    (E) Yes, because of the product rule.

31. Mathematically speaking, casinos and life insurance companies make a profit because of

    (A) their understanding of sampling error and sources of bias.
    (B) their use of well-designed, well-conducted surveys and experiments.
    (C) their use of simulation of probability distributions.
    (D) the central limit theorem.
    (E) the law of large numbers.

32. Consider the following table of ages of U.S. senators:

Age (yr):	<40	40–49	50–59	60–69	70–79	>79
Number of senators:	5	30	36	22	5	2

    What is the probability that a senator is under 70 years old given that he is at least 50 years old?

    (A) .580
    (B) .624
    (C) .643
    (D) .892
    (E) .969

Questions 33–36 refer to the following study: Five hundred people used a home test for HIV, and then all underwent more conclusive hospital testing. The accuracy of the home test was evidenced in the following table.

	HIV	Healthy	
Positive test	35	25	60
Negative test	5	435	440
	40	460	

33. What is the *predictive value* of the test? That is, what is the probability that a person has HIV and tests positive?

    (A) .070
    (B) .130
    (C) .538
    (D) .583
    (E) .875

34. What is the *false-positive* rate? That is, what is the probability of testing positive given that the person does not have HIV?

    (A) .054
    (B) .050
    (C) .130
    (D) .417
    (E) .875

35. What is the *sensitivity* of the test? That is, what is the probability of testing positive given that the person has HIV?

    (A) .070
    (B) .130
    (C) .538
    (D) .583
    (E) .875

36. What is the *specificity* of the test? That is, what is the probability of testing negative given that the person does not have HIV?

    (A) .125
    (B) .583
    (C) .870
    (D) .950
    (E) .946

37. Suppose that 2% of a clinic's patients are known to have cancer. A blood test is developed that is positive in 98% of patients with cancer but is also positive in 3% of patients who do not have cancer. If a person who is chosen at random from the clinic's patients is given the test and it comes out positive, what is the probability that the person actually has cancer?

    (A) .02
    (B) .4
    (C) .5
    (D) .6
    (E) .98

38. Suppose the probability that you will receive an A in AP Statistics is .35, the probability that you will receive A's in both AP Statistics and AP Biology is .19, and the probability that you will receive an A in AP Statistics but not in AP Biology is .17. Which of the following is a proper conclusion?

    (A) The probability that you will receive an A in AP Biology is .36.
    (B) The probability that you didn't take AP Biology is .01.
    (C) The probability that you will receive an A in AP Biology but not in AP Statistics is .18.
    (D) The given probabilities are impossible.
    (E) None of the above

39. Suppose you are one of 7.5 million people who send in their name for a drawing with 1 top prize of $1 million, 5 second-place prizes of $10,000, and 20 third-place prizes of $100. Is it worth the $0.32 postage it cost you to send in your name?

    (A) Yes, because $\frac{1,000,000}{0.32} = 3,125,000$, which is less than 7,500,000.
    (B) No, because your expected winnings are only $0.14.
    (C) Yes, because $\frac{7,500,000}{(1+5+20)} = 288,462$.
    (D) No, because 1,052,000 < 7,500,000.
    (E) Yes, because $\frac{1,052,000}{26} = 40,462$.

40. You have a choice of three investments, the first of which gives you a 10% chance of making $1 million, otherwise you lose $25,000; the second of which gives you a 50% chance of making $500,000, otherwise you lose $345,000; and the third of which gives you an 80% chance of making $50,000, otherwise you make only $1,000. Assuming you will go bankrupt if you don't show a profit, which option should you choose for the best chance of avoiding bankruptcy?

    (A) First choice
    (B) Second choice
    (C) Third choice
    (D) Either the first or the second choice
    (E) All the choices give an equal chance of avoiding bankruptcy.

41. Can the function $f(x) = \frac{x+6}{24}$, for $x = 1, 2,$ and 3, be the probability distribution for some random variable?

    (A) Yes.
    (B) No, because probabilities cannot be negative.
    (C) No, because probabilities cannot be greater than 1.
    (D) No, because the probabilities do not sum to 1.
    (E) Not enough information is given to answer the question.

42. Sixty-five percent of all divorce cases cite incompatibility as the underlying reason. If four couples file for a divorce, what is the probability that exactly two will state incompatibility as the reason?

    (A) .104
    (B) .207
    (C) .254
    (D) .311
    (E) .423

43. Suppose we have a random variable X where the probability associated with the value $\binom{10}{k}(.37)^k(.63)^{10-k}$ for $k = 0, \ldots, 10$. What is the mean of X?

    (A) 0.37
    (B) 0.63
    (C) 3.7
    (D) 6.3
    (E) None of the above

44. A computer technician notes that 40% of computers fail because of the hard drive, 25% because of the monitor, 20% because of a disk drive, and 15% because of the microprocessor. If the problem is not in the monitor, what is the probability that it is in the hard drive?

    (A) .150
    (B) .400
    (C) .417
    (D) .533
    (E) .650

45. Suppose you toss a coin ten times and it comes up heads every time. Which of the following is a true statement?

    (A) By the law of large numbers, the next toss is more likely to be tails than another heads.
    (B) By the properties of conditional probability, the next toss is more likely to be heads given that ten tosses in a row have been heads.
    (C) Coins actually do have memories, and thus what comes up on the next toss is influenced by the past tosses.
    (D) The law of large numbers tells how many tosses will be necessary before the percentages of heads and tails are again in balance.
    (E) The probability that the next toss will again be heads is .5.

46. Which of the following lead to binomial distributions?

    I. An inspection procedure at an automobile manufacturing plant involves selecting a sample of cars from the assembly line and noting for each car whether there are no defects, at least one major defect, or only minor defects.

    II. As students study more and more during their AP Statistics class, their chances of getting an A on any given test continue to improve. The teacher is interested in the probability of any given student receiving various numbers of A's on the class exams.

    III. A committee of two is to be selected from among the five teachers and ten students attending a meeting. What are the probabilities that the committee will consist of two teachers, of two students, or of exactly one teacher and one student?

    (A) I only
    (B) II only
    (C) III only
    (D) II and III
    (E) None of the above gives the complete set of true responses.

47. Which of the following are true statements?

    I. The histogram of a binomial distribution with $p = .5$ is always symmetric no matter what n, the number of trials, is.

    II. The histogram of a binomial distribution with $p = .9$ is skewed to the right.

    III. The histogram of a binomial distribution with $p = .9$ is almost symmetric if n is very large.

    (A) I and II
    (B) I and III
    (C) II and III
    (D) I, II, and III
    (E) None of the above gives the complete set of true responses.

48. Suppose that 60% of students who take the AP Statistics exam score 4 or 5, 25% score 3, and the rest score 1 or 2. Suppose further that 95% of those scoring 4 or 5 receive college credit, 50% of those scoring 3 receive such credit, and 4% of those scoring 1 or 2 receive credit. If a student who is chosen at random from among those taking the exam receives college credit, what is the probability that she received a 3 on the exam?

    (A) .125
    (B) .178
    (C) .701
    (D) .813
    (E) .822

49. Given the probabilities $P(A) = .4$ and $P(A \cup B) = .6$, what is the probability $P(B)$ if A and B are mutually exclusive? If A and B are independent?

    (A) .2, .4
    (B) .2, .33
    (C) .33, .2
    (D) .6, .33
    (E) .6, .4

50. With regard to a particular gene, the percentages of genotypes *AA, Aa,* and *aa* in a particular population are, respectively, 60%, 30%, and 10%. Furthermore, the percentages of these genotypes that contract a certain disease are, respectively, 1%, 5%, and 20%. If a person does contract the disease, what is the probability that the person is of genotype *AA*?

    (A) .006
    (B) .010
    (C) .041
    (D) .146
    (E) .600

Answer Key

1. **B**	11. **E**	21. **D**	31. **E**	41. **A**
2. **E**	12. **D**	22. **C**	32. **D**	42. **D**
3. **A**	13. **B**	23. **E**	33. **A**	43. **C**
4. **A**	14. **B**	24. **D**	34. **A**	44. **D**
5. **A**	15. **D**	25. **C**	35. **E**	45. **E**
6. **B**	16. **B**	26. **D**	36. **E**	46. **E**
7. **D**	17. **D**	27. **A**	37. **B**	47. **B**
8. **A**	18. **D**	28. **C**	38. **D**	48. **B**
9. **C**	19. **E**	29. **E**	39. **B**	49. **B**
10. **E**	20. **A**	30. **A**	40. **C**	50. **D**

Answers Explained

1. **(B)** $\mu = 25(.73) = 18.25$, $\sigma^2 = 25(.73)(.27) = 4.9275$, and $\sigma = 2.22$.

2. **(E)** The probability that *both* events will occur is the product of their separate probabilities only if the events are independent, that is, only if the chance that one event will happen is not influenced by whether or not the second event happens. In this case the probability of different surgeries failing are probably closely related.

3. **(A)** If $p = \frac{1}{9,000,000,000}$, then $P(0 \text{ errors}) = (1 - p)^{20,000,000} = .99778$ and $P(\text{at least 1 error}) = 1 - .99778 = .00222$.

4. **(A)** $10(.15)^2(.85)^3 = .138$ or binompdf(5, .15, 2) = .138

5. **(A)** The probability of the next child being a girl is independent of the gender of the previous children. The probability of having eight girls in a row is $(.5)^8 = .0039$.

6. **(B)** $1 - (.99975)^{100} = .0247$

7. **(D)** $.001(1,500,000) = 1500$

8. **(A)** $\mu = 0(.6) + 1000(.3) + 10,000(.1) = \$1300$, $\sigma^2 = (0 - 1300)^2(.6) + (1000 - 1300)^2(.3) + (10,000 - 1300)^2(.1) = 8,610,000$, and $\sigma = \$2934$.

9. **(C)** $\frac{1500}{6000} = .25$ are honors students, and $\frac{1800}{6000} = .3$ prefer basketball. Because of independence, their intersection is $(.25)(.3) = .075$ of the students, and $.075(6000) = 450$.

10. **(E)** $\frac{1}{4} + \frac{1}{7} - \left(\frac{1}{4}\right)\left(\frac{1}{7}\right) = \frac{5}{14}$. If X is a subset of Y, then $P(X \cup Y) = P(Y)$.

11. **(E)** $(.9)^3 + 3(.9)^2(.1) = .972$

12. **(D)** $P\left(\text{famine}|\text{plague}\right) = \dfrac{P\left(\text{famine} \cap \text{plague}\right)}{P\left(\text{plague}\right)} = \dfrac{.15}{.39} = .385$

13. **(B)** If E and F are independent, then $P(E \cap F) = P(E)P(F)$; however, in this problem, $(.4)(.35) \neq .3$.

14. **(B)** Because A and B are independent, we have $P(A \cap B) = P(A)P(B)$, and thus $P(A \cup B) = .2 + .1 - (.2)(.1) = .28$.

15. **(D)** $\dfrac{5,592,012}{9,664,994} = .579$

16. **(B)** $E(X) = \mu_x = \Sigma x_i p_i = 700(.05) + 540(.25) + 260(.7) = 352$

17. **(D)** The probability of throwing heads is .5. By the law of large numbers, the more times you flip the coin, the more the relative frequency tends to become closer to this probability. With fewer tosses there is a greater chance of wide swings in the relative frequency.

18. **(D)** These are all true facts about probability. Note that the more general statement $P(X \cup Y) = P(X) + P(Y) - P(X \cap Y)$ reduces to $P(X \cup Y) = P(X) + P(Y)$ when the two events cannot occur simultaneously.

19. **(E)** Probabilities are never less than 0.

20. **(A)** $E(X) = \mu_x = \Sigma x_i p_i = 21,000(.8) + 35,000(.2) = 23,800$

21. **(D)** $1000(.2) + 5000(.05) = 450$, and $800 - 450 = 350$.

22. **(C)** The mean is an unchanging, fixed number; the law of large numbers gives that the average of observed *X* values tends to be closer and closer to this mean after more and more repetitions.

23. **(E)** $5\left(\frac{4}{5}\right) + 100\left(\frac{1}{5}\right) = 24$, and $5\left(\frac{400}{500}\right) + 100\left(\frac{100}{500}\right) = 24$

24. **(D)** $P(E \cap F) = P(E)P(F)$ only if the events are independent. In this case, women live longer than men and so the events are not independent.

25. **(C)** $\mu = 1000(.4) + 10,000(.3) + 50,000(.2) + 100,000(.1) = 23,400$, $\sigma^2 = (1000 - 23,400)^2(.4) + (10,000 - 23,400)^2(.3) + (50,000 - 23,400)^2(.2) + (100,000 - 23,400)^2(.1) = 982,840,000$, and $\sigma = \$31,350$.

For questions 26–30, we first sum the rows and columns:

	GPA			
	<2.0	2.0–3.0	>3.0	
Many skipped classes	80	25	5	110
Few skipped classes	175	450	265	890
	255	475	270	1000

26. **(D)** $\frac{25 + 450}{1000} = .475$

27. **(A)** $\frac{80}{1000} = .080$ (probability of an intersection)

28. **(C)** $\frac{80 + 25 + 5 + 175}{1000} = .285$ (probability of a union)

29. **(E)** $\frac{80}{80 + 25 + 5} = .727$ (conditional probability)

30. **(A)** $P(2.0 - 3.0\,\text{GPA}) = \frac{475}{1000} = .475$; however, $P(2.0 - 3.0\,\text{GPA}|\text{few skips}) = \frac{450}{890} = .506$. If independent, these would have been equal.

31. **(E)** While the outcome of any single play on a roulette wheel or the age at death of any particular person is uncertain, the law of large numbers gives that the relative frequencies of specific outcomes in the long run tend to become closer to numbers called probabilities.

32. **(D)** $\frac{36 + 22}{36 + 22 + 5 + 2} = .892$

33. **(A)** $\frac{35}{500} = .070$

34. **(A)** $\frac{25}{460} = .054$

35. **(E)** $\frac{35}{40} = .875$

36. **(E)** $\frac{435}{460} = .946$

37. **(B)**

$$P(\text{positive test}) = P(\text{cancer} \cap \text{positive test})$$
$$+ P(\text{no cancer} \cap \text{positive test})$$
$$= .0196 + .0294$$
$$= .0490$$

$$P\left(\text{cancer}\,\middle|\,\text{positive test}\right) = \frac{P\left(\text{cancer} \cap \text{positive test}\right)}{P\left(\text{positive test}\right)}$$
$$= \frac{.0196}{.0490}$$
$$= .4$$

38. **(D)** The probability that you will receive an A in AP Statistics but not in AP Biology must be $.35 - .19 = .16$, not $.17$.

39. **(B)** Your expected winnings are only

$$\frac{1}{7,500,000}(1,000,000) + \frac{5}{7,500,000}(10,000) + \frac{20}{7,500,000}(100) = 0.14$$

40. **(C)** Even though the first two choices have a higher expected value than the third choice, the third choice gives a 100% chance of avoiding bankruptcy.

41. **(A)** The probabilities $\frac{7}{24}$, $\frac{8}{24}$, and $\frac{9}{24}$ are all nonnegative, and they sum to 1.

42. **(D)** $6(.65)^2(.35)^2 = .311$ or binompdf(4, .65, 2) = .311

43. **(C)** This is a binomial with $n = 10$ and $p = .37$, and so the mean is $np = 10(.37) = 3.7$.

44. **(D)** $\frac{.40}{.40+.20+.15} = .533$

45. **(E)** Coins have no memory, and so the probability that the next toss will be heads is .5 and the probability that it will be tails is .5. The law of large numbers says that as the number of tosses becomes larger, the percentage of heads tends to become closer to .5.

46. **(E)** None of them lead to binomial distributions. The first has three outcomes, not two. In the second, p increases from test to test. In the third, there is a lack of independence in choosing the first and second people to be on the committee; for example, if the first person chosen is a teacher, the probability that the second will also be a teacher is $\frac{4}{14}$, not $\frac{4}{15}$.

47. **(B)** If $p > .5$, the more likely numbers of successes are to the right and the lower numbers of successes have small probabilities, and so the histogram is skewed to the left. No matter what p is, if n is sufficiently large, the histogram will look almost symmetric.

48. **(B)**

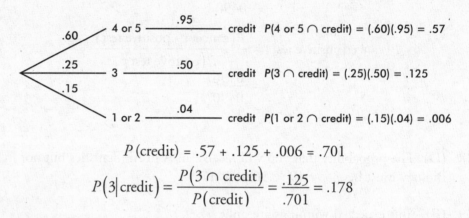

$$P(\text{credit}) = .57 + .125 + .006 = .701$$

$$P(3 \mid \text{credit}) = \frac{P(3 \cap \text{credit})}{P(\text{credit})} = \frac{.125}{.701} = .178$$

49. **(B)** If A and B are mutually exclusive, $P(A \cap B) = 0$. Thus $.6 = .4 + P(B) - 0$, and so $P(B) = .2$. If A and B are independent, then $P(A \cap B) = P(A)P(B)$. Thus $.6 = .4 + P(B) - .4P(B)$, and so $P(B) = \frac{1}{3}$.

50. **(D)**

$$
\begin{array}{l}
.6 \quad AA \xrightarrow{\ .01\ } \text{contract disease} \quad P\!\left(AA \cap \text{contract disease}\right) = (.6)(.01) = .006 \\[2ex]
.3 \quad Aa \xrightarrow{\ .05\ } \text{contract disease} \quad P\!\left(Aa \cap \text{contract disease}\right) = (.3)(.05) = .015 \\[2ex]
.1 \quad aa \xrightarrow{\ .20\ } \text{contract disease} \quad P\!\left(aa \cap \text{contract disease}\right) = (.1)(.20) = .020
\end{array}
$$

$$P\!\left(\text{contract disease}\right) = .006 + .015 + .020 = .041$$

$$P\!\left(AA \mid \text{contract disease}\right) = \frac{P\!\left(AA \cap \text{contract disease}\right)}{P\!\left(\text{contract disease}\right)} = \frac{.006}{.041} = .146$$

Free-Response Questions

Seven Open-Ended Questions

1. Suppose you will be taking exams in English, statistics, and chemistry tomorrow, and from past experience you know that for each exam you have a 50% chance of receiving an A. List all eight possibilities for receiving A's on the various exams. Letting X represent the number of A's you will receive, show the distribution of X, that is, the values and associated probabilities.

2. A sample of applicants for a management position yields the following numbers with regard to age and experience:

	Years of experience		
	0–5	6–10	>10
Less than 50 years old	80	125	20
More than 50 years old	10	75	50

   a. What is the probability that an applicant is less than 50 years old? Has more than 10 years' experience? Is more than 50 years old and has five or fewer years' experience?
   b. What is the probability that an applicant is less than 50 years old given that she has between 6 and 10 years' experience?
   c. Are the two events "less than 50 years old" and "more than 10 years' experience" independent events? How about the two events "more than 50 years old" and "between 6 and 10 years' experience"? Explain.

3. *The New York Times* (September 21, 1994, p. C10) reported that an American woman diagnosed with ovarian cancer has a 37.5% chance of survival and that approximately 20,000 American women are diagnosed with ovarian cancer each year. Discuss these observations with regard to expected number of deaths and also reference some measure of variability.

4. Suppose USAir accounts for 20% of all U.S. domestic flights. As of mid-1994, USAir was involved in four of the previous seven major disasters. "That's enough to begin getting suspicious but not enough to hang them," said Dr. Brad Efton in *The New York Times* (September 11, 1994, Sec. 4, p. 4). Using a binomial distribution, comment on Dr. Efton's remark.

5. Assume there is no overlap between the 56% of the population who wear glasses and the 4% who wear contacts. If 55% of those who wear glasses are women and 63% of those who wear contacts are women, what is the probability that the next person you encounter on the street will be a woman with

glasses? A woman with contacts? A man with glasses? A man with contacts? A person not wearing glasses or contacts? Explain your reasoning.

6. Explain what is wrong with each of the following statements:

   a. The probability that a student will score high on the AP Statistics exam is .43, while the probability that she will not score high is .47.

   b. The probability that a student plays tennis is .18, while the probability that he plays basketball is six times as great.

   c. The probability that a student enjoys her English class is .64, while the probability that she enjoys both her English and her social studies classes is .71.

   d. The probability that a student will be accepted by his first choice for college is .38, while the probability that he will be accepted by his first or second choice is .32.

   e. The probability that a student fails AP Statistics and will still be accepted by an Ivy League school is −.17.

7. Player A rolls a die with 7 on four sides and 11 on two sides. Player B flips a coin with 6 on one side and 10 on the other. Assume a fair die and a fair coin.

   (a) Suppose the winner is the player with the higher number showing. Explain whom you would rather be. If player A receives $0.25 for every win, what should player B receive for every win to make the game fair?

   (b) Suppose player A receives $0.70 from player B when the 7 shows and $1.10 when the 11 shows, while player B receives $0.60 from player A when the 6 shows and $1.00 when the 10 shows. Explain whom you would rather be.

Answers Explained

1. The eight possible ways of scoring A's are

English	A	A	A	A	—	—	—	—
Statistics	A	A	—	—	A	A	—	—
Chemistry	A	—	A	—	A	—	A	—

   The values and probabilities for X, the number of A's, are found by noting there is one way of receiving no A's, three ways of receiving one A, three ways of receiving two A's, and one way of receiving all A's. Each of the eight ways has probability $\frac{1}{8} = .125$.

   | Value of X: | 0 | 1 | 2 | 3 |
   | Probability: | .125 | .375 | .375 | .125 |

2. It's easiest to first sum the rows and columns:

	Years of experience			
	0–5	6–10	>10	
Less than 50 years old	80	125	20	225
More than 50 years old	10	75	50	135
	90	200	70	360

a. $P(\text{age} < 50) = \frac{225}{360} = .625$

$P(\text{experience} > 10) = \frac{70}{360} = .194$

$P(\text{age} > 50 \cap \text{experience } 6–10) = \frac{75}{360} = .208$

b. $P(\text{age} < 50 | \text{experience } 6–10) = \frac{125}{200} = .625$

c. $P(\text{age} < 50) = \frac{225}{360} = .625$; however, $P(\text{age} < 50 | \text{experience} > 10) = \frac{20}{70}$

= .286 and so they are not independent.

$P(\text{age} > 50) = \frac{135}{360} = .375$ and also $P(\text{age} > 50 | \text{experience } 6–10) = \frac{75}{200}$

= .375 and so these two are independent.

3. The probability of death for a woman diagnosed with ovarian cancer is 1 − .375 = .625. Thus we are considering a binomial with n = 20,000 and p = .625. For this we can calculate:

$$\mu_x = 20,000(.625) = 12,500$$

$$\sigma_x = \sqrt{20,000(.625)(.375)} = 68.5$$

Thus one can expect an average of 12,500 deaths per year with a standard deviation of 68.5 deaths among American women diagnosed with ovarian cancer.

4. If USAir accounted for 20% of the major disasters, the chance that it would be involved in at least four of seven such disasters is

$$\binom{7}{4}(.2)^4(.8)^3 + \binom{7}{5}(.2)^5(.8)^2 + \binom{7}{6}(.2)^6(.8)^1 + (.2)^7$$
$$= .029 + .004 + .000 + 000$$
$$= .033$$

[Or binomcdf(7, .8, 3) = .033.]

Mathematically, if USAir accounted for only 20% of the major disasters, there is only a .033 chance of it being involved in four of seven such disasters. This seems more than enough evidence to be suspicious!

5.

$$P(E|F) = \frac{P(E \cap F)}{P(F)} \quad \text{or} \quad P(E \cap F) = P(E|F)P(F),$$

and so

$$P(\text{woman} \cap \text{glasses}) = (.55)(.56)$$
$$= .308$$

$$P(\text{woman} \cap \text{contacts}) = (.63)(.04)$$
$$= .0252$$

If 55% of those who wear glasses are women, 45% of those who wear glasses must be men, and if 63% of those who wear contacts are women, 37% of those who wear contacts must be men. Thus we have

$$P(\text{man} \cap \text{glasses}) = (.45)(.56)$$
$$= .252$$

$$P(\text{man} \cap \text{contacts}) = (.37)(.04)$$
$$= .0148$$

The probability that you will encounter a person not wearing glasses or contacts is $1 - (.308 + .0252 + .252 + .0148) = .4$.

6. *a.* The probability of the complement is 1 minus the probability of the event, but $1 - .43 \neq .47$.
b. Probabilities are never greater than 1, but $6(.18) = 1.08$.
c. The probability of an intersection cannot be greater than the probability of one of the separate events.
d. The probability of a union cannot be less than the probability of one of the separate events.
e. Probabilities are never negative.

7. (a) Player B wins only if both a 10 shows on the coin (a probability of $\frac{1}{2}$) and a 7 shows on the die (a probability of $\frac{2}{3}$). These events will both happen $\left(\frac{1}{2}\right)\left(\frac{2}{3}\right) = \frac{1}{3}$ of the time, and so player A wins $\frac{2}{3}$ of the time or twice as often as player B. Thus, to make this a fair game, player B should receive $0.50 each time he wins.
(b) Player A's expected payoff is

$$\sum xP(x) = (0.70)\left(\frac{2}{3}\right) + (1.10)\left(\frac{1}{3}\right) \approx \$0.83$$

while player B's expected payoff is

$$\sum xP(x) = (0.60)\left(\frac{1}{2}\right) + (1.00)\left(\frac{1}{2}\right) \approx \$0.80$$

Five Investigative Tasks

1. Suppose the probability that someone will make a major mistake on an income tax return is .23. One day, an Internal Revenue Service (IRS) agent plans to audit as many returns as necessary until she finds one with a major mistake.

   a. Use simulation to estimate the probability that a major mistake will be found before the fifth return.
   b. Use probability to calculate the probability that a major mistake will be found before the fifth return.

2. A new antibiotic is tested on six patients in each of 125 hospitals. The number of positive responses among the patients in each hospital is noted with the following results:

Number of positive responses:	0	1	2	3	4	5	6
Actual number of hospitals:	0	0	5	36	51	20	13

   a. How many patients had a positive response?
   b. What is the probability that a patient will have a positive response?
   c. What is the binomial probability distribution with this probability and with $n = 6$?
   d. What is the expected number of hospitals reporting each number of positive responses?
   e. Do the expected and actual numbers of hospitals seem to be close? (If the actual numbers are close to the numbers predicted by the binomial distribution, we can be more confident in concluding that the results are due solely to the antibiotic and not to differences in the hospitals and other factors.)

3. A well-known paradox runs as follows: Suppose every couple continues having children until they have a boy and then they stop. What percentage of male births will result? No couple can have more than one boy, while many can have numerous girls, so it appears that the population will become overwhelmingly female. Test using simulation. What conclusion do you reach? In looking at your simulation outcomes, what do you see is wrong with the reasoning above that led to the conclusion about an increasingly female population?

4. In a championship chess match suppose the probability that player A will win a game is .2, the probability that player B will win is .4, and the probability of a draw is .4. Players receive one point for a win and one-half point for a draw, and the first to reach three points is the winner. If both reach three points at the same time, they keep playing until one wins another game.

   a. Run a simulation to predict the probability of each player winning the championship.
   b. Will the probabilities be the same if no points are given for a draw? Test using simulation.

5. If a car is stopped for speeding, the probability that the driver has illegal drugs hidden in the car is .14. A police officer manning a speed trap plans to thoroughly search cars he stops for speeding until he finds one with illegal drugs.

   (a) Use simulation to estimate the probability that illegal drugs will turn up before he stops the tenth car. Explain your work. Use the following random number table:

8417706757	1761315582	5150681435	4105092031	0644905059
5988431180	5311584469	9486857967	0581184514	7501113006
6339555041	1586606589	1311971020	8594091932	0648874987
5435552704	9035902649	4749671567	9426808844	2629464759
0898957024	9728400637	8928303514	5919507635	0330972605
2935723737	6788103668	3387635841	5286923114	1586438942

   (b) The results of a 100-trial simulation are shown below. Each trial ends when a car with illegal drugs is found.

   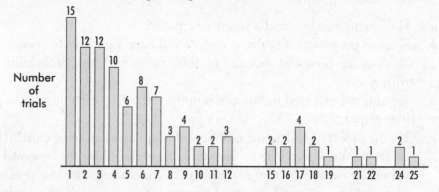

   First car with illegal drugs

   Use the simulation to estimate the mean of the distribution.

   (c) Use probability to calculate the probability that illegal drugs will be found before he stops the tenth car. Show your work.

Answers Explained

1. *a.* Let the numbers 01–23 represent a major mistake, while 24–99 and 00 represent no major mistake. Read off pairs of digits from the random number table until a major mistake is found and note whether it is found before the fifth return. Underlining major mistakes gives

84 <u>17</u> 706757	<u>17</u> 61315582	5150 68 <u>14</u> 35	41 <u>05 09 20</u> 31	<u>06</u> 44 90 50 59
59 88 43 <u>11</u> 80	53 <u>11</u> 584469	9486857967	<u>05</u> 81 <u>18</u> 45 <u>14</u>	75 <u>01</u> 11 30 <u>06</u>
63 39 55 50 41	<u>15</u> 86606589	<u>13 11</u> 97 <u>10</u> 20	85 94 <u>09 19</u> 32	<u>06</u> 48 87 49 87
54 35 55 27 <u>04</u>	9035902649	4749671567	9426808844	2629464759
<u>08</u> 98957024	97 28 40 <u>06</u> 37	89 28 30 35 <u>14</u>	59 <u>19</u> 507635	<u>03</u> 30 97 26 <u>05</u>
29 35 72 37 37	67 88 <u>10</u> 3668	3387635841	5286923114	<u>15</u> 86438942

The first major mistake (17) is found on the second return, the next major mistake (again 17) is found on the fourth return, the next (14) on the eighth return, and so on. Tabulating gives

Return on Which Major Mistake Is Found	Frequency
1	7
2	11
3	3
4	3
5 or beyond	12

The probability that a major mistake will be found before the fifth return is estimated to be $\frac{7+11+3+3}{36} = \frac{24}{36} = .67$.

b. The probability of the first return having a major mistake is .23. The probability of the first return having no major mistake but the second one having one is $(.23)(.77) = .1771$. The probability of the first return with a major mistake being the third return is $(.23)(.77)^2 = .1364$, and being the fourth return is $(.23)(.77)^3 = .1050$. The probability of the first major mistake being found before the fifth return is thus $.23 + .1771 + .1364 + .1050 = .6485$.

2. *a.* $5(2) + 36(3) + 51(4) + 20(5) + 13(6) = 500$

   b. There were $125(6) = 750$ patients; $\frac{500}{750} = .667$.

   c.

$$P(0) = 1(.667)^0(.333)^6 = .001$$
$$P(1) = 6(.667)^1(.333)^5 = .016$$
$$P(2) = 15(.667)^2(.333)^4 = .082$$
$$P(3) = 20(.667)^3(.333)^3 = .219$$
$$P(4) = 15(.667)^4(.333)^2 = .330$$
$$P(5) = 6(.667)^5(.333)^1 = .264$$
$$P(6) = 1(.667)^6(.333)^0 = \underline{.088}$$
$$1.000$$

[Or use binompdf$(6, \frac{2}{3})$.]

d. $.001(125) = 1.25$, $.016(125) = 2$, $.082(125) = 10.25$, $.219(125) = 27.375$, $.330(125) = 41.25$, $.264(125) = 33$, and $.088(125) = 11$, and so we have

Number of positive responses:	0	1	2	3	4	5	6
Expected number of hospitals:	1	2	10	27	41	33	11

e. (0, 0, 5, 36, 51, 20, 13) and (1, 2, 10, 27, 41, 33, 11) seem to differ significantly, indicating that the results are not due solely to the antibiotic. (A chi-square test of significance is indicated—see Topic 14.)

3. Let even numbers represent male births and odd numbers represent female births. Read off digits from the random number table, recording the results every time there is a male birth (even number). We have

8 4 1770 6 757176 131558 2 5150 6 8 14 354 10 50 92 0 310 6 4
4 90 50 59598 8 4 3118 0 531158 4 4 6 994 8 6 8 5796 70 58 118
4 514 750 11130 0 6 6 3395550 4 1158 6 6 0 6 58 913119710 2 0
8 594 0 91932 0 6 4 8 8 74 98 754 35552 70 4 90 3590 2 6 4 94
74 96 7156 794 2 6 8 0 8 8 4 4 2 6 2 94 6 4 7590 8 98 9570 2
4 972 8 4 0 0 6 378 92 8 30 3514 591950 76 350 330 972 6 0 52
93572 37376 78 8 10 36 6 8 338 76 358 4 152 8 6 92 3114 158 6
4 38 94 2

Tabulating the numbers of each size family:

1 boy, 0 girls:	74
1 boy, 1 girl:	29
1 boy, 2 girls:	18
1 boy, 3 girls:	11
1 boy, 4 girls:	6
1 boy, 5 girls:	4
1 boy, 6 girls:	1
1 boy, 8 girls:	1

Totaling the numbers of boys and girls yields 144 boys (48%), 156 girls (52%). The results appear to be close to 50% for each. The reasoning that concludes there will be an increasingly female population fails to take into account the largest group, which is the many families that have no girls.

4. *a.* Let 0 and 1 represent a game win for player A; 2, 3, 4, and 5 represent a game win for player B; and 6, 7, 8, and 9 represent a drawn game. Reading off digits from the random number table, awarding points for wins and draws, and stopping when a player reaches three points (or as soon as either player wins if they reach three at the same time) yields

841770 67571761 3155 82515 0681 435 41050 920310 6449
050595 98843 1180 53115 8446 99486 85796 70581 184514
75011 1300 6633 9555 0411 5866065 89131 1971 0208594 0919
32064 88749 8754 355 52704 90359 02649 4749 67156794
2680884 4262 9464 759089895 70249 7284 00637 89283 03514
59195 07635 0330972 60529 3572 3737 67881 036683 38763
58415 28692 31141 5864 3894 2

Summing wins gives P(A wins) $= \frac{16}{60} = .27$, while P(B wins) $= \frac{44}{60} = .73$.

b. Repeating the procedure giving no points for a draw yields

84177067571 7613155 82515 0681435 41050 920310 644905
05959884 31180 53115 8446994 8685796705811 84514 75011
1300 663395 5504 1158660 65891311 971020 859409193
20648874 987543 555 2704903 590264 947496715 67942680884
4262 946475 90898957024 972840063 78928303 5145 919507635
0330972 605293 5723 73767881036683 387635 84152 86923114
158643 8942

Summing wins gives $P(\text{A wins}) = \frac{10}{43} = .23$, while $P(\text{B wins}) = \frac{33}{43} = .77$.

5. (a) Let the numbers 01 through 14 represent finding an illegal drug, while 15 through 99 and 00 represent no drugs. Read off pairs of digits from the random number table until a number representing illegal drugs is found or until nine clean cars are allowed to pass. Note whether a car with illegal drugs is found before nine clean cars pass. Repeat this procedure. Underlining numbers representing the presence of illegal drugs gives

84<u>17</u>706757 1761315582 5150681435 41<u>05</u><u>09</u>2031 <u>06</u>44905059 598843<u>11</u>180
53<u>11</u>584469 9486857967 <u>05</u>81184514 75<u>01</u>113006 6339555041 1586606589
13<u>11</u>97<u>10</u>20 85940<u>09</u>1932 <u>06</u>48874987 5435552<u>04</u> 9035902649 4749671567
9426808844 2629464759 <u>08</u>98957024 9728400<u>06</u>37 8928303<u>14</u> 5919507635
<u>03</u>30972<u>05</u> 2935723737 6788<u>10</u>3668 3387635841 5286923<u>14</u> 1586438942

The first nine cars are clean. Then 14 is found after four clean cars, then 05 after two clean cars, then 09 before any clean cars, then 06 after two clean cars, then 11 after seven free cars, and so on. Tabulating gives

First nine cars clean: 5
Illegal drugs found before tenth car: 24

The probability that illegal drugs will be found before the tenth car is estimated to be $\frac{24}{5+24} = .83$.

(b) $\Sigma x P(x) = 1(.15) + 2(.12) + \cdots + 25(.01) = 6.81$

(c) The probability that the first stopped car will have illegal drugs is .14. The probability that the first will be clean but that the second will have drugs is $(.86)(.14)$. The probability that two clean cars will be followed by one with drugs is $(.86)^2(.14)$. The probability that three clean ones will be followed by one with drugs is $(.86)^3(.14)$, and so on. The probability of drugs being found before the tenth car is

$$.14 + (.86)(.14) + (.86)^2(.14) + \cdots + (.86)^8(.14) = .7427$$

Or more simply we could solve by subtracting the complementary probability from 1, that is, $1 - (.86)^9 = .7427$.

Combining Independent Random Variables

- Independence vs. Dependence
- Mean and Standard Deviation for Sums and Differences of Independent Random Variables

In many experiments a pair of numbers is associated with each outcome. This situation leads to a study of pairs of random variables. When the random variables are independent, there is an easy calculation for finding both the mean and standard deviation of the sum (and difference) of the two random variables.

INDEPENDENCE VERSUS DEPENDENCE

EXAMPLE 10.1

An automobile salesperson sells three models of vans with selling prices of $20,000, $25,000, and $30,000, respectively. For each sale, the salesperson receives a bonus of either $500 or $750. The probabilities of the various outcomes are given by the following table:

	Bonus $500	Bonus $750
$20,000	.30	.05
Amount of sale $25,000	.20	.20
$30,000	.05	.20

If X is the "amount of sale" random variable, what is the probability distribution of X?

Answer: Summing each row gives

$x(\$)$:	20,000	25,000	30,000
$P(x)$:	.35	.40	.25

If Y is the "bonus" random variable, what is the probability distribution of Y?

(continued)

Answer: Summing each column gives

y($):	500	750
P(y):	.55	.45

What is the probability of a $500 bonus given that the sale was $30,000?
Answer:

$$P\left(\$500\,\text{bonus}\mid\$30,000\,\text{sale}\right) = \frac{P\left(\$500\,\text{bonus}\cap\$30,000\,\text{sale}\right)}{P\left(\$30,000\,\text{sale}\right)}$$

$$= \frac{.05}{.25} = .2$$

However, *P*($500 bonus) = .55, and so *P*($500 bonus|$30,000 sale) ≠ *P*($500 bonus). We say that the amount of sale random variable and the bonus random variable are not independent.

We say that the two random variables *X* and *Y* are independent if $P(x|y) = P(x)$ for all values of *x* and *y*. Equivalently, *X* and *Y* are independent if $P(x \cap y) = P(x)P(y)$ for all values of *x* and *y*. [This expression is also written $P(X = x, Y = y) = P(X = x)P(Y = y)$.]

EXAMPLE 10.2

In a study on vitamin C and the common cold, people were randomly assigned to take either no vitamin C or 500 milligrams daily, and it was noted whether or not they came down with a cold. Letting *X*, the "vitamin C" random variable, take the values 0 and 500, and *Y*, the "colds" random variable, take the values 0 (no colds) and 1 (at least one cold), the following table gives the observed probabilities:

		Colds (Y)	
		0	**1**
Vitamin C (X)	**0**	.10	.30
	500	.15	.45

Are the vitamin C random variable and the colds random variable independent?
Answer: Summing rows and columns gives

$$P(X = 0) \quad = .10 + .30 = .40$$
$$P(X = 500) = .15 + .45 = .60$$
$$P(Y = 0) \quad = .10 + .15 = .25$$
$$P(Y = 1) \quad = .30 + .45 = .75$$

Now checking the conditional probabilities,

$$P\left(X = 0 \mid Y = 0\right) \quad = \frac{.10}{.25} = .4 = P\left(X = 0\right)$$

$$P\left(X = 0 \mid Y = 1\right) \quad = \frac{.30}{.75} = .4 = P\left(X = 0\right)$$

$$P\left(X = 500 \mid Y = 0\right) = \frac{.15}{.25} = .6 = P\left(X = 500\right)$$

$$P\left(X = 500 \mid Y = 1\right) = \frac{.45}{.75} = .6 = P\left(X = 500\right)$$

(continued)

Thus we conclude that X and Y are independent.

Note that an alternative check for independence is the following: $P(X = 0, Y = 0) = .10$ and $P(X = 0)P(Y = 0) = (.40)(.25) = .10$. Similarly, $P(X = 0, Y = 1) = P(X = 0)P(Y = 1) = .30$, $P(X = 500, Y = 0) = P(X = 500)P(Y = 0) = .15$, and $P(X = 500, Y = 1) = P(X = 500)P(Y = 1) = .45$.

MEAN AND STANDARD DEVIATION FOR SUMS AND DIFFERENCES OF INDEPENDENT RANDOM VARIABLES

One way of comparing two populations is to analyze their set of differences.

EXAMPLE 10.3

Suppose set $X = \{2, 9, 11, 22\}$ and set $Y = \{5, 7, 15\}$. Note that the mean of set X is $\mu_x = \frac{2+9+11+22}{4} = 11$ and the mean of set Y is $\mu_y = \frac{5+7+15}{3} = 9$. Form the set Z of differences by subtracting each element of Y from each element of X:

$$Z = \{2-5, 2-7, 2-15, 9-5, 9-7, 9-15, 11-5, 11-7,$$
$$11-15, 22-5, 22-7, 22-15\}$$
$$= \{-3, -5, -13, 4, 2, -6, 6, 4, -4, 17, 15, 7\}$$

What is the mean of Z?
Answer:

$$\mu = \frac{-3-5-13+4+2-6+6+4-4+17+15+7}{12} = \frac{24}{12} = 2$$

Note that $\mu_z = \mu_x - \mu_y$.

In general, the mean of a set of differences is equal to the difference of the means of the two original sets. Even more generally, if a sum is formed by picking one element from each of several sets, the mean of such sums is simply the sum of the means of the various sets.

How is the variance of Z related to the variances of the original sets?
Answer:

$$\mu_x = 11 \qquad \sigma_x^2 = \frac{(2-11)^2 + (9-11)^2 + (11-11)^2 + (22-11)^2}{4}$$
$$= \frac{206}{4} = 51.5$$

$$\mu_y = 9 \qquad \sigma_y^2 = \frac{(5-9)^2 + (7-9)^2 + (15-9)^2}{3}$$
$$= \frac{56}{3} = 18.67$$

$$\mu_z = 2 \qquad \sigma_z^2 = \frac{(-3-2)^2 + (-5-2)^2 + (-13-2)^2 + \cdots + (7-2)^2}{12}$$
$$= \frac{842}{12} = 70.17$$

Note that in the above example $\sigma_z^2 = \sigma_x^2 + \sigma_y^2$. This is true for the variance of any set of differences. More generally, if a total is formed by a procedure that adds or subtracts one element from each of several independent sets, the variance of the resulting totals is simply the sum of the variances of the several sets.

Not only can we sum variances as shown above to calculate total variance, but we can also reverse the process and determine how the total variance is split up among its various sources. For example, we can find what portion of the variance in numbers of sales made by a company's sales representatives is due to the individual salesperson, what portion is due to the territory, what portion is due to the particular products sold by each salesperson, and so on.

The above principles extend more broadly to random variables.

EXAMPLE 10.4

Suppose two random variables X and Y have the following joint probability distribution table:

		Y		
		5	8	9
	3	.10	.05	.15
X	4	.20	.10	.00
	6	.05	.20	.15

What are the expected values (means) of the random variables X, Y, and $X + Y$?

Answer:

$$E(X) = \sum x_i p_i$$
$$= 3(.10 + .05 + .15) + 4(.20 + .10 + .00) + 6(.05 + .20 + .15)$$
$$= 4.50$$

$$E(Y) = \sum y_i p_i$$
$$= 5(.10 + .20 + .15) + 8(.05 + .10 + .20) + 9(.15 + .00 + .15)$$
$$= 7.25$$

$$E(X + Y) = (3 + 5)(.10) + (3 + 8)(.05) + (3 + 9)(.15) + (4 + 5)(.20)$$
$$+ (4 + 8)(.10) + (4 + 9)(.00) + (6 + 5)(.05) + (6 + 8)(.20)$$
$$+ (6 + 9)(.15)$$
$$= 11.75$$

Note that $E(X + Y) = E(X) + E(Y)$. This is true in general for expected values of sums of random variables. It is also true that $E(X - Y) = E(X) - E(Y)$.

(continued)

What are the variances of the random variables X, Y, and X + Y?
Answer:

$$var(X) = \sum(x_i - \mu_x)^2 p_i$$
$$= (3 - 4.50)^2(.30) + (4 - 4.50)^2(.30) + (6 - 4.50)^2(.40)$$
$$= 1.65$$

$$var(Y) = (5 - 7.25)^2(.35) + (8 - 7.25)^2(.35) + (9 - 7.25)^2(.30)$$
$$= 2.8875$$

$$var(X + Y) = (8 - 11.75)^2(.10) + (11 - 11.75)^2(.05) + (12 - 11.75)^2(.15)$$
$$+ (9 - 11.75)^2(.20) + (12 - 11.75)^2(.10) + (13 - 11.75)^2(.00)$$
$$+ (11 - 11.75)^2(.05) + (14 - 11.75)^2(.20) + (15 - 11.75)^2(.15)$$
$$= 5.5875$$

Note that var(X + Y) ≠ var(X) + var(Y).

Do the variances ever sum like the means do? Consider the following example of independent random variables X and Y.

EXAMPLE 10.5

Suppose that two random variables X and Y have the following joint probability distribution table:

	Y			
	2	5	10	
X 5	.07	.05	.08	.20
X 7	.28	.20	.32	.80
	.35	.25	.40	

If one computes all the conditional probabilities, it can be shown that $P(x|y) = P(x)$ for all values of x and y, and thus X and Y are independent. What are the variances of the random variables X, Y, and X + Y?
Answer: First we calculate

$$E(X) = 5(.20) + 7(.80) = 6.60$$

$$E(Y) = 2(.35) + 5(.25) + 10(.40) = 5.95$$

$$E(X + Y) = E(X) + E(Y) = 12.55$$

(continued)

Then we calculate the variances

$$var(X) = (5 - 6.60)^2(.20) + (7 - 6.60)^2(.80) = 0.64$$

$$var(Y) = (2 - 5.95)^2(.35) + (5 - 5.95)^2(.25)$$
$$+ (10 - 5.95)^2(.40)$$
$$= 12.2475$$

$$var(X + Y) = (7 - 12.55)^2(.07) + (10 - 12.55)^2(.05)$$
$$+ (15 - 12.55)^2(.08) + (9 - 12.55)^2(.28)$$
$$+ (12 - 12.55)^2(.20) + (17 - 12.55)^2(.32)$$
$$= 12.8875$$

Note that 0.64 + 12.2475 = 12.8875; that is, var(X + Y) = var(X) + var(Y).

More generally, *if two random variables are independent*, the variance of the sum (or the *difference*) of the two random variables is equal to the *sum* of the two individual variances.

EXAMPLE 10.6

Suppose the mean SAT verbal score is 425 with standard deviation 100, while the mean SAT math score is 475 with standard deviation 100. What can be said about the mean and standard deviation of the combined math and verbal scores?

Answer: The mean is 425 + 475 = 900. Can we add the two variances to obtain $100^2 + 100^2 = 20,000$ and then take the square root to obtain 141.4? No, because we don't have independence—students with high math scores tend to have high verbal scores, and those with low math scores tend to have low verbal scores. The standard deviation of the combined scores cannot be calculated from the given information.

For random variables X and Y the generalizations for transforming X and for combining X and Y are sometimes written as

Transforming X: $E(X \pm a) = E(X) \pm a$ $var(X \pm a) = var(X)$
                     $E(bX) = bE(X)$ $var(bX) = b^2 var(X)$

Combining X and Y: $E(X \pm Y) = E(X) \pm E(Y)$
                     $var(X \pm Y) = var(X) + var(Y)$ if X and Y are independent

Does it make sense that variances *add* for both sums and differences of independent random variables? *Ranges* may illustrate this point more clearly: suppose prices for primary residences in a community go from $210,000 to $315,000, while prices for summer cottages go from $50,000 to $120,000. Someone wants to purchase a primary residence and a summer cottage. The minimum *total* spent will be $260,000, the maximum possible is $435,000, and thus the range for the total is $435,000 − $260,000 = $175,000. The minimum *difference* spent between the two homes will be $90,000, the maximum difference will be $265,000, and thus the range for the difference is $265,000 − $90,000 = $175,000, the same as the range for the total. (Variances and ranges are both measures of spread and behave the same way with regard to the above rule.)

Questions on Topic Ten: Combining Independent Random Variables

Multiple-Choice Questions

> **Directions:** The questions or incomplete statements that follow are each followed by five suggested answers or completions. Choose the response that best answers the question or completes the statement.

Questions 1–4 refer to the following: Students are classified by television usage (unending, average, and infrequent) and how often they exercise (regular, occasional, and never), resulting in the following joint probability table.

TV usage	Exercise		
	Regular (Y = 1)	Occasional (Y = 2)	Never (Y = 3)
Unending (X = 1)	.05	.05	.10
Average (X = 2)	.20	.15	.10
Infrequent (X = 3)	.15	.15	.05

1. What is the probability distribution for X?

   (A) $P(X = 1) = .05$, $P(X = 2) = .20$, $P(X = 3) = .15$
   (B) $P(X = 1) = .10$, $P(X = 2) = .35$, $P(X = 3) = .30$
   (C) $P(X = 1) = .20$, $P(X = 2) = .45$, $P(X = 3) = .35$
   (D) $P(X = 1) = .40$, $P(X = 2) = .35$, $P(X = 3) = .25$
   (E) It cannot be determined from the given information.

2. What is the probability $P(X = 2, Y = 3)$, that is, the probability that a student has average television usage but never exercises?

   (A) .10
   (B) .1125
   (C) .22
   (D) .40
   (E) .60

3. What is the probability $P(X = 2 | Y = 3)$, that is, the probability that a student has average television usage given that he never exercises?

   (A) .10
   (B) .17
   (C) .22
   (D) .40
   (E) .60

4. Are X and Y independent?

   (A) Yes, because the conditional probabilities $P(X = x | Y = y)$ equal the corresponding unconditional probabilities $P(X = x)$.

   (B) Yes, because the joint probabilities are equal to the product of the respective probabilities.

   (C) Yes, because of either of the above answers.

   (D) No.

   (E) The answer cannot be determined from the given information.

5. Following are parts of the probability distributions for the random variables X and Y.

x	$P(x)$		y	$P(y)$
1	?		1	.4
2	.2		2	?
3	.3		3	.1
4	?			

   If X and Y are independent and the joint probability $P(X = 1, Y = 2) = .1$, what is $P(X = 4)$?

   (A) .1

   (B) .2

   (C) .3

   (D) .4

   (E) .5

6. Suppose X and Y are random variables with $E(X) = 25$, var$(X) = 3$, $E(Y) = 30$, and var$(Y) = 4$. What are the expected value and variance of the random variable $X + Y$?

   (A) $E(X + Y) = 55$, var$(X + Y) = 3.5$

   (B) $E(X + Y) = 55$, var$(X + Y) = 5$

   (C) $E(X + Y) = 55$, var$(X + Y) = 7$

   (D) $E(X + Y) = 27.5$, var$(X + Y) = 7$

   (E) There is insufficient information to answer this question.

7. Suppose X and Y are random variables with $\mu_x = 10$, $\sigma_x = 3$, $\mu_y = 15$, and $\sigma_y = 4$. Given that X and Y are independent, what are the mean and standard deviation of the random variable $X + Y$?

   (A) $\mu_{x+y} = 25$, $\sigma_{x+y} = 3.5$

   (B) $\mu_{x+y} = 25$, $\sigma_{x+y} = 5$

   (C) $\mu_{x+y} = 25$, $\sigma_{x+y} = 7$

   (D) $\mu_{x+y} = 12.5$, $\sigma_{x+y} = 7$

   (E) There is insufficient information to answer this question.

8. Suppose X and Y are random variables with $E(X) = 500$, var$(X) = 50$, $E(Y) = 400$, and var$(Y) = 30$. Given that X and Y are independent, what are the expected value and variance of the random variable $X - Y$?

   (A) $E(X - Y) = 100$, var$(X - Y) = 20$
   (B) $E(X - Y) = 100$, var$(X - Y) = 80$
   (C) $E(X - Y) = 900$, var$(X - Y) = 20$
   (D) $E(X - Y) = 900$, var$(X - Y) = 80$
   (E) There is insufficient information to answer this question.

9. Suppose the average height of policemen is 71 inches with a standard deviation of 4 inches, while the average for policewomen is 66 inches with a standard deviation of 3 inches. If a committee looks at all ways of pairing up one male with one female officer, what will be the mean and standard deviation for the difference in heights for the set of possible partners?

   (A) Mean of 5 inches with a standard deviation of 1 inch
   (B) Mean of 5 inches with a standard deviation of 3.5 inches
   (C) Mean of 5 inches with a standard deviation of 5 inches
   (D) Mean of 68.5 inches with a standard deviation of 1 inch
   (E) Mean of 68.5 inches with a standard deviation of 3.5 inches

10. Consider the set of scores of all students taking the AP Statistics exam. What is true about the variance of this set?

    (A) Given a boxplot of this distribution, if the whisker lengths are equal, the variance will equal the interquartile range.
    (B) If the distribution of scores is bell-shaped, the variance will be between -3 and $+3$.
    (C) If there are a few high scores with the bulk of the scores less than the mean, the variance will be skewed to the right.
    (D) If the distribution is symmetric, the variance will equal the standard deviation.
    (E) The variance is the sum of variances coming from a variety of factors such as study time, course grade, IQ, and so on.

11. Following are parts of the probability distributions for the random variables X and Y.

x	$P(x)$		y	$P(y)$
1	.1		1	?
2	?		2	?
3	?		3	.2

    If X and Y are independent and two joint probabilities are $P(X = 1, Y = 1) = .025$ and $P(X = 3, Y = 3) = .08$, what is $P(X = 2, Y = 2)$?

    (A) .115
    (B) .275
    (C) .333
    (D) .725
    (E) .895

12. Suppose X and Y are random variables with $\mu_x = 720$, $\sigma_x = 6$, $\mu_y = 240$, and $\sigma_y = 8$. Given that X and Y are independent, what are the mean and standard deviation of the random variable $X + Y$?

   (A) $\mu_{x+y} = 960$, $\sigma_{x+y} = 7$
   (B) $\mu_{x+y} = 960$, $\sigma_{x+y} = 10$
   (C) $\mu_{x+y} = 960$, $\sigma_{x+y} = 14$
   (D) $\mu_{x+y} = 480$, $\sigma_{x+y} = 14$
   (E) There is insufficient information to answer this question.

Answer Key

1. **C**	4. **D**	7. **B**	10. **E**
2. **A**	5. **C**	8. **B**	11. **B**
3. **D**	6. **E**	9. **C**	12. **B**

Answers Explained

1. **(C)** Summing each row gives $P(X = 1) = .20$, $P(X = 2) = .45$, and $P(X = 3) = .35$.

2. **(A)** This question asks for the probability of an intersection, and the answer is found in the second row, third column.

3. **(D)** $P\left(X = 2 \middle| Y = 3\right) = \frac{P(X=2, Y=3)}{P(Y=3)} = \frac{.10}{.10+.10+.05} = .40$

4. **(D)** For example, $P(X = 2 | Y = 3) = .40$, but $P(X = 2) = .45$. For independence this has to hold for all values of x and y.

5. **(C)** $P(Y = 2) = 1 - (.4 + .1) = .5$. By independence, $P(X = 1, Y = 2) = P(X = 1)P(Y = 2)$, and so $.1 = P(X = 1)(.5)$ and $P(X = 1) = .2$. Then $P(X = 4) = 1 - (.2 + .2 + .3) = .3$.

6. **(E)** Without independence we cannot determine $\text{var}(X + Y)$ from the information given.

7. **(B)** The means and the variances add. Thus the new variance is $3^2 + 4^2 = 25$, and the new standard deviation is 5.

8. **(B)** If two random variables are independent, the mean of the difference of the two random variables is equal to the difference of the two individual means; however, the variance of the difference of the two random variables is equal to the sum of the two individual variances.

9. **(C)** The mean of a set of differences is the difference of the means of the two sets, while the variance is the sum of the two variances. Thus $71 - 66 = 5$, while $4^2 + 3^2 = 25$ and $\sqrt{25} = 5$.

10. **(E)** The total variance can be split up among its various sources. Also, the alternative choices are all nonsense.

11. **(B)** By independence, $(.1)P(Y = 1) = .025$, so $P(Y = 1) = .25$. Similarly, $P(X = 3)(.2) = .08$, so $P(X = 3) = .4$. Now $P(X = 2) = 1 - (.1 + .4) = .5$ and $P(Y = 2) = 1 - (.25 + .2) = .55$. Finally, $P(X = 2, Y = 2) = (.5)(.55) = .275$.

12. **(B)** The means and the variances can be added. Thus the new variance is $6^2 + 8^2 = 100$ and the new standard deviation is 10.

Free-Response Questions

> *Directions:* You must show all work and indicate the methods you use. You will be graded on the correctness of your methods and on the accuracy of your final answers.

An Open-Ended Question

1. Following are the probability distributions for the random variables X and Y.

x	$P(x)$
1	.3
2	.7

y	$P(y)$
1	.6
2	.4

   a. If X and Y are independent random variables, what is their joint probability table?

   b. Give a possible joint probability table for which X and Y are not independent.

Answers Explained

1. *a.* If X and Y are independent random variables, $P(X = x, Y = y) = P(X = x) P(Y = y)$ for all values of x and y. Multiplying probabilities thus leads to

	Y = 1	Y = 2
X = 1	.18	.12
X = 2	.42	.28

   b. There are many possible answers as long as the first row sums to .3, the second row to .7, the first column to .6, and the second column to .4. So start with anything less than .3 (but not .18) in the upper left corner. For example, if $P(X = 1, Y = 1) = .1$, the complete table is

	Y = 1	Y = 2
X = 1	.1	.2
X = 2	.5	.2

The Normal Distribution

- Properties
- Using Tables
- A Model for Measurement
- Commonly Used Probabilities and z-scores

- Finding Means and Standard Deviations
- Normal Approximation to the Binomial
- Checking Normality

Some of the most useful probability distributions are symmetric and bell-shaped:

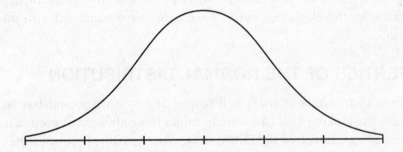

In this topic we study one such distribution called the *normal distribution* (not *all* symmetric, unimodal distributions are normal). The normal distribution is valuable in describing various natural phenomena, especially those involving growth or decay. However, the real importance of the normal distribution in statistics is that it can be used to describe the results of sampling procedures.

The normal distribution can be viewed as a limiting case of the binomial. More specifically, we start with any fixed probability of success p and consider what happens as n becomes arbitrarily large. To obtain a visual representation of the limit, we draw histograms using an increasingly large n for each fixed p.

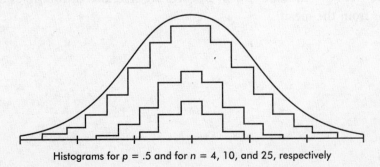

Histograms for $p = .5$ and for $n = 4$, 10, and 25, respectively

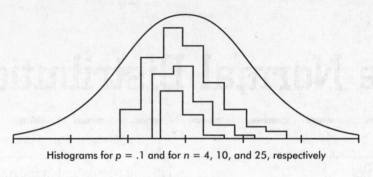

Histograms for $p = .1$ and for $n = 4, 10,$ and 25, respectively

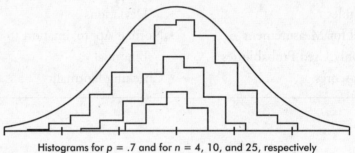

Histograms for $p = .7$ and for $n = 4, 10,$ and 25, respectively

For greater clarity, different scales have been used for the histograms associated with different values of n. From the diagrams above it is reasonable to accept that as n becomes larger without bound, the resulting histograms approach a smooth bell-shaped curve. This is the curve associated with the normal distribution.

PROPERTIES OF THE NORMAL DISTRIBUTION

The normal distribution curve is bell-shaped and symmetric and has an infinite base. Long, flat-looking tails cover many values but only a small proportion of the area. The flat appearance of the tails is deceptive. Actually, far out in the tails, the curve drops proportionately at an ever-increasing rate. In other words, when two intervals of equal length are compared, the one closer to the center may experience a greater numerical drop, but the one further out in the tail experiences a greater drop when measured as a proportion of the height at the beginning of the interval.

The mean here is the same as the median and is located at the center. We want a unit of measurement that applies equally well to any normal distribution, and we choose a unit that arises naturally out of the curve's shape. There is a point on each side where the slope is steepest. These two points are called *points of inflection*, and the distance from the mean to either point is precisely equal to one standard deviation. Thus, it is convenient to measure distances under the normal curve in terms of z-scores (recall from Topic 2 that z-scores are fractions or multiples of standard deviations from the mean).

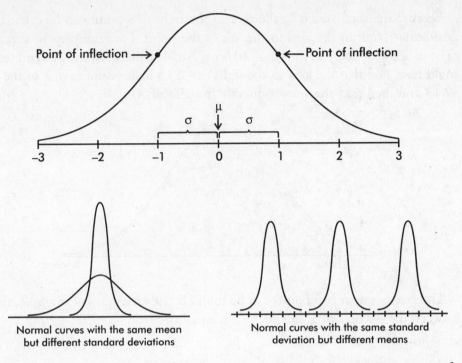

Normal curves with the same mean but different standard deviations

Normal curves with the same standard deviation but different means

Mathematically, the formula for the normal curve turns out to be $y = e^{-z^2/2}$, where y is the relative height above z-score. (By *relative height* we mean proportion of the height above the mean). However, our interest is not so much in relative heights under the normal curve as it is in *proportionate areas*. The probability associated with any interval under the normal curve is equal to the proportionate area found under the curve and above the interval.

A useful property of the normal curve is that approximately 68% of the area (and thus 68% of the observations) falls within one standard deviation of the mean, approximately 95% of the area falls within two standard deviations of the mean, and approximately 99.7% of the area falls within three standard deviations of the mean.

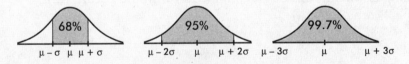

USING TABLES OF THE NORMAL DISTRIBUTION

Table A in the Appendix shows areas under the normal curve. Specifically, this table shows the area to the left of a given point. It shows, for example, that to the left of a z-score of 1.2 there is .8849 of the area:

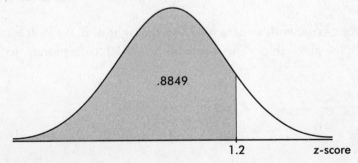

Because the total area is 1, the area to the right of a point can be calculated by calculating 1 minus the area to the left of the point. For example, the area to the right of a *z*-score of − 2.13 is 1 − .0166 = .9834. (Alternatively, by symmetry, we could have said that the area to the right of −2.13 is the same as that to the left of +2.13 and then read the answer directly from Table A.)

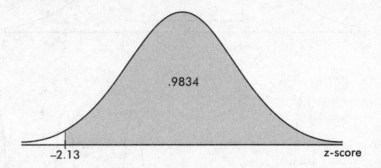

The area between two points can be found by subtraction. For example, the area between the *z*-scores of 1.23 and 2.71 is equal to .9966 − .8907 = .1059.

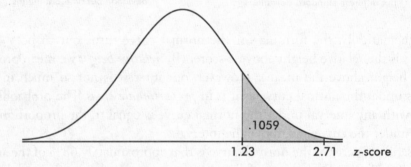

Given a probability, we can use Table A in reverse to find the appropriate *z*-score. For example, to find the *z*-score that has an area of .8982 to the left of it, we search for the probability closest to .8982 in the bulk of the table and read off the corresponding *z*-score. In this case we find .8980 with the *z*-score 1.27.

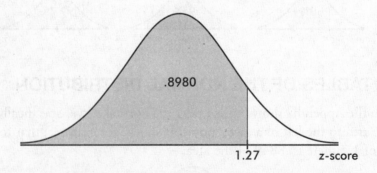

To find the *z*-score with an area of .72 to the right of it, we look for 1− .72 = .28 in the bulk of the table. The probability .2810 corresponds to the *z*-score −0.58.

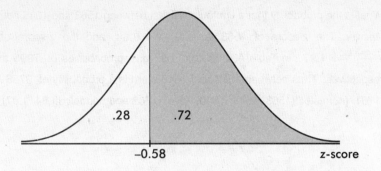

THE NORMAL DISTRIBUTION AS A MODEL FOR MEASUREMENT

In solving a problem using the normal distribution as a model, drawing a picture of a normal curve with horizontal lines showing raw scores and z-scores is usually helpful.

EXAMPLE 11.1

The life expectancy of a particular brand of lightbulb is normally distributed with a mean of 1500 hours and a standard deviation of 75 hours.

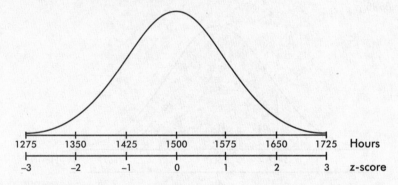

a. What is the probability that a lightbulb will last less than 1410 hours?

Answer: The z-score of 1410 is $\frac{1410-1500}{75} = -1.2$, and from Table A the probability to the left of -1.2 is .1151. [On the TI-84, normalcdf(0, 1410, 1500, 75) = .1151 and normalcdf(-10, -1.2) = .1151.]

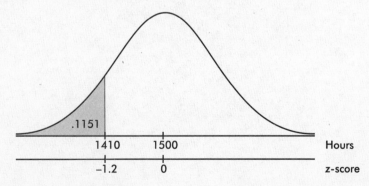

(continued)

b.   What is the probability that a lightbulb will last between 1563 and 1648 hours?

Answer: The *z*-score of 1563 is $\frac{1563-1500}{75} = 0.84$, and the *z*-score of 1648 is $\frac{1648-1500}{75} = 1.97$. In Table A, 0.84 and 1.97 give probabilities of .7995 and .9756, respectively. Thus between 1563 and 1648 there is a probability of .9756 − .7995 = .1761. [normalcdf(1563, 1648, 1500, 75) = .1762 and normalcdf(.84, 1.97) = .1760.]

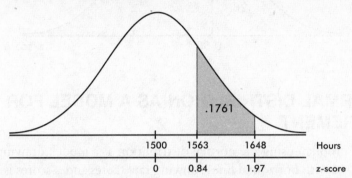

c.   What is the probability that a lightbulb will last between 1416 and 1677 hours?

Answer: The *z*-score of 1416 is $\frac{1416-1500}{75} = -1.12$, and the *z*-score of 1677 is $\frac{1677-1500}{75} = 2.36$. In Table A, −1.12 gives a probability of .1314, and 2.36 gives a probability of .9909. Thus between 1416 and 1677 there is a probability of .9909 − .1314 = .8595. [normalcdf(1416, 1677, 1500, 75) = .8595 and normalcdf(−1.12, 2.36) = .8595.]

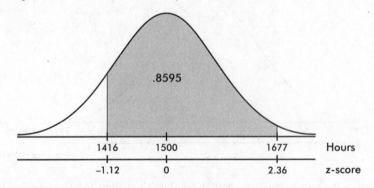

EXAMPLE 11.2

A packing machine is set to fill a cardboard box with a mean average of 16.1 ounces of cereal. Suppose the amounts per box form a normal distribution with a standard deviation equal to 0.04 ounce.

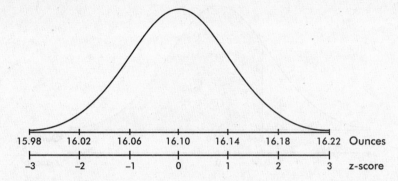

a. What percentage of the boxes will end up with at least 1 pound of cereal?

Answer: The *z*-score of 16 is $\frac{16-16.1}{0.04} = -2.5$, and -2.5 in Table A gives a probability of .0062. Thus the probability of having more than 1 pound in a box is $1 - .0062 = .9938$ or 99.38%. [normalcdf(16, 1000, 16.10, .04) = .9938 and normalcdf(−2.5, 10) = .9938.]

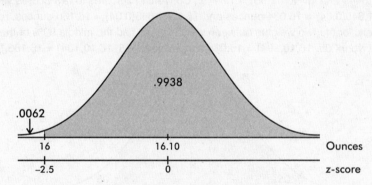

b. Ten percent of the boxes will contain more than what number of ounces?

Answer: In Table A, we note that .1 area (actually .1003) is found to the left of a -1.28 *z*-score. [On the TI-84, invNorm(.1) = −1.2816.] Converting the *z*-score of -1.28 into a raw score yields $-1.28 = \frac{x-16.1}{0.04}$ or $16.1 - 1.28(0.04) = 16.049$ ounces. [Or directly on the TI-84, invNorm(.1, 16.10, .04) = 16.049.]

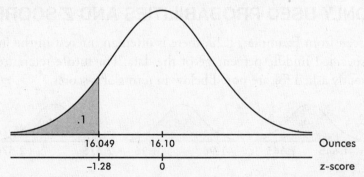

(continued)

c. Eighty percent of the boxes will contain more than what number of ounces?
Answer: In Table A, we note that $1 - .8 = .2$ area (actually .2005) is found to the left of a z-score of -0.84, and so to the right of a -0.84 z-score must be 80% of the area. [Or invNorm(.2) = .8416.] Converting the z-score of -0.84 into a raw score yields $16.1 - 0.84(0.04) = 16.066$ ounces. [Or directly on the TI-84, invNorm(.2, 16.10, .04) = 16.066.]

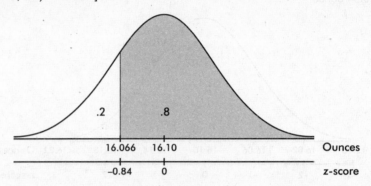

d. The middle 90% of the boxes will be between what two weights?
Answer: Ninety percent in the middle leaves five percent in each tail. In Table A, we note that .0505 area is found to the left of a z-score of -1.64, while 0.495 area is found to the left of a z-score of -1.65, and so we use the z-score of -1.645. Then 90% of the area is between z-scores of -1.645 and 1.645. [Or invNorm(.05) = -1.6449 and invNorm(.95) = 1.6449.] Converting z-scores to raw scores yields $16.1 - 1.645(0.04) = 16.034$ ounces and $16.1 + 1.645(0.04) = 16.166$ ounces, respectively, for the two weights between which we will find the middle 90% of the boxes. [invNorm(.05, 16.10, .04) = 16.034 and invNorm(.95, 16.10, .04) = 16.166.]

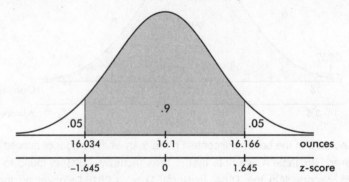

COMMONLY USED PROBABILITIES AND Z-SCORES

As can be seen from Example 11.2*d*, there is often an interest in the limits enclosing some specified middle percentage of the data. For future reference, the limits most frequently asked for are noted below in terms of z-scores.

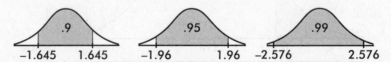

Ninety percent of the values are between z-scores of -1.645 and $+1.645$, 95% of the values are between z-scores of -1.96 and $+1.96$, and 99% of the values are between z-scores of -2.576 and $+2.576$.

Sometimes the interest is in values with particular percentile rankings. For example,

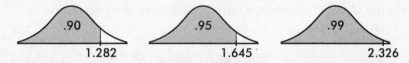

Ninety percent of the values are below a *z*-score of 1.282, 95% of the values are below a *z*-score of 1.645, and 99% of the values are below a *z*-score of 2.326.

There are corresponding conclusions for negative *z*-scores:

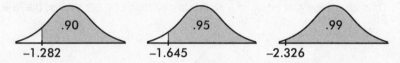

Ninety percent of the values are above a *z*-score of −1.282, 95% of the values are above a *z*-score of −1.645, and 99% of the values are above a *z*-score of −2.326.

It is also useful to note the percentages corresponding to values falling between integer *z*-scores. For example,

Note that 68.26% of the values are between *z*-scores of −1 and +1, 95.44% of the values are between *z*-scores of −2 and +2, and 99.74% of the values are between *z*-scores of −3 and +3.

EXAMPLE 11.3

Suppose that the average height of adult males in a particular locality is 70 inches with a standard deviation of 2.5 inches.

a. If the distribution is normal, the middle 95% of males are between what two heights?
    Answer: As noted above, the critical *z*-scores in this case are ±1.96, and so the two limiting heights are 1.96 standard deviations from the mean. Therefore, 70 ± 1.96(2.5) = 70 ± 4.9, or from 65.1 to 74.9 inches.
b. Ninety percent of the heights are below what value?
    Answer: The critical *z*-score is 1.282, and so the value in question is 70 + 1.282(2.5) = 73.205 inches.
c. Ninety-nine percent of the heights are above what value?
    Answer: The critical *z*-score is −2.326, and so the value in question is 70 − 2.326(2.5) = 64.185 inches.
d. What percentage of the heights are between *z*-scores of ±1? Of ±2? Of ±3?
    Answer: 68.26%, 95.44%, and 99.74%, respectively.

FINDING MEANS AND STANDARD DEVIATIONS

If we know that a distribution is normal, we can calculate the mean μ and the standard deviation σ using percentage information from the population.

EXAMPLE 11.4

Given a normal distribution with a mean of 25, what is the standard deviation if 18% of the values are above 29?

Answer: Looking for a .82 probability in Table A, we note that the corresponding z-score is 0.92. Thus $29 - 25 = 4$ is equal to a standard deviation of 0.92, that is, $0.92\sigma = 4$, and $\frac{4}{0.92} = 4.35$.

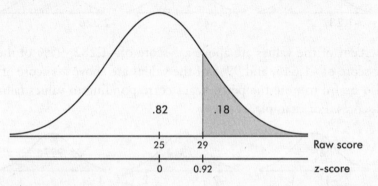

EXAMPLE 11.5

Given a normal distribution with a standard deviation of 10, what is the mean if 21% of the values are below 50?

Answer: Looking for a .21 probability in Table A leads us to a z-score of -0.81. Thus 50 is -0.81 standard deviation from the mean, and so $\mu = 50 + 0.81(10) = 58.1$.

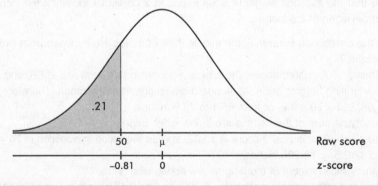

EXAMPLE 11.6

Given a normal distribution with 80% of the values above 125 and 90% of the values above 110, what are the mean and standard deviation?

Answer: Table A gives critical z-scores of –0.84 and –1.28. Thus we have $\frac{125-\mu}{\sigma} = -0.84$ and $\frac{110-\mu}{\sigma} = -1.28$. Solving the system $\{125 - \mu = -0.84\sigma, 110 - \mu = -1.28\sigma\}$ simultaneously gives $\mu = 153.64$ and $\sigma = 34.09$.

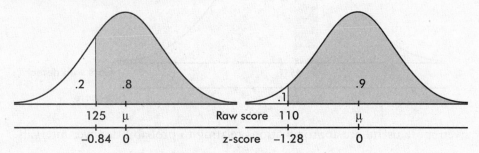

NORMAL APPROXIMATION TO THE BINOMIAL

Many practical applications of the binomial involve examples in which n is large. However, for large n, binomial probabilities can be quite tedious to calculate. Since the normal can be viewed as a limiting case of the binomial, it is natural to use the normal to approximate the binomial in appropriate situations.

The binomial takes values only at integers, while the normal is continuous with probabilities corresponding to areas over intervals. Therefore, we establish a technique for converting from one distribution to the other. For approximation purposes we do as follows. Each binomial probability corresponds to the normal probability over a unit interval centered at the desired value. Thus, for example, to approximate the binomial probability of eight successes we determine the normal probability of being between 7.5 and 8.5.

EXAMPLE 11.7

Suppose that 15% of the cars coming out of an assembly plant have some defect. In a delivery of 40 cars what is the probability that exactly 5 cars have defects?

Answer: The actual answer is $\left(\frac{40!}{35!5!}\right)(.15)^5(.85)^{35}$, but clearly this involves a nontrivial calculation. [If one has a calculator such as the TI-84, then binompdf(40, .15, 5) = .1692.] To approximate the answer using the normal, we first calculate the mean μ and the standard deviation σ as follows:

$$\mu = np = 40(.15) = 6$$
$$\sigma = \sqrt{np(1-p)} = \sqrt{40(.15)(.85)} = 2.258$$

(continued)

We then calculate the appropriate *z*-scores: $\frac{4.5-6}{2.258} = -0.66$, and $\frac{5.5-6}{2.258} = -0.22$. Looking up the corresponding probabilities in Table A, we obtain a final answer of .4129 − .2546 = .1583. (The actual answer is .1692.)

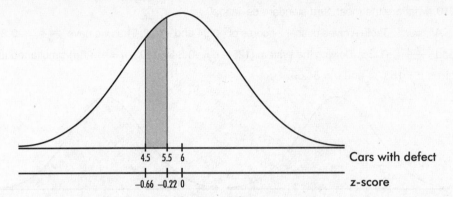

Even more useful are approximations relating to probabilities over intervals.

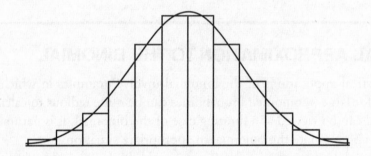

EXAMPLE 11.8

If 60% of the population support massive federal budget cuts, what is the probability that in a survey of 250 people at most 155 people support such cuts?

Answer: The actual answer is the sum of 156 binomial expressions:

$$(.4)^{250} + \cdots + \frac{250!}{155!95!}(.6)^{155}(.4)^{95}$$

However, a good approximation can be obtained quickly and easily by using the normal. We calculate μ and σ:

$$\mu = np = 250(.6) = 150$$

$$\sigma = \sqrt{np(1-p)} = \sqrt{250(.6)(.4)} = 7.746$$

The binomial of at most 155 successes corresponds to the normal probability of ≤155.5. The *z*-score of 155.5 is $\frac{155.5-150}{7.746} = 0.71$. Using Table A leads us to a final answer of .7611. [Or binomcdf(250, .6, 155) = .7605.]

(continued)

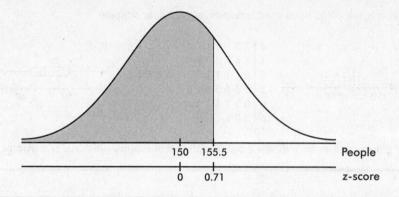

Is the normal a good approximation? The answer, of course, depends on the error tolerances in particular situations. A general rule of thumb is that the normal is a good approximation to the binomial whenever both np and $n(1-p)$ are greater than 10. (Some authors use 5 instead of 10 here.)

CHECKING NORMALITY

In the examples in this Topic we have assumed the population has a normal distribution. When you collect your own data, before you can apply the procedures developed above, you must decide whether it is reasonable to assume the data come from a normal population. In later Topics we will have to check for normality before applying certain inference procedures.

The initial check should be to draw a picture. Dotplots, stemplots, boxplots, and histograms are all useful graphical displays to show that data is unimodal and roughly symmetric.

EXAMPLE 11.9

The ages at inauguration of U.S. presidents from Washington to G. W. Bush were: {57, 61, 57, 57, 58, 57, 61, 54, 68, 51, 49, 64, 50, 48, 65, 52, 56, 46, 54, 49, 51, 47, 55, 55, 54, 42, 51, 56, 55, 51, 54, 51, 60, 61, 43, 55, 56, 61, 52, 69, 64, 46, 54}. Can we conclude that the distribution is roughly normal?

Answer: Entering the 43 data points in a graphing calculator gives the following histogram:

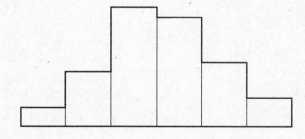

(continued)

Alternatively, we could have used a dotplot, stemplot, or boxplot:

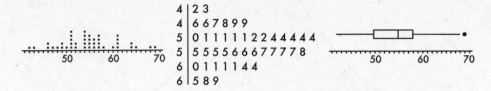

```
4 | 2 3
4 | 6 6 7 8 9 9
5 | 0 1 1 1 1 1 2 2 4 4 4 4 4
5 | 5 5 5 5 6 6 6 7 7 7 7 8
6 | 0 1 1 1 1 4 4
6 | 5 8 9
```

All the graphical displays indicate a distribution that is *roughly* unimodal and symmetric.

A more specialized graphical display to check normality is the *normal probability plot*. When the data distribution is roughly normal, the plot is roughly a diagonal straight line. While this plot more clearly shows deviations from normality, it is not as easy to understand as a histogram. The normal probability plot is difficult to calculate by hand; however, technology such as the TI-84 readily plots the graph. (On the TI-84, in STATPLOT the sixth choice in TYPE gives the normal probability plot.)

The normal probability plot for the data in Exercise 11.9 is:

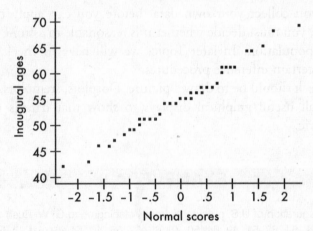

A graph this close to a diagonal straight line indicates that the data have a distribution very close to normal.

Questions on Topic Eleven: The Normal Distribution

Multiple-Choice Questions

Directions: The questions or incomplete statements that follow are each followed by five suggested answers or completions. Choose the response that best answers the question or completes the statement.

1. Which of the following are true statements?

   I. The area under a normal curve is always equal to 1, no matter what the mean and standard deviation are.
   II. The smaller the standard deviation of a normal curve, the higher and narrower the graph.
   III. Normal curves with different means are centered around different numbers.

   (A) I and II
   (B) I and III
   (C) II and III
   (D) I, II, and III
   (E) None of the above gives the complete set of true responses.

2. Which of the following are true statements?

   I. The area under the standard normal curve between 0 and 2 is twice the area between 0 and 1.
   II. The area under the standard normal curve between 0 and 2 is half the area between −2 and 2.
   III. For the standard normal curve, the interquartile range is approximately 3.

   (A) I and II
   (B) I and III
   (C) II and III
   (D) I, II, and III
   (E) None of the above gives the complete set of true responses.

3. Populations P1 and P2 are normally distributed and have identical means. However, the standard deviation of P1 is twice the standard deviation of P2. What can be said about the percentage of observations falling within two standard deviations of the mean for each population?

(A) The percentage for P1 is twice the percentage for P2.
(B) The percentage for P1 is greater, but not twice as great, as the percentage for P2.
(C) The percentage for P2 is twice the percentage for P1.
(D) The percentage for P2 is greater, but not twice as great, as the percentage for P1.
(E) The percentages are identical.

4. Consider the following two normal curves:

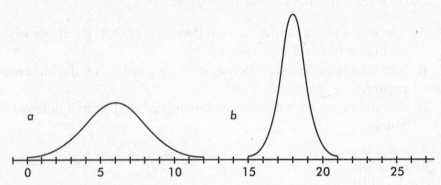

Which has the larger mean and which has the larger standard deviation?

(A) Larger mean, *a*; larger standard deviation, *a*
(B) Larger mean, *a*; larger standard deviation, *b*
(C) Larger mean, *b*; larger standard deviation, *a*
(D) Larger mean, *b*; larger standard deviation, *b*
(E) Larger mean, *b*; same standard deviation

5. Which of the following are true statements?

I. In all normal distributions, the mean and median are equal.
II. All bell-shaped curves are normal distributions for some choice of μ and σ.
III. Virtually all the area under a normal curve is within three standard deviations of the mean, no matter what the particular mean and standard deviation are.

(A) I and II
(B) I and III
(C) II and III
(D) I, II, and III
(E) None of the above gives the complete set of true responses.

In Problems 6–15, assume the given distribution is normal.

6. A trucking firm determines that its fleet of trucks averages a mean of 12.4 miles per gallon with a standard deviation of 1.2 miles per gallon on cross-country hauls. What is the probability that one of the trucks averages fewer than 10 miles per gallon?

   (A) .0082
   (B) .0228
   (C) .4772
   (D) .5228
   (E) .9772

7. A factory dumps an average of 2.43 tons of pollutants into a river every week. If the standard deviation is 0.88 tons, what is the probability that in a week more than 3 tons are dumped?

   (A) .2578
   (B) .2843
   (C) .6500
   (D) .7157
   (E) .7422

8. An electronic product takes an average of 3.4 hours to move through an assembly line. If the standard deviation is 0.5 hour, what is the probability that an item will take between 3 and 4 hours?

   (A) .2119
   (B) .2295
   (C) .3270
   (D) .3811
   (E) .6730

9. The mean score on a college placement exam is 500 with a standard deviation of 100. Ninety-five percent of the test takers score above what?

   (A) 260
   (B) 336
   (C) 405
   (D) 414
   (E) 664

10. The average noise level in a restaurant is 30 decibels with a standard deviation of 4 decibels. Ninety-nine percent of the time it is below what value?

    (A) 20.7
    (B) 32.0
    (C) 33.4
    (D) 37.8
    (E) 39.3

11. The mean income per household in a certain state is $9500 with a standard deviation of $1750. The middle 95% of incomes are between what two values?

    (A) $5422 and $13,578
    (B) $6070 and $12,930
    (C) $6621 and $12,379
    (D) $7260 and $11,740
    (E) $8049 and $10,951

12. Jay Olshansky from the University of Chicago was quoted in *Chance News* as arguing that for the average life expectancy to reach 100, 18% of people would have to live to 120. What standard deviation is he assuming for this statement to make sense?

    (A) 21.7
    (B) 24.4
    (C) 25.2
    (D) 35.0
    (E) 111.1

13. Cucumbers grown on a certain farm have weights with a standard deviation of 2 ounces. What is the mean weight if 85% of the cucumbers weigh less than 16 ounces?

    (A) 13.92
    (B) 14.30
    (C) 14.40
    (D) 14.88
    (E) 15.70

14. If 75% of all families spend more than $75 weekly for food, while 15% spend more than $150, what is the mean weekly expenditure and what is the standard deviation?

    (A) $\mu = 83.33$, $\sigma = 12.44$
    (B) $\mu = 56.26$, $\sigma = 11.85$
    (C) $\mu = 118.52$, $\sigma = 56.26$
    (D) $\mu = 104.39$, $\sigma = 43.86$
    (E) $\mu = 139.45$, $\sigma = 83.33$

15. A coffee machine can be adjusted to deliver any fixed number of ounces of coffee. If the machine has a standard deviation in delivery equal to 0.4 ounce, what should be the mean setting so that an 8-ounce cup will overflow only 0.5% of the time?

    (A) 6.97 ounces
    (B) 7.22 ounces
    (C) 7.34 ounces
    (D) 7.80 ounces
    (E) 9.03 ounces

16. Assume that a baseball team has an average pitcher, that is, one whose probability of winning any decision is .5. If this pitcher has 30 decisions in a season, what is the probability that he will win at least 20 games?

    (A) .0505
    (B) .2514
    (C) .2743
    (D) .3333
    (E) .4300

17. Given that 10% of the nails made using a certain manufacturing process have a length less than 2.48 inches, while 5% have a length greater than 2.54 inches, what are the mean and standard deviation of the lengths of the nails? Assume that the lengths have a normal distribution.

    (A) $\mu = 2.506$, $\sigma = 0.0205$
    (B) $\mu = 2.506$, $\sigma = 0.0410$
    (C) $\mu = 2.516$, $\sigma = 0.0825$
    (D) $\mu = 2.516$, $\sigma = 0.1653$
    (E) The mean and standard deviation cannot be computed from the information given.

Answer Key

1. **D**	5. **B**	9. **B**	13. **A**	17. **A**
2. **E**	6. **B**	10. **E**	14. **D**	
3. **E**	7. **A**	11. **B**	15. **A**	
4. **C**	8. **E**	12. **A**	16. **A**	

Answers Explained

1. **(D)** The area under any probability distribution is equal to 1. The mean of a normal curve determines the value around which the curve is centered, while the standard deviation determines the height and spread.

2. **(E)** Statement II is true by symmetry of the normal curve, but .4772 is not 2 times .3413, and $0.67 - (-0.67)$ is not 3.

3. **(E)** All normal distributions have about 95% of their observations within two standard deviations of the mean.

4. **(C)** Curve *a* has mean 6 and standard deviation 2, while curve *b* has mean 18 and standard deviation 1.

5. **(B)** Because of symmetry, the mean and median are identical for normal distributions. Many bell-shaped curves are *not* normal curves, and we don't *choose* μ and σ. For all normal curves, 99.7% of the area is within three standard deviations of the mean.

6. **(B)** The z-score of 10 is $\frac{10-12.4}{1.2} = -2$. From Table A, to the left of -2 is an area of .0228. [normalcdf(0, 10, 12.4, 1.2) = .0228.]

7. **(A)** The z-score of 3 is $\frac{3-2.43}{0.88} = 0.65$ From Table A, to the left of 0.65 is an area of .7422, and so to the right must be $1 - .7422 = .2578$. [normalcdf(3, 1000, 2.43, .88) = .2589.]

8. **(E)** The z-scores of 3 and 4 are $\frac{3-3.4}{0.5} = -0.8$ and $\frac{4-3.4}{0.5} = 1.2$, respectively. From Table A, to the left of -0.8 is an area of .2119, and to the left of 1.2 is an area of .8849. Thus between 3 and 4 is an area of $.8849 - .2119 = .6730$. [normalcdf(3, 4, 3.4, .5) = .6731.]

9. **(B)** If 95% of the area is to the right of a score, 5% is to the left. Looking for .05 in the body of Table A, we note the z-score of -1.645. Converting this to a raw score gives $500 - 1.645(100) = 336$. [invNorm(.05, 500, 100) = 336.]

10. **(E)** The critical z-score associated with 99% to the left is 2.326, and $30 + 2.326(4) = 39.3$. [invNorm(.99, 30, 4) = 39.3.]

11. **(B)** The critical z-scores associated with the middle 95% are ±1.96, and $9500 \pm 1.96(1750) = 6070$ and 12,930. [invNorm(.025, 9500, 1750) = 6070 and invNorm(.975, 9500, 1750) = 12,930.]

12. **(A)** The critical z-score associated with 18% to the right or 82% to the left is 0.92. Then $100 + 0.92\sigma = 120$ gives $\sigma = 21.7$.

13. **(A)** The critical z-score associated with 85% to the left is 1.04. Then $\mu + 1.04(2) = 16$ gives $\mu = 13.92$.

14. **(D)** The critical z-scores associated with 75% to the right (25% to the left) and with 15% to the right (85% to the left) are -0.67 and 1.04, respectively. Then $\{\mu - 0.67\sigma = 75, \mu + 1.04\sigma = 150\}$ gives $\mu = 104.39$ and $\sigma = 43.86$.

15. **(A)** The critical z-score associated with .5% to the right (99.5% to the left) is 2.576. Then $c + 2.576(0.4) = 8$ gives $c = 6.97$.

16. **(A)** Using the normal as an approximation to the binomial, we have $\mu = 30(.5) = 15$, $\sigma^2 = 30(.5)(.5) = 7.5$, $\sigma = 2.739$, $\frac{19.5-15}{2.739} = 1.64$, and $1 - .9495 = .0505$. [Or binomcdf(30, .5, 10) = .0494.]

17. **(A)** The critical z-scores for 10% to the left and 5% to the right are -1.282 and 1.645, respectively. Then $\{\mu - 1.282\sigma = 2.48, \mu + 1.645\sigma = 2.54\}$ gives $\mu = 2.506$ and $\sigma = 0.0205$.

Free-Response Questions

> ***Directions:*** You must show all work and indicate the methods you use. You will be graded on the correctness of your methods and on the accuracy of your final answers.

Four Open-Ended Questions

1. Explain why each of the following are not normal curves.

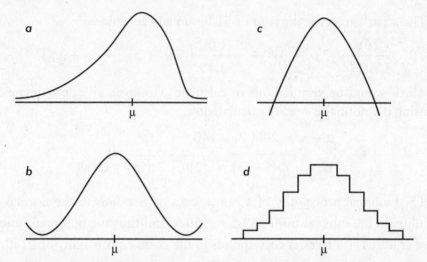

2. Consider two normal curves, one whose mean is larger than the other's mean.
   a. What can be said about which standard deviation is larger?
   b. What can be said about the percentage of area to the left of the mean for the curve with the smaller mean compared to the percentage for the curve with the larger mean?
   c. When drawn above the same axis, must each normal curve extend as far as the mean of the other curve? Explain.

3. A particular form of cancer is fatal within 1 year in 30% of all diagnosed cases. A new drug is tried on 200 patients with this disease. Researchers will judge the new medication effective if at least 150 of the patients survive for longer than 1 year. If the medication has no effect, what is the probability that at least 150 patients will survive for longer than 1 year?

4. A random sample of two players' bowling scores gives:

   Player A: {192, 212, 178, 175, 189, 186, 223, 184, 183, 179, 184, 197, 204}
   Player B: {191, 210, 198, 175, 204, 186, 223, 195, 182, 200, 217, 192, 208}

   Test each distribution for normality. Explain your method.

Answers Explained

1. *a.* Normal curves are symmetric.
   b. The tails of a normal curve continue to become closer and closer to the axis.
   c. Normal curves do not cross the axis.
   d. Normal curves are smooth.

2. *a.* Knowing which mean is larger does not give any information about which standard deviation is larger.
   b. For all normal curves 50% of the area is to the left of the mean.
   c. Yes, because normal curves extend infinitely in both directions.

3. The actual answer is the sum of 51 binomial expressions:

$$\frac{200!}{150!50!}(.7)^{150}(.3)^{50} + \frac{200!}{151!49!}(.7)^{151}(.3)^{49} + \cdots + (.7)^{200}$$

which would be very tedious to calculate. However, an approximate answer using the normal is readily calculated:

$$\mu = np = 200(.7) = 140$$

$$\sigma = \sqrt{np(1-p)} = \sqrt{200(.7)(.3)} = 6.48$$

The binomial probability of 150 successes corresponds to the normal probability on the interval from 149.5 to 150.5, and thus the binomial probability of at least 150 successes corresponds to the normal probability of ≥ 149.5. The z-score of 149.5 is $\frac{149.5-140}{6.48} = 1.47$. Using Table A gives us a final answer of $1 - .9292 = .0708$.

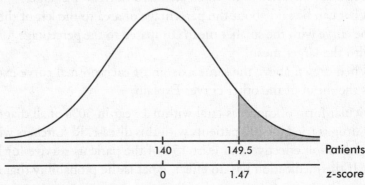

[Or binomcdf(200, .3, 50) = .0695.]

4. Comparing dotplots, stemplots, or normal probability plots shows that while the second distribution is roughly normal, the first is not.

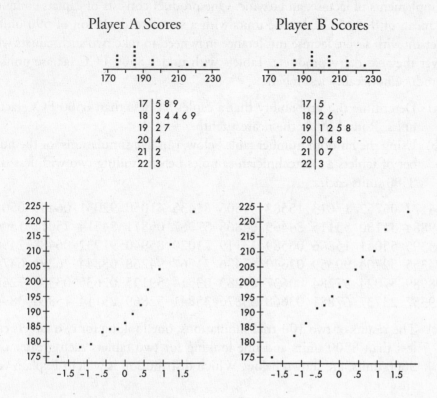

Player A Scores

17	5 8 9
18	3 4 4 6 9
19	2 7
20	4
21	2
22	3

Player B Scores

17	5
18	2 6
19	1 2 5 8
20	0 4 8
21	0 7
22	3

Two Investigative Tasks

1. When Michael Jordan came up to try for a sixth free throw after having made five straight free throws, the announcer commented that the law of averages would be working against Jordan.

   (a) What did the announcer mean? Was this a correct interpretation of probability? Explain.

   (b) Jordan makes 90% of his free throws. What is the probability that he will make six straight free throws? That he will make five straight free throws and then miss the next? That he will make the next free throw given that he has made the last five in a row?

   (c) How could you set up a simulation using a random number table to analyze the situation the announcer was commenting on?

   (d) If Jordan shoots six times from the free throw line, what are the mean and standard deviation for the number of shots he is expected to make?

   (e) If Jordan makes only 35 out of 40 free throw tries during the playoffs, is this sufficient evidence that the probability of his making a free throw is really below .90? Explain.

2. Many people have lactose intolerance leading to cramps and diarrhea when they eat dairy products. Substantial relief can be obtained by taking dietary supplements of lactase, an enzyme. One product consists of caplets containing a mean of 9000 FCC lactase units with a standard deviation of 590 units. A person with severe lactose intolerance may need to take two such tablets whenever they eat dairy products. Tablets with under 8500 FCC lactase units can produce noticeably less relief.

(a) Determine the probability that a caplet has less than 8500 FCC lactase units. Round off to the nearest tenth.

(b) Using the random number table below, run five simulations for the number of tablets a lab technician samples before finding two with less than 8500 units each.

84177 06757 17613 15582 51506 81435 41050 92031 06449 05059
59884 31180 53115 84469 94868 57967 05811 84514 75011 13006
63395 55041 15866 06589 13119 71020 85940 91932 06488 74987
54355 52704 90359 02649 47496 71567 94268 08844 26294 64759
08989 57024 97284 00637 89283 03514 59195 07635 03309 72605
29357 23737 67881 03668 33876 35841 52869 23114 15864 38942

(c) The results of two 100-trial simulations, one looking for two tablets each less than 8500 units and one looking for two tablets each greater than 9000 units, are shown below. Which distribution is which? Explain your answer.

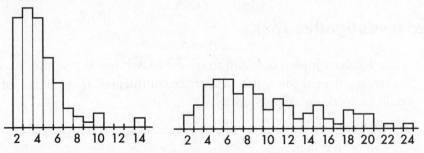

(d) Using the correct barplot above, estimate the expected number of caplets a laboratory will sample before finding two with less than 8500 units each.

Answers Explained

1. (a) The announcer meant that to maintain his tabulated free throw percentage, Jordan would have to soon miss one. This is not a correct use of probability. If Jordan makes a certain percentage of free throws, that probability applies to each throw irrespective of the previous throws. By the law of large numbers, in the long run the relative frequency tends toward the correct probability, but no conclusion is possible about any given outcome.

(b) $P(\text{six in a row}) = (.9)^6 = .531$
$P(\text{five, then a miss}) = (.9)^5(.1) = .059$
$P(\text{next}|\text{previous five}) = P(\text{next}) = .9$

(c) Let the digits 1 through 9 stand for making a free throw, while 0 stands for a miss. Look at blocks of five random numbers. If all stand for "makes" (no 0), look at the very next digit to see if it is a 0 or not. Keep a tally of how many times five makes in a row are followed by a make and how many times five makes in a row are followed by a miss.

(d) This is a binomial distribution with $n = 6$ and $p = .9$, and so the mean is $np = 6(9) = 5.4$ while the standard deviation is $\sqrt{np(1-p)}$ $\sqrt{6(.9)(.1)} = .735$.

(e) Since $nq = 40(1 - .90) = 4$, the normal approximation to the binomial is not recommended. However, a direct binomial calculation yields

$$
\begin{aligned}
P(\le 35) &= 1 - P(\ge 36) \\
&= 1 - \left[\binom{40}{36}(.9)^{36}(.1)^4 + \binom{40}{37}(.9)^{37}(.1)^3 \right. \\
&\quad \left. + \binom{40}{38}(.9)^{38}(.1)^2 + \binom{40}{39}(.9)^{39}(.1) + (.9)^{40} \right] \\
&\approx .37
\end{aligned}
$$

[Or simply use binomcdf(40, .9, 35) on the TI-83.] With such a high probability, there is no evidence to conclude that Jordan's average has dropped!

2. (a) With $z = \frac{8500-9000}{590} = -0.85$, Table A gives a probability of $.1977 \approx .2$.
(b) For example, letting 0 and 1 represent caplets containing less than 8500 units and 2 through 9 represent caplets containing more than 8500 units, we can read off digits until we find two containing less than 8500 caplets. Five simulations would yield

        841770   67571761   55825150   68143541   050

giving 6, 8, 8, 8, and 3 tablets sampled before finding two with less than 8500 units apiece.
(c) The probability of a caplet having more than 9000 units is .5, and thus one would expect to find two caplets with more than 9000 units much quicker than finding two caplets with less than 8500 units. Thus the first histogram results from looking for two caplets with more than 9000 units and the second histogram results from looking for two caplets with less than 8500 units.
(d) An estimate for the expected value is obtained by

$$
\sum xP(x) = 2(.03) + 3(.05) + 4(.09) + 5(.10) + \cdots + 24(.01) = 9.44.
$$

Sampling Distributions

- Sample Proportion
- Sample Mean
- Central Limit Theorem
- Difference Between Two Independent Sample Proportions
- Difference Between Two Independent Sample Means
- The *t*-distribution
- The Chi-square Distribution
- The Standard Error

The *population* is the complete set of items of interest. A *sample* is a part of a population used to represent the population. The population mean μ and population standard deviation σ are examples of *population parameters*. The sample mean $\bar{x}$ and the sample standard deviation s are examples of *statistics*. Statistics are used to make inferences about population parameters. While a population parameter is a fixed quantity, statistics vary depending on the particular sample chosen. The probability distribution showing how a statistic varies is called a *sampling distribution*. The sampling distribution is *unbiased* if its mean is equal to the associated population parameter.

We want to know some important truth about a population, but in practical terms this truth is unknowable. What's the average adult human body weight? What proportion of people have high cholesterol? What we can do is carefully collect data from as large and representative a group of individuals as possible and then use this information to estimate the value of the population parameter. How close are we to the truth? We know that different samples would give different estimates, and so sampling error is unavoidable. What is wonderful, and what we will learn in this Topic, is that we can quantify this sampling error! We can make statements like "the average weight must almost surely be within 5 pounds of 176 pounds" or "the proportion of people with high cholesterol must almost surely be within ±3% of 37%."

SAMPLING DISTRIBUTION OF A SAMPLE PROPORTION

Whereas the mean is basically a quantitative measurement, the proportion represents essentially a qualitative approach. The interest is simply in the presence or absence of some attribute. We count the number of yes responses and form a proportion. For example, what proportion of drivers wear seat belts? What proportion of SCUD missiles can be intercepted? What proportion of new stereo sets have a certain defect?

This separation of the population into "haves" and "have-nots" suggests that we can make use of our earlier work on binomial distributions. We also keep in mind that, when *n* (trials, or in this case sample size) is large enough, the binomial can be approximated by the normal.

In this topic we are interested in estimating a population proportion *p* by considering a single sample proportion $\hat{p}$. This sample proportion is just one of a whole universe of sample proportions, and to judge its significance we must know how sample proportions vary. Consider the set of proportions from all possible samples of a specified size *n*. It seems reasonable that these proportions will cluster around the population proportion (the sample proportion is an unbiased statistic of the population proportion) and that the larger the chosen sample size, the tighter the clustering.

How do we calculate the mean and standard deviation of the set of population proportions? Suppose the sample size is *n* and the actual population proportion is *p*. From our work on binomial distributions, we remember that the mean and standard deviation for the number of successes in a given sample are *pn* and $\sqrt{np(1-p)}$, respectively, and for large *n* the complete distribution begins to look "normal."

Here, however, we are interested in the proportion rather than in the number of successes. From Topic Two we remember that when we multiply or divide every element by a constant, we multiply or divide both the mean and the standard deviation by the same constant. In this case, to change number of successes to proportion of successes, we divide by *n*:

$$\mu_{\hat{p}} = \frac{pn}{n} = p \quad \text{and} \quad \sigma_{\hat{p}} = \frac{\sqrt{np(1-p)}}{n} = \sqrt{\frac{p(1-p)}{n}}$$

Furthermore, if each element in an approximately normal distribution is divided by the same constant, it is reasonable that the result will still be an approximately normal distribution.

Thus the principle forming the basis of the following discussion is

---

Start with a population with a given proportion *p*. Take all samples of size *n*. Compute the proportion in each of these samples. Then

1. the set of all sample proportions is approximately normally distributed.
2. the mean $\mu_{\hat{p}}$ of the set of sample proportions equals p, the population proportion.
3. the standard deviation $\sigma_{\hat{p}}$ of the set of sample proportions is approximately equal to $\sqrt{\frac{p(1-p)}{n}}$.

Alternatively, we say that the sampling distribution of $\hat{p}$ is approximately normal with mean *p* and standard deviation $\sqrt{\frac{p(1-p)}{n}}$.

Since we are using the normal approximation to the binomial, both *np* and $n(1 - p)$ should be at least 10. Furthermore, in making calculations and drawing conclusions from a specific sample, it is important that the sample be a *simple random sample.*

Finally, because sampling is usually done without replacement, the sample cannot be too large; the sample size *n* should be no larger than 10% of the population. (We're actually worried about *independence*, but randomly selecting a relatively small sample allows us to assume independence. Of course, it's always better to have larger samples—it's just that if the sample is large relative to the population, then the proper inference techniques are different from those taught in introductory statistics classes.)

EXAMPLE 12.1

Suppose that 70% of all dialysis patients will survive for at least 5 years. In a simple random sample (SRS) of 100 new dialysis patients, what is the probability that the proportion surviving for at least 5 years will exceed 80%?

Answer: Both $np = (100)(.7) = 70 > 10$ and $n(1 - p) = (100)(.3) = 30 > 10$, and our sample is clearly less than 10% of all dialysis patients. So the set of sample proportions is approximately normally distributed with mean .70 and standard deviation

$$\sigma_p = \sqrt{\frac{(.7)(.3)}{100}} = .0458$$

With a z-score of $\frac{.80-.70}{.0458} = 2.18,$ the probability that the sample proportion will exceed 80% is $1 - .9854 = .0146.$ [normalcdf(.8, 1, .7, .0458) = .0145.]

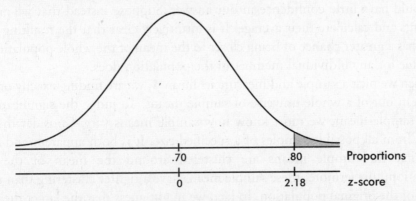

| | .70 | .80 | Proportions |
| | 0 | 2.18 | z-score |

EXAMPLE 12.2

It is estimated that 48% of all motorists use their seat belts. If a police officer observes 400 cars go by in an hour, what is the probability that the proportion of drivers wearing seat belts is between 45% and 55%?

(continued)

Answer: Both $np = (400)(.48) = 192 > 10$ and $n(1 - p) = (400)(.52) = 208 > 10$, and our sample is clearly less than 10% of all motorists. So the set of sample proportions is approximately normally distributed with mean .48 and standard deviation

$$\sigma_p = \sqrt{\frac{(.48)(.52)}{400}} = .0250$$

The *z*-scores of .45 and .55 are $\frac{.45 - .48}{.0250} = -1.2$ and $\frac{.55 - .48}{.0250} = 2.8$, respectively. From Table A, the area between .45 and .55 is $.9974 - .1151 = .8823$. Thus there is a .8823 probability that between 45% and 55% of the drivers are wearing seat belts. [normalcdf(.45, .55, .48, .0250) = .8824.]

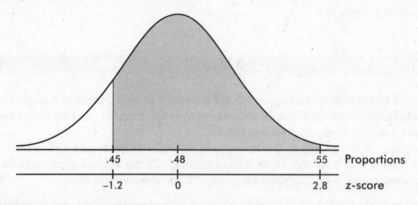

SAMPLING DISTRIBUTION OF A SAMPLE MEAN

Suppose we are interested in estimating the mean μ of a population. For our estimate we could simply randomly pick a single element of the population, but then we would have little confidence in our answer. Suppose instead that we pick 100 elements and calculate their average. It is intuitively clear that the resulting sample mean has a greater chance of being closer to the mean of the whole population than the value for any individual member of the population does.

When we pick a sample and measure its mean $\bar{x}$, we are finding exactly one sample mean out of a whole universe of sample means. To judge the significance of a single sample mean, we must know how sample means vary. Consider the set of means from all possible samples of a specified size. It is both apparent and reasonable that the sample means are clustered around the mean of the whole population; furthermore, these sample means have a tighter clustering than the elements of the original population. In fact, we might guess that the larger the chosen sample size, the tighter the clustering.

How do we calculate the standard deviation $\sigma_{\bar{x}}$ of the set of sample means? Suppose the variance of the population is σ^2 and we are interested in samples of size *n*. Sample means are obtained by first summing together *n* elements and then dividing by *n*. A set of sums has a variance equal to the sum of the variances associated with the original sets. In our case, $\sigma^2_{sums} = \sigma^2 + \cdots + \sigma^2 = n\sigma^2$. When each element of a set is divided by some constant, the new variance is the old one divided by the square of the constant. Since the sample means are obtained by dividing the sums by *n*, the variance of the sample means is obtained by dividing the variance of the sums by n^2. Thus if $\sigma_{\bar{x}}$ symbolizes the standard deviation of the sample means, we find that

$$\sigma_{\bar{x}}^2 = \frac{\sigma_{\text{sums}}^2}{n^2} = \frac{n\sigma^2}{n^2} = \frac{\sigma^2}{n}$$

In terms of standard deviations, we have $\sigma_{\bar{x}} = \frac{\sigma}{\sqrt{n}}$.

We have shown the following:

> Start with a population with a given mean μ and standard deviation σ. Compute the mean of all samples of size n. Then the mean of the set of sample means will equal μ, the mean of the population, and the standard deviation $\sigma_{\bar{x}}$ of the set of sample means will be approximately equal to $\frac{\sigma}{\sqrt{n}}$, that is, the standard deviation of the whole population divided by the square root of the sample size.

Note that the variance of the set of sample means varies directly as the variance of the original population and inversely as the size of the samples, while the standard deviation of the set of sample means varies directly as the standard deviation of the original population and inversely as the square root of the size of the samples.

EXAMPLE 12.3

Suppose that tomatoes weigh an average of 10 ounces with a standard deviation of 3 ounces and a store sells boxes containing 12 tomatoes each. If customers determine the average weight of the tomatoes in each box they buy, what will be the mean and standard deviation of these averages?

Answer: We have samples of size 12. The mean of these sample means will equal the population mean, 10 ounces. The standard deviation of these sample means will equal $\frac{3}{\sqrt{12}} = 0.866$ ounces.

Note that while giving the mean and standard deviation of the set of sample means, we did not describe the shape of the distribution. If we are also given that the original population is normal, then we can conclude that the set of sample means has a normal distribution.

EXAMPLE 12.4

Suppose that the distribution for total amounts spent by students vacationing for a week in Florida is normally distributed with a mean of $650 and a standard deviation of $120. What is the probability that an SRS of 10 students will spend an average of between $600 and $700?

Answer: The mean and standard deviation of the set of all sample means of size 10 are:

$$\mu_{\bar{x}} = 650 \quad \text{and} \quad \sigma_{\bar{x}} = \frac{120}{\sqrt{10}} = 37.95$$

(continued)

The z-scores of 600 and 700 are $\frac{600-650}{37.95} = -1.32$ and $\frac{700-650}{37.95} = 1.32$, respectively. Using Table A, we find the desired probability is $.9066 - .0934 = .8132$. [normalcdf(600, 700, 650, 37.95) = .8123.]

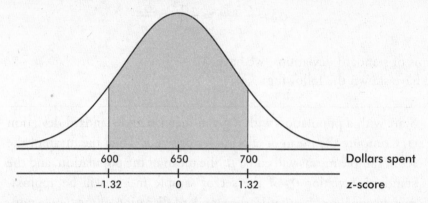

CENTRAL LIMIT THEOREM

We assumed above that the original population had a normal distribution. Unfortunately, few populations are normal, let alone exactly normal. However, it can be shown mathematically that no matter how the original population is distributed, if n is large enough, then the set of sample means is approximately normally distributed. For example, there is no reason to suppose that the amounts of money that different people spend in grocery stores are normally distributed. However, if each day we survey 30 people leaving a store and determine the average grocery bill, these daily averages will have a nearly normal distribution.

The following principle forms the basis of much of what we discuss in this topic and in those following. It is a simplified statement of the *central limit theorem* of statistics.

Start with a population with a given mean μ, a standard deviation σ, and any shape distribution whatsoever. Pick n sufficiently large (at least 30) and take all samples of size n. Compute the mean of each of these samples. Then

1. the set of all sample means is approximately normally distributed.
2. the mean of the set of sample means equals μ, the mean of the population.
3. the standard deviation $\sigma_{\bar{x}}$ of the set of sample means is approximately equal to $\frac{\sigma}{\sqrt{n}}$, that is, equal to the standard deviation of the whole population divided by the square root of the sample size.

Alternatively, we say that the sampling distribution of $\bar{x}$ is approximately normal with mean μ and standard deviation $\frac{\sigma}{\sqrt{n}}$.

While we mention $n \geq 30$ as a rough rule of thumb, $n \geq 40$ is often used, and n should be chosen even larger if more accuracy is required or if the original population is far from normal. As with proportions, we have the assumptions of a simple random sample and of sample size n no larger than 10% of the population.

EXAMPLE 12.5

Suppose that the average outstanding credit card balance for young couples is $650 with a standard deviation of $420. In an SRS of 100 couples, what is the probability that the mean outstanding credit card balance exceeds $700?

Answer: The sample size is over 30, we have an SRS, and our sample is less than 10% of all couples with outstanding balances, and so by the central limit theorem the set of sample means is approximately normally distributed with mean 650 and standard deviation $\frac{420}{\sqrt{100}} = 42$. With a *z*-score of $\frac{700-650}{42} = 1.19$, the probability that the sample mean exceeds 700 is $1 - .8830 = .1170$. [normalcdf(700, 10000, 650, 42) = .1169.]

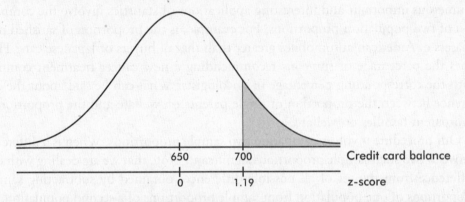

EXAMPLE 12.6

The strength of paper coming from a manufacturing plant is known to be 25 pounds per square inch with a standard deviation of 2.3. In a simple random sample of 40 pieces of paper, what is the probability that the mean strength is between 24.5 and 25.5 pounds per square inch?

Answer: We have a large (*n* = 40) SRS that is still smaller than 10% of all papers coming from the plant. $\mu_{\bar{x}} = 25$ and $\sigma_{\bar{x}} = \frac{2.3}{\sqrt{40}} = 0.364$. The *z*-scores of 24.5 and 25.5 are $\frac{24.5-25}{0.364} = -1.37$ and $\frac{25.5-25}{0.364} = 1.37$, respectively. The probability that the mean strength in the sample is between 24.5 and 25.5 pounds per square inch is $.9147 - .0853 = .8294$. [normalcdf(24.5, 25.5, 25, .364) = .8304.]

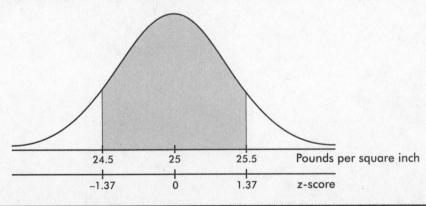

Don't be confused by the several different distributions being discussed! First, there's the distribution of the original population, which may be uniform, bell-shaped, strongly skewed – anything at all. Second, there's the distribution of the data in the sample, and the larger the sample size, the more this will look like the population distribution. Third, there's the distribution of the means of many samples of a given size, and the amazing fact is that this *sampling distribution* can be described by a normal model, regardless of the shape of the original population.

SAMPLING DISTRIBUTION OF A DIFFERENCE BETWEEN TWO INDEPENDENT SAMPLE PROPORTIONS

Numerous important and interesting applications of statistics involve the comparison of two population proportions. For example, is the proportion of satisfied purchasers of American automobiles greater than that of buyers of Japanese cars? How does the percentage of surgeons recommending a new cancer treatment compare with the corresponding percentage of oncologists? What can be said about the difference between the proportion of single parents on welfare and the proportion of two-parent families on welfare?

Our procedure involves comparing two sample proportions. When is a difference between two such sample proportions significant? Note that we are dealing with one difference from the set of all possible differences obtained by subtracting sample proportions of one population from sample proportions of a second population. To judge the significance of one particular difference, we must first determine how the differences vary among themselves. Remember that the variance of a set of differences is equal to the sum of the variances of the individual sets; that is,

$$\sigma_d^2 = \sigma_1^2 + \sigma_2^2$$

Now if

$$\sigma_1 = \sqrt{\frac{p_1(1 - p_1)}{n_1}} \quad \text{and} \quad \sigma_2 = \sqrt{\frac{p_2(1 - p_2)}{n_2}}$$

then

$$\sigma_d^2 = \frac{p_1(1 - p_1)}{n_1} + \frac{p_2(1 - p_2)}{n_2} \quad \text{and} \quad \sigma_d = \sqrt{\frac{p_1(1 - p_1)}{n_1} + \frac{p_2(1 - p_2)}{n_2}}$$

Then we have the following about the sampling distribution of $\hat{p}_1 - \hat{p}_2$:

Start with two populations with given proportions p_1 and p_2. Take all samples of sizes n_1 and n_2, respectively. Compute the difference $\hat{p}_1 - \hat{p}_2$ of the two proportions in each pair of samples. Then

1. the set of all differences of sample proportions is approximately normally distributed.
2. the mean of the set of differences of sample proportions equals $p_1 - p_2$, the difference of population proportions.
3. the standard deviation σ_d of the set of differences of sample proportions is approximately equal to

$$\sqrt{\frac{p_1(1 - p_1)}{n_1} + \frac{p_2(1 - p_2)}{n_2}}$$

Since we are using the normal approximation to the binomial, $n_1 p_1$, $n_1(1 - p_1)$, $n_2 p_2$, and $n_2(1 - p_2)$ should all be at least 10. Furthermore, in making calculations and drawing conclusions from specific samples, it is important both that the samples be *simple random samples* and that they be taken *independently* of each other. Finally, the samples cannot be too large; the sample sizes should be no larger than 10% of the populations.

EXAMPLE 12.7

A promoter knows that 23% of males enjoy watching boxing matches; however, only 12% of females enjoy watching this sport. In an SRS of 100 men and an independent SRS of 125 women, what is the probability that the difference in the percentages of men and women who enjoy watching boxing is more than 10%?

Answer: We note that $n_1 p_1 = 23$, $n_1(1 - p_1) = 77$, $n_2 p_2 = 15$, and $n_2(1 - p_2) = 110$ are all >10; we have independent SRSs, and the samples are less than 10% of their respective populations. For the sampling distribution of $\hat{p}_1 - \hat{p}_2$, the mean is $.23 - .12 = .11$ and the standard deviation is

$$\sqrt{\frac{(.23)(.77)}{100} + \frac{(.12)(.88)}{125}} = .0511$$

The z-score of .10 is $\frac{.10 - .11}{.0511} = -0.20$. Using Table A, we find $1 - .4207 = .5793$ for the probability that the difference is more than 10%. [normalcdf(.10, 1, .11, .0511) = .5776.]

(continued)

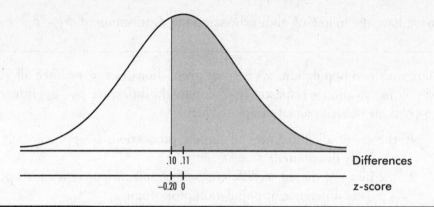

Differences

z-score

EXAMPLE 12.8

In urban America 43% of married couples own their own homes while only 19% of single people own their own homes. In an SRS of 200 married couples and an independent SRS of 180 single people, the probability is .90 that the difference in percentages of married couples and single people who are homeowners is greater than what percentage?

   Answer: We note that $n_1p_1 = 86$, $n_1(1 - p_1) = 114$, $n_2p_2 = 34.2$, and $n_2(1 - p_2) = 145.8$ are all >10; we have independent SRSs, and the samples are less than 10% of their respective populations. For the sampling distribution of $\hat{p}_1 - \hat{p}_2$, the mean is $.43 - .19 = .24$ and the standard deviation is

$$\sqrt{\frac{(.43)(.57)}{200} + \frac{(.19)(.81)}{180}} = .0456$$

The .90 area to the right corresponds to a *z*-score of −1.282 and thus a difference of .24 − 1.282(.0456) = .182. Thus there is a .90 probability that the difference in percentages is greater than 18.2%. [invNorm(.10, .24, .0456) = .1816.]

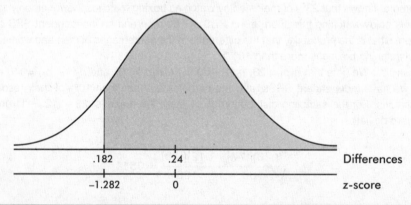

Differences

z-score

SAMPLING DISTRIBUTION OF A DIFFERENCE BETWEEN TWO INDEPENDENT SAMPLE MEANS

Many real-life applications of statistics involve comparisons of two population means. For example, is the average weight of laboratory rabbits receiving a special diet greater than that of rabbits on a standard diet? Which of two accounting firms pays a higher mean starting salary? Is the life expectancy of a coal miner less than that of a school teacher?

First we consider how to compare the means of samples, one from each population. When is a difference between two such sample means significant? The answer is more apparent when we realize that what we are looking at is one difference from a set of differences. That is, there is the set of all possible difference obtained by subtracting sample means from one set from sample means from a second set. To judge the significance of one particular difference we must first determine how the differences vary among themselves. The necessary key is the fact that the variance of a set of differences is equal to the sum of the variances of the individual sets. Thus,

$$\sigma^2_{\bar{x}_1 - \bar{x}_2} = \sigma^2_{\bar{x}_1} + \sigma^2_{\bar{x}_2}$$

Now if

$$\sigma_{\bar{x}_1} = \frac{\sigma_1}{\sqrt{n_1}} \quad \text{and} \quad \sigma_{\bar{x}_2} = \frac{\sigma_2}{\sqrt{n_2}}$$

then

$$\sigma^2_{\bar{x}_1 - \bar{x}_2} = \frac{\sigma_1^2}{n_1} + \frac{\sigma_2^2}{n_2} \quad \text{and} \quad \sigma_{\bar{x}_1 - \bar{x}_2} = \sqrt{\frac{\sigma_1^2}{n_1} + \frac{\sigma_2^2}{n_2}}$$

Then we have the following about the sampling distribution of $\bar{x}_1 - \bar{x}_2$:

Start with two normal populations with means μ_1 and μ_2 and standard deviations σ_1 and σ_2. Take all samples of sizes n_1 and n_2, respectively. Compute the difference $\bar{x}_1 - \bar{x}_2$ of the two means in each pair of these samples. Then

1. the set of all differences of sample means is approximately normally distributed.
2. the mean of the set of differences of sample means equals $\mu_1 - \mu_2$, the difference of population means.
3. the standard deviation $\sigma_{\bar{x}_1 - \bar{x}_2}$ of the set of differences of sample means is approximately equal to $\sqrt{\frac{\sigma_1^2}{n_1} + \frac{\sigma_2^2}{n_2}}$.

The more either population varies from normal, the greater should be the corresponding sample size. We also have the assumptions of independent simple random samples, and of sample sizes no larger than 10% of the populations.

EXAMPLE 12.9

When fertilizer A is used, the vegetable yield is 4.5 tons per acre with a standard deviation of 0.7 tons, while the yield when fertilizer B is used is 4.3 tons per acre with a standard deviation of 0.4 tons. In 45 sample plots using fertilizer A and 50 plots using fertilizer B, what is the probability that the difference in average yields will be negative, that is, that the average yield for the plots using fertilizer A will be less than the average yield for the plots using fertilizer B?

Answer: We have large ($n_1 = 45$ and $n_2 = 50$) SRSs that are still smaller than 10% of all possible plots. The mean of the differences is $4.5 - 4.3 = 0.2$, while the standard deviation is $\sqrt{\frac{(0.7)^2}{45} + \frac{(0.4)^2}{50}} = 0.119$. The *z*-score of a difference of 0 is $\frac{0-0.2}{0.119} = -1.68$. From Table A, the area to the left of −1.68 is .0465. Thus there is a 4.65% chance that the average yield for the sample plots using fertilizer A will be less than the average yield for the sample plots using fertilizer B. [normalcdf(−1000, 0, .2, .119) = .0464.]

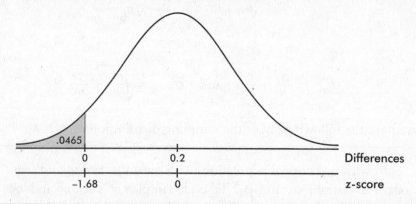

EXAMPLE 12.10

The average number of missed school days for students going to public schools is 8.5 with a standard deviation of 4.1, while students going to private schools miss an average of 5.3 with a standard deviation of 2.9. In an SRS of 200 public school students and an SRS of 150 private school students, with a probability of .95 the difference in average missed days (public average minus private average) is above what number?

Answer: We have large ($n_1 = 200$ and $n_2 = 150$) SRSs that are still smaller than 10% of all public and private school students. The mean of the differences is $8.5 - 5.3 = 3.2$, while the standard deviation is $\sqrt{\frac{(4.1)^2}{200} + \frac{(2.9)^2}{150}} = 0.374$. An area of .95 to the right corresponds to a *z*-score of −1.645. The difference in days is $3.2 - 1.645(0.374) = 2.6$. Thus there is a .95 probability that the difference in average missed days between public and private school samples is over 2.6 days. [invNorm(.05, 3.2, .374) = 2.585.]

(continued)

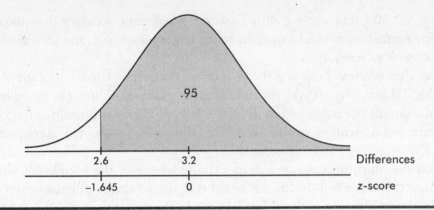

THE *T*-DISTRIBUTION

When the population standard deviation σ is unknown, we use the sample standard deviation s as an estimate for σ. But then $\dfrac{\bar{x} - \mu}{s/\sqrt{n}}$ does not follow a normal distribution. If, however, **the original population is normally distributed**, there is a distribution that can be used when working with the $s/\sqrt{n}$ ratios. This *Student t-distribution* was introduced in 1908 by W. S. Gosset, a British mathematician employed by the Guiness Breweries. (When we are working with small samples from a population that is *not* nearly normal, we must use very different "nonparametric" techniques not discussed in this review book.)

Thus, for a sample from a normally distributed population, we work with the variable

$$t = \frac{\bar{x} - \mu}{s/\sqrt{n}}$$

with a resulting *t*-distribution that is bell-shaped and symmetric, but lower at the mean, higher at the tails, and so more spread out than the normal distribution.

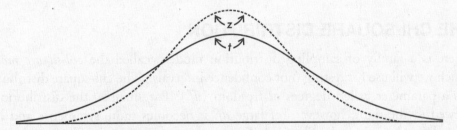

Like the binomial distribution, the *t*-distribution is different for different values of *n*. In the tables these distinct *t*-distributions are associated with the values for degrees of freedom (df). For this discussion the df value is equal to the sample size minus 1. The smaller the df value, the larger the dispersion in the distribution. The larger the df value, that is, the larger the sample size, the closer the distribution to the normal distribution.

Since there is a separate *t*-distribution for each df value, fairly complete tables would involve many pages; therefore, in Table B of the Appendix we list areas and *t*-values for only the more commonly used percentages or probabilities. The last row of Table B is the normal distribution, which is a special case of the *t*-distribution taken when *n* is infinite. For practical purposes, the two distributions are very close for any $n \geq 30$ (some use $n \geq 40$). However, when more accuracy is required, the Student *t*-distribution can be used for much larger values of *n*, and the calculations are easy with technology.

Note that, whereas Table A gives areas under the normal curve to the left of given *z*-values, Table B gives areas to the right of given positive *t*-values. For example, suppose the sample size is 20, and so df = 20 − 1 = 19. Then a probability of .05 in the tail corresponds with a *t*-value of 1.729, while .01 in the tail corresponds to $t = 2.539$.

Thus the *t*-distribution is the proper choice whenever the population standard deviation σ is unknown. In the real world σ is almost always unknown, and so we should almost always use the *t*-distribution. In the past this was difficult because extensive tables were necessary for the various *t*-distributions. However, with calculators such as the TI-84, this problem has diminished. It is no longer necessary to assume that the *t*-distribution is close enough to the *z*-distribution whenever *n* is greater than the arbitrary number 30.

The issue of sample size is refined even further by some statisticians:

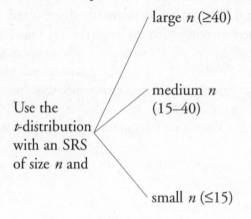

Use the
t-distribution
with an SRS
of size *n* and

large *n* (≥40) — Unnecessary to make any assumptions about parent population.

medium *n* (15–40) — Sample should show no extreme values and little, if any, skewness; or assume parent population is normal.

small *n* (≤15) — Sample should show no outliers and no skewness; or assume parent population is normal.

THE CHI-SQUARE DISTRIBUTION

There is a family of sampling distribution models, called the *chi-square models*, which we will use for testing (not confidence intervals). The chi-square distribution has a parameter called degrees of freedom (*df*). For small *df* the distribution is skewed to the right; however, for large *df* it becomes more symmetric and bell-shaped (as does the *t*-distribution). For one or two degrees of freedom the peak occurs at 0, while for three or more degrees of freedom the peak is at *df* − 2.

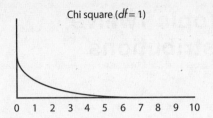

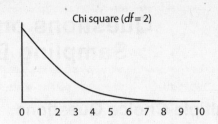

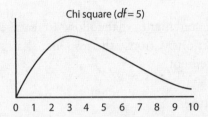

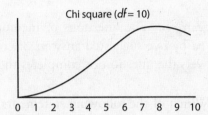

The chi-square distribution, like the *t*-distribution, has a separate curve for each *df* value; therefore, in Table C of the Appendix we list critical chi-square values for only the more commonly used tail probabilities. The chi-square distribution has numerous applications, two of the best known of which are goodness of fit of an observed distribution to a theoretical one, and independence of two criteria of classification of qualitative data. While we will be using chi-square for categorical data, the chi-square distribution as seen above is a *continuous* distribution, and applying it to counting data is just an approximation.

THE STANDARD ERROR

With proportions and means we typically do not know population parameters. So in calculating standard deviations of the sampling models, we actually estimate using sample statistics. In this case, we use the term *standard error*. That is, for proportions,

$$\sigma_{\hat{p}} = \sqrt{\frac{pq}{n}}, \text{ and we have } SE(\hat{p}) = \sqrt{\frac{pq}{n}}.$$

Similarly, for means,

$$\sigma_{\bar{x}} = \frac{\sigma}{\sqrt{n}}, \text{ and we have } SE(\bar{x}) = \frac{s}{\sqrt{n}}.$$

Questions on Topic Twelve: Sampling Distributions

Multiple-Choice Questions

Directions: The questions or incomplete statements that follow are each followed by five suggested answers or completions. Choose the response that best answers the question or completes the statement.

1. Which of the following statements are true?

    I. The larger the sample, the larger the spread in the sampling distribution.

    II. Provided that the population size is significantly greater than the sample size, the spread of a sampling distribution is about the same no matter what the population size.

    III. Bias has to do with the center, not the spread, of a sampling distribution.

    (A) I and II
    (B) I and III
    (C) II and III
    (D) I, II, and III
    (E) None of the above gives the complete set of true responses.

2. Which of the following statements are true?

    I. Sample parameters are used to make inferences about population statistics.

    II. Statistics from smaller samples have more variability.

    III. Parameters are fixed, while statistics vary depending on which sample is chosen.

    (A) I and II
    (B) I and III
    (C) II and III
    (D) I, II, and III
    (E) None of the above gives the complete set of true responses.

3. Which of the following statements are true?

   I. The sampling distribution of $\bar{x}$ has standard deviation $\frac{\sigma}{\sqrt{n}}$ even if the population is not normally distributed.

   II. The sampling distribution of $\bar{x}$ is normal if the population has a normal distribution.

   III. When n is large, the sampling distribution of $\bar{x}$ is approximately normal even if the population is not normally distributed.

   (A) I and II
   (B) I and III
   (C) II and III
   (D) I, II, and III
   (E) None of the above gives the complete set of true responses.

4. Which of the following statements are true?

   I. The mean of the set of sample means varies inversely as the square root of the size of the samples.

   II. The variance of the set of sample means varies directly as the size of the samples and inversely as the variance of the original population.

   III. The standard deviation of the set of sample means varies directly as the standard deviation of the original population and inversely as the square root of the size of the samples.

   (A) I only
   (B) II only
   (C) III only
   (D) I and II
   (E) I and III

5. Which of the following statements are true?

   I. The sampling distribution of $\hat{p}$ has a mean equal to the population proportion p.

   II. The sampling distribution of $\hat{p}$ has a standard deviation equal to
   $$\sqrt{np(1-p)}$$

   III. The sampling distribution of $\hat{p}$ is considered close to normal provided that $n \geq 30$.

   (A) I and II
   (B) I and III
   (C) II and III
   (D) I, II, and III
   (E) None of the above gives the complete set of true responses.

6. Which of the following statements are true?

   I. The sampling distribution of the difference $\bar{x}_1 - \bar{x}_2$ has a mean equal to the difference of the population means.

   II. The sampling distribution of the difference $\bar{x}_1 - \bar{x}_2$ has a standard deviation equal to the sum of the population standard deviations.

   III. The sampling distribution of the difference $\bar{x}_1 - \bar{x}_2$ has a standard deviation equal to the difference of the population standard deviations.

(A) I only
(B) II only
(C) III only
(D) I and II
(E) I and III

7. Which of the following are unbiased estimators for the corresponding population parameters?

   I. Sample means
   II. Sample proportions
   III. Difference of sample means
   IV. Difference of sample proportions

(A) None are unbiased.
(B) I and II
(C) I and III
(D) III and IV
(E) All are unbiased.

8. Suppose that 35% of all business executives are willing to switch companies if offered a higher salary. If a headhunter randomly contacts an SRS of 100 executives, what is the probability that over 40% will be willing to switch companies if offered a higher salary?

(A) .1469
(B) .1977
(C) .4207
(D) .8023
(E) .8531

9. Given that 58% of all gold dealers believe next year will be a good one to speculate in South African gold coins, in a simple random sample of 150 dealers, what is the probability that between 55% and 60% believe that it will be a good year to speculate?

(A) .0500
(B) .1192
(C) .3099
(D) .4619
(E) .9215

10. The average outstanding bill for delinquent customer accounts for a national department store chain is \$187.50 with a standard deviation of \$54.50. In a simple random sample of 50 delinquent accounts, what is the probability that the mean outstanding bill is over \$200?

    (A) .0526
    (B) .0667
    (C) .4090
    (D) .5910
    (E) .9474

11. The average number of daily emergency room admissions at a hospital is 85 with a standard deviation of 37. In a simple random sample of 30 days, what is the probability that the mean number of daily emergency admissions is between 75 and 95?

    (A) .1388
    (B) .2128
    (C) .8612
    (D) .8990
    (E) .9970

12. Two companies offer classes designed to improve students' SAT scores. Suppose that 83% of students enrolling in the first program improve their scores, while 74% of those signing up for the second program also raise their scores. In an SRS of 60 students taking the first program and an independent SRS of 50 taking the second, what is the probability that the difference between the percentages of students improving their scores (first program minus second) is more than 15%?

    (A) .0281
    (B) .2236
    (C) .2764
    (D) .3632
    (E) .7764

13. Pepper plants watered lightly every day for a month show an average growth of 27 centimeters with a standard deviation of 8.3 centimeters, while pepper plants watered heavily once a week for a month show an average growth of 29 centimeters with a standard deviation of 7.9 centimeters. In a simple random sample of 60 plants, half of which are given each of the watering treatments, what is the probability that the difference in average growth between the two halves is between −3 and +3 centimeters?

    (A) .3156
    (B) .3240
    (C) .4639
    (D) .6760
    (E) .6844

14. Which of the following statements are true?

    I. Like the normal, *t*-distributions are always symmetric.

    II. Like the normal, *t*-distributions are always mound-shaped.

    III. The *t*-distributions have less spread than the normal, that is, they have less probability in the tails and more in the center than the normal.

(A) II only
(B) I and II
(C) I and III
(D) II and III
(E) I, II, and III

15. Which of the following statements about *t*-distributions are true?

    I. The greater the number of degrees of freedom, the narrower the tails.

    II. The smaller the number of degrees of freedom, the closer the curve is to the normal curve.

    III. Thirty degrees of freedom gives the normal curve.

(A) I only
(B) I and II
(C) I and III
(D) II and III
(E) I, II, and III

Answer Key

1. **C**	4. **C**	7. **E**	10. **A**	13. **D**
2. **C**	5. **E**	8. **A**	11. **C**	14. **B**
3. **D**	6. **A**	9. **D**	12. **B**	15. **A**

Answers Explained

1. **(C)** The larger the sample, the smaller the spread in the sampling distribution.

2. **(C)** Sample statistics are used to make inferences about population proportions.

3. **(D)** It is always true that the sampling distribution of $\bar{x}$ has mean μ and standard deviation $\frac{\sigma}{\sqrt{n}}$. In addition, the sampling distribution will be normal if the population is normal or if n is large.

4. **(C)** The mean of the set of sample means is equal to the mean of the population; it does not vary with the size of the samples. The variance of the set of sample means varies inversely as the size of the samples and directly as the variance of the original population.

5. **(E)** Only I is true. The sampling distribution of $\hat{p}$ has a standard deviation equal to $\sqrt{\frac{p(1-p)}{n}}$. The sampling distribution of $\hat{p}$ is considered close to

normal provided that both np and $n(1 - p)$ are large enough (greater than either 5 or 10 are standard guidelines).

6. **(A)** Variances can be added; standard deviations cannot.

7. **(E)** All are unbiased estimators for the corresponding population parameters; that is, the means of their sampling distributions are equal to the population parameters.

8. **(A)** Both $np = (100)(.35) = 35 > 10$ and $n(1 - p) = (100)(.65) = 65 > 10$. The set of sample proportions is approximately normally distributed with mean .35 and standard deviation $\sigma_{\hat{p}} = \sqrt{\frac{(.35)(.65)}{100}} = .0477$. With a z-score of $\frac{.40-.35}{.0477} = 1.05$, the probability that the sample proportion exceeds 40% is $1 - .8531 = .1469$. [normalcdf(.40, 1, .35, .0477) = .1473.]

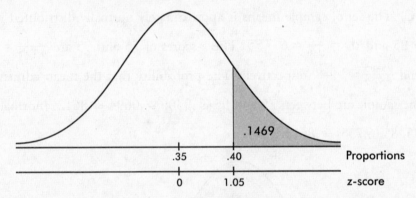

9. **(D)** Both $np = (150)(.58) = 87 > 10$ and $n(1 - p) = (150)(.42) = 63 > 10$. The set of sample proportions is approximately normally distributed with mean .58 and standard deviation $\sigma_{\hat{p}} = \sqrt{\frac{(.58)(.42)}{150}} = .0403$. With z-scores of $\frac{.55-.58}{.0403} = -0.74$ and $\frac{.60-.58}{.0403} = 0.50$, the probability the sample proportion is between .55 and .60 is $.6915 - .2296 = .4619$. [normalcdf(.55, .60, .58, .0403) = .4618.]

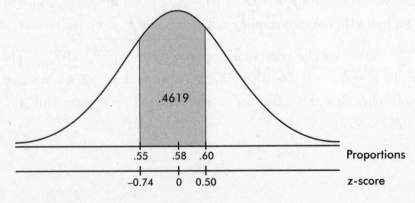

10. **(A)** The set of sample means is approximately normally distributed with $\mu_{\bar{x}} = 187.50$ and $\sigma_{\bar{x}} = \frac{54.50}{\sqrt{50}} = 7.707$. The z-score of 200 is $\frac{200 - 187.50}{7.707} = 1.62$, and the probability that the mean outstanding bill is over \$200 is $1 - .9474 = .0526$. [normalcdf(200, 10000, 187.5, 7.707) = .0524.]

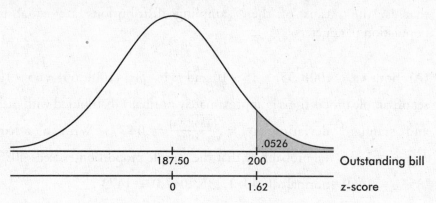

11. **(C)** The set of sample means is approximately normally distributed with $\mu_{\bar{x}} = 85$ and $\sigma_{\bar{x}} = \frac{37}{\sqrt{30}} = 6.755$. The z-scores of 75 and 95 are $\frac{75-85}{6.755} = -1.48$ and $\frac{95-85}{6.755} = 1.48$, respectively. The probability that the mean admissions in the sample are between 75 and 95 is $.9306 - .0694 = .8612$. [normalcdf(75, 95, 85, 6.755) = .8612.]

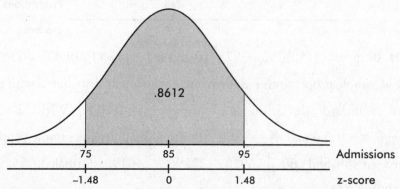

12. **(B)** We note that $n_1 p_1 = 49.8$, $n_1(1 - p_1) = 10.2$, $n_2 p_2 = 37$, and $n_2(1 - p_2) = 13$ are all >10. For the sampling distribution of $\hat{p}_1 - \hat{p}_2$, the mean is $.83 - .74 = .09$ and the standard deviation is $\sqrt{\frac{(.83)(.17)}{60} + \frac{(.74)(.26)}{50}} = .0787$. The z-score of .15 is $\frac{.15-.09}{.0787} = 0.76$. Using Table A, we find $1 - .7764 = .2236$ for the probability that the difference is more than 15%. [normalcdf(.15, 1, .09, .0787) = .2229.]

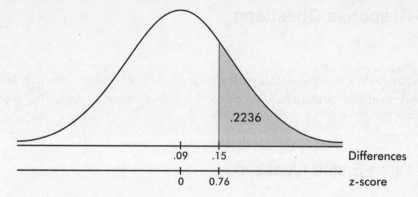

13. **(D)** The mean of the differences equals $27 - 29 = -2$, while the standard deviation is $\sqrt{\frac{(8.3)^2}{30} + \frac{(7.9)^2}{30}} = 2.092$. The z-scores of $+3$ and -3 are $\frac{3-(-2)}{2.092} = 2.39$ and $\frac{-3-(-2)}{2.092} = -0.48$, respectively. The probability that the difference is between these two values is $.9916 - .3156 = .6760$. [normalcdf(-3, 3, -2, 2.092) = .6753.]

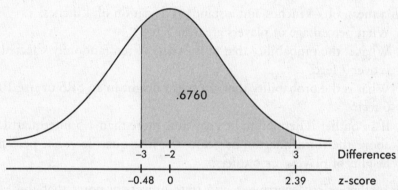

14. **(B)** The t-distributions are symmetric and mound-shaped, and they have more, not less, spread than the normal distribution.

15. **(A)** The larger the number of degrees of freedom, the closer the curve to the normal curve. While around the 30 level is often considered a reasonable approximation to the normal curve, it is not the normal curve.

Free-Response Questions

> ***Directions:*** You must show all work and indicate the methods you use. You will be graded on the correctness of your methods and on the accuracy of your final answers.

Four Open-Ended Questions

1. Give an example of a population and an associated population parameter and sample statistic, and then explain what is meant by the sampling distribution.

2. Heights, weights, and like measurements tend to result in normal distributions, but normally distributed natural phenomena occur much less frequently than might be supposed. Explain why the normal curve has an importance in statistics that is independent of whether or not it appears in nature.

3. Suppose that the heights of college basketball players are normally distributed with a mean of 74 inches and a standard deviation of 4 inches.
   (a) What percentage of players are over 7 feet?
   (b) What is the probability that at least one of ten randomly selected players is over 7 feet?
   (c) What is the probability that the mean height in an SRS of size 10 is over 6 feet?
   (d) If an outlier is defined to be any value more than 1.5 interquartile ranges above the third quartile or below the first quartile, what percentage of heights of players are outliers?

4. The mathematics department at a state university notes that the SAT math scores of high school seniors applying for admission into their program are normally distributed with a mean of 610 and standard deviation of 50.
   (a) What is the probability that a randomly chosen applicant to the department has an SAT math score above 700?
   (b) What is the shape, mean, and standard deviation of the sampling distribution of the mean of a sample of 40 randomly selected applicants?
   (c) What is the probability that the mean SAT math score in an SRS of 40 applicants is above 625?
   (d) Would your answers to (a), (b), or (c) be affected if the original population of SAT math scores were highly skewed instead of normal? Explain.

Answers Explained

1. The population parameter can be something like a mean μ or a proportion p, and then the sample statistic is a sample mean $\bar{x}$ or a sample proportion $\hat{p}$. For example, if the population is the set of salaries earned by adults in the United States, the population parameter is the mean of all these salaries, and the sample statistic can be the mean of the salaries in a randomly chosen sample of 100

adult salaries. The sampling distribution would then be the distribution of sample means from all possible samples of 100 adult salaries.

2. What is significant is that the results of many types of sampling experiments can be analyzed using the normal curve. For example, there is no reason to suppose that the amounts of money that different people spend in grocery stores are normally distributed. However, if every day we survey 30 people leaving a store and determine the average grocery bill, these daily averages will have a nearly normal distribution. The central limit theorem of statistics says that for any population, no matter what its distribution, if n is sufficiently large, then the set of sample means for all samples of size n will be approximately normally distributed.

3. (a) The z-score of 84 is $\frac{84-74}{4} = 2.5$, which gives a probability of .0062 or 0.62%. [normalcdf(84, 1000, 74, 4) = .0062.]
(b) $1 - (.9938)^{10} = .06$
(c) Checking conditions: the original population is given to be normal, we have an SRS, and our sample is less than 10% of all college basketball players. The z-score of 72 is $\dfrac{72-74}{\frac{4}{\sqrt{10}}} = -1.58$ with a resulting probability of .9429. [normalcdf(72, 1000, 74, 1.265) = .9431.]
(d) We have $Q_1 = 74 - 0.674(4) = 71.3$, $Q_3 = 74 + 0.674(4) = 76.7$, and $1.5(\text{IQR}) = 1.5(Q_3 - Q_1) = 8.1$. Then $71.3 - 8.1 = 63.2$ has a z-score of -2.7, while $76.7 + 8.1 = 84.8$ has a z-score of 2.7. The corresponding probabilities give $.0035 + .0035 = .007$ or 0.7%. The problem could also have been worked simply in terms of z-scores: $1.5(0.674 + 0.674) = 2.02$ and $0.674 + 2.02 \approx 2.7$, and so on.

4. (a)

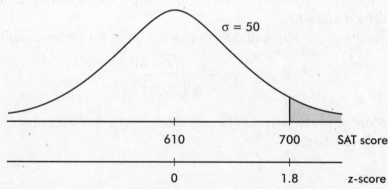

The critical z-score is $\frac{700-610}{50} = 1.8$, and so from Table A the asked-for probability is $1 - .9641 = .0359$. [normalcdf(700, 10000, 610, 50) = .0359.]
(b) It is roughly normal with a mean equal to the population mean 610 and a standard deviation equal to the population standard deviation divided by the square root of the sample size, that is, $\frac{50}{\sqrt{40}} = 7.91$.

(c) Checking conditions: the original population is given to be normal, we have an SRS, and our sample comes from less than 10% of all applicants.

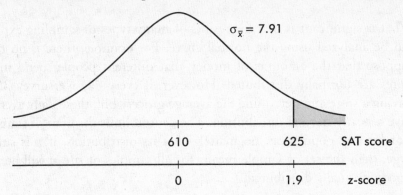

The critical z-score is $\frac{625-610}{7.91} = 1.90$, and from Table A the asked-for probability is $1 - .9713 = .0287$. [normalcdf(625, 10000, 610, 7.91) = .0290.]
(d) The answer to part *a* would be affected because it assumes a normal population. The other answers would not be affected because for large enough *n*, the central limit theorem gives that the sampling distribution will be roughly normal regardless of the distribution of the original population.

Two Investigative Tasks

1. Consider a population of size $N = 5$ consisting of the elements 3, 6, 8, 10, and 18.

   a. Determine the population mean μ and the population standard deviation σ.

   b. List all possible samples of size $n = 2$ [there are $C(5, 2) = 10$ of them] and determine the mean $\bar{x}$ of each.

   c. Show that the mean of the set of ten sample means is equal to the population mean μ.

   d. Show that the standard deviation of the set of ten sample means is

   $$\sigma_{\bar{x}} = \frac{\sigma}{\sqrt{n}} \sqrt{\frac{N-n}{N-1}}$$

   e. More generally, what can be said about $\frac{\sigma}{\sqrt{n}} \sqrt{\frac{N-n}{N-1}}$ if the population size N is very large?

2. Suppose that, of a group of six ($N = 6$) people, A, B, C, D, E, and F, two (B and E) own their own houses. Thus the proportion of the group owning their own homes is $p = \frac{2}{6} = \frac{1}{3}$. Now consider all possible samples of size $n = 4$.

   a. List the samples [there are $C(6, 4) = 15$] and determine the proportion $\hat{p}$ of homeowners for each sample.

   b. Calculate the mean and standard deviation for this set of 15 proportions. Show that the mean is equal to p and that the standard deviation is equal to $\sqrt{\frac{p(1-p)}{n}}\sqrt{\frac{N-n}{N-1}}$.

   c. What happens to $\sqrt{\frac{p(1-p)}{n}}\sqrt{\frac{N-n}{N-1}}$ when the population size N is very large?

Answers Explained

1. *a.* Using a calculator, find $\mu = 9$ and $\sigma = 5.0596$.

   b.

Sample	Mean
{3, 6}	4.5
{3, 8}	5.5
{3, 10}	6.5
{3, 18}	10.5
{6, 8}	7
{6, 10}	8
{6, 18}	12
{8, 10}	9
{8, 18}	13
{10, 18}	14

   c. $\dfrac{(4.5 + 5.5 + 6.5 + 10.5 + 7 + 8 + 12 + 9 + 13 + 14)}{10} = 9$.

   d. Using a calculator, find $\sigma_{\bar{x}} = 3.0984$ and then note that

$$\frac{5.0596}{\sqrt{2}}\sqrt{\frac{5-2}{5-1}} = 3.0984$$

   e. If N is very large, $\frac{N-n}{N-1}$ is approximately equal to 1 and so the expression simplifies to $\frac{\sigma}{\sqrt{n}}$.

2. *a.*

Sample	Proportion
A, B, C, D	.25
A, B, C, E	.50
A, B, C, F	.25
A, B, D, E	.50
A, B, D, F	.25
A, B, E, F	.50
A, C, D, E	.25
A, C, D, F	.00
A, C, E, F	.25
A, D, E, F	.25
B, C, D, E	.50
B, C, D, F	.25
B, C, E, F	.50
B, D, E, F	.50
C, D, E, F	.25

b. Using a calculator, find $\mu_{\hat{p}} = .333$ and $\sigma_{\hat{p}} = .149$ and note that

$$\sqrt{\frac{(.333)(.667)}{4}}\sqrt{\frac{6-4}{6-1}} = .149$$

c. If N is very large, $\frac{N-n}{N-1}$ is approximately equal to 1 and so the expression simplifies to $\sqrt{\frac{p(1-p)}{n}}$.

Confidence Intervals

- Definition
- For a Proportion
- For a Difference of Two Proportions
- For a Mean
- For a Difference Between Two Means
- For the Slope of a Least Squares Regression Line

Using a measurement from a sample, we are never able to say *exactly* what a population proportion or mean is; rather we always say we have a certain *confidence* that the population proportion or mean lies in a particular *interval*. The particular interval is centered around a sample proportion or mean (or other statistic) and can be expressed as the sample estimate plus or minus an associated *margin of error*.

THE MEANING OF A CONFIDENCE INTERVAL

Using what we know about sampling distributions, we are able to establish a certain confidence that a sample proportion or mean lies within a specified interval around the population proportion or mean. However, we then have the same confidence that the population proportion or mean lies within a specified interval around the sample proportion or mean (e.g., the distance from Missoula to Whitefish is the same as the distance from Whitefish to Missoula).

Typically we consider 90%, 95%, and 99% confidence interval estimates, but any percentage is possible. The percentage is the percentage of samples that would pinpoint the unknown p or μ within plus or minus a certain margin or error. We do *not* say there is a .90, .95, or .99 probability that p or μ is within a certain margin of error of a given sample proportion or mean. For a given sample proportion or mean, p or μ either is or isn't within the specified interval, and so the probability is either 1 or 0.

CONFIDENCE INTERVAL FOR A PROPORTION

We are interested in estimating a population proportion p by considering a single sample proportion $\hat{p}$. This sample proportion is just one of a whole universe of sample proportions, and from Topic 12 we remember the following:

1. The set of all sample proportions is approximately normally distributed.
2. The mean $\mu_{\hat{p}}$ of the set of sample proportions equals p, the population proportion.

3. The standard deviation $\sigma_{\hat{p}}$ of the set of sample proportions is approximately equal to $\sqrt{\frac{p(1-p)}{n}}$.

In finding confidence interval estimates of the population proportion p, how do we find $\sqrt{\frac{p(1-p)}{n}}$ since p is unknown? The reasonable procedure is to use the sample proportion $\hat{p}$:

$$\sigma_{\hat{p}} \approx \sqrt{\frac{\hat{p}(1-\hat{p})}{n}}$$

When the standard deviation is estimated in this way (using the sample), we use the term *standard error*. That is,

$$SE_{\hat{p}} = \sqrt{\frac{\hat{p}(1-\hat{p})}{n}}$$

Remember that we are really using a normal approximation to the binomial, so $n\hat{p}$ and $n(1-\hat{p})$ should both be at least 10. Furthermore, in making calculations and drawing conclusions from a specific sample, it is important that the sample be a *simple random sample*. Finally, the population should be large, typically checked by the assumption that the sample is less than 10% of the population. (If the population is small and the sample exceeds 10% of the population, then models other than the normal are more appropriate.)

EXAMPLE 13.1

If 64% of an SRS of 550 people leaving a shopping mall claim to have spent over \$25, determine a 99% confidence interval estimate for the proportion of shopping mall customers who spend over \$25.

Answer: We check that $n\hat{p} = 550(.64) = 352 > 10$ and $n(1 - \hat{p}) = 550(.36) = 198 > 10$, we are given that the sample is an SRS, and it is reasonable to assume that 550 is less than 10% of all mall shoppers. Since $\hat{p} = .64$, the standard deviation of the set of sample proportions is

$$\sigma_p \approx \sqrt{\frac{(.64)(.36)}{550}} = .0205$$

From Topic 11 we know that 99% of the sample proportions should be within 2.576 standard deviations of the population proportion. Equivalently, we are 99% certain that the population proportion is within 2.576 standard deviations of any sample proportion.[1] Thus the 99% confidence interval estimate for the population proportion is $.64 \pm 2.576(.0205) = .64 \pm .053$. We say that the *margin of error* is $\pm.053$. We are 99% certain that the proportion of shoppers spending over \$25 is between .587 and .693.

We can also say, using the definition of confidence *level*, that if the interviewing procedure were repeated many times, about 99% of the resulting confidence intervals would contain the true proportion (thus we're 99% confident that the method worked for the interval we got).

---

[1] Note that we cannot say there is a .99 probability that the population proportion is within 2.576 standard deviations of a given sample proportion. For a given sample proportion, the population proportion either is or isn't within the specified interval, and so the probability is either 1 or 0.

EXAMPLE 13.2

In a simple random sample of machine parts, 18 out of 225 were found to have been damaged in shipment. Establish a 95% confidence interval estimate for the proportion of machine parts that are damaged in shipment.

Answer: We check that $n\hat{p} = 18 > 10$ and $n(1 - \hat{p}) = 225 - 18 = 207 > 10$, we are given that the sample is an SRS, and it is reasonable to assume that 225 is less than 10% of all shipped machine parts. The sample proportion is $\hat{p} = \frac{18}{225} = .08$, and the standard deviation of the set of sample proportions is

$$\sigma_p \approx \sqrt{\frac{(.08)(.92)}{225}} = .0181$$

The 95% confidence interval estimate for the population proportion is $.08 \pm 1.96(.0181) = .08 \pm .035$. Thus we are 95% certain that the proportion of machine parts damaged in shipment is between .045 and .115.

[On the TI-84, under STAT and then TESTS, go to 1-PropZInt. With x:18, n:225, and C-Level:.95, Calculate gives (.04455, .11545).]

Suppose there are 50,000 parts in the entire shipment. We can translate from proportions to actual numbers:

$$.045(50,000) = 2250 \quad \text{and} \quad .115(50,000) = 5750$$

and so we can be 95% confident that there are between 2250 and 5750 defective parts in the whole shipment.

EXAMPLE 13.3

A telephone survey of 1000 adults was taken shortly after the United States began bombing Iraq.

a.   If 832 voiced their support for this action, with what confidence can it be asserted that 83.2% ± 3% of the adult U.S. population supported the decision to go to war?
Answer: We check that $n\hat{p} = 832 > 10$ and $n(1 - \hat{p}) = 1000 - 832 = 168 > 10$, we assume an SRS, and clearly 1000 is <10% of the adult U.S. population.

$$\hat{p} = \frac{832}{1000} = .832 \quad \text{and so} \quad \sigma_{\hat{p}} \approx \sqrt{\frac{(.832)(.168)}{1000}} = .0118$$

The relevant z-scores are $\pm\frac{.03}{.0118} = \pm2.54$. Table A gives probabilities of .0055 and .9945, and so our answer is $.9945 - .0055 = .9890$. In other words, 83.2% ± 3% is a 98.90% confidence interval estimate for U.S. adult support of the war decision.

b.   If the adult U.S. population is 191 million, estimate the actual numerical support.
Answer: Since

$$.802(191,000,000) \approx 153,000,000$$

while

$$.862(191,000,000) \approx 165,000,000$$

we can be 98.90% confident that between 153 and 165 million adults supported the initial bombing decision.

EXAMPLE 13.4

A U.S. Department of Labor survey of 6230 unemployed adults classified people by marital status, gender, and race. The raw numbers are as follows:

| White, 16 yr and older | | | |
	Married	Widow/Div.	Single
Men	1090	337	1168
Women	952	423	632

| Nonwhite, 16 yr and older | | | |
	Married	Widow/Div.	Single
Men	266	135	503
Women	189	186	349

a. Find a 90% confidence interval estimate for the proportion of unemployed men who are married.

Answer: Totaling the first row across both tables, we find that there are 3499 men in the survey; 1090 + 266 = 1356 of them are married. We check that $n\hat{p}$ = 1356 > 10 and $n(1 - \hat{p})$ = 3499 − 1356 = 2143 > 10, we assume an SRS, and clearly the sample is <10% of the total unemployed adult population. Therefore $\hat{p} = \frac{1356}{3499} = .3875$ and

$$\sigma_{\hat{p}} \approx \sqrt{\frac{(.3875)(.6125)}{3499}} = .008236$$

Thus the 90% confidence interval estimate is .3875 ± 1.645(.008236) = .3875 ± .0135. [On the T1-83, 1-PropZInt gives (.37399, .40109).]

b. Find a 98% confidence interval estimate of the proportion of unemployed single persons who are women.

Answer: There are 1168 + 632 + 503 + 349 = 2652 singles in the survey. Of these, 632 + 349 = 981 are women [$n\hat{p}$ = 981 > 10 and $n(1 - \hat{p})$ = 2652 − 981 = 1671 > 10], so $\hat{p} = \frac{981}{2652} = .3699$ and

$$\sigma_{\hat{p}} \approx \sqrt{\frac{(.3699)(.6301)}{2652}} = .009375$$

The 98% confidence interval estimate is .3699 ± 2.326(.009375) = .3699 ± .0218. [On the TI-84, 1-PropZInt gives (.3481, .39172).]

As we have seen, there are two types of statements that come out of confidence intervals. First, we can interpret the confidence *interval* and say we are 90% confident that between 60% and 66% of all voters favor a bond issue. Second, we can interpret the confidence *level* and say that if this survey were conducted many times, then about 90% of the resulting confidence intervals would contain the true proportion of voters who favor the bond issue. *Incorrect* statements include "The percentage of all voters who support the bond issue is between 60% and 66%," "There is a .90 probability that the true percentage of all voters who favor the bond issue is between 60% and 66%," and "If this survey were conducted many times, then about 90% of the sample proportions would be in the interval (.60, .66)."

(continued)

One important consideration in setting up a survey is the choice of sample size. To obtain a smaller, more precise interval estimate of the population proportion, we must either decrease the degree of confidence or increase the sample size. Similarly, if we want to increase the degree of confidence, we can either accept a wider interval estimate or increase the sample size. Again, while choosing a larger sample size may seem desirable, in the real world this decision involves time and cost considerations.

In setting up a survey to obtain a confidence interval estimate of the population proportion, what should we use for $\sigma_{\hat{p}}$? To answer this question, we first must consider how large $\sqrt{p(1-p)}$ can be. We plot various values of p:

p:	.1	.2	.3	.4	.5	.6	.7	.8	.9
$\sqrt{p(1-p)}$:	.3	.4	.458	.490	.5	.490	.458	.4	.3

which gives .5 as the intuitive answer. Thus $\sqrt{\frac{p(1-p)}{n}}$ is at most $\frac{.5}{\sqrt{n}}$. We make use of this fact to determine sample sizes in problems such as Examples 13.5 and 13.6.

EXAMPLE 13.5

An Environmental Protection Agency (EPA) investigator wants to know the proportion of fish that are inedible because of chemical pollution downstream of an offending factory. If the answer must be within ±.03 at the 96% confidence level, how many fish should be in the sample tested?

Answer: We want $2.05\sigma_{\hat{p}} \le .03$. From the above remark, $\sigma_{\hat{p}}$ is at most $\frac{.5}{\sqrt{n}}$ and so it is sufficient to consider $2.05\left(\frac{.5}{\sqrt{n}}\right) \le .03$. Algebraically, we have $\sqrt{n} \ge \frac{2.05(.5)}{.03} = 34.17$ and $n \ge 1167.4$. Therefore, choosing a sample of 1168 fish gives the inedible proportion to within ±.03 at the 96% level.

Note that the accuracy of the estimate does *not* depend on what fraction of the whole population we have sampled. What is critical is the *absolute size* of the sample. Is some minimal value of n necessary for the procedures we are using to be meaningful? Since we are using the normal approximation to the binomial, both np and $n(1-p)$ should be at least 10 (see Topic 11).

EXAMPLE 13.6

A study is undertaken to determine the proportion of industry executives who believe that workers' pay should be based on individual performance. How many executives should be interviewed if an estimate is desired at the 99% confidence level to within ±.06? To within ±.03? To within ±.02?

Answer: Algebraically, $2.576\left(\frac{.5}{\sqrt{n}}\right) \le .06$ gives $\sqrt{n} \ge \frac{2.576(.5)}{.06} = 21.5$, so $n \ge 462.25$. Similarly, $2.576\left(\frac{.5}{\sqrt{n}}\right) \le .03$ gives $\sqrt{n} \ge \frac{2.576(.5)}{.03} = 42.9$, so $n \ge 1840.4$. Finally, $2.576\left(\frac{.5}{\sqrt{n}}\right) \le .02$ gives $\sqrt{n} \ge \frac{2.576(.5)}{.02} = 64.4$, so $n \ge 4147.4$. Thus 463, 1841, or 4148 executives should be interviewed, depending on the accuracy desired.

Note that to cut the interval estimate in half (from ±.06 to ±.03), we would have to increase the sample size fourfold, and to cut the interval estimate to a third (from ±.06 to ±.02), a nine-fold increase in the sample size would be required (answers are not exact because of round-off error.)

More generally, to divide the interval estimate by d without affecting the confidence level, we must increase the sample size by a multiple of d^2.

A formula for the calculations in Examples 13.5 and 13.6 is

$$n = \left[\frac{z}{2(\text{error})} \right]^2$$

(Note: this formula is not given on the AP exam formula page.)

CONFIDENCE INTERVAL FOR A DIFFERENCE OF TWO PROPORTIONS

From Topic 12, we have the following information about the sampling distribution of $\hat{p}_1 - \hat{p}_2$:

1. The set of all differences of sample proportions is approximately normally distributed.
2. The mean of the set of differences of sample proportions equals $p_1 - p_2$, the difference of population proportions.
3. The standard deviation σ_d of the set of differences of sample proportions is approximately equal to

$$\sqrt{\frac{p_1(1-p_1)}{n_1} + \frac{p_2(1-p_2)}{n_2}}$$

Remember that we are using the normal approximation to the binomial, so $n_1\hat{p}_1$, $n_1(1 - \hat{p}_1)$, $n_2\hat{p}_2$, and $n_2(1 - \hat{p}_2)$ should all be at least 10. In making calculations and drawing conclusions from specific samples, it is important both that the samples be *simple random samples* and that they be taken *independently* of each other. Finally, the original populations should be large compared to the sample sizes.

EXAMPLE 13.7

Suppose that 84% of an SRS of 125 nurses working 7 a.m. to 3 p.m. shifts in city hospitals express positive job satisfaction, while only 72% of an SRS of 150 nurses on 11 p.m. to 7 a.m. shifts express similar fulfillment. Establish a 90% confidence interval estimate for the difference.

Answer: Note that $n_1\hat{p}_1 = (125)(.84) = 105$, $n_1(1 - \hat{p}_1) = (125)(.16) = 20$, $n_2\hat{p}_2 = (150)(.72) = 108$, and $n_2(1 - \hat{p}_2) = (150)(.28) = 42$ are all >10, we are given SRSs, and the population of city hospital nurses is assumed to be large.

$$n_1 = 125 \quad n_2 = 150$$
$$\hat{p}_1 = .84 \quad \hat{p}_2 = .72$$

$$\sigma_d \approx \sqrt{\frac{(.84)(.16)}{125} + \frac{(.72)(.28)}{150}} = .0492$$

The observed difference is $.84 - .72 = .12$, and the critical z-scores are ± 1.645. The confidence interval estimate is $.12 \pm 1.645(.0492) = .12 \pm .081$. We can be 90% certain that the proportion of satisfied nurses on 7 a.m. to 3 p.m. shifts is between .039 and .201 higher than the proportion for nurses on 11 p.m. to 7 a.m. shifts.

EXAMPLE 13.8

A grocery store manager notes that in an SRS of 85 people going through the express checkout line, only 10 paid with checks, whereas, in an SRS of 92 customers passing through the regular line, 37 paid with checks. Find a 95% confidence interval estimate for the difference between the proportion of customers going through the two different lines who use checks.

Answer: Note that $n_1\hat{p}_1 = 10$, $n_1(1 - \hat{p}_1) = 75$, $n_2\hat{p}_2 = 37$, and $n_2(1 - \hat{p}_2) = 55$ are all at least 10, we are given SRSs, and the total number of customers is large.

$$n_1 = 85 \qquad n_2 = 92$$
$$\hat{p}_1 = \frac{10}{85} = .118 \quad \hat{p}_2 = \frac{37}{92} = .402$$

$$\sigma_d \approx \sqrt{\frac{(.118)(.882)}{85} + \frac{(.402)(.598)}{92}} = .0619$$

The observed difference is $.118 - .402 = -.284$, and the critical z-scores are ± 1.96. Thus, the confidence interval estimate is $-.284 \pm 1.96(.0619) = -.284 \pm .121$. The manager can be 90% sure that the proportion of customers passing through the express line who use checks is between .163 and .405 lower than the proportion going through the regular line who use checks.

With regard to choosing a sample size, $\sqrt{p(1 - p)}$ is at most .5. Thus

$$\sqrt{p(1 - p)\left(\frac{1}{n_1} + \frac{1}{n_2}\right)} \leq (.5)\sqrt{\frac{1}{n_1} + \frac{1}{n_2}}$$

Now, if we simplify by insisting that $n_1 = n_2 = n$, the above statement can be reduced as follows:

$$(.5)\sqrt{\frac{1}{n} + \frac{1}{n}} = (.5)\sqrt{\frac{2}{n}} = \frac{.5\sqrt{2}}{\sqrt{n}}$$

(Note: this formula is not given on the AP exam formula page.)

EXAMPLE 13.9

A pollster wants to determine the difference between the proportions of high-income voters and low-income voters who support a decrease in the capital gains tax. If the answer must be known to within ±.02 at the 95% confidence level, what size samples should be taken?

Answer: Assuming we will pick the same size samples for the two sample proportions, we have $\sigma_d \leq \frac{.5\sqrt{2}}{\sqrt{n}}$ and $1.96\sigma_d \leq .02$. Thus $\frac{1.96(.5)\sqrt{2}}{\sqrt{n}} \leq .02$. Algebraically we find that $\sqrt{n} \geq \frac{1.96(.5)\sqrt{2}}{.02} = 69.3$. Therefore, $n \geq 69.3^2 = 4802.5$, and the pollster should use 4803 people for each sample.

CONFIDENCE INTERVAL FOR A MEAN

We are interested in estimating a population mean μ by considering a single sample mean $\bar{x}$. This sample mean is just one of a whole universe of sample means, and from Topic 12 we remember that if n is sufficiently large,

1. the set of all sample means is approximately normally distributed.
2. the mean of the set of sample means equals μ, the mean of the population.
3. the standard deviation $\sigma_{\bar{x}}$ of the set of sample means is approximately equal to $\frac{\sigma}{\sqrt{n}}$, that is, equal to the standard deviation of the whole population divided by the square root of the sample size.

Frequently we do not know σ, the population standard deviation. In such cases, we must use s, the *standard deviation of the sample*, as an estimate of σ. In this case $\frac{s}{\sqrt{n}}$ is called the *standard error*, $SE_{\bar{x}}$, and is used as an estimate for $\sigma_{\bar{x}} = \frac{\sigma}{\sqrt{n}}$. (Note that we use *t*-distributions instead of the standard normal curve whenever σ is unknown, no matter what the sample size, and the population must be assumed approximately normal.)

Remember that in making calculations and drawing conclusions from a specific sample, it is important that the sample be a *simple random sample* and be no more than 10% of the population.

EXAMPLE 13.10

A bottling machine is operating with a standard deviation of 0.12 ounce. Suppose that in an SRS of 36 bottles the machine inserted an average of 16.1 ounces into each bottle.

a.   Estimate the mean number of ounces in all the bottles this machine fills. More specifically, give an interval within which we are 95% certain that the mean lies.
Answer: For samples of size 36, the sample means are approximately normally distributed with a standard deviation of $\sigma_{\bar{x}} = \frac{\sigma}{\sqrt{n}} = \frac{0.12}{\sqrt{36}} = 0.02$. From Topic 11 we know that 95% of the sample means should be within 1.96 standard deviations of the population mean. Equivalently, we are 95% certain that the population mean is within 1.96 standard deviations of any sample mean. In our case, $16.1 \pm 1.96(0.02)$ = 16.1 ± 0.0392, and we are 95% sure that the mean number of ounces in all bottles is between 16.0608 and 16.1392. This is called a *95% confidence interval estimate*. [On the TI-84, under STAT and then TESTS, go to ZInterval. With Inpt:Stats, σ:.12, $\bar{x}$:16.1, n:36, and C-Level:.95, Calculate gives (16.061, 16.139).]

b.   How about a 99% confidence interval estimate?
Answer: Here, $16.1 \pm 2.576(0.02) = 16.1 \pm 0.0515$, and we are 99% sure that the mean number of ounces in all bottles is between 16.0485 and 16.1515. [On the TI-84, ZInterval gives (16.048, 16.152).] We can also say, using the definition of confidence *level*, that if the sampling procedure were repeated many times, about 99% of the resulting confidence intervals would contain the true population mean (thus we're 99% confident that the method worked for the interval we got).

Note that when we wanted a higher certainty (99% instead of 95%), we had to settle for a larger, less specific interval (± 0.0515 instead of ± 0.0392).

EXAMPLE 13.11

At a certain plant, batteries are being produced with a life expectancy that has a variance of 5.76 months squared. Suppose the mean life expectancy in an SRS of 64 batteries is 12.35 months.

a. Find a 90% confidence interval estimate of life expectancy for all the batteries produced at this plant.
Answer: The standard deviation of the population is $\sigma = \sqrt{5.76} = 2.4$, and the standard deviation of the sample means is $\sigma_{\bar{x}} = \frac{\sigma}{\sqrt{n}} = \frac{2.4}{\sqrt{64}} = 0.3$. The 90% confidence interval estimate for the population mean is $12.35 \pm 1.645(0.3) = 12.35 \pm 0.4935$. Thus we are 90% certain that the mean life expectancy of the batteries is between 11.8565 and 12.8435 months. [The TI-84 gives (11.857, 12.843).]

b. What would the 90% confidence interval estimate be if the sample mean of 12.35 had come from a sample of 100 batteries?
Answer: The standard deviation of the sample means would then have been $\sigma_{\bar{x}} = \frac{\sigma}{\sqrt{n}} = \frac{2.4}{\sqrt{100}} = 0.24$, and the 90% confidence interval estimate would be $12.35 \pm 1.645(0.24) = 12.35 \pm 0.3948$. [The TI-84 gives (11.955, 12.745).]

Note that when the sample size increased (from 64 to 100), the same sample mean resulted in a narrower, more specific interval (± 0.3948 versus ± 0.4935).

EXAMPLE 13.12

A new drug results in lowering the heart rate by varying amounts with a standard deviation of 2.49 beats per minute.

a. Find a 95% confidence interval estimate for the mean lowering of the heart rate in all patients if a 50-person SRS averages a drop of 5.32 beats per minute.
Answer: The standard deviation of sample means is $\sigma_{\bar{x}} = \frac{\sigma}{\sqrt{n}} = \frac{2.49}{\sqrt{50}} = 0.352$. We are 95% certain that the mean lowering of the heart rate is in the range $5.32 \pm 1.96(0.352) = 5.32 \pm 0.69$ or between 4.63 and 6.01 heartbeats per minute. [The TI-84 gives (4.6298, 6.0102).]

b. With what certainty can we assert that the new drug lowers the heart rate by a mean of 5.32 ± 0.75 beats per minute?
Answer: Converting ± 0.75 to z-scores yields $\frac{\pm 0.75}{0.352} = \pm 2.13$. From Table A, these z-scores give probabilities of .0166 and .9834, respectively, so our answer is .9834 − .0166 = .9668. In other words, 5.32 ± 0.75 beats per minute is a 96.68% confidence interval estimate of the mean lowering of the heart rate effected by this drug.

Remember, when σ is unknown (which is almost always the case), we use the t-distribution instead of the z-distribution. Furthermore, we must have that the parent population is normal, or at least nearly normal, which we can roughly check using a dotplot, stemplot, boxplot, histogram, or normal probability plot of the sample data. If we are not given that the parent population is normal, then for small samples ($n \leq 15$) the sample data should be unimodal and symmetric with no outliers and no skewness; for medium samples ($15 < n < 40$) the sample data should be unimodal and reasonably symmetric with no extreme values and little, if any, skewness; while for large samples ($n \geq 40$) the t-methods can be used no matter what the sample data show.

EXAMPLE 13.13

When ten cars of a new model were tested for gas mileage, the results showed a mean of 27.2 miles per gallon with a standard deviation of 1.8 miles per gallon. What is a 95% confidence interval estimate for the gas mileage achieved by this model? (Assume that the population of mpg results for all the new model cars is approximately normally distributed.)

Answer: The population standard is unknown, and so we use the *t*-distribution. The standard deviation of the sample means is $\sigma_{\bar{x}} = \frac{1.8}{\sqrt{10}} = 0.569$. With $10 - 1 = 9$ degrees of freedom and 2.5% in each tail, the appropriate *t*-scores are ±2.262. Thus we can be 95% certain that the gas mileage of the new model is in the range $27.2 \pm 2.262(0.569) = 27.2 \pm 1.3$ or between 25.9 and 28.5 miles per gallon.

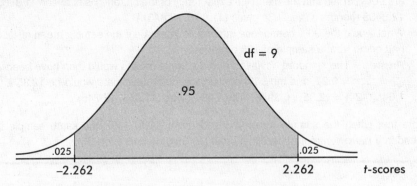

[On the TI-84, go to STAT, then TESTS, then TInterval. Using Stats, put in $\bar{x}$:27.2, *Sx*:1.8, n:10, and C-level:.95. Then Calculate gives (25.912, 28.488).]

EXAMPLE 13.14

A new process for producing synthetic gems yielded six stones weighing 0.43, 0.52, 0.46, 0.49, 0.60, and 0.56 carats, respectively, in its first run. Find a 90% confidence interval estimate for the mean carat weight from this process. (Assume that the population of carats of all gems produced by this new process is approximately normally distributed.)

Answer: Using the statistical package on your calculator or, much more laboriously, by hand, calculate the sample mean $\bar{x}$ and standard deviation *s*, and then find $\sigma_{\bar{x}}$ the standard deviation of the set of sample means:

$$\bar{x} = \frac{\sum x}{n} = \frac{0.43 + 0.52 + 0.46 + 0.49 + 0.60 + 0.56}{6} = \frac{3.06}{6} = 0.51$$

$$s = \sqrt{\frac{\sum (x - \bar{x})^2}{n - 1}}$$
$$= \sqrt{\frac{(0.08)^2 + (0.01)^2 + (0.05)^2 + (0.02)^2 + (0.09)^2 + (0.05)^2}{5}}$$
$$= 0.0632$$

and

$$\sigma_{\bar{x}} \approx \frac{s}{\sqrt{n}} = \frac{0.0632}{\sqrt{6}} = 0.0258$$

With df = $6 - 1 = 5$ and 5% in each tail, the *t*-scores are ±2.015. Thus we can be 90% sure that the new process will yield stones weighing $0.51 \pm 2.015(0.0258) = 0.51 \pm 0.052$ or between 0.458 and 0.562 carats.

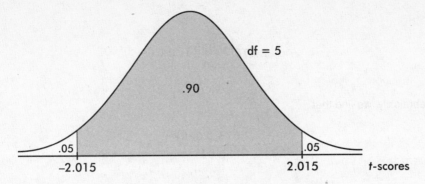

[On the TI-84, put the data in a list and then use Data under TInterval to obtain (.45797, .56203).]

EXAMPLE 13.15

A survey was conducted involving 250 out of 125,000 families living in a city. The average amount of income tax paid per family in the sample was $3540 with a standard deviation of $1150. Establish a 99% confidence interval estimate for the total taxes paid by all the families in the city. (Assume that the population of all income taxes paid is approximately normally distributed.)

Answer: We are given $\bar{x} = 3540$ and $s = 1150$. We use s as an estimate for σ and calculate the standard deviation of the sample means to be $\sigma_{\bar{x}} \approx \frac{s}{\sqrt{n}} = \frac{1150}{\sqrt{250}} = 72.73$.

Since σ is unknown, we use a t-distribution. In this case we obtain $3540 \pm 2.596(72.73) = 3540 \pm 188.8$ and then $125{,}000(3540 \pm 188.8) = 442{,}500{,}000 \pm 23{,}600{,}000$. That is, we are 99% confident that the total tax paid by all families in the city is between $418,900,000 and $466,100,000.

Statistical principles are useful not only in analyzing data but also in setting up experiments, and one important consideration is the choice of a sample size. In making interval estimates of population means, we have seen that each inference must go hand in hand with an associated confidence level statement. Generally, if we want a smaller, more precise interval estimate, we either decrease the degree of confidence or increase the sample size. Similarly, if we want to increase the degree of confidence, we either accept a wider interval estimate or increase the sample size. Thus, choosing a larger sample size always seems desirable; in the real world, however, time and cost considerations are involved.

EXAMPLE 13.16

Ball bearings are manufactured by a process that results in a standard deviation in diameter of 0.025 inch. What sample size should be chosen if we wish to be 99% sure of knowing the diameter to within ±0.01 inch.

Answer: We have

$$\sigma_{\bar{x}} = \frac{\sigma}{\sqrt{n}} = \frac{0.025}{\sqrt{n}} \quad \text{and} \quad 2.576\sigma_{\bar{x}} \leq 0.01$$

(continued)

Thus

$$2.576\left(\frac{0.025}{\sqrt{n}}\right) \leq 0.01$$

Algebraically, we find that

$$\sqrt{n} \geq \frac{2.576(0.025)}{0.01} = 6.44$$

so $n \geq 41.5$. We choose a sample size of 42.
A formula for the above calculation is

$$n = \left(\frac{z\sigma}{\text{error}}\right)^2$$

(Note: this formula is not given on the AP exam formula page.)

CONFIDENCE INTERVAL FOR A DIFFERENCE BETWEEN TWO MEANS

We have the following information about the sampling distribution of $\bar{x}_1 - \bar{x}_2$:

1. The set of all differences of sample means is approximately normally distributed.
2. The mean of the set of differences of sample means equals $\mu_1 - \mu_2$, the difference of population means.
3. The standard deviation $\sigma_{\bar{x}_1 - \bar{x}_2}$ of the set of differences of sample means is approximately equal to $\sqrt{\frac{\sigma_1^2}{n_1} + \frac{\sigma_2^2}{n_2}}$.

In making calculations and drawing conclusions from specific samples, it is important both that the samples be *simple random samples* and that they be taken *independently* of each other.

EXAMPLE 13.17

A 30-month study is conducted to determine the difference in the numbers of accidents per month occurring in two departments in an assembly plant. Suppose the first department averages 12.3 accidents per month with a standard deviation of 3.5, while the second averages 7.6 accidents with a standard deviation of 3.4. Determine a 95% confidence interval estimate for the difference in the numbers of accidents per month. (Assume that the two populations are independent and approximately normally distributed.)
Answer:

$$n_1 = 30 \qquad n_2 = 30$$
$$\bar{x}_1 = 12.3 \qquad \bar{x}_2 = 7.6$$
$$s_1 = 3.5 \qquad s_2 = 3.4$$

(continued)

$$\sigma_{\bar{x}_1 - \bar{x}_2} = \sqrt{\frac{\sigma_1^2}{n_1} + \frac{\sigma_2^2}{n_2}} \approx \sqrt{\frac{s_1^2}{n_1} + \frac{s_2^2}{n_2}} = \sqrt{\frac{(3.5)^2}{30} + \frac{(3.4)^2}{30}} = 0.89$$

The observed difference is 12.3 − 7.6 = 4.7, and with df = $(n_1 - 1) + (n_2 - 1)$ = 58, the critical *t*-scores are ±2.00. Thus the confidence interval estimate is 4.7 ± 2.00(0.89) = 4.7 ± 1.78. We are 95% confident that the first department has between 2.92 and 6.48 more accidents per month than the second department. (Using the conservative *df* = min($x_1 - 1$, $x_2 - 1$) = 29 gives 4.7 ± 2.045(0.89) = 4.7 ± 1.82.)

[On the TI-84, 2-SampTInt gives (2.9167, 6.4833).]

EXAMPLE 13.18

A survey is run to determine the difference in the cost of groceries in suburban stores versus inner city stores. A preselected group of items is purchased in a sample of 45 suburban and 35 inner city stores, and the following data are obtained.

Suburban stores	Inner city stores
$n_1 = 45$	$n_2 = 35$
$\bar{x}_1 = \$36.52$	$\bar{x}_2 = \$39.40$
$s_1 = \$1.10$	$s_2 = \$1.23$

Find a 90% confidence interval estimate for the difference in the cost of groceries. (Assume that the two populations are independent and approximately normally distributed.)
Answer:

$$\sigma_{\bar{x}_1 - \bar{x}_2} \approx \sqrt{\frac{(1.10)^2}{45} + \frac{(1.23)^2}{35}} = 0.265$$

The observed difference is 36.52 − 39.40 = −2.88, the critical *t*-scores are ±1.664, and the confidence interval estimate is −2.88 ± 1.664(0.265) = −2.88 ± 0.44. Thus we are 90% certain that the selected group of items costs between $2.44 and $3.32 *less* in suburban stores than in inner city stores.

[On the TI-84, 2-SampTInt gives (−3.321, −2.439).]

EXAMPLE 13.19

A hardware store owner wishes to determine the difference between the drying times of two brands of paint. Suppose the standard deviation between cans in each population is 2.5 minutes. How large a sample (same number) of each must the store owner use if he wishes to be 98% sure of knowing the difference to within 1 minute?
Answer:

$$\sigma_{\bar{x}_1 - \bar{x}_2} = \sqrt{\frac{(2.5)^2}{n} + \frac{(2.5)^2}{n}} = \frac{3.536}{\sqrt{n}}$$

With a critical *z*-score of 2.326, we have $2.326\left(\frac{3.536}{\sqrt{n}}\right) \le 1$, so $\sqrt{n} \ge 8.22$ and $n \ge 67.6$. Thus the owner should test samples of 68 paint patches from each brand.

In the case of independent samples, there is also a procedure based on the assumption that both original populations not only are normally distributed but also have equal variances. Then we can get a better estimate of the common population variance by *pooling* the two sample variances. But it's never wrong not to pool.

The analysis and procedure discussed in this section require that the two samples being compared be independent of each other. However, many experiments and tests involve comparing two populations for which the data naturally occur in pairs. In this case of *paired differences*, the proper procedure is to run a one-sample analysis on the single variable consisting of the differences from the paired data.

EXAMPLE 13.20

An SAT preparation class of 30 students produces the following improvement in scores:

Student	First Score	Second Score	Improvement
1	912	1025	113
2	1025	1085	60
3	1295	1350	55
4	1123	1202	79
5	875	982	107
6	890	950	60
7	1002	1089	87
8	998	1159	161
9	1235	1246	11
10	1045	1135	90
11	956	1005	49
12	987	1010	23
13	1028	1015	-13
14	954	1032	78
15	1152	1310	158
16	1215	1302	87
17	948	1010	62
18	1190	1235	45
19	1077	1103	26
20	1223	1200	-23
21	1100	1187	87
22	842	910	68
23	985	1049	64
24	1107	1123	16
25	847	901	54
26	987	1086	99
27	1228	1276	48
28	1005	1029	24
29	1166	1221	55
30	808	874	66

Find a 90% confidence interval estimate of the average improvement in test scores.

(continued)

Answer: The two sets of test scores (before and after the course) are not independent; they are associated in matched pairs, with one pair of scores for each student. The proper procedure is to run a one-sample analysis on the single variable consisting of the set of improvements. In this case we obtain $\bar{x} = 63.2$, $s = 41.89$, $\sigma_{\bar{x}} \approx \frac{41.89}{\sqrt{30}} = 7.65$, and a 90% confidence interval estimate of $63.2 \pm 1.699(7.648) = 63.2 \pm 13.0$. We can be 90% confident that the average improvement in test scores is between 50.2 and 76.2.

When to do a two-sample analysis versus a matched pair analysis can be confusing. The key idea is whether or not the two groups are independent. The difference has to do with design. What is the average reaction time of people who are sober compared to that of people who have had two beers? If our experimental design calls for randomly assigning half of a group of volunteers to drink two beers and then comparing reaction times with the remaining sober volunteers, then a two-sample analysis is called for. However, if our experimental design calls for testing reaction times of all the volunteers and then giving all the volunteers two beers and retesting, then a matched pair analysis is called for. Matching may come about as above because you have made two measurements on the same person, or measurements might be made on sets of twins, or between salaries of president and provost at a number of universities, etc. The key is whether the two sets of measurements are independent, or related in some way relevant to the question under consideration.

CONFIDENCE INTERVAL FOR THE SLOPE OF A LEAST SQUARES REGRESSION LINE

In Topic 4 we discussed the least squares line

$$\hat{y} = \bar{y} + b_1(x - \bar{x})$$

where b_1 is the slope of the line. This slope is readily found using the statistical software on a calculator. It is an estimate of the slope β of the true regression line. A confidence interval estimate for β can be found using t-scores, where df $= n - 2$, and the appropriate standard deviation is

$$s_{b_1} = \frac{\sqrt{\frac{\Sigma(y_i - \hat{y}_i)^2}{n-2}}}{\sqrt{\Sigma(x_i - \bar{x})^2}}$$

where the sum in the numerator is the sum of the squared residuals (see Topic 4) and the sum in the denominator is the sum of the squared deviations from the mean. The value of s_{b_1} is found at the same time that the statistical software on your calculator finds the value of b_1. Some calculators calculate the numerator $s = \sqrt{\frac{\Sigma(y_i - \hat{y}_i)^2}{n-2}}$, and then the denominator can be calculated by $\sqrt{\Sigma(x_i - \bar{x})^2} = s_x\sqrt{n - 1}$.

Assumptions here include: (1) the sample must be randomly selected; (2) the scatterplot should be approximately linear; (3) there should be no apparent pattern in the residuals plot (On the TI-84, after Stat $\rightarrow$ Calc $\rightarrow$ LinReg, the list of residuals is stored in RESID under LIST NAMES); and (4) the distribution of the residuals should be approximately normal. (The fourth point can be checked by a histogram, dotplot, stemplot, or a normal probability plot of the residuals.)

EXAMPLE 13.21

A random sample of ten high school students produced the following results for number of hours of television watched per week and GPA.

TV hours	12	21	8	20	16	16	24	0	11	18
GPA	3.1	2.3	3.5	2.5	3.0	2.6	2.1	3.8	2.9	2.6

Determine the least squares line and give a 95% confidence interval estimate for the true slope.

Answer: Checking the assumptions:

We are told that the data come from a *random* sample of students.

Scatterplot is approximately linear.　　No apparent pattern is evident in residuals plot.

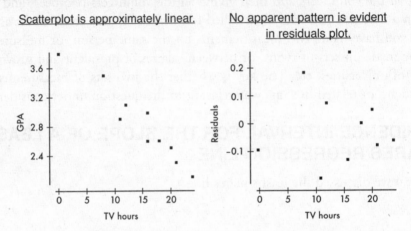

Distribution of residuals is approximately normal.

Checked by histogram.　　　　　Checked by normal probability plot.

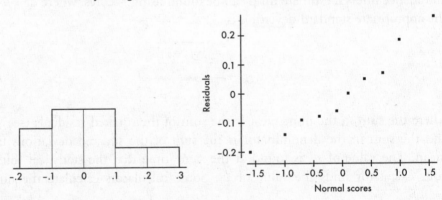

Now using the statistics software on a calculator gives

$$\hat{y} = 3.892 - 0.07202x$$

Using, for example, the TI-84 (LinRegTTest and 1-Var Stats) one also obtains

$$s = 0.1531 \quad \text{and} \quad s_x = 7.074$$

Thus

$$s_{b_1} = \frac{0.1531}{7.074\sqrt{10-1}} = 0.0072$$

[Note: On the TI-84, LinRegTTest gives both b_1 and t, and you can also calculate

$s_{b_1} = \dfrac{b_1}{t} = \dfrac{-0.07202}{-9.984} = 0.0072$, which comes from $t = \dfrac{b_1}{s_{b_1}}$.]

With ten data points, the degrees of freedom are $df = 10 - 2 = 8$, and the critical t-values (with .025 in each tail) are ± 2.306. The 95% confidence interval of the true slope is

$$b_1 \pm t\, s_{b_1} = -0.07202 \pm 2.306(0.0072) = -0.07202 \pm 0.0166$$

We are thus 95% confident that each additional hour before the television each week is associated with between a 0.055 and a 0.089 drop in GPA. (A TI-84 with LinRegTInt will give this result more directly.) Remember we must be careful about drawing conclusions concerning cause and effect; it is possible that students who have lower GPAs would have these averages no matter how much television they watched per week.

Note that $s = \sqrt{\dfrac{\sum(y-\hat{y})^2}{n-2}}$, called the *standard deviation of the residuals*, is sometimes written s_e. It is a measure of the spread about the regression line, and is almost always in the regression output (usually simply labeled as s or S). Note that S_{b_1} is sometimes written $\text{SE}(b_1)$ and is also referred to as the *standard error of the slope*. So we can also write $\text{SE}(b_1) = \dfrac{s_e}{s_x\sqrt{n-1}}$. We see that increasing s_e (that is, increasing spread about the line) *increases* the slope's standard error, while increasing s_x (that is, increasing the range of x-values) or increasing n (the sample size) *decreases* the slope's standard error.

EXAMPLE 13.22

Information concerning SAT verbal scores and SAT math scores was collected from 15 randomly selected subjects. A linear regression performed on the data using a statistical software package produced the following printout:

```
Dependent variable: Math
Variable          Coef          SE Coef          T          Prob
Constant          92.5724        31.75           2.92        0.012
Verbal            0.763604       0.05597         13.6        0.000
S = 16.69     R-Sq = 93.5%     R-Sq(adj) = 93.0%
```

(continued)

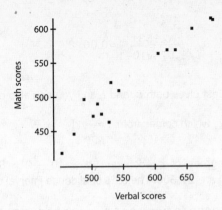

What is the regression equation?

Answer. Assuming that all assumptions for regression are met (we are given that the sample is random, and the scatterplot is approximately linear), the *y*-intercept and slope of the equation are found in the `Coef` column of the above printout.

$$\text{Math} = 92.57 + 0.764 \text{ Verbal}$$

What is a 95% confidence interval estimate for the slope of the regression line?

Answer. The standard deviation of the residuals is $S = 16.69$ and the standard error of the slope is $S_{b_1} = 0.05597$. With 15 data points, $df = 15 - 2 = 13$, and the critical *t*-values are ± 2.160. The 95% confidence interval of the true slope is:

$$b_1 \pm t\, s_{b_1} = 0.764 \pm 2.160(0.05597) = 0.764 \pm 0.121$$

We are 95% confident that for every 1-point increase in verbal SAT score, the average increase in math SAT score is between .64 and .89

Questions on Topic Thirteen: Confidence Intervals

Multiple-Choice Questions

Directions: The questions or incomplete statements that follow are each followed by five suggested answers or completions. Choose the response that best answers the question or completes the statement.

1. Changing from a 95% confidence interval estimate for a population proportion to a 99% confidence interval estimate, with all other things being equal,

   (A) increases the interval size by 4%.
   (B) decreases the interval size by 4%.
   (C) increases the interval size by 31%.
   (D) decreases the interval size by 31%.
   (E) This question cannot be answered without knowing the sample size.

2. In general, how does doubling the sample size change the confidence interval size?

   (A) Doubles the interval size
   (B) Halves the interval size
   (C) Multiplies the interval size by 1.414
   (D) Divides the interval size by 1.414
   (E) This question cannot be answered without knowing the sample size.

3. A confidence interval estimate is determined from the GPAs of a simple random sample of n students. All other things being equal, which of the following will result in a smaller margin of error?

   I. A smaller confidence level
   II. A smaller sample standard deviation
   III. A smaller sample size

   (A) I and II
   (B) I and III
   (C) II and III
   (D) I, II, and III
   (E) None of the above gives the complete set of true responses.

4. A survey was conducted to determine the percentage of high school students who planned to go to college. The results were stated as 82% with a margin of error of ±5%. What is meant by ±5%?

   (A)   Five percent of the population were not surveyed.
   (B)   In the sample, the percentage of students who plan to go to college was between 77% and 87%.
   (C)   The percentage of the entire population of students who plan to go to college is between 77% and 87%.
   (D)   It is unlikely that the given sample proportion result would be obtained unless the true percentage was between 77% and 87%.
   (E)   Between 77% and 87% of the population were surveyed.

5. One month the actual unemployment rate in France was 13.4%. If during that month you took an SRS of 100 Frenchmen and constructed a confidence interval estimate of the unemployment rate, which of the following would have been true?

   I. The center of the interval was 13.4.
   II. The interval contained 13.4.
   III. A 99% confidence interval estimate contained 13.4.

   (A) I and II
   (B) I and III
   (C) II and III
   (D) I, II, and III
   (E) None of the above gives the complete set of true responses.

6. The margin of error in a confidence interval estimate using z-scores covers which of the following?

   I. Random sampling errors
   II. Errors due to undercoverage and nonresponse in obtaining sample surveys
   III. Errors due to using sample standard deviations as estimates for population standard deviations

   (A) I only
   (B) II only
   (C) III only
   (D) I and III
   (E) None of the above gives the complete set of true responses.

7. A *USA Today* "Lifeline" column reported that in a survey of 500 people, 39% said they watch their bread while it's being toasted. Establish a 90% confidence interval estimate for the percentage of people who watch their bread being toasted.

   (A) 39% ± .078%
   (B) 39% ± 2.2%
   (C) 39% ± 2.8%
   (D) 39% ± 3.6%
   (E) 39% ± 4.3%

8. In a survey funded by Burroughs-Welcome, 750 of 1000 adult Americans said they didn't believe they could come down with a sexually transmitted disease (STD). Construct a 95% confidence interval estimate of the proportion of adult Americans who don't believe they can contract an STD.

   (A) (.728, .772)
   (B) (.723, .777)
   (C) (.718, .782)
   (D) (.713, .787)
   (E) (.665, .835)

9. A 1993 *Los Angeles Times* poll of 1703 adults revealed that only 17% thought the media was doing a "very good" job. With what degree of confidence can the newspaper say that 17% ± 2% of adults believe the media is doing a "very good" job?

   (A) 72.9%
   (B) 90.0%
   (C) 95.0%
   (D) 97.2%
   (E) 98.6%

10. A politician wants to know what percentage of the voters support her position on the issue of forced busing for integration. What size voter sample should be obtained to determine with 90% confidence the support level to within 4%?

    (A) 21
    (B) 25
    (C) 423
    (D) 600
    (E) 1691

11. In a 1994 *New York Times* poll measuring a candidate's popularity, the newspaper claimed that in 19 of 20 cases its poll results should be no more than three percentage points off in either direction. What confidence level are the pollsters working with, and what size sample should they have obtained?

    (A) 3%, 20
    (B) 6%, 20
    (C) 6%, 100
    (D) 95%, 33
    (E) 95%, 1068

12. In a test for acid rain, an SRS of 49 water samples showed a mean pH level of 4.4 with a standard deviation of 0.35. Find a 90% confidence interval estimate for the mean pH level.

    (A) 4.4 ± 0.01
    (B) 4.4 ± 0.08
    (C) 4.4 ± 0.32
    (D) 4.4 ± 0.35
    (E) 4.4 ± 0.58

13. One gallon of gasoline is put in each of 30 test autos, and the resulting mileage figures are tabulated with $\bar{x} = 28.5$ and $s = 1.2$. Determine a 95% confidence interval estimate of the mean mileage.

    (A) (28.46, 28.54)
    (B) (28.42, 28.58)
    (C) (28.1, 28.9)
    (D) (27.36, 29.64)
    (E) (27.3, 29.7)

14. The number of accidents per day at a large factory is noted for each of 64 days with $\bar{x} = 3.58$ and $s = 1.52$. With what degree of confidence can we assert that the mean number of accidents per day at the factory is between 3.20 and 3.96?

    (A) 48%
    (B) 63%
    (C) 90%
    (D) 95%
    (E) 99%

15. A company owns 335 trucks. For an SRS of 30 of these trucks, the average yearly road tax paid is $9540 with a standard deviation of $1205. What is a 99% confidence interval estimate for the total yearly road taxes paid for the 335 trucks?

    (A) $9540 ± $103
    (B) $9540 ± $567
    (C) $3,196,000 ± $606
    (D) $3,196,000 ± $35,000
    (E) $3,196,000 ± $203,000

16. What sample size should be chosen to find the mean number of absences per month for school children to within ±.2 at a 95% confidence level if it is known that the standard deviation is 1.1?

    (A) 11
    (B) 29
    (C) 82
    (D) 96
    (E) 117

17. Hospital administrators wish to learn the average length of stay of all surgical patients. A statistician determines that, for a 95% confidence level estimate of the average length of stay to within ±0.5 days, 50 surgical patients' records will have to be examined. How many records should be looked at to obtain a 95% confidence level estimate to within ±0.25 days?

    (A) 25
    (B) 50
    (C) 100
    (D) 150
    (E) 200

18. The National Research Council of the Philippines reported that 210 of 361 members in biology are women, but only 34 of 86 members in mathematics are women (*Science*, March 11, 1994, page 1491). Establish a 95% confidence interval estimate of the difference in proportions of women in biology and women in mathematics in the Philippines.

    (A) .187 ± .115
    (B) .187 ± .154
    (C) .395 ± .103
    (D) .543 ± .154
    (E) .582 ± .051

19. In a simple random sample of 300 elderly men, 65% were married, while in an independent simple random sample of 400 elderly women, 48% were married. Determine a 99% confidence interval estimate for the difference between the percentages of elderly men and women who are married.

    (A) 17% ± 0.36%
    (B) 17% ± 9.6%
    (C) 55% ± 6.7%
    (D) 56.5% ± 6.7%
    (E) 56.5% ± 9.6%

20. A researcher plans to investigate the difference between the proportion of psychiatrists and the proportion of psychologists who believe that most emotional problems have their root causes in childhood. How large a sample should be taken (same number for each group) to be 90% certain of knowing the difference to within ±.03?

    (A) 39
    (B) 376
    (C) 752
    (D) 1504
    (E) 3007

21. In a study aimed at reducing developmental problems in low-birth-weight (under 2500 grams) babies (*Journal of the American Medical Association*, June 13, 1990, page 3040), 347 infants were exposed to a special educational curriculum while 561 did not receive any special help. After 3 years the children exposed to the special curriculum showed a mean IQ of 93.5 with a standard deviation of 19.1; the other children had a mean IQ of 84.5 with a standard deviation of 19.9. Find a 95% confidence interval estimate for the difference in mean IQs of low-birth-weight babies who receive special intervention and those who do not.

    (A) 9.0 ± 2.60
    (B) 9.0 ± 4.42
    (C) 9.0 ± 6.24
    (D) 89.0 ± 19.5
    (E) 89.0 ± 39.0

22. Does socioeconomic status relate to age at time of HIV infection? For 274 high-income HIV-positive individuals the average age of infection was 33.0 years with a standard deviation of 6.3, while for 90 low-income individuals the average age was 28.6 years with a standard deviation of 6.3 (*The Lancet*, October 22, 1994, page 1121). Find a 90% confidence interval estimate for the difference in ages of high- and low-income people at the time of HIV infection.

    (A) 4.4 ± 0.963
    (B) 4.4 ± 1.26
    (C) 4.4 ± 2.51
    (D) 30.8 ± 2.51
    (E) 30.8 ± 6.3

23. An engineer wishes to determine the difference in life expectancies of two brands of batteries. Suppose the standard deviation of each brand is 4.5 hours. How large a sample (same number) of each type of battery should be taken if the engineer wishes to be 90% certain of knowing the difference in life expectancies to within 1 hour?

    (A) 10
    (B) 55
    (C) 110
    (D) 156
    (E) 202

24. Two confidence interval estimates from the same sample are (16.4, 29.8) and (14.3, 31.9). What is the sample mean, and if one estimate is at the 95% level while the other is at the 99% level, which is which?

    (A) $\bar{x} = 23.1$; (16.4, 29.8) is the 95% level.
    (B) $\bar{x} = 23.1$; (16.4, 29.8) is the 99% level.
    (C) It is impossible to completely answer this question without knowing the sample size.
    (D) It is impossible to completely answer this question without knowing the sample standard deviation.
    (E) It is impossible to completely answer this question without knowing both the sample size and standard deviation.

25. Two 90% confidence interval estimates are obtained: I (28.5, 34.5) and II (30.3, 38.2).

    a. If the sample sizes are the same, which has the larger standard deviation?
    b. If the sample standard deviations are the same, which has the larger size?

    (A) *a.* I   *b.* I
    (B) *a.* I   *b.* II
    (C) *a.* II   *b.* I
    (D) *a.* II   *b.* II
    (E) More information is needed to answer these questions.

26. Suppose (25, 30) is a 90% confidence interval estimate for a population mean μ. Which of the following are true statements?

    I. There is a .90 probability that μ is between 25 and 30.

    II. If 100 random samples of the given size are picked and a 90% confidence interval estimate is calculated from each, then μ will be in 90 of the resulting intervals.

    III. If 90% confidence intervals are calculated from all possible samples of the given size, μ will be in 90% of these intervals.

    (A) I and II
    (B) I and III
    (C) II and III
    (D) I, II, and III
    (E) None of the above gives the complete set of true responses.

27. Under what conditions would it be meaningful to construct a confidence interval estimate when the data consist of the entire population?

    (A) If the population size is small ($n < 30$)
    (B) If the population size is large ($n \geq 30$)
    (C) If a higher level of confidence is desired
    (D) If the population is truly random
    (E) Never

28. A social scientist wishes to determine the difference between the percentage of Los Angeles marriages and the percentage of New York marriages that end in divorce in the first year. How large a sample (same for each group) should be taken to estimate the difference to within ±.07 at the 94% confidence level?

    (A) 181
    (B) 361
    (C) 722
    (D) 1083
    (E) 1443

29. A sample of 57 laboratory rabbits fed a special carrot juice diet for a month showed an average weight loss of 242 grams with a standard deviation of 39 grams. What is a 95% confidence interval estimate for the average weight loss?

    (A) 242 ± 1.3
    (B) 242 ± 5.2
    (C) 242 ± 8.5
    (D) 242 ± 10.3
    (E) 242 ± 76.4

30. What is the critical t-value for finding a 90% confidence interval estimate from a sample of 15 observations?

    (A) 1.341
    (B) 1.345
    (C) 1.350
    (D) 1.753
    (E) 1.761

31. In a sleep laboratory experiment, 16 volunteers sleep an average of 7.4 hours with a standard deviation of 1.3 hours. What is a 99% confidence interval estimate for the mean number of hours that people sleep at night? Assume the population has a normal distribution.

    (A) 7.4 ± 0.24
    (B) 7.4 ± 0.84
    (C) 7.4 ± 0.85
    (D) 7.4 ± 0.96
    (E) 7.4 ± 3.83

32. Acute renal graft rejection can occur years after the graft. In one study (*The Lancet*, December 24, 1994, page 1737), 21 patients showed such late acute rejection when the ages of their grafts (in years) were 9, 2, 7, 1, 4, 7, 9, 6, 2, 3, 7, 6, 2, 3, 1, 2, 3, 1, 1, 2, and 7, respectively. Establish a 90% confidence interval estimate for the ages of renal grafts that undergo late acute rejection.

    (A) 2.024 ± 0.799
    (B) 2.024 ± 1.725
    (C) 4.048 ± 0.799
    (D) 4.048 ± 1.041
    (E) 4.048 ± 1.725

33. Nine subjects, 87 to 96 years old, were given 8 weeks of progressive resistance weight training (*Journal of the American Medical Association*, June 13, 1990, page 3032). Strength before and after training for each individual was measured as maximum weight (in kilograms) lifted by left knee extension:

Before:	3	3.5	4	6	7	8	8.5	12.5	15
After:	7	17	19	12	19	22	28	20	28

    Find a 95% confidence interval estimate for the strength gain.

    (A) 11.61 ± 3.03
    (B) 11.61 ± 3.69
    (C) 11.61 ± 3.76
    (D) 19.11 ± 1.25
    (E) 19.11 ± 3.69

34. Six capsules of drug A took an average of 75 seconds with a standard deviation of 1.4 seconds to dissolve, while the average time for six capsules of drug B was 71 seconds with a standard deviation of 1.7 seconds. Establish a 99% confidence interval estimate for the difference in dissolving time between the two brands (assuming we have normal distributions).

    (A) 4 ± 1.55
    (B) 4 ± 2.48
    (C) 4 ± 2.56
    (D) 4 ± 2.85
    (E) 4 ± 3.10

35. A simple random sample of 20 batteries are tested and show a mean life expectancy of 218 hours with a standard deviation of 11 hours. Determine a 90% confidence interval estimate for this mean life expectancy.

    (A) 218 ± 0.95
    (B) 218 ± 3.27
    (C) 218 ± 4.05
    (D) 218 ± 4.25
    (E) 218 ± 19.02

36. A catch of five fish of a certain species yielded the following ounces of protein per pound of fish: 3.1, 3.5, 3.2, 2.8, and 3.4. What is a 90% confidence interval estimate for ounces of protein per pound of this species of fish?

    (A) 3.2 ± 0.202
    (B) 3.2 ± 0.247
    (C) 3.2 ± 0.261
    (D) 4.0 ± 0.202
    (E) 4.0 ± 0.247

Answer Key

1. **C**	7. **D**	13. **C**	19. **B**	25. **C**	31. **D**
2. **D**	8. **B**	14. **D**	20. **D**	26. **E**	32. **D**
3. **A**	9. **D**	15. **E**	21. **A**	27. **E**	33. **C**
4. **D**	10. **C**	16. **E**	22. **B**	28. **B**	34. **D**
5. **E**	11. **E**	17. **E**	23. **C**	29. **D**	35. **D**
6. **A**	12. **B**	18. **A**	24. **A**	30. **E**	36. **C**

Answers Explained

1. **(C)** The critical z-scores will go from ± 1.96 to ± 2.576, resulting in an increase in the interval size: $\frac{2.576}{1.96} = 1.31$ or an increase of 31%.

2. **(D)** Increasing the sample size by a multiple of d divides the interval estimate by $\sqrt{d}$.

3. **(A)** The margin of error varies directly with the critical z-value and directly with the standard deviation of the sample, but inversely with the square root of the sample size.

4. **(D)** Although the sample proportion *is* between 77% and 87% (more specifically, it is 82%), this is not the meaning of $\pm 5\%$. Although the percentage of the entire population is likely to be between 77% and 87%, this is not known for certain.

5. **(E)** There is no guarantee that 13.4 is anywhere near the interval, so none of the statements are true.

6. **(A)** The margin of error has to do with measuring chance variation but has nothing to do with faulty survey design. As long as *n* is large, *s* is a reasonable estimate of σ; however, again this is not measured by the margin of error. (With *t*-scores, there is a correction for using *s* as an estimate of σ.)

7. **(D)** $\sigma_{\hat{p}} = \sqrt{\frac{(.39)(.61)}{500}} = .0218$

   $.39 \pm 1.645(.0218) = .39 \pm .036$

8. **(B)** $\sigma_{\hat{p}} = \sqrt{\frac{(.75)(.25)}{1000}} = .0137$

   $.75 \pm 1.96(.0137) = .75 \pm .027$ or $(.723, .777)$

9. **(D)** $\sigma_{\hat{p}} = \sqrt{\frac{(.17)(.83)}{1703}} = .0091$

   $z(.0091) = .02 \quad z = 2.20, \quad .9861 - .0139 = 97.2\%$

10. **(C)** $1.645\left(\frac{.5}{\sqrt{n}}\right) \le .04, \sqrt{n} \ge 20.563, n \ge 422.8$, so choose $n = 423$.

11. **(E)** $\frac{19}{20} = 95\%; 1.96\left(\frac{.5}{\sqrt{n}}\right) \le .03, \sqrt{n} \ge 32.67, n \ge 1067.1$, and so the pollsters should have obtained a sample size of at least 1068. (They actually interviewed 1148 people.)

12. **(B)** Using *t*-scores: $4.4 \pm 1.677\left(\frac{0.35}{\sqrt{49}}\right) = 4.4 \pm 0.084$.

13. **(C)** Using *t*-scores: $28.5 \pm 2.045\left(\frac{1.2}{\sqrt{30}}\right) = 28.5 + 0.45$.

14. **(D)** $\sigma_{\bar{x}} = \frac{1.52}{\sqrt{64}} = 0.19, \frac{0.38}{0.19} = 2$, and $.4772 + .4772 = .9544 \approx 95\%$.

15. **(E)** Using *t*-scores:

$335\left[9540 \pm 2.756\left(\frac{1205}{\sqrt{30}}\right)\right] = 335(9540 \pm 606.3) = \$3,196,000 \pm \$203,000.$

16. **(E)** $1.96\left(\frac{1.1}{\sqrt{n}}\right) \le 0.2, \sqrt{n} \ge 10.78$, and $n \ge 116.2$; choose $n = 117$.

17. **(E)** To divide the interval estimate by *d* without affecting the confidence level, multiply the sample size by a multiple of d^2. In this case, $4(50) = 200$.

18. **(A)**

$$n_1 = 361 \qquad n_2 = 86$$
$$\hat{p}_1 = \frac{210}{361} = .582 \quad \hat{p}_2 = \frac{34}{86} = .395$$
$$\sigma_d = \sqrt{\frac{(.582)(.418)}{361} + \frac{(.395)(.605)}{86}} = .0588$$
$$(.582 - .395) \pm 1.96(0.0588) = .187 \pm .115$$

19. **(B)**

$$n_1 = 300 \quad n_2 = 400$$
$$\hat{p}_1 = .65 \quad \hat{p}_2 = .48$$

$$\sigma_d = \sqrt{\frac{(.65)(.35)}{300} + \frac{(.48)(.52)}{400}} = .0372$$

$$(.65 - .48) \pm 2.576(.0372) = .17 \pm .096$$

20. **(D)** $1.645\left(\frac{.5\sqrt{2}}{\sqrt{n}}\right) \leq .03$, $\sqrt{n} \geq 38.77$, and $n \geq 1503.3$; the researcher should choose a sample size of at least 1504.

21. **(A)** $\sigma_{\bar{x}_1 - \bar{x}_2} = \sqrt{\frac{(19.1)^2}{347} + \frac{(19.9)^2}{561}} = 1.326$

$$(93.5 - 84.5) \pm 1.96(1.326) = 9.0 \pm 2.60$$

22. **(B)** $\sigma_{\bar{x}_1 - \bar{x}_2} = \sqrt{\frac{(6.3)^2}{274} + \frac{(6.3)^2}{90}} = 0.765$

$$(33.0 - 28.6) \pm 1.645(0.765) = 4.4 \pm 1.26$$

23. **(C)** $\sigma_{\bar{x}_1 - \bar{x}_2} = \sqrt{\frac{(4.5)^2}{n} + \frac{(4.5)^2}{n}} = \frac{6.364}{\sqrt{n}}$

$$1.645\left(\frac{6.364}{\sqrt{n}}\right) \leq 1, \sqrt{n} \geq 10.47, n \geq 109.6; \text{choose } n = 110.$$

24. **(A)** The sample mean is at the center of the confidence interval; the lower confidence level corresponds to the narrower interval.

25. **(C)** Narrower intervals result from smaller standard deviations and from larger sample sizes.

26. **(E)** Only III is true. The 90% refers to the method; 90% of all intervals obtained by this method will capture μ. Nothing is sure about any particular set of 100 intervals. For any particular interval, the probability that it captures μ is either 1 or 0 depending on whether μ is or isn't in it.

27. **(E)** In determining confidence intervals, one uses sample statistics to estimate population parameters. If the data are actually the whole population, making an estimate has no meaning.

28. **(B)** $\frac{1.88(.5)\sqrt{2}}{\sqrt{n}} \leq .07$ gives $\sqrt{n} \geq 18.99$ and $n \geq 360.7$.

29. **(D)** Using a t-distribution, a calculator like the TI-83 gives 242 ± 10.35.

30. **(E)** With df = 15 − 1 = 14 and .05 in each tail, the critical *t*-value is 1.761.

31. **(D)** df = 15, and $7.4 \pm 2.947\left(\frac{1.3}{\sqrt{16}}\right) = 7.4 \pm 0.96$.

32. **(D)** $\bar{x} = 4.048$, $s = 2.765$, df = 20, and

$$4.048 \pm 1.725\left(\frac{2.765}{\sqrt{21}}\right) = 4.048 \pm 1.041.$$

33. **(C)** The confidence interval estimate of the set of nine differences is

$$11.61 \pm 2.306\left(\frac{4.891}{\sqrt{9}}\right) = 11.61 \pm 3.76.$$

34. **(D)** $\sigma_{\bar{x}_1 - \bar{x}_2} = \sqrt{\frac{(1.4)^2}{6} + \frac{(1.7)^2}{6}} = 0.899$

with df = 6 + 6 − 2 = 10, $(75 - 71) \pm 3.169(0.899) = 4 \pm 2.85$.

35. **(D)** df = 19, and $218 \pm 1.729\left(\frac{11}{\sqrt{20}}\right) = 218 \pm 4.25$.

36. **(C)** df = 4, and $3.2 \pm 2.132\left(\frac{0.274}{\sqrt{5}}\right) = 3.2 \pm 0.261$.

Free-Response Questions

> ***Directions:*** You must show all work and indicate the methods you use. You will be graded on the correctness of your methods and on the accuracy of your final answers.

Eight Open-Ended Questions

1. An SRS of 1000 voters finds that 57% believe that competence is more important than character in voting for President of the United States.

   a. Determine a 95% confidence interval estimate for the percentage of voters who believe competence is more important than character.
   b. If your parents know nothing about statistics, how would you explain to them why you can't simply say that 57% of voters believe that competence is more important.
   c. Also explain to your parents what is meant by 95% confidence level.

2. Pollsters sometimes simply give a percentage result such as "40% of the voters support Candidate Smith," while at other times they express results in a form such as "40% plus or minus 5%." Comment on the missing pieces of information in such statements and on what a pollster might add so that readers can better understand the strength of such statements?

3. An SRS of 40 inner city gas stations shows a mean price for regular unleaded gasoline to be $1.45 with a standard deviation of $0.05, while an SRS of 120 suburban stations shows a mean of $1.38 with a standard deviation of $0.08.

   a. Construct 95% confidence interval estimates for the mean price of regular gas in inner city and in suburban stations.

   b. The confidence interval for the inner city stations is wider than the interval for the suburban stations even though the standard deviation for inner city stations is less than that for suburban stations. Explain why this happened.

   c. Based on your answer in part *a*, are you confident that the mean price of inner city gasoline is less than $1.50? Explain.

4. In a simple random sample of 30 subway cars during rush hour, the average number of riders per car was 83.5 with a standard deviation of 5.9. Assume the sample data are unimodal and reasonably symmetric with no extreme values and little, if any, skewness.

   (a) Establish a 90% confidence interval estimate for the average number of riders per car during rush hour. Show your work.

   (b) Assuming the same standard deviation of 5.9, how large a sample of cars would be necessary to determine the average number of riders to within ±1 at the 90% confidence level? Show your work.

5. In a sample of ten basketball players the mean income was $196,000 with a standard deviation of $315,000.

   (a) Assuming all necessary assumptions are met, find a 95% confidence interval estimate of the mean salary of basketball players.

   (b) What assumptions are necessary for the above estimate? Do they seem reasonable here?

6. An SRS of ten brands of breakfast cereals is tested for the number of calories per serving. The following data result: 185, 190, 195, 200, 205, 205, 210, 210, 225, 230.

   Establish a 95% confidence interval estimate for the mean number of calories for servings of breakfast cereals. Be sure to check assumptions.

7. A new drug is tested for relief of allergy symptoms. In a double-blind experiment, 12 patients are given varying doses of the drug and report back on the number of hours of relief. The following table summarizes the results:

Dosage (mg)	2	3	4	4	5	6	6	7	8	8	9	10
Duration of relief (hrs)	3	7	6	8	10	8	13	16	15	21	23	24

   a. Determine the equation of the least squares regression line.
   b. Construct a 90% confidence interval estimate for the slope of the regression line, and interpret this in context.

8. Information with regard to the assessed values (in $1000) and the selling prices (in $1000) of a random sample of homes sold in an NE market yields the following computer output:

   Dependent variable is   **Price**

   R squared = 89.8%   R squared (adjusted) = 89.2%
   s = 9.508   with   20 – 2 = 18   degrees of freedom

Source	SS	df	MS	F-ratio
Regression	14271.2	1	14271.2	158
Residual	1627.31	18	90.4063	

Variable	Coeff	s.e. of coeff	t-ratio	prob
Constant	0.890087	16.16	0.055	0.956
Assessed	1.0292	0.08192	12.6	0.000

   a. Determine the equation of the least squares regression line.
   b. Construct a 99% confidence interval estimate for the slope of the regression line, and interpret this in context.

Answers Explained

1. *a.* $\hat{p} = .57$ and $\sigma_{\hat{p}} \approx \sqrt{\frac{(.57)(.43)}{1000}} = .0157$. [Note that $np = 1000(.57) = 570$ and $nq = 1000(.43) = 430$ are both greater than 10, we are given an SRS, and $1000 < 10\%$ of all voters.] The critical z-scores associated with the 95% level are ± 1.96. Thus the confidence interval estimate is $.57 \pm 1.96(.0157) = .57 \pm .031$, or between 54% and 60%. (We are 95% confident that between 54% and 60% of the voters believe competence is more important than character.)
   b. Explain to your parents that by using a measurement from a sample we are never able to say *exactly* what a population proportion is; rather we are only able to say we are confident that it is within some range of values, in this case between 54% and 60%.
   c. In 95% of all possible samples of 1000 voters, the method used gives an estimate that is within three percentage points of the true answer.

2. The first statement is close to meaningless. The reader has no idea whether 1000 voters were interviewed and 400 expressed support for the candidate or whether only 5 people were polled with 2 of them expressing support. The second statement is better; however, it is still missing a critical piece of information, the confidence level. The reader does not know if the calculation

resulting in the ±5% figure is associated with 99%, 90%, or even 50% confidence that the resulting interval contains the actual support percentage. While the general public might not understand the term *confidence level,* they should be given something like: "In 9 out of 10 cases, our poll results are no more than five percentage points off in either direction from the true figure."

3. *a.* Checking conditions: We are given random samples and must assume that the parent populations are nearly normal. Then a calculator like the TI-84 gives $\bar{x} \pm t\frac{s}{\sqrt{n}} = 1.45 \pm .016$ for inner city stations and

$\bar{x} \pm t\frac{s}{\sqrt{n}} = 1.38 \pm .015$ for suburban stations. That is, we are 95% confident

that the mean price for gas in inner city stations is between \$1.434 and \$1.466, and are 95% confident that the mean price for gas in suburban stations is between \$1.365 and \$1.395.

b. Because the sample size of inner city stations is smaller.

c. Yes, because the standard deviation of the set of sample means is $\frac{.05}{\sqrt{140}} =$ .0079, and so \$1.50 is more than three standard deviations away from \$1.45. Thus the probability that the true mean is this far from the sample mean is extremely small.

4. (a) Checking conditions: We are given a simple random sample and must assume that the numbers of riders per car during rush hour are normally distributed. Then $\bar{x} \pm t\frac{s}{\sqrt{n}} = 83.5 \pm 1.699\left(\frac{5.9}{\sqrt{30}}\right) = 83.5 \pm 1.83$. That is, we are 90% confident that the average number of riders per car during rush hour is between 81.67 and 85.33.

(b) $1.645\left(\frac{5.9}{\sqrt{n}}\right) \leq 1$ gives $\sqrt{n} \geq 9.7055$, and $n \geq 94.2$, so you must choose a random sample of at least 95 subway cars.

5. (a) With df = 9, $\bar{x} \pm t\frac{s}{\sqrt{n}} = 196,000 \pm 2.262\frac{315,000}{\sqrt{10}} = 196,000 \pm$ 225,000.

(b) We must assume the sample is an SRS and that basketball salaries are normally distributed. This does *not* seem reasonable—the salaries are probably strongly skewed to the right by a few high ones.

6. We are given that we have a random sample, and the nearly normal condition seems reasonable from, for example, either a stemplot or a normal probability plot:

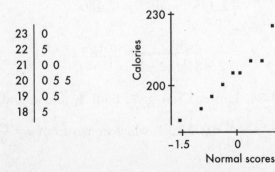

Under these conditions the mean calories can be modeled by a *t*-distribution with $n - 1 = 10 - 1 = 9$ degrees of freedom.

We calculate $\bar{x} = 205.5$ and $s = 14.23$, and look up $t = 2.262$.

A one-sample *t*-interval for the mean gives:

$$\bar{x} \pm t \frac{s}{\sqrt{n}} = 205.5 \pm 2.262 \frac{14.23}{\sqrt{10}} = 205.5 \pm 10.2$$

We are 95% confident that the true mean number of calories of all breakfast cereals is between 195.3 and 215.7.

7. We first check conditions: we must assume use of randomization. The scatterplot is roughly linear, and there is no apparent pattern in the residuals plot.

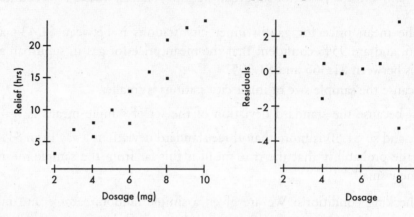

The distribution of the residuals is very roughly normal.

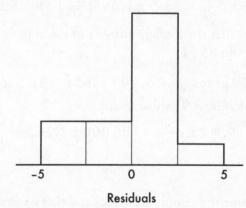

a. Now using the statistics software on a calculator gives

$$Hours = -3.225 + 2.676\, Dosage$$

b. Using, for example, the TI-84 (LinRegTTest and 1-Var Stats) one also obtains

$$s = 2.335 \quad \text{and} \quad s_x = 2.486$$

Thus

$$s_{b_1} = \frac{2.335}{2.486\sqrt{12-1}} = 0.2832$$

[Note: On the TI-84, LinRegTTest gives both b_1 and t, and you can also calculate $s_{b_1} = \frac{b_1}{t} = \frac{2.676}{9.450} = 0.2832$, which comes from $t = \frac{b_1}{s_{b_1}}$.]

With 12 data points, the degrees of freedom are df $= 12 - 2 = 10$, and the critical t-values (with .05 in each tail) are ± 1.812. The 99% confidence interval of the true slope is

$$b_1 \pm t\, s_{b_1} = 2.676 \pm 1.812(0.2832) = 2.676 \pm 0.513$$

We are thus 99% confident that each additional gram of the new drug is associated with an average of between 2.2 and 3.2 more hours of allergy relief. (A TI-84+ with `LinRegTInt` will give this result more directly.) We should be careful about extrapolating; note how the scatterplot of Relief versus Dosage seems to be leveling off at the higher end.

8. *a.* Assuming that all assumptions for regression are met, the y-intercept and slope of the equation are found in the `Coeff` column of the computer printout.

$$Selling\ price = 0.890 + 1.029\ (Assessed\ value)$$

where both the selling price and the assessed value are in $1000.

b. From the printout, the standard error of the slope is $s_{b_1} = 0.08192$. With df $= 18$ and .005 in each tail, the critical t-values are ± 2.878. The 99% confidence interval of the true slope is:

$$b_1 \pm t\, s_{b_1} = 1.029 \pm 2.878(0.08192) = 1.029 \pm 0.236$$

We are 99% confident that for every $1 increase in assessed value, the average increase in selling price is between $0.79 and $1.27 (or for every $1000 increase in assessed value, the average increase in selling price is between $790 and $1270).

Tests of Significance— Proportions and Means

- Logic of Significance Testing
- Null and Alternative Hypotheses
- *P*-values
- Type I and Type II Errors
- Concept of Power
- Hypothesis Testing

Closely related to the problem of estimating a population proportion or mean is the problem of testing a hypothesis about a population proportion or mean. For example, a travel agency might determine an interval estimate for the proportion of sunny days in the Virgin Islands or, alternatively, might test a tourist bureau's claim about the proportion of sunny days. A major stockholder in a construction company might ascertain an interval estimate for the proportion of successful contract bids or, alternatively, might test a company spokesperson's claim about the proportion of successful bids. A consumer protection agency might determine an interval estimate for the mean nicotine content of a particular brand of cigarettes or, alternatively, might test a manufacturer's claim about the mean nicotine content of its cigarettes. An agricultural researcher could find an interval estimate for the mean productivity gain caused by a specific fertilizer or, alternatively, might test the developer's claimed mean productivity gain. In each of these cases, the experimenter must decide whether the interest lies in an interval estimate of a population proportion or mean or in a hypothesis test of a claimed proportion or mean.

LOGIC OF SIGNIFICANCE TESTING, NULL AND ALTERNATIVE HYPOTHESIS, *P*-VALUES, ONE- AND TWO-SIDED TESTS, TYPE I AND TYPE II ERRORS, AND THE CONCEPT OF POWER

The general testing procedure is to choose a specific hypothesis to be tested, called the *null hypothesis*, pick an appropriate random sample, and then use measurements from the sample to determine the likelihood of the null hypothesis. If the sample statistic is far enough away from the claimed population parameter, we say that there is sufficient evidence to reject the null hypothesis. We attempt to show that the null hypothesis is unacceptable by showing that it is improbable.

Consider the context of the population proportion. The null hypothesis H_0 is stated in the form of an equality statement about the *population* proportion (for example, H_0: $p = .37$). There is an *alternative hypothesis*, stated in the form of an inequality (for example, H_a: $p < .37$ or H_a: $p > .37$ or H_a: $p \neq .37$). The strength of the sample statistic $\hat{p}$ can be gauged through its associated *P-value*, which is the probability of obtaining a sample statistic as extreme as the one obtained if the null hypothesis is assumed to be true. The smaller the *P*-value, the more significant the difference between the null hypothesis and the sample results.

Note that only population parameter symbols appear in H_0 and H_a. Sample statistics like $\bar{x}$ and $\hat{p}$ do NOT appear in H_0 or H_a.

There are two types of possible errors: the error of mistakenly rejecting a true null hypothesis and the error of mistakenly failing to reject a false null hypothesis. The α-risk, also called the *significance level* of the test, is the probability of committing a *Type I error* and mistakenly rejecting a true null hypothesis. A *Type II error*, a mistaken failure to reject a false null hypothesis, has associated probability β. There is a different value of β for each possible correct value for the population parameter p. For each β, $1 - \beta$ is called the power of the test against the associated correct value. The *power* of a hypothesis test is the probability that a Type II error is not committed. That is, given a true alternative, the power is the probability of rejecting the false null hypothesis. Increasing the sample size or increasing the significance level are both ways of increasing the power.

A simple illustration of the difference between a Type I and a Type II error is as follows: Suppose the null hypothesis is that all systems are operating satisfactory with regard to a NASA liftoff. A Type I error would be to delay the liftoff mistakenly thinking that something was malfunctioning when everything was actually OK. A Type II error would be to fail to delay the liftoff mistakenly thinking everything was OK when something was actually malfunctioning. The power is the probability of recognizing a particular malfunction. (Note the complimentary aspect of power, a "good" thing, with Type II error, a "bad" thing.)

Our justice system provides another often quoted illustration. If the null hypothesis is that a person is innocent, then a Type I error results when an innocent person is found guilty, while a Type II error results when a guilty person is not convicted. We try to minimize Type I errors in criminal trials by demanding unanimous jury guilty verdicts. In civil suits, however, many states try to minimize Type II errors by accepting simple majority verdicts.

It should be emphasized that with regard to calculations, questions like "What is the *power* of this test?" and "What is the probability of a *Type II error* in this test?" cannot be answered without reference to a specific alternative hypothesis. Furthermore, AP students are not required to know how to calculate these probabilities. However, they are required to understand the concepts and interactions among the concepts.

HYPOTHESIS TEST FOR A PROPORTION

We assume that we have a simple random sample, that both np_0 and $n(1 - p_0)$ are at least 10, and that the sample size is less than 10% of the population.

EXAMPLE 14.1

A union spokesperson claims that 75% of union members will support a strike if their basic demands are not met. A company negotiator believes the true percentage is lower and runs a hypothesis test at the 10% significance level. What is the conclusion if 87 out of an SRS of 125 union members say they will strike?

Answer: We note that $np_0 = (125)(.75) = 93.75$ and $n(1 - p_0) = (125)(.25) = 31.25$ are both >10 and that we have an SRS, and we assume that 125 is less than 10% of the total union membership.

$$H_0: \ p = .75$$
$$H_a: \ p < .75$$

We use the claimed proportion to calculate the standard deviation of the sample proportions:

$$\sigma_p = \sqrt{\frac{(.75)(.25)}{125}} = .03873$$

The observed sample proportion is $\hat{p} = \frac{87}{125} = .696$.

To measure the strength of the disagreement between the sample proportion and the claimed proportion, we calculate the *P*-value (also called the *attained significance level*). The z-score for .696 is $\frac{.696 - .75}{.03873} = -1.39$, with a resulting *P*-value of .0823. [On the TI-84, normalcdf(−10, −1.39) = .0823 and normalcdf(0, .696, .75, .03873) = .0816.] Since .0823 < .10, there is sufficient evidence to reject H_0 at the 10% significance level. The company negotiator should challenge the union claim at this level. Note, however, that there is not sufficient evidence to reject H_0 at the 5%, or even at the 8%, significance level because .0823 > .05 and .0823 > .08. When .05 < *P* < .10, we usually say that there is *some* evidence against H_0.

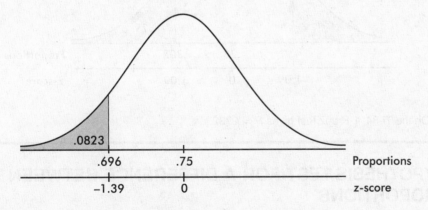

[On the TI-84, under STAT and then TESTS, go to 1-PropZTest. With Po:.75, x:87, n:125, and prop < Po, Calculate gives *P* = .0816.]

EXAMPLE 14.2

A cancer research group surveys 500 women more than 40 years old to test the hypothesis that 28% of women in this age group have regularly scheduled mammograms. Should the hypothesis be rejected at the 5% significance level if 151 of the women respond affirmatively?

Answer: Note that $np_0 = (500)(.28) = 140$ and $n(1 - p_0) = (500)(.72) = 360$ are both >10, we assume an SRS, and clearly 500 < 10% of all women over 40 years old. Since no suspicion is voiced that the 28% claim is low or high, we run a two-sided *z*-test for proportions:

$$H_0: \ p = .28$$
$$H_a: \ p \neq .28$$
$$\alpha = .05$$
$$\sigma_p = \sqrt{\frac{(.28)(.72)}{500}} = .0201$$

The observed $\hat{p} = \frac{151}{500} = .302$. The *z*-score for .302 is $\frac{.302 - .28}{.0201} = 1.09$, which corresponds to a probability of $1 - .8621 = .1379$ in the tail. Doubling this value (because the test is two-sided), we obtain a *P*-value of $2(.1379) = .2758$. Since $.2758 > .05$, there is not sufficient evidence to reject H_0; that is, the cancer research group should not dispute the 28% claim. Note, for example, that the null hypothesis should not be rejected even if α is a relatively large .25. When $P > .10$, we usually say there is little or no evidence to reject H_0.

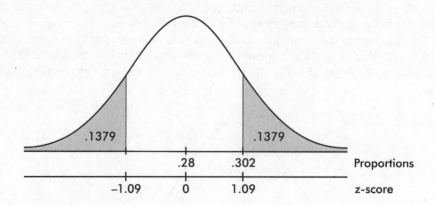

[On the TI-84, 1-PropZTest gives $P = .2732$.]

HYPOTHESIS TEST FOR A DIFFERENCE BETWEEN TWO PROPORTIONS

As with confidence intervals for the difference of two proportions, $n_1\hat{p}_1$, $n_1(1 - \hat{p}_1)$, $n_2\hat{p}_2$, and $n_2(1 - \hat{p}_2)$ should all be at least 10, it is important both that the samples be *simple random samples* and that they be taken *independently* of each other, and the original populations should be large compared to the sample sizes.

The fact that a sample proportion from one population is greater than a sample proportion from a second population does not automatically justify a similar conclusion about the population proportions themselves. Two points need to be stressed. First, sample proportions from the same population can vary from each other. Second, what we are really comparing are confidence interval estimates, not just single points.

For many problems the null hypothesis states that the population proportions are equal or, equivalently, that their difference is 0:

$$H_0: p_1 - p_2 = 0$$

The alternative hypothesis is then

$$H_a: p_1 - p_2 < 0 \quad H_a: p_1 - p_2 > 0 \quad \text{or} \quad H_a: p_1 - p_2 \neq 0$$

where the first two possibilities lead to one-sided tests and the third possibility leads to two-sided tests.

Since the null hypothesis is that $p_1 = p_2$, we call this common value p and use it in calculating σ_d:

$$\sigma_d = \sqrt{\frac{p(1-p)}{n_1} + \frac{p(1-p)}{n_2}} = \sqrt{p(1-p)\left(\frac{1}{n_1} + \frac{1}{n_2}\right)}$$

In practice, if $\hat{p}_1 = \frac{x_1}{n_1}$ and $\hat{p}_2 = \frac{x_2}{n_2}$, we use $\hat{p} = \frac{x_1 + x_2}{n_1 + n_2}$ as an estimate of p in calculating σ_d.

EXAMPLE 14.3

Suppose that early in an election campaign a telephone poll of 800 registered voters shows 460 in favor of a particular candidate. Just before election day, a second poll shows only 520 of 1000 registered voters expressing the same preference. At the 10% significance level is there sufficient evidence that the candidate's popularity has decreased?

Answer: Note that $n_1\hat{p}_1 = 460$, $n_1(1 - \hat{p}_1) = 340$, $n_2\hat{p}_2 = 520$, and $n_2(1 - \hat{p}_2) = 480$ are all at least 10, we assume the samples are independent SRSs, and clearly the total population of registered voters is large.

$$H_0 : p_1 - p_2 = 0 \quad \hat{p}_1 = \frac{460}{800} = .575$$
$$H_a : p_1 - p_2 > 0 \quad \hat{p}_2 = \frac{520}{1000} = .520$$
$$\alpha = .10 \quad \hat{p} = \frac{460+520}{800+1000} = .544$$

$$\sigma_d \approx \sqrt{(.544)(.456)\left(\frac{1}{800} + \frac{1}{1000}\right)} = .0236$$

The observed difference is $.575 - .520 = .055$. The *z*-score for .055 is $\frac{.055-0}{.0236} = 2.33$, and so the *P*-value is $1 - .9901 = .0099$. [normalcdf(2.33, 100) = .00990 and normalcdf(.055, 1, 0, .0236) = .00989.] Since $.0099 < .10$, we conclude that at the 10% significance level the candidate's popularity *has* dropped. We note that $.0099 < .01$, so that the observed difference is statistically significant even at the 1% level.

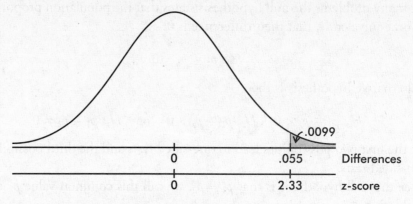

[On the TI-84, 2-PropZTest gives $P = .00995$.]

EXAMPLE 14.4

An automobile manufacturer tries two distinct assembly procedures. In a sample of 350 cars coming off the line using the first procedure there are 28 with major defects, while a sample of 500 autos from the second line shows 32 with defects. Is the difference significant at the 10% significance level?

Answer: Note that $n_1\hat{p}_1 = 28$, $n_1(1 - \hat{p}_1) = 322$, $n_2\hat{p}_2 = 32$, and $n_2(1 - \hat{p}_2) = 468$ are all at least 10, we assume the samples are independent SRSs, and the total population of autos manufactured at the plant is large. Since there is no mention that one procedure is believed to be better or worse than the other, this is a two-sided test.

$$H_0 : p_1 - p_2 = 0 \quad \hat{p}_1 = \frac{28}{350} = .080$$
$$H_a : p_1 - p_2 \neq 0 \quad \hat{p}_2 = \frac{32}{500} = .064$$
$$\alpha = .10 \quad \hat{p} = \frac{28+32}{350+500} = .0706$$

$$\sigma_d \approx \sqrt{(.0706)(.9294)\left(\frac{1}{350} + \frac{1}{500}\right)} = .0179$$

The observed difference is $.080 - .064 = .016$. The *z*-score for $.016$ is $\frac{.016-0}{.0179} = 0.89$, which corresponds to a tail probability of $1 - .8133 = .1867$. [normalcdf(.89, 100) = .1867 and normalcdf(.016, 1, 0, .0179) = .1857.] Doubling this value because the test is two-sided results in a *P*-value of $2(.1867) = .3734$. Since $.3734 > .10$, we conclude that the observed difference is *not* significant at the 10% level. We note that the smallest significance level for which the observed difference is significant is more than 37.34%. When *P* is so large, we can safely say that there is no evidence against H_0. That is, the observed difference in proportions of cars with major defects coming from the two assembly procedures is not significant.

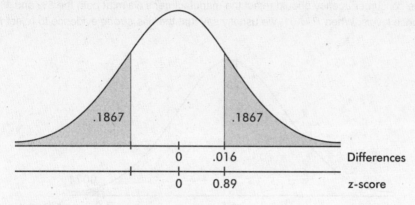

.1867   .1867

| | 0 | .016 | Differences |
| | 0 | 0.89 | z-score |

[On the TI-84, 2-PropZTest gives $P = .3701$.]

HYPOTHESIS TEST FOR A MEAN

In the following examples we assume that we have simple random samples from approximately normally distributed populations. (If we are given the sample data we check the normality assumption with a graph such as a histogram or a normal probability plot; the histogram should be unimodal and reasonably symmetric, although this condition can be relaxed the larger the sample size.)

EXAMPLE 14.5

A manufacturer claims that a new brand of air-conditioning unit uses only 6.5 kilowatts of electricity per day. A consumer agency believes the true figure is higher and runs a test on a sample of size 50. If the sample mean is 7.0 kilowatts with a standard deviation of 1.4, should the manufacturer's claim be rejected at a significance level of 5%? Of 1%?

Answer: Assuming an SRS (the normality assumption is less important because $n \geq 40$) we have:

$$H_0: \quad \mu = 6.5$$
$$H_a: \quad \mu > 6.5$$
$$\alpha = .05 \text{ or } .01$$

Here $\sigma_{\bar{x}} = \frac{\sigma}{\sqrt{n}} \approx \frac{s}{\sqrt{n}} = \frac{1.4}{\sqrt{50}} = 0.198$ The t-score of 7.0 is $\frac{7.0-6.5}{0.198} = 2.53$. The critical t-values for the 5% and 1% tests are 1.676 and 2.403, respectively, and we have both $2.53 > 1.676$ and $2.53 > 2.403$. [On the TI-84, STAT → TESTS → T-Test, then Inpt:Stats, μ_0:6.5, $\bar{x}$:7, Sx:1.4, n:50, μ:>μ_0, Calculate yields $t = 2.525$ and $P = .0074$.]

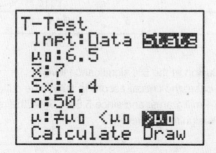

The consumer agency should reject the manufacturer's claim at both the 5% and 1% significance levels. When $P < .01$, we usually say that there is *strong* evidence to reject H_0.

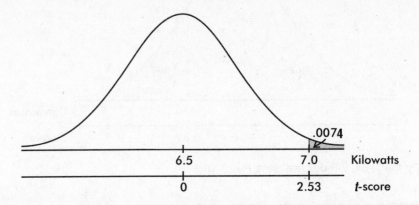

EXAMPLE 14.6

A cigarette industry spokesperson remarks that current levels of tar are no more than 5 milligrams per cigarette. A reporter does a quick check on 15 cigarettes representing a cross section of the market.

a.. What conclusion is reached if the sample mean is 5.63 milligrams of tar with a standard deviation of 1.61? Assume a 10% significance level.

Answer: Assuming an SRS from an approximately normally distributed population, we have $H_0: \mu = 5$, $H_a: \mu > 5$, $\alpha = .10$, and

$$\sigma_{\bar{x}} \approx \frac{s}{\sqrt{n}} = \frac{1.61}{\sqrt{15}} = 0.42$$

With df $= 15 - 1 = 14$ and $\alpha = .10$, the critical *t*-score is 1.345. The critical number of milligrams of tar is $5 + 1.345(0.42) = 5.56$. Since $5.63 > 5.56$, the industry spokesperson's remarks should be rejected at the 10% significance level.

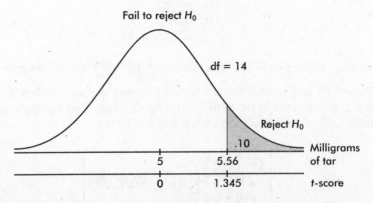

b. What is the conclusion at the 5% significance level?

Answer: In this case, the critical *t*-score is 1.761, the critical tar level is $5 + 1.761(0.42) = 5.74$ milligrams, and since $5.63 < 5.74$, the remarks cannot be rejected at the 5% significance level.

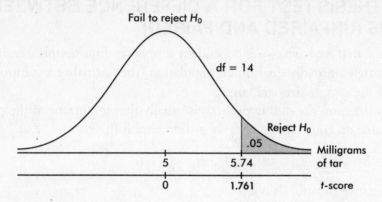

c. What is the *P*-value?

Answer: Without a software program we can't calculate the *P*-value exactly. However, the *t*-score for 5.63 is $\frac{5.63-5}{0.42} = 1.5$, and since 1.5 is between 1.345 and 1.761, we can conclude that *P* is between .05 and .10. [On the TI-84, use STAT and then TESTS; then T-Test results in *P* = .0759. Note that .05 < .0759, while .0759 < .10.]

EXAMPLE 14.7

A local chamber of commerce claims that the mean sale price for homes in the city is $90,000. A real estate salesperson notes eight recent sales of $75,000, $102,000, $82,000, $87,000, $77,000, $93,000, $98,000, and $68,000. How strong is the evidence to reject the chamber of commerce claim?

Answer: Assuming an SRS, we check for normality with a histogram, which yields the roughly unimodal, roughly symmetric:

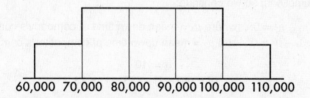

$$H_0: \mu = 90{,}000 \qquad H_a: \mu \neq 90{,}000$$

A calculator gives $\bar{x} = 85{,}250$ and $s = 11{,}877$, and then one calculates $\sigma_{\bar{x}} \approx \frac{s}{\sqrt{n}} = \frac{11877}{\sqrt{8}} = 4199$. The *t*-score of 85,250 is $\frac{85{,}250-90{,}000}{4199} = -1.13$. We note that 1.13 is between 1.119 and 1.415, corresponding to .15 and .10, respectively. Doubling these values because this is a two-sided test, we see that the *P*-value is between .20 and .30. With such a high *P*, we conclude that there is no evidence to reject the chamber of commerce claim. [The TI-84 gives *P* = .2953.]

HYPOTHESIS TEST FOR A DIFFERENCE BETWEEN TWO MEANS (UNPAIRED AND PAIRED)

We assume that we have two independent simple random samples, each from an approximately normally distributed population (the normality assumptions can be relaxed if the sample sizes are large).

In this situation the null hypothesis is usually that the means of the populations are the same or, equivalently, that their difference is 0:

$$H_0: \mu_1 - \mu_2 = 0$$

The alternative hypothesis is then

$$H_a: \mu_1 - \mu_2 < 0 \quad H_a: \mu_1 - \mu_2 > 0 \quad \text{or} \quad H_a: \mu_1 - \mu_2 \neq 0$$

The first two possibilities lead to one-sided tests, and the third possibility leads to two-sided tests.

EXAMPLE 14.8

A sales representative believes that his company's computer has more average downtime per week than a similar computer sold by a competitor. Before taking this concern to his director, the sales representative gathers data and runs a hypothesis test. He determines that in a simple random sample of 40 week-long periods at different firms using his company's product, the average downtime was 125 minutes with a standard deviation of 37 minutes. However, 35 week-long periods involving the competitor's computer yield an average downtime of only 115 minutes with a standard deviation of 43 minutes. What conclusion should the sales representative draw assuming a 10% significance level?

Answer: We are given independent SRSs and large enough sample sizes to relax the normality assumptions, so we continue:

$$H_0: \mu_1 - \mu_2 = 0 \quad \text{(where } \mu_1 = \text{mean down time of company's computers}$$
$$H_a: \mu_1 - \mu_2 > 0 \quad \text{and } \mu_2 = \text{mean down time of competitor's computers)}$$

$$\alpha = .10$$
$$df = (40 - 1) + (35 - 1) = 73$$

(or the more conservative df = min(40–1, 35–1) = 34)

$$\sigma_{\bar{x}_1 - \bar{x}_2} \approx \sqrt{\frac{(37)^2}{40} + \frac{(43)^2}{35}} = 9.33$$

The difference in sample means is 125 − 115 = 10, and the *t*-score is $\frac{10}{9.33} = 1.07$. Using Table B, we find the *P*-value to be between .10 and .15. Thus, the observed difference is not significant at the 10% significance level. The sales representative does not have sufficient evidence that his company's computer has more downtime. While the observed difference would be significant at, for example, the 15% level, still, with *P* > .1, we say that the evidence is weak that the salesman's company's computer has more downtime than that of the competitor's.

Remark: On the TI-84, STAT → TESTS → 2-SampTTest, then Inpt : Stats, $\bar{x}1$: 125, Sx1 : 37, n1 : 40, $\bar{x}2$: 115, Sx2 : 43, n2 : 35, $\mu1$: > $\mu2$ and Calculate yields *t* = 1.0718 and *P* = .1438.

EXAMPLE 14.9

A store manager wishes to determine whether there is a significant difference between two trucking firms with regard to the handling of egg cartons. In a simple random sample of 200 cartons on one firm's truck there was an average of 0.7 broken eggs per carton with a standard deviation of 0.31, while a sample of 300 cartons on the second firm's truck showed an average of 0.775 broken eggs per carton with a standard deviation of 0.42. Is the difference between the averages significant at a significance level of 5%? At a significance level of 1%?

Answer: We are given independent SRSs and large enough sample sizes to relax the normality assumptions, so we continue:

$H_0: \mu_1 - \mu_2 = 0$ (where μ_1 = mean number of broken eggs per carton
$H_a: \mu_1 - \mu_2 \neq 0$ for the first firm and μ_2 = mean number of broken eggs per carton for
                  the second firm)

$$\alpha = .05$$
$$df = (200 - 1) + (300 - 1) = 498$$

(or the more conservative df = min(200 − 1, 300 − 1) = 199)

$$\sigma_{\bar{x}_1 - \bar{x}_2} \approx \sqrt{\frac{(0.31)^2}{200} + \frac{(0.42)^2}{300}} = 0.0327$$

Since there is no mention that it is believed that one firm does better than the other, this is a two-sided test. The observed difference of 0.7 − 0.775 = −0.075 has a *t*-score of $\frac{-0.075}{0.0327} = -2.29$. Using Table B, we find that this gives a probability between .01 and .02. The test is two-sided, and so the *P*-value is between .02 and .04. So the observed difference in average number of broken eggs per carton is statistically significant at the 5% level but not at the 1% level. With .01 < *P* < .05, we usually say that there is *moderate* evidence to reject H_0. [On the TI-84, 2-SampTTest gives *P* = .0222.]

EXAMPLE 14.10

A city council member claims that male and female officers wait equal times for promotion in the police department. A women's spokesperson, however, believes women must wait longer than men. If five men waited 8, 7, 10, 5, and 7 years, respectively, for promotion while four women waited 9, 5, 12, and 8 years, respectively, what conclusion should be drawn?

Answer: We must assume that we have two independent SRSs, each from an approximately normally distributed population (the samples are too small for histograms to show anything, but at least a quick calculation indicates no outliers and normal probability plots are fairly linear).

We have $H_0: \mu_1 - \mu_2 = 0$ and $H_a: \mu_1 - \mu_2 < 0$. Using a calculator we find $\bar{x}_1 = 7.4$, $s_1 = 1.82$, $\bar{x}_2 = 8.5$, and $s_2 = 2.89$. Then

$$\sigma_{\bar{x}_1 - \bar{x}_2} = \sqrt{\frac{(1.82)^2}{5} + \frac{(2.89)^2}{4}} = 1.66$$

The *t*-score of the observed difference is $\frac{7.4 - 8.5}{1.66} = -0.66$. With df = 5 + 4 − 2 = 7, we note that 0.66 < 0.711, where 0.711 corresponds to a tail probability of .25. Thus the *P*-value is greater than .25, and we conclude there is no evidence to dispute the council member's claim. (Using the conservative df = min(5−1, 4−1) = 3 leads to the same conclusion.) [On the TI-84, inserting the data into lists and using 2-SampTTest with Data gives *P* = .2685.]

The analysis and procedure described above require that the two samples being compared be independent of each other. However, many experiments and tests involve comparing two populations for which the data naturally occur in pairs. In this case, the proper procedure is to run a one-sample test on a single variable consisting of the differences from the paired data.

EXAMPLE 14.11

Does a particular drug slow reaction times? If so, the government might require a warning label concerning driving a car while taking the medication. An SRS of 30 people are tested before and after taking the drug, and their reaction times (in seconds) to a standard testing procedure are noted. The resulting data are as follows:

Person	Before	After	Person	Before	After
1	1.42	1.48	16	1.83	1.75
2	1.87	1.75	17	1.40	1.51
3	1.34	1.31	18	1.75	1.67
4	0.98	1.22	19	1.56	1.72
5	1.51	1.58	20	1.56	1.63
6	1.43	1.57	21	2.03	1.81
7	1.52	1.48	22	1.38	1.48
8	1.61	1.55	23	1.42	1.35
9	1.37	1.54	24	1.69	1.75
10	1.49	1.37	25	1.50	1.39
11	0.95	1.07	26	1.12	1.24
12	1.32	1.35	27	1.38	1.40
13	1.68	1.77	28	1.71	1.65
14	1.44	1.44	29	0.91	1.11
15	1.17	1.27	30	1.59	1.85

We can calculate the mean reaction times before and after, 1.46 seconds and 1.50 seconds, respectively, and ask if this observed rise is significant. If we performed a two-sample test, we would calculate the P-value to be .2708 and would conclude that with such a large P, the observed rise is *not* significant. However, this would not be the proper test or conclusion! The two-sample test works for *independent* sets. However, in this case, there is a clear relationship between the data, in pairs, and this relationship is completely lost in the procedure for the two-sample test. The proper procedure is to form the set of 30 differences, being careful with signs, {−0.06, 0.12, 0.03, −0.24, . . .}, and to perform a single sample hypothesis test as follows:

Name the test:

We are using a *paired t-test*, that is, a single sample hypothesis test on the set of differences.

State the hypothesis:

H_o: The reaction times of individuals to a standard testing procedure are the same before and after they take a particular drug; the mean difference is zero: $\mu_d = 0$.

H_a: The reaction time is greater after they take the drug; the mean difference is less than zero: $\mu_d < 0$.

Check the conditions:

1. The *data are paired* because they are measurements on the same individuals before and after taking the drug.
2. The reaction times of any individual are independent of the reaction times of the others, so the *differences are independent*.
3. A *random sample* of people are tested.
4. The histogram of the differences looks *nearly normal* (roughly unimodal and symmetric):

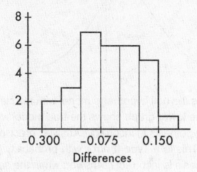

Perform the mechanics:

A calculator readily gives: $n = 30$, $\bar{x} = -0.037667$, and $s = 0.11755$.

Then $\sigma_{\bar{x}} \approx \dfrac{s}{\sqrt{n}} = \dfrac{0.11755}{\sqrt{30}} = 0.021462$ and

$$t = \frac{\bar{x} - 0}{\sigma_{\bar{x}}} = \frac{-0.037667}{0.021462} = -1.755$$

With df $= n - 1 = 29$, the P-value is $P(t < -1.755) = .0449$.

Give a conclusion in context:

The resulting P-value of .0449 is small enough to justify a conclusion (at the 5% significance level) that the observed rise in reaction times after taking the drug is significant.

MORE ON POWER AND TYPE II ERRORS

Given a specific alternative hypothesis, a Type II error is a mistaken failure to reject the false null hypothesis, while the *power* is the probability of rejecting that false null hypothesis.

EXAMPLE 14.12

A candidate claims to have the support of 70% of the people, but you believe that the true figure is lower. You plan to gather an SRS and will reject the 70% claim if in your sample shows 65% or less support. What if in reality only 63% of the people support the candidate?

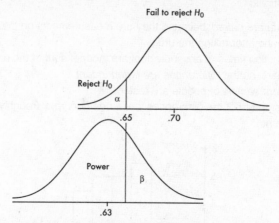

The upper graph shows the null hypothesis model with the claim that $p_0 = 70$, and the plan to reject H_0 if $\hat{p} < .65$. The lower graph shows the true model with $p = .63$. When will we fail to pick up that the null hypothesis is incorrect? Answer: precisely when the sample proportion is greater than .65. This is a Type II error with probability β. When will we rightly conclude that the null hypothesis is incorrect? Answer: when the sample proportion is less than .65. This is the power of the test and has probability $1-\beta$.

The following points should be emphasized:

- Power gives the probability of avoiding a Type II error.

- Power has a different value for different possible correct values of the population parameter; thus it is actually a *function* where the independent variable ranges over specific alternative hypotheses.

- Choosing a smaller α (that is, a tougher standard to reject H_0) results in a higher risk of Type II error and a lower power—observe in the above graphs how making α smaller (in this case moving the critical cutoff value to the left) makes the power less and β more!

- The greater the difference between the null hypothesis p_0 and the true value p, the smaller the risk of a Type II error and the greater the power—observe in the above picture how moving the lower graph to the left makes the power greater and β less. (The difference between p_0 and p is sometimes called the *effect*—thus the greater the effect, the greater is the power to pick it up.)

- A larger sample size n will reduce the standard deviations, making both graphs narrower resulting in smaller α, smaller β, and larger power!

Questions on Topic Fourteen: Tests of Significance—Proportions and Means

Multiple-Choice Questions

Directions: The questions or incomplete statements that follow are each followed by five suggested answers or completions. Choose the response that best answers the question or completes the statement.

1. Which of the following are true statements?

    I. The *P*-value of a test is the probability of obtaining a result as extreme as the one obtained assuming the null hypothesis is true.

    II. If the *P*-value for a test is .015, the probability that the null hypothesis is true is .015.

    III. When the null hypothesis is rejected, it is because it is not true.

    (A) I only
    (B) II only
    (C) III only
    (D) I and III
    (E) None of the above gives the complete set of true responses.

2. A coffee-dispensing machine is supposed to deliver 8 ounces of liquid into each paper cup, but a consumer believes that the actual amount is less. As a test he plans to obtain a sample of 36 cups of the dispensed liquid and, if the mean content is less than 7.75 ounces, to reject the 8-ounce claim. If the machine operates with a standard deviation of 0.9 ounces, what is the probability that the consumer will mistakenly reject the 8-ounce claim even though the claim is true?

    (A) .0475
    (B) .0950
    (C) .1500
    (D) .3897
    (E) .4525

3. Which of the following are true statements?

    I. Tests of significance (hypothesis tests) are designed to measure the strength of evidence against the null hypothesis.

    II. A well-planned test of significance should result in a statement either that the null hypothesis is true or that it is false.

    III. The alternative hypothesis is one-sided if there is interest in deviations from the null hypothesis in only one direction.

    (A) I and II
    (B) I and III
    (C) II and III
    (D) I, II, and III
    (E) None of the above gives the complete set of true responses.

4. A government statistician claims that the mean monthly rainfall along the Liberian coast is 15.0 inches with a standard deviation of 12.0 inches. A meteorologist plans to test this claim with measurements over 3.5 years (42 months). If she finds a sample mean more than 2.0 inches different from the claimed 15.0 inches, she will reject the government statistician's claim. What is the probability that the meteorologist will mistakenly reject a true claim?

    (A) .0675
    (B) .1401
    (C) .2802
    (D) .4325
    (E) .8650

5. Which of the following are true statements?

    I. The alternative hypothesis is stated in terms of a sample statistic.

    II. A large *P*-value indicates strong evidence against the null hypothesis.

    III. If a sample is large enough, the necessity for it to be a simple random sample is diminished.

    (A) I only
    (B) II only
    (C) III only
    (D) Exactly two of the above statements are true.
    (E) None of the above statements are true.

6. An automotive company executive claims that a mean of 48.3 cars per dealership are being sold each month. A major stockholder believes this claim is high and runs a test by sampling 30 dealerships. What conclusion is reached if the sample mean is 45.4 cars with a standard deviation of 15.4?

    (A) There is sufficient evidence to prove the executive's claim is true.
    (B) There is sufficient evidence to prove the executive's claim is false.
    (C) The stockholder has sufficient evidence to reject the executive's claim.
    (D) The stockholder does not have sufficient evidence to reject the executive's claim.
    (E) There is not sufficient data to reach any conclusion.

7. A pharmaceutical company claims that a medication will produce a desired effect for a mean time of 58.4 minutes. A government researcher runs a hypothesis test of 250 patients and calculates a mean of $\bar{x} = 59.5$ with a standard deviation of $s = 8.3$. In which of the following intervals is the P-value located?

   (A) $P < .01$
   (B) $.01 < P < .02$
   (C) $.02 < P < .05$
   (D) $.05 < P < .10$
   (E) $P > .10$

8. A dentist believes that brand C toothpaste is better than brand P toothpaste. She asks 30 of her patients to use each brand and records the numbers of cavities observed over a 3-year period. She plans to reject any equality claim if the average number of cavities for brand C users is at least 1 fewer than the average for brand P users. If the standard deviation in cavities per person for each brand is 2.3, what is the probability the dentist will mistakenly reject a correct null hypothesis of equality?

   (A) .0087
   (B) .0233
   (C) .0465
   (D) .5940
   (E) There is insufficient information given to calculate this probability.

9. To test which of two fuel additives results in better gas mileage, the average miles per gallon is noted when 40 cars are run for 1 week using the first additive, and then the average miles per gallon is calculated for the same 40 cars when they are run for another week. What is the conclusion at a 10% significance level if a two-sample hypothesis test, H_0: $\mu_1 - \mu_2 = 0$, H_a: $\mu_1 - \mu_2 \neq 0$, results in a P-value of .25?

   (A) The observed difference in miles per gallon is significant.
   (B) The observed difference in miles per gallon is not significant.
   (C) A conclusion is not possible without knowing the mean miles per gallon obtained using each additive.
   (D) A conclusion is not possible without knowing both the mean and the standard deviation resulting from the use of each additive.
   (E) A two-sample hypothesis test should not be used in this example.

10. A building inspector believes that the percentage of new construction with serious code violations may be even greater than the previously claimed 7%. She conducts a hypothesis test on 200 new homes and finds 23 with serious code violations. Is this strong evidence against the .07 claim?

    (A) Yes, because the P-value is .0062.
    (B) Yes, because the P-value is 2.5.
    (C) No, because the P-value is only .0062.
    (D) No, because the P-value is over 2.0.
    (E) No, because the P-value is .045.

11. A survey of 1000 Canadians reveals that 585 believe that there is too much violence on television. In a survey of 1500 Americans, 780 believe that there is too much television violence. To test at the 5% significance level whether or not the data are significant evidence that the proportion of Canadians who believe that there is too much violence on television is not equal to the proportion of Americans who believe that there is too much violence on television, a student sets up the following: H_0: $p = .585$ and H_a: $p \neq .585$, where p is the proportion of Americans who believe there is too much violence on television. Which of the following is a true statement?

(A) The student has set up a correct hypothesis test.
(B) Given the sample sizes, a 1% significance level would be more appropriate.
(C) Given that $\frac{780}{1500} = .52$, H_a: $p < .585$ would be more appropriate.
(D) Given that $\frac{585+780}{1000+1500} = .546$, H_0: $p = .546$ would be more appropriate.
(E) A two-population difference in proportions hypothesis test would be more appropriate.

12. In a one-sided hypothesis test for the mean, in a random sample of size 10 the t-score of the sample mean is 2.79. Is this significant at the 5% level? At the 1% level?

(A) Significant at the 1% level but not at the 5% level
(B) Significant at the 5% level but not at the 1% level
(C) Significant at both the 1% and 5% levels
(D) Significant at neither the 1% nor 5% level
(E) Cannot be determined from the given information

13. An IRS representative claims that the average deduction for medical care is $1250. A taxpayer who believes that the real figure is lower samples 12 families and comes up with a mean of $934 and a standard deviation of $616. Where is the P-value?

(A) Below .01
(B) Between .01 and .025
(C) Between .025 and .05
(D) Between .05 and .10
(E) Over .10

14. A city spokesperson claims that the mean response time for arrival of a fire truck at a fire is 12 minutes. A newspaper reporter suspects that the response time is actually longer and runs a test by examining the records of 64 fire emergency situations. What conclusion is reached if the sample mean is 13.1 minutes with a standard deviation of 6 minutes?

    (A) The *P*-value is less than .001, indicating very strong evidence against the 12-minute claim.
    (B) The *P*-value is .01, indicating strong evidence against the 12-minute claim.
    (C) The *P*-value is .07, indicating some evidence against the 12-minute claim.
    (D) The *P*-value is .18, indicating very little evidence against the 12-minute claim.
    (E) The *P*-value is .43, indicating no evidence against the 12-minute claim.

15. It is believed that using a new fertilizer will result in a yield of 1.6 tons per acre. A botanist carries out a two-tailed test on a field of 64 acres. Determine the *P*-value if the mean yield per acre in the sample is 1.72 tons with a standard deviation of 0.4. What is the conclusion at a level of significance of 10%? 5%? 1%?

    (A) *P* = .008, and so the 1.6-ton claim should be disputed at all three of these levels.
    (B) *P* = .02, and so the 1.6-ton claim should be disputed at the 10% and 5% levels but not at the 1% level.
    (C) *P* = .08, and so the 1.6-ton claim should be disputed at the 10% level but not at the 5% and 1% levels.
    (D) *P* = .02, and so the 1.6-ton claim should be disputed at the 1% level but not at the 10% and 5% levels.
    (E) *P* = .008, and so there is not enough evidence to dispute the 1.6-ton claim at any of the three levels.

16. In an effort to curb certain diseases, especially autoimmune (AIDS), San Francisco has a program whereby drug users can exchange used needles for fresh ones. As reported in the *Journal of the American Medical Association* (January 12, 1994, p. 115), 35% of 5644 intravenous drug users in San Francisco admitted to sharing needles. Is this sufficient evidence to say that the rate of sharing needles has dropped from the pre-needle exchange rate of 66%?

    (A) *P* < .001, so this is very strong evidence that the rate has dropped.
    (B) *P* = .0063, so this is strong evidence of a drop in rate.
    (C) *P* is between .01 and .05, so there is moderate evidence of a drop in rate.
    (D) *P* is between .05 and .10, so there is some evidence of a drop in rate.
    (E) *P* = .31, so there is no real evidence of a drop in rate.

17. A recent study of health service costs for coronary angioplasty versus coronary artery bypass surgery at a London hospital (*The Lancet*, October 1, 1994, page 929) showed an average cost of £6176 with a standard deviation of £329 for 231 angioplasties and an average cost of £8164 with a standard deviation of £264 for 221 bypass surgeries. Is this sufficient evidence to say that the average cost of angioplasty is less than the average cost of bypass surgery?

    (A) $P < .001$, so this is very strong evidence that angioplasty costs less.
    (B) P is between .001 and .01, so this is strong evidence that angioplasty costs less.
    (C) P is between .01 and .05, so this is moderate evidence that angioplasty costs less.
    (D) P is between .05 and .10, so there is some evidence that angioplasty costs less.
    (E) $P > .10$, so there is little evidence that angioplasty costs less.

18. Amos Tversky and Thomas Gilovich, in their study on the "Hot Hand" in basketball (*Chance*, Winter 1989, page 20), found that in a random sample of games, Larry Bird hit a second free throw in 48 of 53 attempts after the first free throw was missed, and hit a second free throw in 251 of 285 attempts after the first free throw was made. Is there sufficient evidence to say that the probability that Bird will make a second free throw is different depending on whether or not he made the first free throw?

    (A) $P < .001$, so this is very strong evidence that the probability that Bird will make a second free throw is different depending on whether he made the first.
    (B) P is between .001 and .01, so this is strong evidence that the probabilities are different.
    (C) P is between .01 and .05, so this is moderate evidence that the probabilities are different.
    (D) P is between .05 and .10, so there is some evidence that the probabilities are different.
    (E) $P > .10$, so there is little or no evidence that the probabilities are different.

19. An auditor remarks that the accounts receivable for a company seem to average about $2000. A quick check of 20 accounts gives a mean of $2250 with a standard deviation of $600. Where is the P-value?

    (A) Below .01
    (B) Between .01 and .025
    (C) Between .025 and .05
    (D) Between .05 and .10
    (E) Over .10

20. A researcher believes a new diet should improve weight gain in laboratory mice. If ten control mice on the old diet gain an average of 4 ounces with a standard deviation of 0.3 ounces, while the average gain for ten mice on the new diet is 4.8 ounces with a standard deviation of 0.2 ounces, where is the *P*-value?

    (A) Below .01
    (B) Between .01 and .025
    (C) Between .025 and .05
    (D) Between .05 and .10
    (E) Over .10

21. Which of the following statements are true?

    I. It is helpful to examine your data before deciding whether to use a one-sided or a two-sided hypothesis test.
    II. If the *P*-value is .05, the probability that the null hypothesis is correct is .05.
    III. The larger the *P*-value, the more evidence there is against the null hypothesis.

    (A) I only
    (B) II only
    (C) III only
    (D) II and III
    (E) None of the above gives the complete set of true responses.

22. A test is run to determine whether there is a difference in miles per gallon between two car models. A simple random sample of 40 autos of the first model and an independent simple random sample of 50 of the second model are available for the test. What is the probability of mistakenly claiming there is a difference when there isn't if the cutoff difference scores are ±0.5? Assume standard deviations of 1.2 and 0.65 miles per gallon, respectively, for the two models.

    (A) .0045
    (B) .0089
    (C) .0178
    (D) .4911
    (E) .9822

23. The 26 contestants in the pentathlon event of the 1992 Olympics ran the 200-meter dash in the following times (in seconds): 25.44, 24.39, 25.66, 23.93, 23.34, 25.01, 24.27, 24.54, 25.44, 24.86, 23.95, 23.31, 24.60, 23.12, 25.29, 24.40, 25.24, 24.43, 23.70, 25.20, 25.09, 24.48, 26.13, 25.28, 24.18, and 23.83 (*Journal of the American Statistical Association*, September 1994, page 1101). In testing the null hypothesis that the mean time required for pentathlon contestants to run this race is 25 seconds against the alternative hypothesis that the mean time required is less than 25 seconds, in which of the following intervals is the *P*-value located?

    (A) $P < .005$
    (B) $.005 < P < .01$
    (C) $.01 < P < .05$
    (D) $.05 < P < .10$
    (E) $.10 < P$

24. A local restaurant owner claims that only 15% of visiting tourists stay for more than 2 days. A chamber of commerce volunteer is sure that the real percentage is higher. He plans to survey 100 tourists and intends to speak up if at least 18 of the tourists stay longer than 2 days. What is the probability of mistakenly rejecting the restaurant owner's claim if it is true?

    (A) .0357
    (B) .0714
    (C) .1428
    (D) .2005
    (E) .4010

25. A geologist claims that a particular rock formation will yield a mean amount of 24 pounds of a chemical per ton of excavation. His company, fearful that the true amount will be less, plans to run a test on a random sample of 50 tons. They will reject the 24 pound claim if the sample mean is less than 22. Suppose the standard deviation between tons is 5.8 pounds. If the true mean is 20 pounds of chemical, what is the probability that the test will result in a failure to reject the incorrect 24 pound claim?

    (A) .0073
    (B) .4927
    (C) .5073
    (D) .8200
    (E) .9927

26. A factory manager claims that the plant's smokestacks spew forth only 350 pounds of pollution per day. A government investigator suspects that the true value is higher and plans a hypothesis test with a critical value of 375 pounds. Suppose the standard deviation in daily pollution poundage is 150 and the true mean is 385 pounds. If the sample size is 100 days, what is the probability that the investigator will mistakenly fail to reject the factory manager's false claim?

    (A) .0475
    (B) .2514
    (C) .7486
    (D) .7514
    (E) .9525

Answer Key

1. **A**	7. **C**	12. **B**	17. **A**	22. **C**
2. **A**	8. **C**	13. **D**	18. **E**	23. **B**
3. **B**	9. **E**	14. **C**	19. **D**	24. **D**
4. **C**	10. **A**	15. **B**	20. **A**	25. **A**
5. **E**	11. **E**	16. **A**	21. **E**	26. **B**
6. **D**				

Answers Explained

1. **(A)** The *P*-value is the smallest value of the significance level α for which the null hypothesis would be rejected; or, equivalently, if the null hypothesis is assumed to be true, the *P*-value of a sample statistic is the probability of obtaining a result as extreme as the one obtained. The *P*-value does not give the probability that the null hypothesis is true. Depending on the sample we happen to choose, we may mistakenly reject a true null hypothesis, in which case we commit a Type I error.

2. **(A)** We have $H_0: \mu = 8$ and $H_a: \mu < 8$. The standard deviation of sample means is $\sigma_{\bar{x}} = \frac{\sigma}{\sqrt{n}} = \frac{0.9}{\sqrt{36}} = 0.15$. The *z*-score for 7.75 is $\frac{7.75-8}{0.15} = -1.67$.

    Using Table A, we obtain $\alpha = .0475$. Thus, if the 8-ounce claim is correct, there is a .0475 probability that the consumer will still obtain a sample mean less than 7.75 and will mistakenly reject the claim.

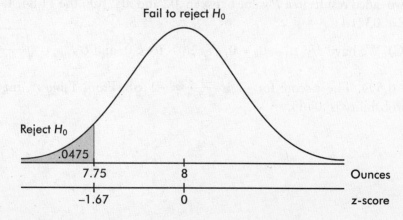

3. **(B)** We attempt to show that the null hypothesis is unacceptable by showing that it is improbable; however, we cannot show that it is definitely true or false.

4. **(C)** We have $H_0: \mu = 15.0$ and $H_a: \mu \neq 15.0$. This is an example of a two-sided test. The standard deviation of the sample means is $\sigma_{\bar{x}} = \frac{12.0}{\sqrt{42}} = 1.85$.

The z-score for 17 is $\frac{17-15}{1.85} = 1.08$; similarly, the z-score for 13 is -1.08. From Table A, we obtain .1401, for a total probability of Type I error (mistakenly rejecting a true null hypothesis) of $.1401 + .1401 = .2802$.

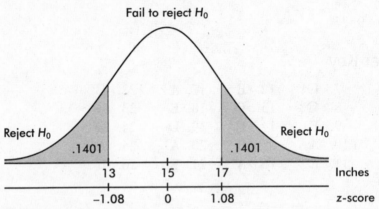

5. **(E)** Both the null and alternative hypotheses are stated in terms of a population parameter, not a sample statistic. Small P-values are evidence against the null hypothesis. These hypothesis tests assume simple random samples.

6. **(D)** We have $H_0: \mu = 48.3$ and $H_a: \mu < 48.3$. Here $\sigma_{\bar{x}} = \frac{\sigma}{\sqrt{n}} \approx \frac{s}{\sqrt{n}} = \frac{15.4}{\sqrt{30}} = 2.81$. The t-score of 45.4 is $\frac{45.4-48.3}{2.81} = -1.03$, and with df $= 30 - 1 = 29$, the P-value is over .15. [On the TI-84, T-Test gives $P = .1554$.] With such a large P-value, there is little evidence against H_0 and the stockholder should not reject the executive's claim.

7. **(C)** We have $H_0: \mu = 58.4$, $H_a: \mu \neq 58.4$, and $\sigma_{\bar{x}} \approx \frac{s}{\sqrt{n}} = \frac{8.3}{\sqrt{250}} = 0.525$. The t-score for 59.5 is $\frac{59.5-58.4}{0.525} = 2.10$. Using Table B, we find the corresponding probability to be between .01 and .02. Doubling this value because the test is two-sided results in a P-value between .02 and .04. [On the TI-84, T-Test gives $P = .0371$.]

8. **(C)** We have $H_0: \mu_1 - \mu_2 = 0$, $H_a: \mu_1 - \mu_2 < 0$, and $\sigma_{\bar{x}_1 - \bar{x}_2} = \sqrt{\frac{(2.3)^2}{30} + \frac{(2.3)^2}{30}}$ $= 0.594$. The z-score for -1 is $\frac{-1-0}{0.594} = -1.68$. From Table A, the resulting probability is .0465.

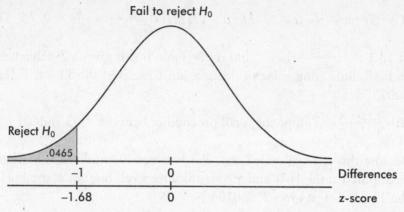

9. **(E)** The two-sample hypothesis test is not the proper one and should be used only when the two sets are independent. In this case, there is a clear relationship between the data, in pairs, and this relationship is completely lost in the procedure for the two-sample test. The proper procedure is to run a one-sample test on the single variable consisting of the differences from the paired data.

10. **(A)** We have H_0: $p = .07$, H_a: $p > .07$, and $\sigma_{\hat{p}} = \sqrt{\frac{(.07)(.93)}{200}} = .018$.

$\hat{p} = \frac{23}{200} = .115$ with a z-score of $\frac{.115-.07}{.018} = 2.5$. Thus the P-value is $1 - .9938 = .0062$. This low a P-value is strong evidence against H_0.

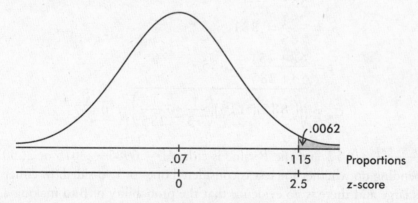

11. **(E)** We are comparing two population proportions and thus the correct test involves H_0: $p_1 - p_2 = 0$ and H_a: $p_1 - p_2 \neq 0$.

12. **(B)** With df $= 10 - 1 = 9$, the critical t-scores for the .05 and .01 tail probabilities are 1.833 and 2.821, respectively. We have $2.79 > 1.833$, but $2.79 < 2.821$, and so 2.79 is significant at the 5% level but not at the 1% level.

13. **(D)** $\frac{934-1250}{616/\sqrt{12}} = -1.777$. Since 1.777 is between 1.363 and 1.769, P is between .05 and .10.

14. **(C)** We have $H_0: \mu = 12$, $H_a: \mu > 12$, and $\sigma_{\bar{x}} \approx \frac{s}{\sqrt{n}} = \frac{6}{\sqrt{64}} = 0.75$. The *t*-score

of 13.1 is $\frac{13.1-12}{0.75} = 1.47$, and using Table B this gives a *P*-value between .05 and .10, indicating some evidence against H_0. [On the TI-84, T-Test gives $P = .0737$.]

15. **(B)** $\frac{1.72-1.6}{0.4/\sqrt{64}} = 2.4$, giving a tail probability between .005 and .01 in Table B.

Because this is a two-tailed test, P is between .01 and .02. The claim should be disputed at the 10% and 5% significance levels but not at the 1% level. [On the TI-83, T-Test gives $P = .0194$.]

16. **(A)** $\sigma_{\hat{p}} = \sqrt{\frac{(.66)(.34)}{5644}} = .0063$

$\frac{.35-.66}{.0063} = -49$, and the *P*-value is .0000.

17. **(A)** $\sigma_{\bar{x}_1-\bar{x}_2} = \sqrt{\frac{(329)^2}{231} + \frac{(264)^2}{221}} = 28.00$

$\frac{6176-8164}{28.00} = -71$, and the *P*-value is .0000. This is very strong evidence that angioplasty costs less than bypass surgery.

18. **(E)**

$$\hat{p}_1 = \frac{48}{53} = .906$$

$$\hat{p}_2 = \frac{251}{285} = .881$$

$$\hat{p} = \frac{48+251}{53+285} = .885$$

$$\sigma_d = \sqrt{(.885)(.115)\left(\frac{1}{53} + \frac{1}{285}\right)} = .0477$$

$\frac{.906-.881}{.0477} = 0.52$ and the *P*-value is either $1 - .6985 = .3015$ or $2(.3015)$ depending on whether the test is considered one- or two-sided. In either case, P is large and there is no evidence that the probability of Bird making a second free throw depends on whether he makes the first free throw.

19. **(D)** We have $\frac{2250-2000}{600/\sqrt{20}} = 1.863$. With df = 20 − 1 = 19, we note that 1.863

is between 1.729 and 2.093, which correspond to .05 and .025, respectively. This is a two-sided test, and so we double these probabilities and find the interval for the *P*-value to be between .10 and .05.

20. **(A)**

$$\sigma_{\bar{x}_1 - \bar{x}_2} = \sqrt{\frac{(0.3)^2}{10} + \frac{(0.2)^2}{10}} = 0.114$$

or if we believe the variances are equal and use pooling, then

$$\sigma_{\bar{x}_1 - \bar{x}_2} = \sqrt{\frac{(10-1)(0.3)^2 + (10-1)(0.2)^2}{10 + 10 - 2}} \sqrt{\frac{1}{10} + \frac{1}{10}} = 0.114$$

21. **(E)** All are false. The null and alternate hypotheses are decided on before the data come in. The *P*-value does not give the probability that the null hypothesis is true; it gives the probability of such an extreme value, assuming the null hypothesis is true. The smaller the *P*-value, the stronger the evidence against the null hypothesis.

22. **(C)** We have H_0: $\mu_1 - \mu_2 = 0$, H_a: $\mu_1 - \mu_2 \neq 0$, and

$$\sigma_{\bar{x}_1 - \bar{x}_2} \approx \sqrt{\frac{(1.2)^2}{40} + \frac{(0.65)^2}{50}} = 0.211$$

The *z*-score for 0.5 is $\frac{0.5 - 0}{0.211} = 2.37$. From Table A, each tail has a probability of $1 - .9911 = .0089$. Adding the two tails gives a probability of $.0089 + .0089 = .0178$.

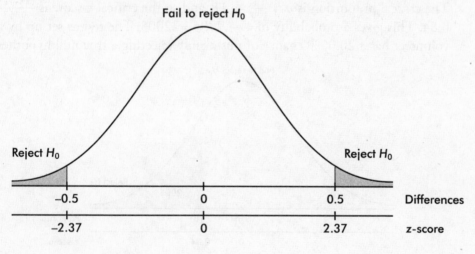

23. **(B)** We have $H_0: \mu = 25$, $H_a: \mu < 25$, $\bar{x} = 24.581$, $s = 0.7793$, and $\sigma_{\bar{x}} =$ $\frac{0.7793}{\sqrt{26}} = 0.153$. The *t*-score of 24.581 is $\frac{24.581-25}{0.153} = -2.74$. In looking at the df = 25 row in Table B, note that 2.74 is between 2.485 and 2.787, so *P* is between .005 and .01. (A calculator like the TI-84 quickly gives the critical *t*-score as −2.74 and *P* = .0056.)

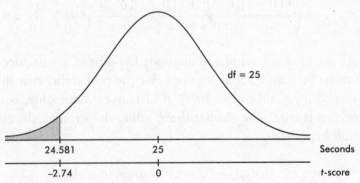

24. **(D)** We have $H_0: p = .15$ and $H_a: p > .15$. Using the claimed 15%, we calculate the standard deviation of sample proportions to be

$$\sigma_{\hat{p}} = \sqrt{\frac{(.15)(.85)}{100}} = .0357$$

The critical proportion is $c = \frac{18}{100} = .18$, and so the critical *z*-score is $\frac{.18-.15}{.0357} = 0.84$. This gives a probability of $1 - .7995 = .2005$. The test as set up by the volunteer has a 20.05% chance of mistakenly rejecting a true null hypothesis.

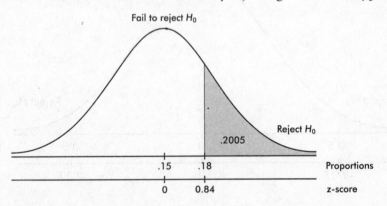

25. **(A)** $\sigma_{\bar{x}} = \frac{\sigma}{\sqrt{n}} = \frac{5.8}{\sqrt{50}} = 0.820$ If the true mean is 20 pounds per ton of rock, the *z*-score for 22 is $\frac{22-20}{0.820} = 2.44$, and so the probability is $1 - .9927 = .0073$.

26. **(B)** $\sigma_{\bar{x}} = \frac{150}{\sqrt{100}} = 15.0$. With a true mean of 385, the *z*-score for 375 is $\frac{375-385}{15.0} = -0.67$ and the probability to the left of 375 is .2514.

Free-Response Questions

> ***Directions:*** You must show all work and indicate the methods you use. You will be graded on the correctness of your methods and on the accuracy of your final answers.

Nine Open-Ended Questions

1. Three in-store surveys are taken: one before, one during, and one after a special promotion for tropical fruits. The proportions of customers expressing interest in these products are 29 out of 157, 37 out of 165, and 40 out of 152 for the surveys before, during, and after the promotion, respectively.

   a. Is the interest increase from before to during the promotion significant? From during to after the promotion? From before to after the promotion?

   b. What does the result obtained in part *a* say about the possibility that $\hat{p}_1$ and $\hat{p}_3$ are significantly different when $\hat{p}_1$ and $\hat{p}_2$ are not and when $\hat{p}_2$ and $\hat{p}_3$ are not?

2. Polychlorinated biphenyl (PCB) contamination of a river by a manufacturer is being measured by amounts of the pollutant found in fish. A company scientist claims that the fish contain only 5 parts per million, but an investigator believes the figure is higher. What is the conclusion if six fish are caught and show the following amounts of PCB (in parts per million): 6.8, 5.6, 5.2, 4.7, 6.3, and 5.4, respectively?

3. Suppose that your rival in the marketing department has what you consider a dumb idea. Your boss, who knows nothing about statistics, is relatively conservative but will follow up on this dumb idea if there is good evidence that more than 25% of the company's customers like it. A survey of 40 customers is taken, and 12 of them favor the idea. Your rival points out that $\frac{12}{40} = 30\%$. You know a little about statistics and perform the usual hypothesis test with H_0: $p = .25$ and H_a: $p > .25$. Write a paragraph explaining to your boss why these data are not strong evidence that more than 25% of the customers like your rival's idea.

4. The weight of an aspirin tablet is 300 milligrams according to the bottle label. Should a Food and Drug Administration (FDA) investigator reject the label if she weighs seven tablets and obtains weights of 299, 300, 303, 302, 301, 301, and 302 milligrams, respectively?

5. An employer wishes to compare typing speeds of graduates from two different study programs. Eight graduates of the first course type at 62, 85, 59, 64, 73, 70, 75, and 72 words per minute, respectively, while six graduates of the second course have typing speeds of 75, 64, 81, 55, 69, and 58 words per minute, respectively. Should the employer conclude that one program is better than the other?

6. One hospital has 1000 live births during the summer, while a second hospital has 500 live births during the same time period.

   a. Assuming there is a 50% chance for a live-born infant to be male, which hospital is more likely to have less than 45% male births? Explain your answer.
   b. Suppose the first hospital has 515 male births, while the second has only 240. Is the difference statistically significant? Show your work.

7. It is reported that the national average cost of an automobile tune-up is $74.35.

   a. Suppose you interview a simple random sample of 40 people in your community who recently have had a tune-up. The sample mean is $76.72 with a standard deviation of $5.22. Find a 95% confidence interval estimate of the cost of a tune-up in your area. Show your work.
   b. What conclusion would you come to after testing the hypothesis that a tune-up costs more in your area than the national average? Explain your conclusion using a *P*-value calculation in your answer.

8. Jeremy, as a point guard on the IHS junior varsity basketball team, connected on 54% of his shots from beyond the three-point line. In his first year on the varsity team he connected on 73 of 120 three-point attempts.

   a. Is there sufficient evidence that he is shooting better than he did as a junior varsity player?
   b. Suppose Jeremy tries 15 three-point shots in the next two games. Could the outcome of these shots change your conclusion above?
   c. Give a 95% confidence interval estimate for Jeremy's varsity three-point shooting proportion.

9. An article ("Undergraduate Marijuana Use and Anger" by Sue B. Stoner) in a 1988 issue of the *Journal of Psychology* (Vol. 122, p. 343) reported that in a sample of 17 marijuana users, the mean and standard deviation on an anger expression scale were $\bar{x} = 42.72$ and $s = 6.05$.

   a. Test whether this result is significantly greater than the established mean of 41.6 for nonusers. Show your work.
   b. What assumptions are necessary for the above test to be valid?

Answers Explained

1. $\hat{p}_1 = \frac{29}{157} = .185$, $\hat{p}_2 = \frac{37}{165} = .224$, $\hat{p}_3 = \frac{40}{152} = .263$. [Note that $n_1 p_1 = 29$, $n_1 q_1 = 128$, $n_2 p_2 = 37$, $n_2 q_2 = 128$, $n_3 p_3 = 40$, and $n_3 q_3 = 112$ are all ≥ 10, we assume the samples are independent SRSs, and the population of all customers is large compared to our sample sizes.] For $\hat{p}_1$ and $\hat{p}_2$ we calculate $\hat{p} = \frac{29+37}{157+165} = .205$ with

$$\sigma_d \approx \sqrt{(.205)(.795)\left(\frac{1}{157} + \frac{1}{165}\right)} = .0450$$

The observed difference is $.224 - .185 = .039$, and the associated z-score is $\frac{.039}{.0450} = 0.87$. Table A gives the P-value as $1 - .8078 = .1922$. (On the TI-84, $P = .1899$.) Such a large P-value indicates that the interest increase from before to during the promotion is not significant.

For $\hat{p}_2$ and $\hat{p}_3$ we have $\hat{p} = \frac{37+40}{165+152} = .243$ with

$$\sigma_d \approx \sqrt{(.243)(.757)\left(\frac{1}{165} + \frac{1}{152}\right)} = .0482$$

The observed difference is $.263 - .224 = .039$, and the associated z-score is $\frac{.039}{.0482} = 0.81$. Table A gives the P-value as $1 - .7910 = .2090$. (On the TI-84, $P = .2098$.) Such a large P-value indicates that the interest increase from during to after the promotion also is not significant.

For $\hat{p}_1$ and $\hat{p}_3$ we have $\hat{p} = \frac{29+40}{157+152} = .223$ with

$$\sigma_d \approx \sqrt{(.223)(.777)\left(\frac{1}{157} + \frac{1}{152}\right)} = .0474$$

The observed difference is $.263 - .185 = .078$, and the associated z-score is $\frac{.078}{.0474} = 1.65$. Table A gives the P-value as $1 - .9505 = .0495$. (On the TI-84, $P = .0489$.) This small P-value indicates that the interest increase from before to after the promotion is significant.

Thus we see that it is possible for $\hat{p}_1$ and $\hat{p}_3$ to be significantly different when $\hat{p}_1$ and $\hat{p}_2$ are not and when $\hat{p}_2$ and $\hat{p}_3$ are not.

2. Name the test:

We are using a *one-sample t-test for the mean.*

State the hypothesis:
H_o: Mean contamination $\mu = 5$ ppm
H_a: Mean contamination $\mu > 5$ ppm.

Check the conditions:

1. *Randomization:* We must assume that the six fish that were caught are a representative sample.
2. *Nearly normal population:* A boxplot of the sample shows no outliers and is very roughly normal:

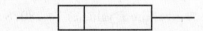

Perform the mechanics:

A calculator readily gives: $n = 6$, $\bar{x} = 5.6667$, and $s = 0.76333$.

$$\text{Then } \sigma_{\bar{x}} \approx \frac{s}{\sqrt{n}} = \frac{0.76333}{\sqrt{6}} = 0.31163 \text{ and}$$

$$t = \frac{\bar{x} - 5}{\sigma_{\bar{x}}} = \frac{5.6667 - 5}{0.31163} = 2.139.$$

With $df = n - 1 = 5$, the P-value is $P(t > 2.139) = .0427$

Give a conclusion in context:

The resulting P-value of .0427 shows moderate evidence against H_0, and so (at the 5% significance level) there is sufficient evidence to dispute the company scientist's claim that the mean contamination per fish is only 5 ppm PCB.

3. Dear Boss:

The 40 customers surveyed represent just one possible sample out of many. Each sample has its own proportion of customers who like the idea, and these sample proportions are bunched around the true proportion. The question is whether this one sample proportion of 30% is so far away from the 25% figure that it is improbable that 25% is the true figure. I calculated a rough measure of the spread of the sample proportions to be

$\sqrt{\frac{(.25)(.75)}{40}} = .0685$. Note that .30 is not even a whole one of these measure-

ments away from .25. In fact it is only $\frac{.05}{.0685} = 0.73$ of one of these measurements of spread. Using a table that converts these figures to probabilities, I see that if the 25% figure is correct, there is more than a .23 chance of stumbling on a sample with a proportion as high as 30%. Statisticians call this .23 a P-value, and this high a P-value is no evidence that the original claim of 25% is wrong.

4. Name the test:

   We are using a *one-sample t-test for the mean.*

   State the hypothesis:
   H_0: Mean weight of aspirin tablets is $\mu = 300$ mg
   H_a: Mean weight is $\mu \neq 300$ mg

   Check the conditions:

   1. *Randomization:* We must assume that the seven aspirin tablets are a representative sample.
   2. *Nearly normal population:* A histogram of the sample shows no outliers and is roughly unimodal and symmetric:

   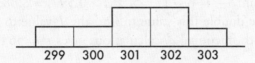

   Perform the mechanics:

   A calculator readily gives: $n = 7$, $\bar{x} = 301.14$, and $s = 1.345$.

   $$\text{Then } \sigma_{\bar{x}} \approx \frac{s}{\sqrt{n}} = \frac{1.345}{\sqrt{7}} = 0.5084 \text{ and}$$

   $$t = \frac{\bar{x} - 300}{\sigma_{\bar{x}}} = \frac{301.14 - 300}{0.5084} = 2.24.$$

   With $df = n - 1 = 6$, $P(t > 2.24) = .033$. Since this is a two-sided test, we double this value to find the P-value to be $P = .066$. (On the T1-84, `T-Test` with `Data` gives $P = .062$.)

   Give a conclusion in context:

   The resulting P-value of .066 (or .062) shows some evidence against H_0, and so (at the 10%, but not 5%, significance level) there is sufficient evidence for the FDA to dispute the label claim of 300 mg.

5. Name the test:

   We are using a *two-sample t-test for means.*

   State the hypothesis:
   H_0: The difference in the mean typing speeds of graduates from the two schools is zero: $\mu_1 - \mu_2 = 0$
   H_a: The difference in mean typing speeds is not zero: $\mu_1 - \mu_2 \neq 0$

   Check the conditions:

   1. *Randomization:* We must assume that the graduates from the two schools are representative samples.
   2. *Independent groups:* It is reasonable to assume that the graduates from one school have no relationship to the graduates from the other school.
   3. *Nearly normal populations:* Boxplots of the samples shows no outliers and are very roughly normal:

Perform the mechanics:

A calculator readily gives: $n_1 = 8$, $\overline{x}_1 = 70$, $s_1 = 8.315$, $n_2 = 6$, $\overline{x}_2 = 67$, and $s_2 = 9.980$.

Then $\sigma_{\overline{x}_1 - \overline{x}_2} \approx \sqrt{\dfrac{(8.315)^2}{8} + \dfrac{(9.980)^2}{6}} = 5.024$.

The *t*-score of the observed difference is $\dfrac{70 - 67}{5.024} = 0.597$.

With $df = \min(8 - 1, 6 - 1) = 5$, $P(t > 0.597) = .288$. Since this is a two-sided test, we double this value to find the *P*-value to be $P = .576$. [Using the TI-84 with fractional *df* calculations gives $P = .564$.]

Give a conclusion in context:

With such a large *P*, there is *no* evidence that one program produces graduates with faster typing speeds than the other.

6. (a) By the law of large numbers, if the probability of a male birth is actually .5, the more births there are, the closer the relative frequency tends toward this probability. With fewer births there is a greater chance for wide swings in the relative frequency, and so the hospital with only 500 births has a greater likelihood of having less than 45% male births.

(b) Note that $n_1\hat{p}_1 = 515$, $n_1(1 - \hat{p}_1) = 485$, $n_2\hat{p}_2 = 240$, and $n_2(1 - \hat{p}_2) = 260$ are all at least 10. We have H_0: $p_1 - p_2 = 0$ and H_a: $p_1 - p_2 \neq 0$, where $p_1 =$ the probability of a male birth at the first hospital, and $p_2 =$ the probability of a male birth at the second hospital.

$$\hat{p} = \frac{515 + 240}{1000 + 500} = .503$$

$$\sigma_d = \sqrt{(.503)(.497)}\sqrt{\frac{1}{1000} + \frac{1}{500}} = .0274$$

The observed difference is $.515 - .48 = .035$ with a *z*-score of $\frac{.035}{.0274} = 1.28$, giving a tail probability of .1003. This is a two-sided test, and so the *P*-value is $2(.1003) = .2006$. With a *P* this large, there is no evidence to reject H_0. Thus, the observed difference in the proportions of male births between the two hospitals is not statistically significant.

7. We are given that we have an SRS, and the sample size, 40, is large enough to relax the population normality assumption.

   (a) The population standard deviation is unknown, and so we use a

   t-distribution to obtain $76.72 \pm 2.022\left(\frac{5.22}{\sqrt{40}}\right) = 76.72 \pm 1.67$. Conclusion

   in context: we are 95% confident that the mean cost of a tune-up in your area is between $75.05 and $78.39.

   (b) We have H_0: $\mu = 74.35$ and H_a: $\mu > 74.35$. Running a t-test on the TI-83 gives a P-value of .0033. With such a small P, there is strong evidence to reject H_0 and conclude that the cost of a tune-up in your area is higher than the national average.

8. (a) Perform a one-sample hypothesis test concerning the population proportion. We are testing the null hypothesis H_0: $p = .54$ against the alternative hypothesis H_a: $p > .54$. Note that both

$$np = (120)(.54) = 64.8 > 10$$

   and

$$n(1 - p) = (120)(.46) = 55.2 > 10.$$

   With

$$\sigma_{\hat{p}} = \sqrt{\frac{(.54)(.46)}{120}} = .04550$$

   the sample proportion $\hat{p} = \frac{73}{120} = .6083$ has a z-score of $\frac{.6083 - .54}{.04550} = 1.50$ and a P-value from Table A of .0668. (A 1-PropZTest on the TI-84 gives $P = .0666$.) With $.05 < P < .10$ we conclude there is *some* evidence to reject H_0 and conclude that Jeremy is shooting better. In particular there is sufficient evidence to reject H_0 at the 10% significance level (with $\alpha = .10$), but not at the 5% significance level (with $\alpha = .05$).

   (b) We would then have

$$\sigma_{\hat{p}} = \sqrt{\frac{(.54)(.46)}{135}} = .04290.$$

   If Jeremy missed all these 15 shots, $\hat{p} = \frac{73}{135} = .5407$ would give $z = \frac{.5407 - .54}{.0429} = 0.02$ and $P = .4920$ and thus *no* evidence to reject H_0. Alternatively, if Jeremy made all those 15 shots, $\hat{p} = \frac{38}{135} = .6519$ would give $z = \frac{.6519 - .54}{.0429} = 2.61$ and $P = .0045$ and thus *strong* evidence to reject H_0. (The TI-84 1-PropZTest would have given $P = .4931$ and $P = .0046$ respectively.)

(c) $\hat{p} = \frac{73}{120} = .6083$. We check that $n\hat{p} = (120)(.6083) = 73.0 > 10$ and $n(1 - \hat{p}) = (120)(.3917) = 47.0 > 10$. With $\sigma_{\hat{p}} = \sqrt{\frac{(.6083)(.3917)}{120}} = .0446$, we get $.6083 \pm 1.96(.0446) = .6083 \pm .0874$ for a 95% confidence interval estimate. [1-PropZInt on the TI-84 gives (.521, .696).]

9. (a) We have $H_0: \mu = 41.6$, $H_a: \mu > 41.6$, and $\sigma_{\bar{x}} = \frac{6.05}{\sqrt{17}} = 1.47$. The *t*-score of 42.72 is $\frac{42.72-41.6}{1.47} = 0.762$. Looking across the df = 16 row in Table B, note that 0.762 is between .690 and .865. Thus the *P*-value is between .20 and .25. With such a large *P*, there is no evidence for rejecting H_0; that is, there is no evidence indicating that marijuana users score higher on the anger expression scale than do nonusers. [On the TI-84, T-Test gives *P* = .2282.]

(b) In addition to the usual assumption that we are dealing with a simple random sample, we must assume that we are also working with a normal population.

Five Investigative Tasks

1. An efficiency expert wishes to analyze the difference in productivity between workers exposed to two different lighting arrangements. She chooses a procedure in which the same 30 employees work under each arrangement for 1 week in turn. The outputs of each employee under the two lighting arrangements, LA1 and LA2, are summarized as follows.

Employee	1	2	3	4	5	6	7	8	9	10
LA1	15	23	18	19	27	13	22	20	20	19
LA2	12	22	18	20	24	12	20	20	19	17

Employee	11	12	13	14	15	16	17	18	19	20
LA1	25	26	15	19	19	21	20	30	24	16
LA2	26	20	14	17	15	16	22	26	24	19

Employee	21	22	23	24	25	26	27	28	29	30
LA1	15	20	19	20	17	26	12	10	23	26
LA2	19	18	19	15	20	22	14	13	20	25

   a. Find the mean and standard deviation of each of the two 30-element sets corresponding to LA1 and LA2 separately.
   b. Run a two-sample hypothesis test with $H_0: \mu_1 - \mu_2 = 0$, $H_a: \mu_1 - \mu_2 \neq 0$, and $\alpha = .05$.
   c. Find the *P*-value.
   d. From the set of 30 differences (be careful with signs), find the mean and standard deviation.
   e. Run a single sample hypothesis test with $H_0: \mu = 0$, $H_a: \mu \neq 0$, and $\alpha = .05$ on this set.
   f. Find the *P*-value.

g. Are the observed differences in lighting arrangements significant? Explain the contrasting conclusions between the two methods of analysis indicated above.

2. An accounting firm measured the blood pressures of ten of its certified public accountants (CPAs) before and during the spring 2007 tax season. The systolic pressures for the ten individuals, designated as A through J, were as follows:

	A	B	C	D	E	F	G	H	I	J
Before:	110	124	98	105	115	120	118	110	123	95
During:	115	126	97	108	115	124	119	113	121	96

Is there sufficient evidence that blood pressure rises during tax season?

a. Assuming normal populations, perform a two-population, small-sample hypothesis test on H_0: $\mu_1 - \mu_2 = 0$, and H_a: $\mu_1 - \mu_2 < 0$. Show that this test does not indicate that the observed difference is significant even at $\alpha = .10$.

b. Which is the more appropriate test? Explain why. What conclusion can be drawn from this test?

3. One of your friends knows that you are taking AP Statistics and asks for your help with her chemistry lab report. She has come up with five measurements of the melting point of a compound: 122.4, 121.8, 122.0, 123.0, and 122.3 degrees Celsius.

(a) The lab manual asks for a 95% confidence interval estimate for the melting point. Show her how to find this estimate.

(b) Explain to your friend in simple language what 95% confidence means.

(c) Would a 90% confidence interval be narrower or wider than the 95% interval. Explain your answer to your friend.

(d) The lab manual asks whether the data show sufficient evidence to call into question the established melting point of 122 degrees Celsius. State the null and alternative hypotheses and find the P-value.

(e) Explain the meaning of the specific P-value to your friend.

4. A pharmaceutical firm would like to show that taking a particular mixture of vitamins and natural herbs will raise the IQ of a teenager. A sample of 15 teenagers is chosen, and their IQs are measured before and after taking the vitamin-herb mixture for 1 year. The individual results are as follows:

IQ before Vitamin-Herb Mixture	IQ after Vitamin-Herb Mixture
103	107
98	100
95	96
107	104
112	110
125	129
111	110
89	87
100	97
121	121
114	117
107	105
95	98
106	108
115	119

(a) The pharmaceutical firm claims that after taking the vitamin-herb mixture, teenagers have a higher than average IQ. (Average IQ is 100.) Do the data support their conclusion? Explain.

(b) The firm also claims that teenagers can improve their IQs by taking the mixture. Do the data support this conclusion? Explain.

(c) Can after-mixture IQs be predicted by before-mixture IQs? Explain.

(d) Can the change in IQs be predicted by before-mixture IQs? Explain.

5. As reported in *Chances: Risks and Odds in Everyday Life* by James Burke, a research team at Cornell University conducted a study in which they concluded that 10% of all businessmen who wear ties wear them so tight that they reduce blood flow to the brain, diminishing cerebral functions.

   (a) Assuming the 10% figure is correct, use simulation to determine the approximate probability that at a board meeting of ten businessmen, all of whom wear ties, at least two are wearing their ties too tight. Show your work.

8417706757	1761315582	5150681435	4105092031	0644905059	5988431180
5311584469	9486857967	0581184514	7501113006	6339555041	1586606589
1311971020	8594091932	0648874987	5435552704	9035902649	4749671567
9426808844	2629464759	0898957024	9728400637	8928303514	5919507635
0330972605	2935723737	6788103668	3387635841	5286923114	1586438942

   (b) Calculate the above probability exactly. Show your work.

   (c) Suppose you believe the Cornell University claim is too high and run a hypothesis test. If in a simple random sample of 200 businessmen who wear ties you find only 17 wearing them too tight, what will be your conclusion? Explain your answer.

   (d) A 100-trial simulation is performed to determine the number of businessmen sampled before finding one with a tight tie. The results are as follows:

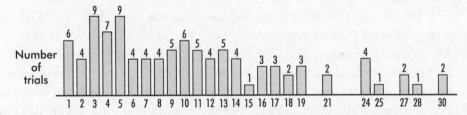

   Number of trials

   Number sampled before finding tight tie

   Use the above to estimate the mean number of businessmen to be sampled before finding one with diminished cerebral function due to an overly tight tie.

Answers Explained

1. *a.* For LA1 we calculate $\bar{x}_1 = 19.97$ and $s_1 = 4.7086$, while for LA2 we calculate $\bar{x}_2 = 18.93$ and $s_2 = 3.9474$.

   b.

   $$\sigma_{\bar{x}_1 - \bar{x}_2} \approx \sqrt{\frac{(4.7086)^2}{30} + \frac{(3.9474)^2}{30}} = 1.122$$

   and with df = $(30 - 1) + (30 - 1) = 58$, the critical scores are $0 \pm 2.000(1.122) = \pm 2.244$. Since the observed difference, $19.97 - 18.93 = 1.04$, is between these scores, the observed difference is not significant.

   c. The *t*-score of $1.04/1.122 = .93$ gives a probability between .15 and .20 and a resulting *P*-value between .30 and .40. [On the TI-84, 2-SampT Test gives

$P = .3578.$] With such a large P-value, there is no evidence that the observed difference in sample means is significant.

d. $\bar{x} = 1.03$, $s = 2.6455$

e. $\sigma_{\bar{x}} \approx \frac{2.6455}{\sqrt{30}} = 0.483$, and with df $= 30 - 1 = 29$, the critical scores are $0 \pm 2.045(0.483) = \pm 0.988$.

f. Since 1.03 is outside these scores, the observed difference is significant. The t-score of $1.03/0.483 = 2.13$ gives a probability between .02 and .025 and a resulting P-value between .04 and .05. [On the TI-84, T-Test gives $P = .0416.$] With a P-value this small, there *is* evidence that employees have different productivity under different lighting arrangements.

g. The observed difference *is* significant. The reason for the apparent inconsistency is that the first test used is not the proper one. This two-sample test should be used only when the two sets are independent. In this case, there is a clear relationship between the data, in pairs, and this relationship is completely lost in the procedure for the two-sample test.

2. *a.* A calculator gives $\bar{x}_1 = 111.8$, $\bar{x}_2 = 113.4$, $s_1 = 10.09$, $s_2 = 10.36$, and

$$\sigma_{\bar{x}_1 - \bar{x}_2} = \sqrt{\frac{(10.09)^2}{10} + \frac{(10.36)^2}{10}} = 4.57$$

The t-score of the observed difference is then $\frac{111.8 - 113.4}{4.57} = -0.35$. With df $= 18$, we note that $.35 < .688$, and the P-value is more than .25. With such a large value of P, there is no evidence of a rise in blood pressure. [On the TI-84, 2-SampT Test gives $P = .365.$]

b. Looking at the data, however, it is apparent that a change in blood pressure occurred. The test used in part *a* does not apply the knowledge of what happened to each individual. In fact, since the before and during sets are not independent but rather are related in pairs, that test is not the proper one to use. The appropriate test is a one-population, small-sample hypothesis test on the set of differences.

Form the set of differences, being careful of sign, and then proceed:

Name the test:

We are using a *paired t-test*, that is, a single sample hypothesis test on the set of differences.

State the hypothesis:

H_0: The systolic blood pressures of CPAs were the same before and during the spring 2007 tax season: $\mu_d = 0$.

H_a: The blood pressures of CPAs were greater during the tax season than before; the mean difference is less than zero: $\mu_d < 0$.

Check the conditions:

1. The *data are paired* because they are measurements on the same individuals before and during tax season.
2. The blood pressures of any individual are assumed independent of the blood pressures of the others, so the *differences are independent.*

3. *Randomization*: We must assume that the CPAs are a representative sample.
4. The boxplot of the differences has no outliers, is symmetric, and *roughly normal*.

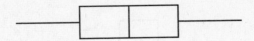

Perform the mechanics:

A calculator readily gives: $n = 10$, $\bar{x} = -1.6$, and $s = 2.221$.

Then $\sigma_{\bar{x}} \approx \dfrac{s}{\sqrt{n}} = \dfrac{2.221}{\sqrt{10}} = 0.7023$ and $t = \dfrac{\bar{x} - 0}{\sigma_{\bar{x}}} = \dfrac{-1.6}{0.7023} = -2.278$.

With df $= n - 1 = 9$, the *P*-value is $P(t < -2.278) = .0244$.

Give a conclusion in context:

The resulting *P*-value of .0244 is small enough to justify a conclusion (at the 5% significance level) that the observed rise in blood pressure during tax season is significant.

3. (a) A calculator gives $\bar{x} = 122.3$ and $s = 0.458$. Then $\sigma_x = \dfrac{0.458}{\sqrt{5}} = 0.205$.

With df $= 5 - 1 = 4$, Table B gives critical *t*-scores of ± 2.776. Thus the 95% confidence interval is $122.3 \pm 2.776(0.205) = 122.3 \pm 0.57$ degrees Celsius.
(b) We are 95% confident that the true mean lies in the interval [121.73, 122.87]. Or, in terms of confidence level, 95% of the time intervals so constructed contain the true mean.
(c) A 90% confidence interval would be narrower than the 95% interval. If we care only about being 90% confident in our answer, we can use a tighter interval.
(d) We have H_0: $\mu = 122$, and H_a: $\mu \neq 122$, and the *t*-score of the observed mean of 122.3 is $\frac{122.3 - 122}{0.205} = 1.463$. In the df $= 4$ row in Table B, 1.463 lies between 1.190 and 1.533, corresponding to tail probabilities of .15 and .10, respectively. Since this is a two-sided test, we must double these probabilities and find that the *P*-value is between .20 and .30.
(e) If the claimed 122 value is correct, there is still a greater than .20 probability that a sample mean will be as extreme as ours. With such a large *P*, there is not sufficient evidence to question the 122-degree-Celsius claim.

4. (a) If there is any reason to suspect that the pharmaceutical firm did not choose a simple random sample of teenagers, then their claim is suspect. Assuming they did pick a *random sample*, and noting that the after-mixture IQ sample has a *roughly normal* histogram:

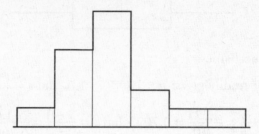

we can run a *t*-test on the after-mixture IQs with H_0: $\mu = 100$ and H_a: $\mu > 100$. This test yields a small *P*-value of .012. Thus there *is* sufficient evidence to reject the null hypothesis, and the data do support the pharmaceutical firm's claim that after taking the vitamin-herb mixture, teenagers have a higher than average IQ.

(b) Running a *t*-test on the difference (a matched pairs test) between after-mixture IQs and before-mixture IQs with H_0: $\mu = 0$ and H_a: $\mu > 0$ yields a large *P*-value of .174. Thus there is *not* sufficient evidence to reject the null hypothesis, and the data do *not* support the pharmaceutical firm's claim that the vitamin-herb mixture improves teenage IQs.

(c) A least square regression line using the after-mixture and before-mixture IQs has a high correlation of $r = .972$, indicating a strong association. Thus after-mixture IQs can be predicted from before-mixture IQs. (The reason for this is that after-mixture IQs are almost identical to before-mixture IQs.)

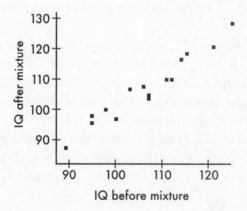

(d) A least square regression line using the before-mixture IQs and the difference IQs yields a very low correlation of $r = .235$, indicating almost no association. Thus the change in IQs *cannot* be predicted from before-mixture IQs.

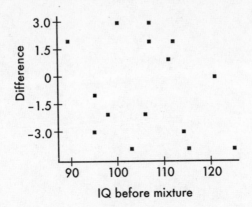

5. (a) Let the digit 9 represent a tie that is too tight, while the digits 0 through 8 represent ties that are not too tight. Read off groups of ten digits, checking for the number of 9s in each group. Tabulating from the table gives

No tight ties (no 9s): 8
One tight tie (one 9): 15
Two tight ties (two 9s): 5
Three tight ties (three 9s): 2

The estimated probability of at least two tight ties among the ten businessmen is thus $\frac{5+2}{30} = .23$.

(b) P (at least 2) $= 1 - [P(0) + P(1)]$
$$= 1 - [(.9)^{10} + 10(.1)(.9)^9]$$
$$= .26$$

(c) Note that $np_0 = (200)(.10) = 20$ and $n(1 - p_0) = (200)(.90) = 180$ are both >10, we are given an SRS, and clearly 200 < 10% of all businessmen who wear

ties. We have H_0: $p = .1$, H_a: $p < .1$, and $\sigma_{\hat{p}} = \sqrt{\frac{(.1)(.9)}{200}}$
$= .0212$. $\hat{p} = \frac{17}{200} = .085$ with z-score $\frac{.085-.1}{.0212} = -0.71$ with a corresponding P-value of .2389 from Table A. With such a large P, there is not sufficient evidence to dispute the Cornell University claim.

(d) $\sum xP(x) = 1(.06) + 2(.04) + 3(.09) + \cdots + 30(.02) = 10.39$

Tests of Significance—
Chi Square and Slope of Least
Squares Line

- Chi-square Test for Goodness of Fit
- Chi-square Test for Independence
- Chi-square Test for Homogeneity of Proportions
- Hypothesis Test for Slope of Least Squares Line

In this topic we continue our development of tools to analyze data. We learn about inference on distributions of counts using chi-square models. This can be used to solve such problems as "Do test results support Mendel's genetic principles?" (goodness-of-fit test); "Was surviving the Titanic sinking independent of a passenger's status?" (independence test); and "Do students, teachers, and staff show the same distributions in types of cars driven?" (homogeneity test). We then learn about inference with regard to linear association of two variables. This can be used to solve such problems as "Is there a linear relationship between the grade received on a term paper and the number of pages turned in?"

CHI-SQUARE TEST FOR GOODNESS OF FIT

A critical question is often whether or not an observed pattern of data fits some given distribution. A perfect fit cannot be expected, and so we must look at discrepancies and make judgments as to the *goodness of fit*.

One approach is similar to that developed earlier. There is the null hypothesis of a good fit, that is, the hypothesis that a given theoretical distribution correctly describes the situation, problem, or activity under consideration. Our observed data consist of one possible sample from a whole universe of possible samples. We ask about the chance of obtaining a sample with the observed discrepancies if the null hypothesis is really true. Finally, if the chance is too small, we reject the null hypothesis and say that the fit is not a good one.

How do we decide about the significance of observed discrepancies? It should come as no surprise that the best information is obtained from squaring the discrepancy values, as this has been our technique for studying variances from the beginning. Furthermore, since, for example, an observed difference of 23 is more significant if the original values are 105 and 128 than if they are 10,602 and 10,625, we must appropriately *weight* each difference. Such weighting is accomplished by dividing each difference by the expected values. The sum of these weighted differences or discrepancies is called *chi-square* and is denoted as χ^2 (χ is the lowercase Greek letter chi):

$$\chi^2 = \sum \frac{(\text{obs} - \text{exp})^2}{\text{exp}}$$

The smaller the resulting χ^2-value, the better the fit. The *P*-value is the probability of obtaining a χ^2 value as extreme as the one obtained if the null hypothesis is assumed true. If the χ^2 value is large enough, that is, if the *P*-value is small enough, we say there is sufficient evidence to reject the null hypothesis and to claim that the fit is poor.

To decide how large a calculated χ^2-value must be to be significant, that is, to choose a critical value, we must understand how χ^2-values are distributed. A χ^2-distribution is not symmetric and is always skewed to the right. There are distinct χ^2-distributions, each with an associated number of degrees of freedom (df). The larger the df value, the closer the χ^2-distribution to a normal distribution. Note, for example, that squaring the often-used *z*-scores 1.645, 1.96, and 2.576 results in 2.71, 3.84, and 6.63, respectively, which are entries found in the first row of the χ^2-distribution table.

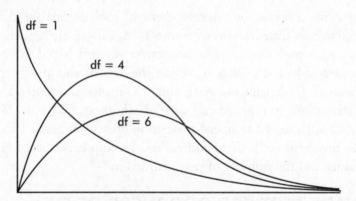

A large χ^2-value may or may not be significant—the answer depends on which χ^2-distribution we are using. A table is given of critical χ^2-values for the more commonly used percentages or probabilities. To use the χ^2-distribution for approximations in goodness-of-fit problems, the individual expected values cannot be too small. An often-used rule of thumb is that no expected value should be less than 5. Finally, as in all hypothesis tests we've looked at, the sample should be randomly chosen from the given population.

EXAMPLE 15.1

In a recent year, at the 6 p.m. time slot, television channels 2, 3, 4, and 5 captured the entire audience with 30%, 25%, 20%, and 25%, respectively. During the first week of the next season, 500 viewers are interviewed.

a. If viewer preferences have not changed, what number of persons is expected to watch each channel?
Answer: .30(500) = 150, .25(500) = 125, .20(500) = 100, and .25(500) = 125, so we have

Channel

	2	3	4	5
Expected number	150	125	100	125

b. Suppose that the actual observed numbers are as follows:

Channel

	2	3	4	5
Observed number	139	138	112	111

Do these numbers indicate a change? Are the differences significant?
Answer: Check the conditions:

1. *Randomization*: We must assume that the 500 viewers are a representative sample.
2. We note that the expected values (150, 125, 100, 125) are all ≥ 5.

H_0: The television audience is distributed over channels 2, 3, 4, and 5 with percentages 30%, 25%, 20%, and 25%, respectively.
H_a: The audience distribution is not 30%, 25%, 20%, and 25%, respectively.

We calculate

$$\chi^2 = \sum \frac{(obs - exp)^2}{exp}$$

$$= \frac{(139 - 150)^2}{150} + \frac{(138 - 125)^2}{125} + \frac{(112 - 100)^2}{100}$$

$$+ \frac{(111 - 125)^2}{125}$$

$$= 5.167$$

Then the *P*-value is $P = P(\chi^2 > 5.167) = .1600$. [With df $= n - 1 = 3$, the TI-84 gives χ^2cdf(5.167,1000,3) = .1600, or a direct test for goodness of fit can be downloaded onto TI-84+ calculators and comes preloaded on some!]

Conclusion:

With this large a *P*-value (.1600) there is not sufficient evidence to reject H_0. That is, there is not sufficient evidence that viewer preferences have changed.

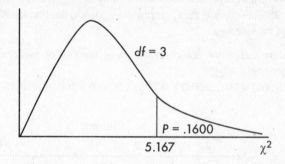

Note: While the TI-83 does not have the goodness of fit download of the TI-84, one can still use a TI-83 to calculate the above χ^2 by putting the observed values in list L1 and the expected values in L2, then calculating (L1 − L2)²/L2 → L3 and χ^2 = sum(L3), where "sum" is found under LIST → MATH.

EXAMPLE 15.2

A grocery store manager wishes to determine whether a certain product will sell equally well in any of five locations in the store. Five displays are set up, one in each location, and the resulting numbers of the product sold are noted.

	Location				
	1	2	3	4	5
Actual number sold	43	29	52	34	48

Is there enough evidence that location makes a difference? Test at both the 5% and 10% significance levels.

Answer:

H_0: Sales of the product are uniformly distributed over the five locations.
H_a: Sales are not uniformly distributed over the five locations.

A total of 43 + 29 + 52 + 34 + 48 = 206 units were sold. If location doesn't matter, we would expect $\frac{206}{5}$ = 41.2 units sold per location (uniform distribution).

	Location				
	1	2	3	4	5
Expected number sold	41.2	41.2	41.2	41.2	41.2

Check the conditions:

1. *Randomization:* We must assume that the 206 units sold are a representative sample.
2. We note that the expected values (all 41.2) are all ≥ 5.

Thus

$$\chi^2 = \frac{(43 - 41.2)^2}{41.2} + \frac{(29 - 41.2)^2}{41.2} + \frac{(52 - 41.2)^2}{41.2}$$
$$+ \frac{(34 - 41.2)^2}{41.2} + \frac{(48 - 41.2)^2}{41.2}$$
$$= 8.903$$

The number of degrees of freedom is the number of classes minus 1; that is, df = 5 − 1 = 4.

The *P*-value is $P = P(\chi^2 > 8.903) = .0636$. [On the TI-84: χ^2cdf(8.903,1000,4).]

With $P = .0636$ there is sufficient evidence to reject H_0 at the 10% level but not at the 5% level. If the grocery store manager is willing to accept a 10% chance of committing a Type I error, there is enough evidence to claim location makes a difference.

CHI-SQUARE TEST FOR INDEPENDENCE

In the goodness-of-fit problems above, a set of expectations was based on an assumption about how the distribution should turn out. We then tested whether an observed sample distribution could reasonably have come from a larger set based on the assumed distribution.

In many real-world problems we want to compare two or more observed samples without any prior assumptions about an expected distribution. In what is called a *test of independence*, we ask whether the two or more samples might reasonably have come from some larger set. For example, do nonsmokers, light smokers, and heavy smokers all have the same likelihood of being eventually diagnosed with cancer, heart disease, or emphysema? Is there a relationship (association) between smoking status and being diagnosed with one of these diseases?

We classify our observations in two ways and then ask whether the two ways are independent of each other. For example, we might consider several age groups and within each group ask how many employees show various levels of job satisfaction. The null hypothesis is that age and job satisfaction are independent, that is, that the proportion of employees expressing a given level of job satisfaction is the same no matter which age group is considered.

Analysis involves calculating a table of *expected* values, assuming the null hypothesis about independence is true. We then compare these expected values with the observed values and ask whether the differences are reasonable if H_0 is true. The significance of the differences is gauged by the same χ^2-value of weighted squared differences. The smaller the resulting χ^2-value, the more reasonable the null hypothesis of independence. If the χ^2-value is large enough, that is, if the *P*-value is small enough, we can say that the evidence is sufficient to reject the null hypothesis and to claim that there *is* some relationship between the two variables or methods of classification.

In this type of problem,

$$df = (r - 1)(c - 1)$$

where df is the number of degrees of freedom, *r* is the number of rows, and *c* is the number of columns.

A point worth noting is that even if there is sufficient evidence to reject the null hypothesis of independence, we cannot necessarily claim any direct *causal* relationship. In other words, although we can make a statement about some link or relationship between two variables, we are *not* justified in claiming that one causes the other. For example, we may demonstrate a relationship between salary level and job satisfaction, but our methods would not show that higher salaries cause higher job satisfaction. Perhaps an employee's higher job satisfaction impresses his superiors and thus leads to larger increases in pay. Or perhaps there is a third variable, such as training, education, or personality, that has a direct causal relationship to both salary level and job satisfaction.

EXAMPLE 15.3

In a nationwide telephone poll of 1000 adults representing Democrats, Republicans, and Independents, respondents were asked two questions: their party affiliation and if their confidence in the U.S. banking system had been shaken by the savings and loan crisis. The answers, cross-classified by party affiliation, are given in the following *contingency table*.

Confidence Shaken

Observed	Yes	No	No opinion
Democrats	175	220	55
Republicans	150	165	35
Independents	75	105	20

Test the null hypothesis that shaken confidence in the banking system is independent of party affiliation. Use a 10% significance level.

Answer:

H_0: Party affiliation and shaken confidence in the banking system are independent.
H_a: Party affiliation and shaken confidence in the banking system are not independent.

The above table gives the observed results. To determine the expected values, we must first determine the row and column totals:

Row totals: 175 + 220 + 55 = 450,
150 + 165 + 35 = 350,
75 + 105 + 20 = 200.
Column totals: 175 + 150 + 75 = 400,
220 + 165 + 105 = 490,
55 + 35 + 20 = 110.

	Yes	No	No opinion	
Democrats				450
Republicans				350
Independents				200
	400	490	110	

To calculate, for example, the expected value in the upper left box, we can proceed in any of several equivalent ways. First, we could note that the proportion of Democrats is

$\frac{450}{1000} = .45$; and so, if independent, the expected number of Democrat *yes* responses is .45(400) = 180. Instead, we could note that the proportion of *yes* responses is $\frac{400}{1000} = .4$; and so, if independent, the expected number of Democrat *yes* responses is .4(450) = 180. Finally, we could note that both these calculations simply involve $\frac{(450)(400)}{1000} = 180$.

In other words, *the expected value of any box can be calculated by multiplying the corresponding row total by the appropriate column total and then dividing by the grand total.* Thus, for example, the expected value for the middle box, which corresponds to Republican *no* responses, is $\frac{(350)(490)}{1000} = 171.5$.

Continuing in this manner, we fill in the table as follows:

	Expected			
	Yes	No	No opinion	
Democrats	180	220.5	49.5	450
Republicans	140	171.5	38.5	350
Independents	80	98	22	200
	400	490	110	

[An appropriate check at this point is that each expected cell count is at least 5.]

Next we calculate the value of chi-square:

$$\chi^2 = \frac{(175-180)^2}{180} + \frac{(220-220.5)^2}{220.5} + \frac{(55-49.5)^2}{49.5}$$
$$+ \frac{(150-140)^2}{140} + \frac{(165-171.5)^2}{171.5} + \frac{(35-38.5)^2}{38.5}$$
$$+ \frac{(75-80)^2}{80} + \frac{(100-98)^2}{98} + \frac{(20-22)^2}{22}$$
$$= 3.024$$

[On the TI-84, go to MATRIX and EDIT. Put the data into a matrix. Then STAT, TESTS, χ^2-Test, will give $\chi^2 = 3.0243$. Note also that the expected values are automatically stored in a second matrix.]

Note that, once the 180, 220.5, 140, and 171.5 boxes are calculated, the other expected values can be found by using the row and column totals. Thus the number of degrees of freedom here is 4. Or we calculate

$$\text{df} = (r-1)(c-1) = (3-1)(3-1) = 4$$

The *P*-value is calculated to be $P = P(\chi^2 > 3.024) = .5538$.

With such a large *P*-value, there is *no* evidence of any relationship between party affiliation and shaken confidence in the banking system.

As for conditions to check for chi-square tests for independence, we should check that the sample is randomly chosen and that the expected values for all cells are at least 5. If a category has one of its expected cell count less than 5, we can combine categories that are logically similar (for example "disagree" and "strongly disagree"), or combine numerically small categories collectively as "other."

EXAMPLE 15.4

To determine whether men with a combination of childhood abuse and a certain abnormal gene are more likely to commit violent crimes, a study is run on a simple random sample of 575 males in the 25 to 35 age group. The data are summarized in the following table:

	Not abused, normal gene	Abused, normal gene	Not abused, abnormal gene	Abused, abnormal gene
Criminal behavior	48	21	32	26
Normal behavior	201	79	118	50

a. Is there evidence of a relationship between the four categories (based on childhood abuse and abnormal genetics) and behavior (criminal versus normal)? Explain.

b. Is there evidence that among men with the normal gene, the proportion of abused men who commit violent crimes is greater than the proportion of nonabused men who commit violent crimes? Explain.

c. Is there evidence that among men who were not abused as children, the proportion of men with the abnormal gene who commit violent crimes is greater than the proportion of men with the normal gene who commit violent crimes? Explain.

d. Is there a contradiction in the above results? Explain.

Answers:

a. A chi-square test for independence is indicated. The expected cell counts are as follows:

55.0	22.1	33.1	16.8
194.0	77.9	116.9	59.2

The condition that all cell counts are greater than 5 is met.

H_0: The four categories (based on childhood abuse and abnormal genetics) and behavior (criminal versus normal) are independent.

H_a: The four categories (based on childhood abuse and abnormal genetics) and behavior (criminal versus normal) are not independent (there is a relationship).

Running a chi-square test gives $\chi^2 = 7.752$. With df $= (r-1)(c-1) = 3$, we get $P = .0514$. Since $.0514 < .10$, the data do provide some evidence (at least at the 10% significance level) to reject H_0 and conclude that there is evidence of a relationship between the four categories (based on childhood abuse and abnormal genetics) and behavior (criminal versus normal).

b. A two-proportion z-test is indicated. We must check that n is large enough: $n_1\hat{p}_1 = 21 > 10$, $n_1(1-\hat{p}_1) = 79 > 10$, $n_2\hat{p}_2 = 48 > 10$, $n_2(1-\hat{p}_2) = 201 > 10$. We must assume simple random samples from the target population, since this is not given.

H_0: $p_1 - p_2 = 0$ (where p_1 is the proportion of abused men with normal genetics who commit violent crimes and p_2 is the proportion of nonabused men with normal genetics who commit violent crimes)

H_a: $p_1 - p_2 > 0$ (the proportion of abused men with normal genetics who commit violent crimes is greater than the proportion of nonabused men with normal genetics who commit violent crimes)

$$\hat{p}_1 = \frac{21}{100} = .21, \ \hat{p}_2 = \frac{48}{249} = .193, \ \text{and} \ \hat{p} = \frac{21+48}{100+249} = .198$$

$$\sigma_d = \sqrt{(.198)(.802)\left(\frac{1}{100}+\frac{1}{249}\right)} = .0472$$

$$\text{So} \ z = \frac{.21-.193}{.0472} = 0.360 \ \text{and} \ P = .359.$$

With such a large P-value there is no evidence to reject H_0, and thus we conclude that there is *no* evidence that the proportion of abused men with normal genetics who commit violent crimes is greater than the proportion of nonabused men with normal genetics who commit violent crimes.

c. A two-proportion z-test is indicated. We must check that n is large enough: $n_1\hat{p}_1 = 32 > 10$, $n_1(1 - \hat{p}_1) = 118 > 10$, $n_2\hat{p}_2 = 48 > 10$, $n_2(1 - \hat{p}_2) = 201 > 10$. We must assume simple random samples from the target population, since this is not given.

H_0: $p_1 - p_2 = 0$ (where p_1 is the proportion of nonabused men with abnormal genetics who commit violent crimes and p_2 is the proportion of nonabused men with normal genetics who commit violent crimes)

H_a: $p_1 - p_2 > 0$ (the proportion of nonabused men with abnormal genetics who commit violent crimes is greater than the proportion of nonabused men with normal genetics who commit violent crimes)

$$\hat{p}_1 = \frac{32}{150} = .213, \ \hat{p}_2 = \frac{48}{249} = .193, \ \text{and} \ \hat{p} = \frac{32+48}{150+249} = .2005$$

$$\sigma_d = \sqrt{(.2005)(.7995)\left(\frac{1}{150}+\frac{1}{249}\right)} = .0414$$

$$\text{So} \ z = \frac{.213-.193}{.0414} = 0.483 \ \text{and} \ P = .315.$$

With such a large P-value there is no evidence to reject H_0, and thus we conclude that there is *no* evidence that the proportion of nonabused men with the abnormal gene who commit violent crimes is greater than the proportion of nonabused men with the normal gene who commit violent crimes.

d. There is no contradiction. It is possible to have evidence of an overall relationship without significant evidence showing in a subset of the categories.

CHI-SQUARE TEST FOR HOMOGENEITY OF PROPORTIONS

In chi-square goodness-of-fit tests we work with a single variable in comparing a single sample to a population model. In chi-square independence tests we work with a single sample classified on two variables. Chi-square procedures can also be used with a single variable to compare samples from two or more populations. It is important that the samples be *simple random samples*, that they be taken *independently* of each other, that the original populations be large compared to the sample sizes, and that the expected values for all cells be at least 5. The contingency table used has a row for each sample.

EXAMPLE 15.5

In a large city, a group of AP Statistics students work together on a project to determine which group of school employees has the greatest proportion who are satisfied with their jobs. In independent simple random samples of 100 teachers, 60 administrators, 45 custodians, and 55 secretaries, the numbers satisfied with their jobs were found to be 82, 38, 34, and 36, respectively. Is there evidence that the proportion of employees satisfied with their jobs is different in different school system job categories?

Answer:

H_0: The proportion of employees satisfied with their jobs is the same across the various school system job categories.

H_a: At least two of the job categories differ in the proportion of employees satisfied with their jobs.

The observed counts are as follows:

	Satisfied	Not satisfied
Teachers	82	18
Administrators	38	22
Custodians	34	11
Secretaries	36	19

Just as we did in the previous section, we can calculate the expected value of any cell by multiplying the corresponding row total by the appropriate column total and then dividing by the grand total. In this case, this results in the following expected counts:

	Satisfied	Not satisfied	
Teachers	73.1	26.9	100
Administrators	43.8	16.2	60
Custodians	32.9	12.1	45
Secretaries	40.2	14.8	55
	190	70	260

We note that all expected cell counts are >5, and then calculate chi-square:

$$\chi^2 = \sum \frac{(obs - exp)^2}{exp} = \frac{(82 - 73.1)^2}{73.1} + \cdots + \frac{(19 - 14.8)^2}{14.8} = 8.640$$

With $4 - 1 = 3$ degrees of freedom, we calculate the *P*-value to be $P(\chi^2 > 8.640) = .0345$. [On the TI-84: χ^2cdf(8.640,1000,3).] With this small a *P*-value, there is sufficient evidence to reject H_0, and we can conclude that there is evidence that the proportion of employees satisfied with their jobs is *not* the same across all the school system job categories.

Note: On the TI-84 we could also have put the observed data into a matrix and used χ^2-Test, resulting in $\chi^2 = 8.707$ and $P = .0335$.

The difference between the test for independence and the test for homogeneity can be confusing. When we're simply given a two-way table, it's not obvious which test is being performed. The crucial difference is in the *design* of the study. Did we pick samples from each of two or more populations to compare the distribution of some variable among the different populations? If so, we are doing a test for homogeneity. Did we pick one sample from a single population and cross-categorize on two variables to see if there is an association between the variables? If so, we are doing a test for independence.

For example, if we separately sample Democrats, Republicans, and Independents to determine whether they are for, against, or have no opinion with regard to stem cell research (several samples, one variable), then we do a test for homogeneity with a null hypothesis that the distribution of opinions on stem cell research is the same among Democrats, Republicans, and Independents. However, if we sample the general population, noting the political preference and opinions on stem cell research of the respondents (one sample, two variables), then we do a test for independence with a null hypothesis that political preference is independent of opinion on stem cell research.

Finally, it should be remembered that while we used χ^2 for categorical data, the χ^2 distribution is a *continuous* distribution, and applying it to counting data is just an approximation.

HYPOTHESIS TEST FOR SLOPE OF LEAST SQUARES LINE

In addition to finding a confidence interval for the true slope, one can also perform a hypothesis test for the value of the slope. Often one uses the null hypothesis H_0: $\beta = 0$, that is, that there is no linear relationship between the two variables. Assumptions for inference for the slope of the least squares line include the following: (1) the sample must be randomly selected; (2) The scatterplot should be approximately linear; (3) there should be no apparent pattern in the residuals plot; and (4) the distribution of the residuals should be approximately normal. (This third point can be checked by a histogram, dotplot, stemplot, or normal probability plot of the residuals.)

Note that a low *P*-value tells us that if the two variables did not have some linear relationship, it would be highly unlikely to find such a random sample; however, strong evidence that there is some association does not mean the association is strong.

EXAMPLE 15.6

The following table gives serving speeds in mph (using a flat or "cannonball" serve) of ten randomly selected professional tennis players before and after using a newly developed tennis racket.

With old racquet	125	133	108	128	115	135	125	117	130	121
With new racket	133	134	112	139	123	142	140	129	139	126

a. Is there evidence of a *straight line* relationship with positive slope between serving speeds of professionals using their old and the new racquets?

(continued)

Answer. Checking the assumptions:

We are told that the data come from a *random* sample of professional players.

<u>Scatterplot is approximately linear.</u>   <u>No apparent pattern in residuals plot.</u>

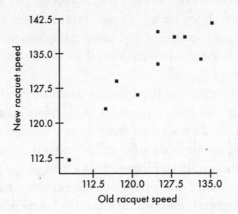
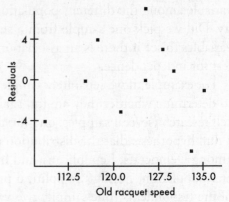

<u>Distribution of residuals is approximately normal.</u>

Checked by histogram.   *or*   Checked by normal probability plot.

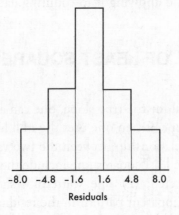
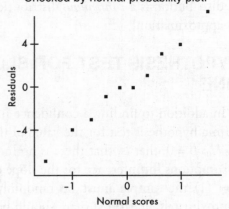

We proceed with the hypothesis test for the slope of the regression line.

$$H_0: \beta_1 = 0 \qquad H_a: \beta_1 > 0$$

Using the statistics software on a calculator (for example, `LinRegTTest` on the TI-84), gives:

New speed = 8.76 + 0.99 (Old speed)   with $t = 5.853$ and $P = .00019$

With such a small *P* value, there is very strong evidence of a straight line relationship with positive slope between serving speeds of professionals using their old and the new racquets.

b.   Interpret in context the least squares line.

Answer. With a slope of approximately 1 and a *y*-intercept of 8.76, the regression line indicates that use of the new racquet increases serving speed on the average by 8.76 mph regardless of the old racquet speed. That is, players with lower and higher old racquet speeds experience on the average the same numerical (rather than percentage) increase when using the new racquet.

It is important to be able to interpret computer output, and appropriate computer output is helpful in checking the assumptions for inference.

EXAMPLE 15.7

A company offers a 10-lesson program of study to improve students' SAT scores. A survey is made of a random sampling of 25 students. A scatterplot of improvement in total (math + verbal) SAT score versus number of lessons taken is as follows:

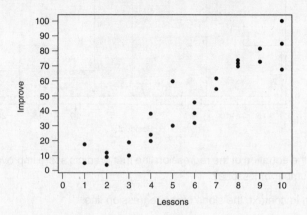

Some computer output of the regression analysis follows, along with a plot and a histogram of the residuals.

Regression Analysis: Improvement Versus Lessons

```
Dependent variable: Improvement

Predictor      Coef      SE Coef        T          P
Constant      -7.718       4.272     -1.81      0.084
Lessons       9.2854      0.6789     13.68      0.000

S = 9.744          R-Sq = 89.1%

Analysis of Variance

Source          DF        SS         MS         F        P
Regression       1      17761      17761     187.06    0.000
Residual Error  23       2184         95
Total           24      19945
```

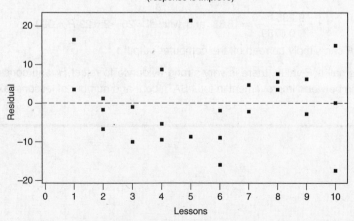

Residuals Versus Lessons
(response is Improve)

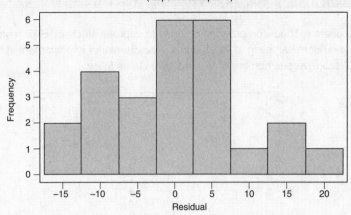

Histogram of the Residuals
(response is Improve)

a. What is the equation of the regression line that predicts score improvement as a function of number of lessons?

b. Interpret, in context, the slope of the regression line.

c. What is the meaning of R–Sq in context of this study?

d. Give the value of the correlation coefficient.

e. Perform a test of significance for the slope of the regression line.

Answers:

a. Improvement = −7.718 + 9.2854 (Lessons)

b. The slope of the regression line is 9.2854, meaning that on the average, each additional lesson is predicted to improve one's total SAT score by 9.2854.

c. R-Sq = 89.1% means that 89.1% of the variation in total SAT score improvement can be explained by variation in the number of lessons taken.

d. $r = +\sqrt{.891} = .949$, where the sign is positive because the slope is positive.

e. A test of significance of the slope of the regression line with H_0: $\beta = 0$, H_a: $\beta \neq 0$ (test of significance of correlation) is as follows:

Assumptions: We are told the data came from a random sample, the scatterplot appears to be approximately linear, there is no apparent pattern in the residuals plot, and the histogram of residuals appears to be approximately normal.

$$t = \frac{9.2854 - 0}{0.6789} = 13.68, \text{ and with df} = 25 - 2 = 23, P = .000$$

(Or *t* and *P* can simply be read off the computer output.)

With this small a *P*-value, there is very strong evidence to reject H_0 and conclude that the relationship between improvement in total SAT score and number of lessons taken is significant.

Questions on Topic Fifteen: Tests of Significance—Chi-square and Slope of Least Squares Line

Multiple-Choice Questions

Directions: The questions or incomplete statements that follow are each followed by five suggested answers or completions. Choose the response that best answers the question or completes the statement.

1. A geneticist claims that four species of fruit flies should appear in the ratio 1 : 3 : 3 : 9. Suppose that a sample of 4000 flies contained 226, 764, 733, and 2277 flies of each species, respectively. Is there sufficient evidence to reject the geneticist's claim?

   (A) The test proves the geneticist's claim.
   (B) The test proves the geneticist's claim is false.
   (C) The test does not give sufficient evidence to reject the geneticist's claim.
   (D) The test gives sufficient evidence to reject the geneticist's claim.
   (E) The test is inconclusive.

2. A medical researcher tests 640 heart attack victims for the presence of a certain antibody in their blood and cross-classifies against the severity of the attack. The results are reported in the following table:

		Severity of attack		
		Severe	Medium	Mild
Antibody test	Positive test	85	125	150
	Negative test	40	95	145

   Is there evidence of a relationship between presence of the antibody and severity of the heart attack? Test at the 5% significance level.

   (A) There is evidence that presence of the antibody causes more severe heart attacks.
   (B) There is evidence that heart attacks cause the antibody to form.
   (C) There is evidence of a relationship between presence of the antibody and occurrence of a heart attack.
   (D) There is evidence of a relationship between presence of the antibody and severity of the heart attack.
   (E) At the 5% significance level there is not sufficient evidence of any relationship.

3. A study of accident records at a large engineering company in England (*The Lancet*, October 22, 1994, page 1137) reported the following number of injuries on each shift for 1 year:

Shift:	Morning	Afternoon	Night
Number of injuries:	1372	1578	1686

Is there sufficient evidence to say that the number of accidents on the three shifts are not the same? Test at the .05, .01, and .001 levels.

(A) There is sufficient evidence at all three levels to say that the number of accidents on each shift are not the same.

(B) There is sufficient evidence at both the .05 and .01 levels but not at the .001 level.

(C) There is sufficient evidence at the .05 level but not at the .01 and .001 levels.

(D) There is sufficient evidence at the .001 level but not at the .01 and .05 levels.

(E) There is not sufficient evidence at any of these levels.

4. A 4-year study analyzing blood cholesterol in men more than 70 years old, reported in *The New York Times*, noted how many men with different cholesterol levels suffered a nonfatal or fatal heart attack.

	Low cholesterol	Medium cholesterol	High cholesterol
Nonfatal heart attacks	29	17	18
Fatal heart attacks	19	20	9

Is there evidence of a relationship? Does level of cholesterol seem to predict risk of heart disease for men more than 70 years old? Test at the 10% significance level.

(A) At the 10% significance level, the given data prove that level of cholesterol predicts risk of heart disease for men more than 70 years old.

(B) At the 10% significance level, the given data prove that level of cholesterol does not predict risk of heart disease for men more than 70 years old.

(C) Since .221 > .10, there is sufficient evidence at the 10% significance level of a relationship between cholesterol level and risk of heart disease for men more than 70 years old.

(D) Since .221 > .10, there is not sufficient evidence at the 10% significance level of a relationship between cholesterol level and risk of heart disease for men more than 70 years old.

(E) Relationship and causation give contradictory conclusions in this problem.

5. A surprising study of 1437 male hospital admissions reported in *The New York Times* (February 24, 1993, page C12) found that, of 665 patients admitted with heart attacks, 214 had vertex baldness, while of the remaining 772 non-heart-related admissions, 175 had vertex baldness. Is this evidence sufficient to say that there is a relationship between heart attacks and vertex baldness?

(A) Yes, because χ^2 = 4.05 with *P* = .044
(B) Yes, because χ^2 = 16.39 with *P* = .000052
(C) No, because χ^2 = 16.39 with *P* = .000052
(D) No, because χ^2 = 4.05 with *P* = .0025
(E) Yes, because χ^2 = 16.39 with *P* = .0025

6. Is hair loss pattern related to body mass index? One study on 769 men (*Journal of the American Medical Association*, February 24, 1993, page 1000) showed the following numbers:

Hair loss pattern

		None	Frontal	Vertex
	<25	137	22	40
Body mass index	25–28	218	34	67
	>28	153	30	68

The χ^2-value is 4.18. At the 5% and 10% significance levels is there a relationship between hair loss pattern and body mass index?

(A) There is sufficient evidence of a relationship at both the 5% and 10% levels.
(B) There is sufficient evidence of a relationship at the 5% level but not at the 10% level.
(C) There is sufficient evidence of a relationship at the 10% level but not at the 5% level.
(D) There is not sufficient evidence of a relationship at either the 5% or the 10% level.
(E) There is not enough information to answer this question.

7. A highway superintendent states that five bridges into a city are used in the ratio 2 : 3 : 3 : 4 : 6 during the morning rush hour. A highway study of a simple random sample of 6000 cars indicates that 720, 970, 1013, 1380, and 1917 cars use the five bridges, respectively. Can the superintendent's claim be rejected at the 2.5% or 5% level of significance?

(A) There is sufficient evidence to reject the claim at either of these two levels.
(B) There is sufficient evidence to reject the claim at the 2.5% level but not at the 5% level.
(C) There is sufficient evidence to reject the claim at the 5% level but not at the 2.5% level.
(D) There is not sufficient evidence to reject the claim at either of these two levels.
(E) There is not sufficient information to answer this question.

8. Is there a relationship between fitness level and smoking habits? A study reported in the *New England Journal of Medicine* (February 25, 1993, page 535) cross-classified individuals in four categories of fitness and four categories of smoking habits:

	Fitness level		
	Low	Medium	High	
Never smoked	113	113	110	159
Former smokers	119	135	172	190
1 to 9 cigarettes daily	77	91	86	65
≥ 10 cigarettes daily	181	152	124	73

The χ^2-value is 87.6. Is there a relationship between fitness level and smoking habits?

(A) The data prove that smoking causes lower fitness levels.

(B) The data prove that lower fitness levels cause people to smoke.

(C) There is sufficient evidence at the 0.1% significance level of a relationship between fitness level and smoking habits.

(D) There is sufficient evidence at the 1% level, but not at the 0.1% level, of a relationship between fitness level and smoking habits.

(E) There is sufficient evidence at the 10% level, but not at the 1% level, of a relationship between fitness level and smoking habits.

9. A building inspector inspects four major construction sites every day. In a sample 100 days, the number of times each site passed inspection was

Site:	A	B	C	D
Passed:	45	33	67	83

Test the hypothesis that sites C and D pass inspection twice as often as sites A and B, that is, that the frequency of passing inspection is in the ratio 1 : 1 : 2 : 2.

(A) There is sufficient evidence at both the 10% and 5% significance levels that there is not a good fit with the indicated ratio.

(B) There is sufficient evidence at the 10% level, but not at the 5% level, that there is not a good fit with the indicated ratio.

(C) There is sufficient evidence at the 5% level, but not at the 10% level, that there is not a good fit with the indicated ratio.

(D) There is not sufficient evidence at either the 10% or 5% significance levels that there is not a good fit with the indicated ratio.

(E) There is not sufficient information to answer the question.

Answer Key

1. **C**	4. **D**	7. **C**
2. **D**	5. **B**	8. **C**
3. **A**	6. **D**	9. **D**

Answers Explained

1. **(C)** Since $1 + 3 + 3 + 9 = 16$, according to the geneticist the expected number of fruit flies of each species is

$$\tfrac{1}{16}\big(4000\big) = 250, \tfrac{3}{16}\big(4000\big) = 750, \tfrac{3}{16}\big(4000\big)$$

$= 750$, and $\tfrac{9}{16}\big(4000\big) = 2$. We calculate chi-square:

$$\chi^2 = \frac{\big(226 - 250\big)^2}{250} + \frac{\big(764 - 750\big)^2}{750} + \frac{\big(733 - 750\big)^2}{750} + \frac{\big(2277 - 2250\big)^2}{2250} = 3.27$$

With df $= 4 - 1 = 3$, the *P*-value is $P(\chi^2 > 3.27) = .352$. With such a large *P*-value, there is not sufficient evidence to reject H_0. The geneticist's claim should not be rejected.

2. **(D)** We have H_0: independence (no relation between presence of antibody and severity of attack), H_a: dependence (severity of attack related to presence of antibody), $\alpha = .05$, and df $= (2 - 1)(3 - 1) = 2$. Totaling rows and columns yields

	Severe	Medium	Mild	
Positive				360
Negative				280
	125	220	295	

The expected value for each box is calculated by multiplying the row total by the column total and dividing by 640 as follows: $\frac{(360)(125)}{640} = 70.3$,

$\frac{(360)(220)}{640} = 123.8$, $\frac{(360)(295)}{640} = 165.9$, $\frac{(280)(125)}{640} = 54.7$, $\frac{(280)(220)}{640} = 94.2$, and

$\frac{(280)(295)}{640} = 129.1$.

Expected

	Severe	Medium	Mild	
Positive	70.3	123.8	165.9	360
Negative	54.7	96.2	129.1	280
	125	220	295	

Then

$$\chi^2 = \frac{(85-70.3)^2}{70.3} + \frac{(125-123.8)^2}{123.8} + \frac{(150-165.9)^2}{165.9}$$
$$+ \frac{(40-54.7)^2}{54.7} + \frac{(95-96.2)^2}{96.2} + \frac{(145-129.1)^2}{129.1}$$
$$= 10.533$$

With $\alpha = .05$ and df = 2, the critical χ^2-value is 5.99. Since $10.533 > 5.99$, there is sufficient evidence to reject the null hypothesis of independence. Thus, at the 5% significance level, there is evidence of a relationship between presence of antibody and severity of heart attack.

3. **(A)** $\frac{1372+1578+1686}{3} = 1545$

$$\chi^2 = \frac{(1372-1545)^2}{1545} + \frac{(1578-1545)^2}{1545} + \frac{(1686-1545)^2}{1545} = 32.94$$

With df = 3 − 1 = 2, the P-value is $P(\chi^2 > 32.94) = .000$. With such a small P-value, there is very strong evidence to reject H_0. Since .000 is less than .05, .01, and .001, there is sufficient evidence at all these levels to say that the number of accidents on each shift is not the same.

4. **(D)**

	Low cholesterol	Medium cholesterol	High cholesterol	
Nonfatal heart attacks	27.4	21.1	15.5	64
Fatal heart attacks	20.6	15.9	11.5	48
	48	37	27	

$$\chi^2 = \frac{(29-27.4)^2}{27.4} + \frac{(17-21.1)^2}{21.1} + \frac{(18-15.5)^2}{15.5}$$
$$+ \frac{(19-20.6)^2}{20.6} + \frac{(20-15.9)^2}{15.9} + \frac{(9-11.5)^2}{11.5}$$
$$= 3.018$$

With df = (2 − 1)(3 − 1) = 2, the P-value is $P(\chi^2 > 3.018) = .221$. With such a large P-value, there is not sufficient evidence to reject H_0. Since .221 is greater than .10, there is not sufficient evidence at the 10% significance level of a relationship between cholesterol level and risk of heart disease for men more than 70 years old.

5. **(B)**

	Vertex baldness	No vertex baldness	
Heart attacks	214	451	665
Other admissions	175	597	772
	389	1048	1437

Expected	
180	485
209	563

$$\frac{(665)(389)}{1437} = 180, \frac{(665)(1048)}{1437} = 485, \frac{(772)(389)}{1437} = 209, \text{ and } \frac{(772)(1048)}{1437} = 563.$$

$$\chi^2 = \frac{(214-180)^2}{180} + \frac{(451-485)^2}{485} + \frac{(175-209)^2}{209} + \frac{(597-563)^2}{563} = 16.39$$

(Input of the original matrix into a calculator like the TI-83 quickly gives χ^2 = 16.37.) With df = $(2-1)(2-1) = 1$ the *P*-value is $P = P(\chi^2 > 16.39) = .000052$. With such a small *P*-value, there is sufficient evidence to say that a relationship exists between heart attacks and vertex baldness.

6. **(D)** With df = $(3-1)(3-1) = 4$, the *P*-value is $P = P(\chi^2 > 4.18) = .3822$. Since .3822 is greater than both .05 and .10, there is not sufficient evidence at either level of a relationship between hair loss and body mass index.

7. **(C)** We have $2 + 3 + 3 + 4 + 6 = 18$, and thus the expected numbers according to the superintendent's claim are

$$\frac{2}{18}(6000) = 667 \qquad \frac{3}{18}(6000) = 1000$$
$$\frac{4}{18}(6000) = 1333 \qquad \frac{6}{18}(6000) = 2000$$

Using chi-square to test for goodness of fit, we calculate

$$\chi^2 = \frac{(720-667)^2}{667} + \frac{(970-1000)^2}{1000} + \frac{(1013-1000)^2}{1000}$$
$$+ \frac{(1380-1333)^2}{1333} + \frac{(1917-2000)^2}{2000}$$
$$= 10.38$$

With df = $5 - 1 = 4$, the *P*-value is $P = P(\chi^2 > 10.38) = .0345$. Note that .0345 < .05, but .0345 > .025. Thus there is sufficient evidence to reject the superintendent's claim at the 5% significance level but not at the 2.5% level.

8. **(C)** Such hypothesis tests suggest relationships but do not prove cause and effect. With df = $(4-1)(4-1) = 9$, the *P*-value is $P = P(\chi^2 > 87.6) = .0000$. Thus there is sufficient evidence at the 0.1% significance level of a relationship between fitness level and smoking habits.

9. **(D)** We have H_0: good fit with a 1 : 1 : 2 : 2 ratio. Noting that $45 + 33 + 74 + 76 = 228$, $\left(\frac{1}{6}\right)228 = 38$, and $\left(\frac{2}{6}\right)228 = 76$, we have the following table of expected values:

Site:	A	B	C	D
Passed inspection:	38	38	76	76

The difference between the observed and expected values is measured by chi-square:

$$\chi^2 = \frac{7^2}{38} + \frac{5^2}{38} + \frac{9^2}{76} + \frac{7^2}{76} = 3.65$$

With df = $4 - 1 = 3$, the *P*-value is $P = P(\chi^2 > 3.65) = .3018$. With this high a *P*-value, there is not sufficient evidence to reject H_0 at either the 5% or the 10% significance level. There is no evidence that the fit with the indicated ratios is not good.

Free-Response Questions

Directions: You must show all work and indicate the methods you use. You will be graded on the correctness of your methods and on the accuracy of your final answers.

Five Open-Ended Questions

1. Suppose your mathematics instructor gives as a homework assignment the problem of testing the fairness of a certain die. You are asked to roll the die 6000 times and to note how often it comes up 1, 2, 3, 4, 5, or 6. Tossing the die soon becomes tiresome, so you decide to invent some "reasonable" data. Being careful to make the total 6000, you write

	Number on die					
	1	2	3	4	5	6
Observed	988	991	1010	990	1013	1008

Your instructor gives you an F, claiming that your results are too good to be true! Explain.

2. A restaurant owner surveys a random sample of 385 customers to determine whether customer satisfaction is related to gender and age. The numerical data are presented in the following table:

	Young male	Young female	Older male	Older female
Satisfied	25	30	135	112
Not satisfied	8	16	22	37

Is there evidence of a relationship between customer satisfaction and age and gender?

3. Five partners in a law firm brought in the following numbers of new clients during the past year:

Partner:	Jones	Smith	Brown	Allen	Cross
Number of new clients:	35	42	22	41	30

Is there sufficient evidence at the 5% or 10% significance level that the partners do not bring in equal numbers of new clients? Show your work.

4. A mass transportation study concludes that in large cities 40% of commuters drive to work by themselves, 15% drive in car pools, 20% take trains, and 25% take buses. A city council wishes to examine whether commuters in their city follow this pattern. A random survey of 600 commuters yields the following breakdown: 263 solo drivers, 101 in car pools, 106 train riders, and 130 bus riders. Is there evidence that the city's commuters do not follow the national pattern? Show your work.

5. A study is conducted to determine if there is a linear relationship between the number of hours of sleep a student receives the night before taking the SAT math exam and the score on that exam. The data from 15 randomly interviewed students follow:

Sleep (hrs)	4	2	9	8	14	2	11	14	7	4	1	9	9	10	5
Math score	423	520	550	309	690	401	470	582	284	440	452	568	339	355	472

Is there evidence of a linear relationship?

Answers Explained

1. We have H_0: good fit to uniform distribution; that is, the die is fair. The expected values are all 1000, so

$$\chi^2 = \frac{12^2}{1000} + \frac{9^2}{1000} + \frac{10^2}{1000} + \frac{10^2}{1000} + \frac{13^2}{1000} + \frac{8^2}{1000} = 0.658$$

What is the conclusion if $\alpha = .01$? $.05$? $.10$?

With df = $6 - 1 = 5$, the critical χ^2-values are 15.09, 11.07, and 9.24. Since 0.658 is less than each of these, there is not enough evidence to claim that the die is unfair. What is the conclusion if we're willing to accept $\alpha = .15$? $.20$? $.25$? (.25 is the limit of Table C.)

These numbers give critical χ^2-values of 8.12, 7.29, and 6.63, respectively, which are all far above our χ^2-value of 0.658. Even if the die is fair, the probability of obtaining such a good-fitting sample, with $\chi^2 = 0.658$, is extremely small. Your instructor reasonably concluded that you simply made up the data. Your results were too good to be true!

2. Total the rows and columns and then find the expected numbers if customer satisfaction is independent of gender and age by multiplying row totals by column totals and dividing by 385:

25.9	36.1	123.1	116.9	302
7.1	9.9	33.9	32.1	83
33	46	157	149	385

[Note: we are given that the sample is random, and we calculate that all expected cells are ≥ 5.]

The difference between observed and expected values is measured by chi-square:

$$\chi^2 = \frac{(25 - 25.9)^2}{25.9} + \frac{(34 - 36.1)^2}{36.1} + \frac{(129 - 123.1)^2}{123.1} + \frac{(112 - 116.9)^2}{116.9}$$

$$+ \frac{(8 - 7.1)^2}{7.1} + \frac{(12 - 9.9)^2}{9.9} + \frac{(28 - 33.9)^2}{33.9} + \frac{(37 - 32.1)^2}{32.1}$$

$$= 11.1$$

With df = $(2 - 1)(4 - 1) = 3$, the P-value is $P = P(\chi^2 > 11.1) = .0112$. With this small a P-value, there is sufficient evidence to reject independence and conclude that customer satisfaction is related to age and gender.

3. If the $35 + 42 + 22 + 41 + 30 = 170$ new clients were uniformly distributed, each partner would have brought in $\frac{170}{5} = 34$ new clients. The difference between the expected and observed numbers is measured by chi-square:

$$\chi^2 = \frac{1^2}{34} + \frac{8^2}{34} + \frac{12^2}{34} + \frac{7^2}{34} + \frac{4^2}{34} = 8.06$$

With df = $5 - 1 = 4$, the *P*-value is $P = P(\chi^2 > 8.06) = .0894$. Thus there is sufficient evidence at the 10% level, but not at the 5% level, that the partners bring in different numbers of new clients.

4. If there is a good fit with the transportation study pattern, the expected numbers are $.40(600) = 240$, $.15(600) = 90$, $.20(600) = 120$, and $.25(600) = 150$. Note that all expected values are >5. The difference between the observed and expected values is measured by chi-square:

$$\chi^2 = \frac{(263 - 240)^2}{240} + \frac{(101 - 90)^2}{90} + \frac{(106 - 120)^2}{120} + \frac{(130 - 150)^2}{150} = 7.85$$

With df = $4 - 1 = 3$, the *P*-value is $P = P(\chi^2 > 7.85) = .0492$. Thus there is sufficient evidence to reject the null hypothesis of a good fit and say that the city's commuters do not follow the national pattern.

5. Checking the assumptions:

We are told that the data come from a *random* sample of students.

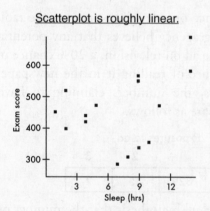

Scatterplot is roughly linear.

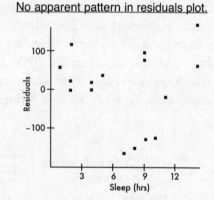

No apparent pattern in residuals plot.

<u>Distribution of residuals is approximately normal.</u>
Histogram is unimodal, roughly symmetric, with no outliers.

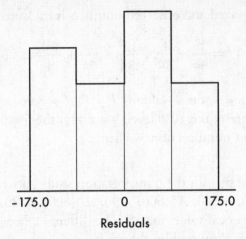

Residuals

We proceed with the hypothesis test for the slope of the regression line.

$$H_0: \beta_1 = 0 \qquad H_a: \beta_1 \neq 0$$

Using the statistics software on a calculator (for example, `LinRegTTest` on the TI-84) gives:

Test score = 385.4 + 9.85 (Hours of sleep)   with $t = 1.418$ and $P = .180$

Conclusion in context: With such a large P value, there is no evidence of a straight line relationship between hours of sleep and SAT math scores.

Six Investigative Tasks

1. Suppose that a commercial is run once on television, once on the radio, and once in a newspaper. The advertising agency believes that any potential consumer has a 20% chance of seeing the ad on television, a 20% chance of hearing it on the radio, and a 20% chance of reading it in the newspaper. In a telephone survey of 800 consumers, the numbers claiming to have been exposed to the ad 0, 1, 2, or 3 times are as follows.

	Exposures to ad			
	0	1	2	3
Observed number of people	434	329	35	2

At the 1% significance level, test the null hypothesis that the number of times any consumer saw the ad follows a binomial distribution with $\pi = .2$.

2. Suppose that in a telephone survey of 800 registered voters, the data are cross-classified both by gender of respondent and by respondent's opinion on an environmental bond issue.

Bond issue

	For	Against
Men	450	150
Women	160	40

a. Determine the χ^2-value for these data.

b. Now use the approach testing the significance of the difference between the proportions $\hat{p}_1 = \frac{450}{600} = .75$ and $\hat{p}_2 = \frac{160}{200} = .8$ and find the observed z-value.

c. Compare the square of this z-value to the χ^2-value determined above.

(It can be shown more generally that the χ^2-value for a two-by-two table is equal to the squared observed z-value for the difference in proportions.)

3. The number of oil paintings drawn per year by the artist J. K. during his 25-year professional career is shown in the following scatterplot:

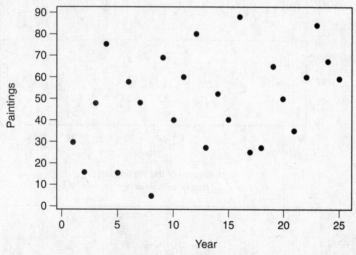

Some computer output of regression analysis follows, along with a plot and a histogram of the residuals.

Regression Analysis: Paintings Versus Year

Response variable: Paintings

Predictor	Coef	SE Coef	T	P
Constant	35.420	8.951	3.96	0.001
Year	1.0385	0.6021	1.72	0.098

S = 21.71   R-Sq = 11.5%

Analysis of Variance

Source	DF	SS	MS	F	P
Regression	1	1401.9	1401.9	2.97	0.098
Residual Error	23	10839.9	471.3		
Total	24	12241.8			

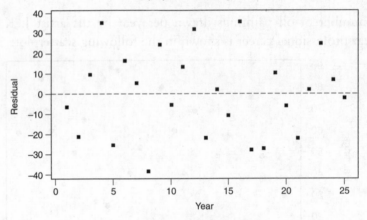

Residuals Versus Year
(response is Painting)

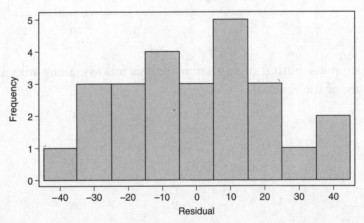

Histogram of the Residuals
(response is Painting)

a. What is the equation of the regression line that predicts the number of paintings as a function of number of years into the artist's career?

b. Interpret the slope of the regression line in context.

c. Interpret R-Sq in context of this study.

d. What is the correlation?

e. Perform a test of significance on the slope of the regression line.

4. A survey of college students focuses on a cross-classification by year in school and positive or negative feelings about the college experience.

Year in school

	Freshman	Sophomore	Junior	Senior
Positive	105	100	90	125
Negative	35	50	30	25

a. Determine the value of χ^2.

b. There are three degrees of freedom in this example. Suppose the table is split into the following three smaller tables, each having a single degree of freedom:

	Freshman	Sophomore
Pos.	105	100
Neg.	35	50

	Junior	Senior
Pos.	90	125
Neg.	30	25

	Underclass	Upperclass
Pos.	205	215
Neg.	85	55

Determine the three χ^2-values, one from each of these tables.

c. Find a relationship between the χ^2-value of the original table and the three χ^2-values calculated above.

5. A random sample of 12 families yields the following data on IQs of father, mother, and child:

Family	1	2	3	4	5	6	7	8	9	10	11	12
Father's IQ	99	117	96	94	102	103	102	109	87	96	99	108
Mother's IQ	85	99	99	111	110	108	109	104	90	85	101	106
Child's IQ	82	115	106	114	115	107	101	113	95	84	103	117

While the mean IQs of the fathers and mothers are both 101, the mean IQ of the children is 104.

The regression analysis of the data follows, along with plots of the residuals.

Regression Analysis: Child IQ Versus Father IQ

Dependent variable: Child IQ

```
Predictor       Coef    SE Coef       T       P
Constant       23.15      41.54    0.56   0.590
Father IQ     0.8038     0.4102    1.96   0.078

S = 10.68    R-Sq = 27.8%    R-Sq (adj) = 20.5%
```

Analysis of Variance

```
Source          DF        SS      MS       F       P
Regression       1     438.1   438.1    3.84   0.078
Residual Error  10    1140.6   114.1
Total           11    1578.7
```

Regression Analysis: Child IQ Versus Mother IQ

Dependent variable: Child IQ

```
Predictor       Coef    SE Coef       T       P
Constant       -3.30      22.30   -0.15   0.885
Mother IQ     1.0701     0.2208    4.85   0.001

S = 6.867    R-Sq = 70.1%    R-Sq (adj) = 67.1%
```

Analysis of Variance

```
Source          DF        SS        MS       F       P
Regression       1     1107.2    1107.2   23.48   0.001
Residual Error  10      471.5      47.2
Total           11     1578.7
```

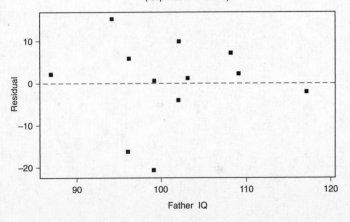

Residuals Versus Father IQ
(response is Child IQ)

Residuals Versus Mother IQ
(response is Child IQ)

a. Do the data support evidence that a child's IQ is greater than his/her father's IQ? Explain.

b. Do the data support evidence that a child's IQ is greater than his/her mother's IQ? Explain.

c. Is the IQ of the father or the IQ of the mother a better predictor of the IQ of the child? Explain.

d. Given the analysis in part c, predict the IQ of a child whose father's IQ is 110 and whose mother's IQ is 100.

e. Given the analysis in part c, what is the approximate range of IQs for children whose father's IQ is 110 and mother's IQ is 100?

6. A purchasing agent weighs a sample of 225 bags of rice labeled as containing 50 pounds each with the following results.

Weight (lb):	Below 49.25	49.25– 49.75	49.75– 50.25	50.25– 50.75	Above 50.75
Observed number of bags:	25	61	70	59	10

Test the null hypothesis that the data follow a normal distribution with $\mu = 50$ and $\sigma = 0.5$. Show your work.

Answers Explained

1. The complete binomial distribution with $\pi = .2$ and $n = 3$ is as follows:

$$P(0) = (.8)^3 = .512$$
$$P(1) = 3(.2)(.8)^2 = .384$$
$$P(2) = 3(.2)^2(.8) = .096$$
$$P(3) = (.2)^3 = .008$$

Multiplying each of these probabilities by 800 gives the expected number of occurrences: $.512(800) = 409.6$, $.384(800) = 307.2$, $.096(800) = 76.8$, and $.008(800) = 6.4$.

	Exposures to ad			
	0	1	2	3
Expected number of people	409.6	307.2	76.8	6.4

Check the assumptions:

1. *Randomization:* We must assume that the 800 consumers who are reached form a representative sample.
2. All the expected cells are ≥ 5.

State the hypotheses:

H_0: The number of times consumers saw the ad follows a binomial with $\pi = .2$

H_a: Consumer ad viewings does not follow a binomial with $\pi = .2$

Show the mechanics:

$$\chi^2 = \frac{(434 - 409.6)^2}{409.6} + \frac{(329 - 307.2)^2}{307.2} + \frac{(35 - 76.8)^2}{76.8}$$
$$+ \frac{(2 - 6.4)^2}{6.4} = 28.776$$

With df $= 4 - 1 = 3$, the P-value is $P = P(\chi^2 > 28.776) = .0000$.

[On the TI-84, $P = \chi^2\text{cdf}(28.776, 1000, 3) = .0000025$.]

Conclusion in context:

With this small a P-value, there is sufficient evidence to reject H_0 and to conclude that the number of ads seen by each consumer does *not* follow a binomial distribution with $\pi = .2$.

2. *a.* The table of expected values is calculated to be

	Bond issue		
	For	Against	
Men	457.5	142.5	600
Women	152.5	47.5	200
	610	190	800

[Note that each expected cell count is ≥ 5.]

$$\chi^2 = \frac{(450 - 457.5)^2}{457.5} + \frac{(150 - 142.5)^2}{142.5} + \frac{(160 - 152.5)^2}{152.5}$$
$$+ \frac{(40 - 47.5)^2}{47.5}$$
$$= 2.071$$

b. Let $\hat{p} = \frac{450+160}{800} = .7625$ and calculate

$$\sigma_d = \sqrt{(.7625)(.2375)\left(\frac{1}{600} + \frac{1}{200}\right)} = .0347$$

The observed z-value is $\frac{.75-.8}{\sigma_d} = 1.439$.

c. $(1.439)^2 = 2.071$. We see that the chi-square test for independence and the two proportion z-test give related test statistics.

3. *a.* Number of paintings = 35.420 + 1.0385 (Year)
b. The slope of the regression line is 1.0385, meaning that the predicted increase in oil paintings by this artist was 1.0385 per year.
c. R-Sq = 11.5% means that 11.5% of the variation in number of paintings each year can be explained by variation in the years into the career.
d. $r = +\sqrt{.115} = .3$ where the sign is positive because the slope is positive.
e. A test of significance of the slope of the regression line with $H_0\colon \beta = 0$, $H_a\colon \beta \neq 0$ is as follows:

Assumptions: The scatterplot appears very roughly linear, there is no apparent pattern in the residuals plot, and the histogram of residuals appears to be approximately normal.

$$t = \frac{1.0385-0}{0.6021} = 1.72, \text{ and with df} = 25 - 2 = 23, P = .099$$

(Or t and P can simply be read off the computer output.)

Since $P = .099 < .10$, there is some evidence to reject H_0 and conclude that at the 10% significance level the relationship between number of paintings and years into career is significant. (Although, with $r^2 = .115$, years into career explains only a small portion of the differences in number of paintings per year.)

4. *a.*

	Freshman	Sophomore	Junior	Senior	
Positive	105	112.5	90	112.5	420
Negative	35	37.5	30	37.5	140
	140	150	120	150	560

[Note that each expected cell count is ≥ 5.]

$$\chi^2 = 0 + \frac{12.5^2}{112.5} + 0 + \frac{12.5^2}{112.5} + 0 + \frac{12.5^2}{37.5} + 0 + \frac{12.5^2}{37.5} = 11.11$$

b.

	Fresh-man	Soph-omore			Junior	Senior			Under-class	Upper-class	
Pos.	99	106	205	Pos.	95.6	119.4	215	Pos.	217.5	202.5	420
Neg.	41	44	85	Neg.	24.4	30.6	55	Neg.	72.5	67.5	140
	140	150	290		120	150	270		290	270	560

[Note that each expected cell count is ≥ 5.]

$$\chi^2 = \frac{6^2}{99} + \frac{6^2}{106} + \frac{6^2}{41} + \frac{6^2}{44} = 2.40$$

$$\chi^2 = \frac{5.6^2}{95.6} + \frac{5.6^2}{119.4} + \frac{5.6^2}{24.4} + \frac{5.6^2}{30.6} = 2.90$$

$$\chi^2 = \frac{12.5^2}{217.5} + \frac{12.5^2}{202.5} + \frac{12.5^2}{72.5} + \frac{12.5^2}{67.5} = 5.96$$

c. Note that the first χ^2-value is equal to the sum of the other three χ^2-values (up to round-off error), illustrating the *additive property* of the χ^2-statistic.

5. *a.* The proper hypothesis test is a matched-pairs *t*-test on the set of differences, $\{-17, -2, 10, 20, 13, 4, -1, 4, 8, -12, 4, 9\}$, with H_0: $\mu_D = 0$, H_a: $\mu_D > 0$. (Or use "father" minus "child," changing the signs of the elements in the set and switching to H_a: $\mu_D < 0$.)

Assumptions to be checked are that samples are random (given) and that the population of differences is approximately normal (use dotplot, stemplot, boxplot, or histogram of differences).

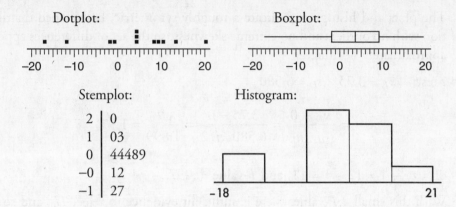

Dotplot:

Boxplot:

Stemplot:

Histogram:

The plots and histogram indicate a roughly symmetric, bell-shaped distribution with no outliers and no extreme skewness, so the set of differences appears approximately normal.

t-test: $\bar{x}_D = 3.333$ $s_D = 10.30$

$$t = \frac{\bar{x}_D - 0}{s_D / \sqrt{n_D}} = \frac{3.333 - 0}{10.30 / \sqrt{12}} = \frac{3.333}{2.973} = 1.121$$

df $= n_D - 1 = 12 - 1 = 11$, and *P*-value $= .143$.

With this large a *P*-value, there is not sufficient evidence to reject H_0, and so we conclude that the differences are not significant, and that the data do not suggest that a child's IQ is greater than his/her father's IQ.

b. The proper hypothesis test is a matched pairs *t*-test on the set of differences, {−3, 16, 7, 3, 5, −1, −8, 9, 5, −1, 2, 11}, with $H_0: \mu_D = 0$, $H_a: \mu_D > 0$. (Or use "mother" minus "child," changing the signs of the elements in the set and switching to $H_a: \mu_D < 0$.)

Assumptions to be checked are that samples are random (given) and that the population of differences is approximately normal (use histogram, boxplot, stemplot, or normal probability plot).

Histogram:

Boxplot:

Stemplot:

Normal probability plot:

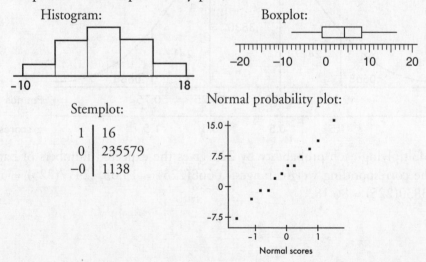

The plots and histogram indicate a roughly symmetric, bell-shaped distribution with no outliers and no extreme skewness, so the set of differences appears approximately normal.

t-test: $\bar{x}_D = 3.75$ $s_D = 6.580$

$$t = \frac{\bar{x}_D - 0}{s_D/\sqrt{n_D}} = \frac{3.75 - 0}{6.580/\sqrt{12}} = \frac{3.75}{1.899} = 1.97$$

df = $n_D - 1 = 12 - 1 = 11$, and P-value = .037.

With this small a P-value, there is sufficient evidence to reject H_0, and so we conclude that the differences are significant, and that the data *do* suggest that a child's IQ is greater than his/her mother's IQ.

c. In both regression analyses the scatterplot of residuals shows no apparent pattern, however, while 70.1% of the variance in child's IQ is explained by variance in mother's IQ, only 27.8% of the variance in child's IQ is explained by variance in father's IQ.

d. Using the regression equation giving child's IQ as a function of mother's IQ gives 1.0701(100) – 3.30 = 103.7.

e. The regression output gives a standard deviation of S = 6.867. Using the 103.7 IQ estimate from part *d* gives a range of IQs for children whose mother's IQ is 100 to be 103.7 ± 3(6.867) = 103.7 ± 20.6. Thus for a mother's IQ of 100, almost all children will have an IQ between 83 and 124.

6. The z-scores for 50.25 and 50.75 are, respectively, $\frac{50.25-50}{0.5} = 0.5$ and $\frac{50.75-50}{0.5} = 1.5$. Similarly, 49.75 and 49.25 have z-scores of –0.5 and –1.5, respectively. Using Table A, we find these values:

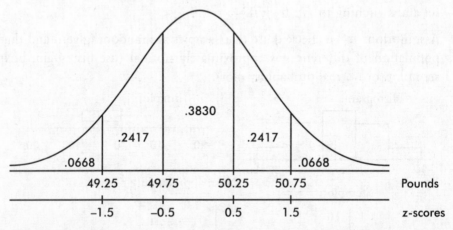

Multiplying each probability by 225 gives the expected numbers of bags for the corresponding weight ranges: .0668(225) = 15.03, .2417(225) = 54.38, .3830(225) = 86.18.

Weight (lb):	Below 49.25	49.25– 49.75	49.75– 50.25	50.25– 50.75	Above 50.75
Expected number of bags:	15.03	54.38	86.18	54.38	15.03

The difference between observed and expected numbers can be measured using chi-square. Thus

$$\chi^2 = \frac{(25 - 15.03)^2}{15.03} + \frac{(61 - 54.38)^2}{54.38}$$
$$+ \frac{(70 - 86.18)^2}{86.18} + \frac{(59 - 54.38)^2}{54.38} + \frac{(10 - 15.03)^2}{15.03}$$
$$= 12.533$$

H_0: the distribution of weights of rice bags is normal with $\mu = 50$ and $\sigma = 0.5$
H_a: the distribution is not normal with $\mu = 50$ and $\sigma = 0.5$

With df = $5 - 1 = 4$, the *P*-value is $P = P(\chi^2 > 12.533) = .0138$. With this small a *P*-value, we have sufficient evidence to reject H_0, and the purchasing agent should conclude that the distribution of weights of rice bags is not normal.

Answer Sheet

PRACTICE EXAMINATION 1

1. A B C D E	11. A B C D E	21. A B C D E	31. A B C D E
2. A B C D E	12. A B C D E	22. A B C D E	32. A B C D E
3. A B C D E	13. A B C D E	23. A B C D E	33. A B C D E
4. A B C D E	14. A B C D E	24. A B C D E	34. A B C D E
5. A B C D E	15. A B C D E	25. A B C D E	35. A B C D E
6. A B C D E	16. A B C D E	26. A B C D E	36. A B C D E
7. A B C D E	17. A B C D E	27. A B C D E	37. A B C D E
8. A B C D E	18. A B C D E	28. A B C D E	38. A B C D E
9. A B C D E	19. A B C D E	29. A B C D E	39. A B C D E
10. A B C D E	20. A B C D E	30. A B C D E	40. A B C D E

Practice Examination 1

Section I

===

Questions 1–40

Spend 90 minutes on this part of the exam.

> **Directions:** The questions or incomplete statements that follow are each followed by five suggested answers or completions. Choose the response that best answers the question or completes the statement.

1. Following is a histogram of home sale prices (in thousands of dollars) in one community:

   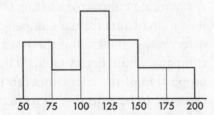

   Which of the following statements are true?

   I. The median price was $125,000.
   II. More homes sold for between $100,000 and $125,000 than for over $125,000.
   III. $10,000 is a reasonable estimate of the standard deviation in selling prices.

   (A) I only
   (B) II only
   (C) III only
   (D) All are true.
   (E) None is true.

2. Which of the following can stemplots show?

   I. Symmetry
   II. Gaps
   III. Clusters
   IV. Outliers

   (A) I, II, and III
   (B) I, II, and IV
   (C) I, III, and IV
   (D) II, III, and IV
   (E) I, II, III, and IV

3. The average yearly snowfall in a city is 55 inches. What is the standard deviation if 15% of the years have snowfalls above 60 inches? Assume yearly snowfalls are normally distributed.

   (A) 4.83
   (B) 5.18
   (C) 6.04
   (D) 8.93
   (E) The standard deviation cannot be computed from the information given.

GO ON TO THE NEXT PAGE ➤

4. To determine the average cost of running for a congressional seat, a simple random sample of 50 politicians is chosen and the politicians' records examined. The cost figures show a mean of $125,000 with a standard deviation of $32,000. Which of the following is the best interpretation of a 90% confidence interval estimate for the average cost of running for office?

(A) 90% of politicians running for a congressional seat spend between $117,500 and $132,500.
(B) 90% of politicians running for a congressional seat spend a mean dollar amount that is between $117,500 and $132,500.
(C) We are 90% confident that politicians running for a congressional seat spend between $117,500 and $132,500.
(D) We are 90% confident that politicians running for a congressional seat spend a mean dollar amount between $117,500 and $132,500.
(E) We are 90% confident that in the chosen sample, the mean dollar amount spent running for a congressional seat is between $117,500 and $132,500.

5. In one study on the effect that eating meat products has on weight level, an SRS of 500 subjects who admitted to eating meat at least once a day had their weights compared with those of an independent SRS of 500 people who claimed to be vegetarians. In a second study, an SRS of 500 subjects were served at least one meat meal per day for 6 months, while an independent SRS of 500 others were chosen to receive a strictly vegetarian diet for 6 months, with weights compared after 6 months.

(A) The first study is a controlled experiment, while the second is an observational study.
(B) The first study is an observational study, while the second is a controlled experiment.
(C) Both studies are controlled experiments.
(D) Both studies are observational studies.
(E) Each study is part controlled experiment and part observational study.

6. A plumbing contractor obtains 60% of her boiler circulators from a company whose defect rate is 0.005, and the rest from a company whose defect rate is 0.010. What proportion of the circulators can be expected to be defective? If a circulator is defective, what is the probability that it came from the first company?

(A) .0070, .429
(B) .0070, .600
(C) .0075, .500
(D) .0075, .600
(E) .0150, .571

GO ON TO THE NEXT PAGE ➤

7. A kidney dialysis center periodically checks a sample of its equipment and performs a major recalibration if readings are sufficiently off target. Similarly, a fabric factory periodically checks the sizes of towels coming off an assembly line and halts production if measurements are sufficiently off target. In both situations, we have the null hypothesis that the equipment is performing satisfactorily. For each situation, which is the *more* serious concern, a Type I or Type II error?

(A) Dialysis center: Type I error, towel manufacturer: Type I error
(B) Dialysis center: Type I error, towel manufacturer: Type II error
(C) Dialysis center: Type II error, towel manufacturer: Type I error
(D) Dialysis center: Type II error, towel manufacturer: Type II error
(E) This is impossible to answer without making an expected value judgment between human life and accurate towel sizes.

8. In the following table, what value for n results in a table showing perfect independence?

40	60
50	n

(A) 30
(B) 50
(C) 70
(D) 75
(E) 100

9. During the years 1886 through 2000 there were an average of 8.7 tropical cyclones per year, of which an average of 5.1 became hurricanes. Assuming that the probability of any cyclone becoming a hurricane is independent of what happens to any other cyclone, if there are five cyclones in one year, what is the probability that at least three become hurricanes?

(A) .313
(B) .345
(C) .586
(D) .658
(E) .686

10. Given the probabilities $P(A) = .3$ and $P(B) = .2$, what is the probability of the union $P(A \cup B)$ if A and B are mutually exclusive? If A and B are independent? If B is a subset of A?

(A) .44, .5, .2
(B) .44, .5, .3
(C) .5, .44, .2
(D) .5, .44, .3
(E) 0, .5, .3

11. A nursery owner claims that a recent drought stunted the growth of 22% of all her evergreens. A botanist tests this claim by examining an SRS of 1100 evergreens and finds that 268 trees in the sample show signs of stunted growth. With H_0: $p = .22$ and H_a: $p \neq .22$, what is the value of the test statistic?

(A) $z = \dfrac{.244 - .22}{\sqrt{1100(.22)(1 - .22)}}$

(B) $z = 2\dfrac{.244 - .22}{\sqrt{1100(.22)(1 - .22)}}$

(C) $z = \dfrac{.244 - .22}{\sqrt{\dfrac{(.22)(1 - .22)}{1100}}}$

(D) $z = 2\dfrac{.244 - .22}{\sqrt{\dfrac{(.22)(1 - .22)}{1100}}}$

(E) $z = 2\dfrac{.244 - .22}{\sqrt{1100(.244)(1 - .244)}}$

12. A medical research team tests for tumor reduction in a sample of patients using three different dosages of an experimental cancer drug. Which of the following is true?

(A) There are three explanatory variables and one response variable.

(B) There is one explanatory variable with three levels of response.

(C) Tumor reduction is the only explanatory variable, but there are three response variables corresponding to the different dosages.

(D) There are three levels of a single explanatory variable.

(E) Each explanatory level has an associated level of response.

13. The graph below shows cumulative proportion plotted against age for a population.

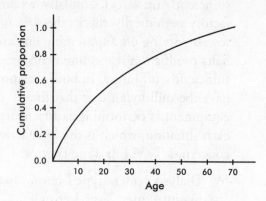

The median is approximately what age?

(A) 17
(B) 30
(C) 35
(D) 40
(E) 45

14. Mr. Bee's statistics class had a standard deviation of 11.2 on a standardized test, while Mr. Em's class had a standard deviation of 5.6 on the same test. Which of the following is the most reasonable conclusion concerning the two classes' performance on the test?

(A) Mr. Bee's class is less heterogeneous than Mr. Em's.

(B) Mr. Em's class is more homogeneous than Mr. Bee's.

(C) Mr. Bee's class performed twice as well as Mr. Em's.

(D) Mr. Em's class did not do as well as Mr. Bee's.

(E) Mr. Bee's class had the higher mean, but this may not be statistically significant.

GO ON TO THE NEXT PAGE ➤

15. A college admissions officer is interested in comparing the SAT math scores of high school applicants who have and have not taken AP Statistics. She randomly pulls the files of five applicants who took AP Statistics and five applicants who did not, and proceeds to run a *t*-test to compare the mean SAT math scores of the two groups. Which of the following is a necessary assumption?

(A) The population variances from each group are known.
(B) The population variances from each group are unknown.
(C) The population variances from the two groups are equal.
(D) The population of SAT scores from each group is normally distributed.
(E) The samples must be independent simple random samples, and for each sample np and $n(1 - p)$ must both be at least 10.

16. The appraised values of houses in a city have a mean of $125,000 with a standard deviation of $23,000. Because of a new teachers' contract, the school district needs an extra 10% in funds compared to the previous year. To raise this additional money, the city instructs the assessment office to raise all appraised house values by $5,000. What will be the new mean and standard deviation of the appraised values of houses in the city?

(A) Mean: $130,000, standard deviation: $23,000
(B) Mean: $130,000, standard deviation: $28,000
(C) Mean: $142,500, standard deviation: $25,300
(D) Mean: $142,500, standard deviation: $30,300
(E) Mean: $143,000, standard deviation: $30,800

17. A psychologist hypothesizes that scores on an aptitude test are normally distributed with a mean of 70 and a standard deviation of 10. In a random sample of size 100, scores are distributed as in the table below. What is the χ^2 statistic for a goodness-of-fit test?

Score:	below 60	60–70	70–80	above 80
Number of people:	10	40	35	15

(A) $\dfrac{(10-16)^2}{10} + \dfrac{(40-34)^2}{40} + \dfrac{(35-34)^2}{35} + \dfrac{(15-16)^2}{15}$

(B) $\dfrac{(10-16)^2}{16} + \dfrac{(40-34)^2}{34} + \dfrac{(35-34)^2}{34} + \dfrac{(15-16)^2}{16}$

(C) $\dfrac{(10-25)^2}{25} + \dfrac{(40-25)^2}{25} + \dfrac{(35-25)^2}{25} + \dfrac{(15-25)^2}{25}$

(D) $\dfrac{(10-25)^2}{10} + \dfrac{(40-25)^2}{40} + \dfrac{(35-25)^2}{35} + \dfrac{(15-25)^2}{15}$

(E) $\dfrac{(10-25)^2}{16} + \dfrac{(40-25)^2}{34} + \dfrac{(35-25)^2}{34} + \dfrac{(15-25)^2}{16}$

18. A talk show host recently reported that in response to his on-air question, 82% of the more than 2500 e-mail messages received through his publicized address supported the death penalty for anyone convicted of selling drugs to children. What does this show?

(A) The survey is meaningless because of voluntary response bias.
(B) No meaningful conclusion is possible without knowing something more about the characteristics of his listeners.
(C) The survey would have been more meaningful if he had picked a random sample of the 2500 listeners who responded.
(D) The survey would have been more meaningful if he had used a control group.
(E) This was a legitimate sample, randomly drawn from his listeners, and of sufficient size to be able to conclude that most of his listeners support the death penalty for such a crime.

GO ON TO THE NEXT PAGE ➤

19. Define a new measurement as the difference between the 60th and 40th percentile scores in a population. This measurement will give information concerning

    (A) central tendency.
    (B) variability.
    (C) symmetry.
    (D) skewness.
    (E) clustering.

20. Suppose X and Y are random variables with $E(X) = 37$, var$(X) = 5$, $E(Y) = 62$, and var$(Y) = 12$. What are the expected value and variance of the random variable $X + Y$?

    (A) $E(X + Y) = 99$, var$(X + Y) = 8.5$
    (B) $E(X + Y) = 99$, var$(X + Y) = 13$
    (C) $E(X + Y) = 99$, var$(X + Y) = 17$
    (D) $E(X + Y) = 49.5$, var$(X + Y) = 17$
    (E) There is insufficient information to answer this question.

21. Which of the following can affect the value of the correlation r?

    (A) A change in measurement units
    (B) A change in which variable is called x and which is called y
    (C) Adding the same constant to all values of the x-variable
    (D) All of the above can affect the r value.
    (E) None of the above can affect the r value.

22. The American Medical Association (AMA) wishes to determine the percentage of obstetricians who are considering leaving the profession because of the rapidly increasing number of lawsuits against obstetricians. How large a sample should be taken to find the answer to within ±3% at the 95% confidence level?

    (A) 6
    (B) 33
    (C) 534
    (D) 752
    (E) 1068

23. Following are parts of the probability distributions for the random variables X and Y.

x	$P(x)$		y	$P(y)$
1	.25		1	.3
2	?		2	.5
3	.35		3	?
4	?			

    If X and Y are independent and the joint probability $P(X = 2, Y = 3) = .03$, what is $P(X = 4)$?

    (A) .10
    (B) .15
    (C) .20
    (D) .25
    (E) .30

24. Which of the following sets has the smallest standard deviation? Which has the largest?

    I. 1, 2, 3, 4, 5, 6, 7
    II. 1, 1, 1, 4, 7, 7, 7
    III. 1, 4, 4, 4, 4, 4, 7

    (A) I, II
    (B) II, III
    (C) III, I
    (D) II, I
    (E) III, II

25. A researcher plans a study to examine long-term confidence in the U.S. economy among the adult population. She obtains a simple random sample of 30 adults as they leave a Wall Street office building one weekday afternoon. All but two of the adults agree to participate in the survey. Which of the following are true statements?

I. Proper use of chance as evidenced by the simple random sample makes this a well-designed survey.
II. The high response rate makes this a well-designed survey.
III. Selection bias makes this a poorly designed survey.

(A) I only
(B) II only
(C) III only
(D) I and II
(E) None of the statements is true.

26. Which among the following would result in the narrowest confidence interval?

(A) Small sample size and 95% confidence
(B) Small sample size and 99% confidence
(C) Large sample size and 95% confidence
(D) Large sample size and 99% confidence
(E) This cannot be answered without knowing an appropriate standard deviation.

27. Consider the following three scatterplots:

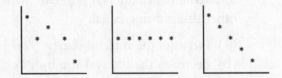

What is the relationship among r_1, r_2, and r_3, the correlations associated with the first, second, and third scatterplots, respectively?

(A) $r_1 < r_2 < r_3$
(B) $r_1 < r_3 < r_2$
(C) $r_2 < r_3 < r_1$
(D) $r_3 < r_1 < r_2$
(E) $r_3 < r_2 < r_1$

28. There are two games involving flipping a coin. In the first game you win a prize if you can throw between 45% and 55% heads. In the second game you win if you can throw more than 80% heads. For each game would you rather flip the coin 30 times or 300 times?

(A) 30 times for each game
(B) 300 times for each game
(C) 30 times for the first game and 300 times for the second
(D) 300 times for the first game and 30 times for the second
(E) The outcomes of the games do not depend on the number of flips.

29. City planners are trying to decide among various parking plan options ranging from more on-street spaces to multilevel facilities to spread-out small lots. Before making a decision, they wish to test the downtown merchants' claim that shoppers park for an average of only 47 minutes in the downtown area. The planners have decided to tabulate parking durations for 225 shoppers and to reject the merchants' claim if the sample mean exceeds 50 minutes. If the merchants' claim is wrong and the true mean is 51 minutes, what is the probability that the random sample will lead to a mistaken failure to reject the merchants' claim? Assume that the standard deviation in parking durations is 27 minutes.

(A) $P\left(z < \dfrac{50-47}{27/\sqrt{225}}\right)$

(B) $P\left(z > \dfrac{50-47}{27/\sqrt{225}}\right)$

(C) $P\left(z < \dfrac{50-51}{27/\sqrt{225}}\right)$

(D) $P\left(z > \dfrac{50-51}{27/\sqrt{225}}\right)$

(E) $P\left(z > \dfrac{51-47}{27/\sqrt{225}}\right)$

GO ON TO THE NEXT PAGE ➤

30. Consider the following parallel boxplots indicating the starting salaries (in thousands of dollars) for blue collar and white collar workers at a particular production plant:

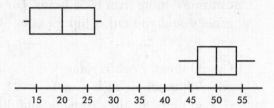

Which of the following are true statements?

   I. The ranges are the same.
  II. The interquartile ranges are the same.
 III. Because of symmetry, their medians are the same.

(A) I only
(B) II only
(C) I and II
(D) I and III
(E) II and III

31. The mean Law School Aptitude Test (LSAT) score for applicants to a particular law school is 650 with a standard deviation of 45. Suppose that only applicants with scores above 700 are considered. What percentage of the applicants considered have scores below 740? (Assume the scores are normally distributed.)

(A) 13.3%
(B) 17.1%
(C) 82.9%
(D) 86.7%
(E) 97.7%

32. If all the other variables remain constant, which of the following will increase the power of a hypothesis test?

   I. Increasing the sample size.
  II. Increasing the significance level.
 III. Increasing the probability of a Type II error.

(A) I only
(B) II only
(C) III only
(D) I and II
(E) All are true.

33. A researcher planning a survey of school principals in a particular state has lists of the school principals employed in each of the 125 school districts. The procedure is to obtain a random sample of principals from each of the districts rather than grouping all the lists together and obtaining a sample from the entire group. Which of the following statements about the resulting stratified sample are true?

   I. It is not a simple random sample.
  II. It is easier and less costly to obtain than a simple random sample.
 III. It gives comparative information that a simple random sample wouldn't give.

(A) I only
(B) I and II
(C) I and III
(D) II and III
(E) I, II, and III

34. A simple random sample is defined by

(A) the method of selection.
(B) examination of the outcome.
(C) both of the above.
(D) how representative the sample is of the population.
(E) the size of the sample versus the size of the population.

35. Changing from a 90% confidence interval estimate for a population proportion to a 99% confidence interval estimate, with all other things being equal,

(A) increases the interval size by 9%.
(B) decreases the interval size by 9%.
(C) increases the interval size by 57%.
(D) decreases the interval size by 57%.
(E) This question cannot be answered without knowing the sample size.

GO ON TO THE NEXT PAGE ➤

36. Which of the following are true statements?

    I. If there is sufficient evidence to reject a null hypothesis at the 10% level, then there is sufficient evidence to reject it at the 5% level.
    II. Whether to use a one- or a two-sided test is typically decided after the data are gathered.
    III. If a hypothesis test is conducted at the 1% level, there is a 1% chance of rejecting the null hypothesis.

    (A) I only
    (B) II only
    (C) III only
    (D) I, II, and III
    (E) None are true.

37. A population is normally distributed with mean 25. Consider all samples of size 10.

    The variable $\dfrac{\bar{x} - 25}{\frac{s}{\sqrt{10}}}$

    (A) has a normal distribution.
    (B) has a t-distribution with df = 10.
    (C) has a t-distribution with df = 9.
    (D) has neither a normal distribution or a t-distribution.
    (E) has either a normal or a t-distribution depending on the characteristics of the population standard deviation.

38. A correlation of .6 indicates that the percentage of variation in y that is explained by the variation in x is how many times the percentage indicated by a correlation of .3?

    (A) 2
    (B) 3
    (C) 4
    (D) 6
    (E) There is insufficient information to answer this question.

39. As reported on CNN, in a May 1999 national poll 43% of high school students expressed fear about going to school. Which of the following best describes what is meant by the poll having a margin of error of 5%?

    (A) It is likely that the true proportion of high school students afraid to go to school is between 38% and 48%.
    (B) Five percent of the students refused to participate in the poll.
    (C) Between 38% and 48% of those surveyed expressed fear about going to school.
    (D) There is a .05 probability that the 43% result is in error.
    (E) If similar size polls were repeatedly taken, they would be wrong about 5% of the time.

40. In a high school of 1650 students, 132 have personal investments in the stock market. To estimate the total stock investment by students in this school, two plans are proposed. Plan I would sample 30 students at random, find a confidence interval estimate of their average investment, and then multiply both ends of this interval by 1650 to get an interval estimate of the total investment. Plan II would sample 30 students at random from among the 132 who have investments in the market, find a confidence interval estimate of their average investment, and then multiply both ends of this interval by 132 to get an interval estimate of the total investment. Which is the better plan for estimating the total stock market investment by students in this school?

    (A) Plan I
    (B) Plan II
    (C) Both plans use random samples and so will produce equivalent results.
    (D) Neither plan will give an accurate estimate.
    (E) The resulting data must be seen to evaluate which is the better plan.

STOP

If there is still time remaining, you may review your answers.

SECTION II

PART A

Questions 1–5

Spend about 65 minutes on this part of the exam.
Percentage of Section II grade—75

You must show all work and indicate the methods you use. You will be graded on the correctness of your methods and on the accuracy of your results and explanations.

1. An AP Statistics student takes a random survey of 100 high school teachers and classifies department affiliation versus music preference. The results are as follows:

	Classical	Country	Jazz
Humanities/Soc. Sci.	15	5	20
Math/Physical Sci.	10	15	10
Fine Arts	15	5	5

(a) What is the probability that a teacher selected at random from this sample is a fine arts teacher who prefers country music?

(b) What is the probability that a teacher selected at random from this sample is a math/physical science teacher given that he/she prefers jazz?

(c) What is the probability that a teacher selected at random from this sample prefers classical music given that he/she is a humanities/social science teacher?

(d) Is there evidence of a relationship between department affiliation and music preference? Explain.

2. A high school principal would like to know whether or not a new integrated math curriculum is working better than the traditional style of separation of topics. He notes that of the six math teachers, three are using the new approach and three the traditional. He asks them to administer a common final exam and plans to compare the average student scores from each of the two groups. If one group does significantly better than the other, he'll ask all of the teachers to adopt the indicated teaching mode.

(a) Comment on the principal's scheme.

(b) Design an appropriate experiment for the principal to use.

(c) How would you modify the experiment if you are concerned that a confounding variable is the ability of a student as measured by the student's scoring above or below 475 on a national exam?

GO ON TO THE NEXT PAGE ➤

3. A printing firm takes on new jobs on a first come, first served basis. Typically, 25% of the orders are small, taking 3 hours to finish, 30% of the orders take 8 hours to finish, 35% take 12 hours, and 10% take 20 hours.

   (a) What is the average number of hours the jobs take to finish?

   (b) Suppose in any given week that when the total number of hours reaches or exceeds 40, no further orders will be taken. Explain how you would use simulation with a random number table to estimate the distribution of the number of new jobs the firm takes on each week and the distribution of the number of job-hours the firm accepts each week.

   (c) Run three trials of your simulation using the random number table below. Give your results, explaining your procedure in action.

   84177 06757 17613 15582 51506 81435 41050 92031 06449 05059
   59884 31180 53115 84469 94868 57967 05811 84514 75011 13006
   63395 55041 15866 06589 13119 71020 85940 91932 06488 74987

4. In a study of migraine headache sufferers, each of 500 subjects flipped a coin to determine who was given either a placebo or an experimental pain medication. After one hour each was asked to judge relief on a scale of 1 to 10 with "no relief" = 1 and "complete relief" = 10. The results appear in the following frequency table:

Pain relief	Frequency of placebo takers	Frequency of experimental medication takers
1	39	18
2	68	21
3	50	20
4	22	33
5	21	49
6	23	40
7	16	31
8	9	13
9	7	6
10	9	5

$n = 264$ $n = 236$
$\bar{x} = 3.76$ $\bar{x} = 4.92$

   (a) Use a graphical display to compare the effectiveness of the placebo versus the experimental medication in treating migraine headaches. Write a few sentences, based on your graph, to compare the results of the subjects receiving placebo versus the subjects receiving the medication.

GO ON TO THE NEXT PAGE ➤

(b) Does there appear to be significantly greater pain relief from taking the medication as opposed to the placebo? Give a statistical justification for your answer.

5. A guidance counselor believes that eating a good breakfast leads to higher grades. She interviews 20 students taking the health course that she also happens to teach and records grade point averages and average calories consumed for breakfast. Some computer graphical output follows:

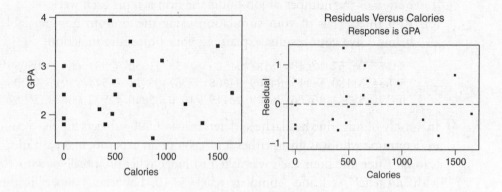

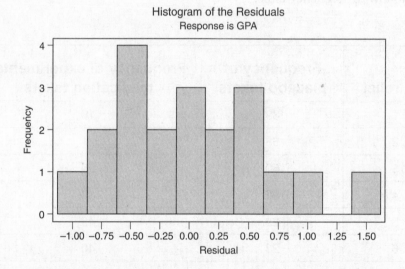

(a) Is this an observational study or a controlled experiment? Explain.
(b) Comment on the design of the study.
(c) The counselor plans to do a test of significance for evidence of an association between GPA and breakfast calories. Comment on whether the necessary assumptions are met for inference on slope for a least squares regression line.

SECTION II

PART B

Question 6

Spend about 25 minutes on this part of the exam.
Percentage of Section II grade—25

6. For a final project, three AP Statistics students plan a study of right-hand versus left-hand grip strengths of male high school students in a large city. From each of the city's five high schools, they pick a simple random sample of 40 male students and measure right-hand and left-hand grip strengths using a gripping device that gives values on a 1 to 10 scale. In their sample, right-hand grip strengths are approximately normally distributed with a mean of 7.3 and a standard deviation of 1.3, while left-hand grip strengths are approximately normally distributed with a mean of 6.1 and a standard deviation of 1.2. There are 58 students whose left-hand grip strength is greater than that of the right-hand, while 142 students show higher grip strength in the right hand.

    (a) Explain why the students did or did not obtain a simple random sample of all male high school students in the city. Explain why your answer would or would not change if you knew that all five high schools had the same number of male students.

    (b) Calculate and interpret a 90% confidence interval estimate for the proportion of male high school students in the city for which the right-hand grip strength is greater than that of the left.

    (c) Suppose that two male students are chosen at random. What is the approximate probability that the right-hand grip strength of the first student is greater than the left-hand grip strength of the second?

    (d) Based on your answers to parts (b) and (c), is there evidence that left-hand and right-hand grip strengths of male students in the city are independent? Explain.

STOP

If there is still time remaining, you may review your answers.

Answer Key

Section I

1. **E**	9. **D**	17. **B**	25. **C**	33. **E**
2. **E**	10. **D**	18. **A**	26. **C**	34. **A**
3. **A**	11. **C**	19. **B**	27. **D**	35. **C**
4. **D**	12. **D**	20. **E**	28. **D**	36. **E**
5. **B**	13. **A**	21. **E**	29. **C**	37. **C**
6. **A**	14. **B**	22. **E**	30. **A**	38. **C**
7. **C**	15. **D**	23. **D**	31. **C**	39. **A**
8. **D**	16. **A**	24. **E**	32. **D**	40. **B**

Answers Explained

Section I

1. **(E)** The median price would have 50% of the prices below and 50% of the prices above; however, looking at areas, it is clear that 60% of the prices are below $125,000. Area considerations also show that 30% of the prices are between $100,000 and $125,000, while 40% are above $125,000. Many values are too far from the mean for the standard deviation to be only $10,000. Between one-sixth and one-fourth of the range is a reasonable estimate for the standard deviation.

2. **(E)** Stemplots are not used for categorical data sets and are unwieldy for use with large data sets; however, they do show symmetry, gaps, clusters, and outliers.

3. **(A)** The critical z-score is 1.036. Thus $60 - 55 = 1.036\sigma$ and $\sigma = 4.83$.

4. **(D)** We are 90% confident that the population mean is within the interval calculated using the data from the sample.

5. **(B)** The first study was observational because the subjects were not chosen for treatment.

6. **(A)**

$$P(\text{def}) = P(1\text{st} \cap \text{def}) + P(2\text{nd} \cap \text{def})$$
$$= (.6)(.005) + (.4)(.010)$$
$$= .0030 + .0040 = .0070$$
$$P\left(1\text{st}\mid\text{def}\right) = \frac{.0030}{.0070} = .429$$

7. **(C)** At the dialysis center the more serious concern would be a Type II error, which is that the equipment is not performing correctly, yet the check does not pick this up; while at the towel manufacturing plant the more serious concern

would be a Type I error, which is that the equipment is performing correctly, yet the check causes a production halt.

8. **(D)** Relative frequencies must be equal. Either looking at rows gives $\frac{40}{100} = \frac{50}{50+n}$ or looking at columns gives $\frac{40}{90} = \frac{60}{60+n}$. We could also set up a proportion $\frac{n}{50} = \frac{60}{40}$ or $\frac{n}{60} = \frac{50}{40}$. Solving any of these equations gives $n = 75$.

9. **(D)** In a binomial distribution with probability of success $p = \frac{5.1}{8.7} = .586$, the probability of at least 3 successes is

$$\binom{5}{3}(.586)^3(.414)^2 + \binom{5}{4}(.586)^4(.414)$$

$+ (.586)^5 = .658$, or using the TI-83, one finds that 1-binomcdf(5, .586, 2) = .658.

10. **(D)** If A and B are mutually exclusive, then $P(A \cap B) = 0$, and so $P(A \cup B) = .3 + .2 - 0 = .5$. If A and B are independent, then $P(A \cap B) = P(A)P(B)$, and so $P(A \cup B) = .3 + .2 - (.3)(.2) = .44$. If B is a subset of A, then $A \cup B = A$, and so $P(A \cup B) = P(A) = .3$.

11. **(C)** The standard deviation of the test statistic is $\sigma_{\hat{p}} = \sqrt{\frac{p(1-p)}{n}}$. Since this is a two-sided test, the P-value will be twice the tail probability of the test statistic; however, the test statistic itself is not doubled.

12. **(D)** Dosage is the only explanatory variable, and it is being tested at three levels. Tumor reduction is the single response variable.

13. **(A)** The median corresponds to a cumulative proportion of 0.5.

14. **(B)** Standard deviation is a measure of variability. The less variability, the more homogeneity.

15. **(D)** Since the sample sizes are small, the samples must come from normally distributed populations. While the samples should be independent simple random samples, np and $n(1 - p)$ refer to conditions for tests involving sample proportions, not means.

16. **(A)** Adding the same constant to all values in a set will increase the mean by that constant, but will leave the standard deviation unchanged.

17. **(B)** With about 68% of the values within 1 standard deviation of the mean, the expected numbers for a normal distribution are as follows:

16	34	34	16

Thus $\chi^2 = \sum \dfrac{(\text{obs} - exp)^2}{exp}$

$= \dfrac{(10 - 16)^2}{16} + \dfrac{(40 - 34)^2}{34} + \dfrac{(35 - 34)^2}{34} + \dfrac{(15 - 16)^2}{16}$.

18. **(A)** This is a good example of voluntary response bias, which often overrepresents strong or negative opinions. The people who chose to respond were very possibly the parents of children facing drug problems or people who had had bad experiences with drugs being sold in their neighborhoods. There is very little chance that the 2500 respondents were representative of the population. Knowing more about his listeners or taking a sample of the sample would not have helped.

19. **(B)** The range (difference between largest and smallest values), the interquartile range ($Q_3 - Q_1$), and this difference between the 60th and 40th percentile scores all are measures of variability, or how spread out is the population or a subset of the population.

20. **(E)** Without independence we cannot determine $\text{var}(X + Y)$ from the information given.

21. **(E)** The correlation coefficient r is not affected by changes in units, by which variable is called x or y, or by adding or multiplying all the values of a variable by the same constant.

22. **(E)** $1.96\left(\dfrac{.5}{\sqrt{n}}\right) \leq .03$ gives $\sqrt{n} \geq 32.67$ and $n \geq 1067.1$.

23. **(D)** $P(Y = 3) = 1 - (.3 + .5) = .2$. By independence, $P(X = 2, Y = 3) = P(X = 2)P(Y = 3)$, and so $.03 = P(X = 2)(.2)$ and $P(X = 2) = .15$. Then $P(X = 4) = 1 - (.25 + .15 + .35) = .25$.

24. **(E)** Note that all three sets have the same mean and the same range. The third set has most of its values concentrated right at the mean, while the second set has most of its values concentrated far from the mean.

25. **(C)** People coming out of a Wall Street office building are a very unrepresentative sample of the adult population, especially given the question under consideration. Using chance and obtaining a high response rate will not change the selection bias and make this into a well-designed survey.

26. **(C)** Larger samples (so $\frac{\sigma}{\sqrt{n}}$ is smaller) and less confidence (so the critical z or t is smaller) both result in smaller intervals.

27. **(D)** The third scatterplot shows perfect negative association, so $r_3 = -1$. The first scatterplot shows strong, but not perfect, negative correlation, so $-1 < r_1 < 0$. The second scatterplot shows no correlation, so $r_2 = 0$.

28. **(D)** The probability of throwing heads is .5. By the law of large numbers, the more times you flip the coin, the more the relative frequency tends to become closer to this probability. With fewer tosses there is more chance for wide swings in the relative frequency.

29. **(C)** $\sigma_{\bar{x}} = \frac{\sigma}{\sqrt{n}} = \frac{27}{\sqrt{225}} = 1.8$. If the true mean parking duration is 51 minutes, the normal curve should be centered at 51. The critical value of 50 has a z-score of $\frac{50-51}{1.8}$.

30. **(A)** The overall lengths (between tips of whiskers) are the same, and so the ranges are the same. The box lengths are different, and so the interquartile ranges are different. The medians are 20 and 50, respectively.

31. **(C)** Critical z-scores are $\frac{700-650}{45} = 1.11$ and $\frac{740-650}{45} = 2$ with right tail probabilities of .1335 and .0228, respectively. The percentage below 740 given that the scores are above 700 is $\frac{.1335-.0228}{.1335} = 82.9\%$.

32. **(D)** There is a different probability of Type II error for each possible correct value of the population parameter, and 1 minus this probability is the power of the test against the associated correct value.

33. **(E)** This is not a simple random sample because all possible sets of the required size do not have the same chance of being picked. For example, a set of principals all from just half the school districts has no chance of being picked to be the sample. Stratified samples are often easier and less costly to obtain and also make comparative data available. In this case responses can be compared among various districts.

34. **(A)** A simple random sample can be any size and may or may not be representative of the population. It is a method of selection in which every possible sample of the desired size has an equal chance of being selected.

35. **(C)** The critical z-scores go from ± 1.645 to ± 2.576, resulting in an increase in the interval size: $\frac{2.576}{1.645} = 1.57$ or an increase of 57%.

36. **(E)** If the P-value is less than .10, it does not follow that it is less than .05. Decisions such as whether a test should be one- or two-sided are made before the data are gathered. If $\alpha = .01$, there is a 1% chance of rejecting the null hypothesis *if* the null hypothesis is true.

37. **(C)** $\dfrac{\bar{x} - \mu}{\frac{s}{\sqrt{n}}}$ has a *t*-distribution with df = $n - 1$.

38. **(C)** A correlation of .6 explains $(.6)^2$ or 36% of the variation in *y*, while a correlation of .3 explains only $(.3)^2$ or 9% of the variation in *y*.

39. **(A)** Using a measurement from a sample, we are never able to say *exactly* what a population proportion is; rather we always say we have a certain *confidence* that the population proportion lies in a particular *interval*. In this case that interval is 43% ± 5% or between 38% and 48%.

40. **(B)** With Plan I the expected number of students with stock investments is only 2.4 out of 30. Plan II allows an estimate to be made using a full 30 investors.

SECTION II

1. (a) $P(\text{fine arts} > \text{country music}) = \dfrac{5}{100} = \dfrac{1}{20}$

   (b) $P(\text{math/phys sci} \mid \text{jazz}) = \dfrac{10}{20+10+5} = \dfrac{10}{35} = \dfrac{2}{7}$

   (c) $P(\text{classical} \mid \text{hum/soc sci}) = \dfrac{15}{15+5+20} = \dfrac{10}{40} = \dfrac{3}{8}$

   (d) Chi-square test for independence. The expected cell counts are as follows:

16	10	14
14	8.75	12.25
10	6.25	8.75

The assumptions that all cell counts are greater than 5 and that the sample is randomly drawn are met.

H_0: Department affiliation and music preference are independent.

H_a: There is a relationship between department affiliation and music preference.

Running a chi-square test gives $\chi^2 = 15.51$. With df = $(r - 1)(c - 1) = 4$, $P = .0037$. With this small a *P*-value ($P < .01$), we say that the data provide strong evidence to reject H_0, and so we conclude that there is evidence of a relationship (or association) between department affiliation and music preference. (That is, department affiliation and music preference are not independent.)

Scoring

Part (d) involves naming the test, checking the assumptions, stating the hypotheses, calculating χ^2 and P, and stating a correct conclusion in context. Part (d) is essentially correct if at least four out of these five steps are complete and accurate. Part (d) is partially correct if at least two out of these five steps are complete and accurate.

4 Complete Answer All four parts essentially correct.

3 Substantial Answer Parts (a), (b), and (c) correct and part (d) partially correct OR part (d) and two out of the other three problems correct.

2 Developing Answer Parts (a), (b), and (c) correct OR part (d) and one of the other parts correct OR part (d) partially correct and two of the other three problems correct.

1 Minimal Answer Part (d) correct OR two of parts (a), (b), and (c) correct OR part (d) partially correct and one of the other three problems correct.

2. (a) The principal's scheme suffers from a lack of randomization of *both* teachers and students. For example, if one group does better, it could be that the three better teachers all chose one teaching method, or it could be that better students chose or were assigned to certain teachers' classes.

   (b) The principal should randomly choose three of the teachers to teach using the traditional style of separation of topics and require the other three teachers to use the new integrated approach. The students should then be randomly distributed into the six classes. A common final can be given and the average scores of the two groups (based on teaching method) compared.

   (c) Three of the teachers should be randomly picked to teach by each method. The students should be separated into two blocks, one block consisting of those who scored under 475 and the other consisting of those who scored over 475. In each of these two blocks the students should be randomly distributed into the six classes. A common final should be given. Among those students who scored under 475, the average score of those being taught by each method should be separately calculated. Similarly, among those students who scored over 475, the average score of those being taught by each method should be separately calculated. The principal can then note which seems to be the best teaching method to reach each of the two blocks of students.

Scoring

(a) An answer is essentially correct if the solution mentions a lack of randomization for both teachers and students, and is partially correct if the solution only talks about a lack of randomization for either teachers or students.

(b) An essentially correct answer gives a randomization scheme, in context, addressing randomization for both student and teacher selection. A partially correct answer addresses randomization for either teachers or students.

(c) An essentially correct answer addresses both blocking and randomization in context. A partially correct answer states the need for both blocking and randomization, but doesn't clearly explain how this is to be done in the context of this problem.

4 Complete Answer   All three parts essentially correct.

3 Substantial Answer Two parts essentially correct and the third partially correct.

2 Developing Answer   Two parts essentially correct OR one part essentially correct and the other two parts partially correct.

1 Minimal Answer   One part essentially correct OR two parts partially correct.

3. (a) The average or expected value for the number of hours per job is given by the following:

$$E(X) = \Sigma x P(x) = (3)(.25) + (8)(.30) + (12)(.35) + (20)(.10) = 9.35 \text{ hours}$$

(b) One possible scheme—from the random number table, read off two digits at a time. If the digits give a number between 01 and 25, this represents a 3-hour job, a number between 26 and 55 represents an 8-hour job, between 56 and 90 represents a 12-hour job, and from 91 to 99 and 00 represents a 20-hour job. Keep a running count of the total number of hours and stop taking further orders when this total reaches or exceeds 40. Note both the number of jobs and the total number of hours reached.

(c) The solution will depend on the scheme and on whether or not a new row is started for each trial. Using the above scheme and not starting a new row for each trial would give the following result:

```
        ....Trial 1....  ....Trial 2....    ......Trial 3.......
Hrs:     12  3  12  12 12  3  12  8   8  12   8  8  12   3  8  8
Rand #'s: 84 17 7   0 67 57 17 61  3   1  55 82  51 50 6   8 14 35 41 050
```

Number of jobs: 5   Number of jobs: 5   Number of jobs: 6
Number of hours: 51   Number of hours: 43   Number of hours: 47

Scoring

There are five components to the answer:

1. Correct calculation of the expected value for number of hours.

2. A clear and correct statement of the scheme to assign two-digit random numbers to the different type (length) jobs together with clear directions on how the random number table is to be used.

3. Statement that a trial ends when a total number of hours of 40 or more is reached.

4. Statement that both number of hours and number of jobs are tabulated for each trial.

5. Correct execution of three trials of the scheme with stopping rule.

While components 2 and 3 must be stated in the answer to part (b), and component 5 is covered in part (c), component 4 may be covered in parts (b) or (c).

4 Complete Answer   All five components correct.

3 Substantial Answer   Four components correct.

2 Developing Answer   Three components (including either 2 or 5) correct.

1 Minimal Answer   Two components (including either 2 or 5) correct.

4. (a) You can use parallel boxplots or two histograms or a double bar chart. With histograms or double bar charts, relative frequency rather than frequency should be used because of unequal sample sizes.

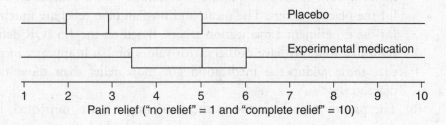

OR:

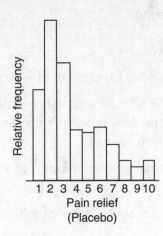

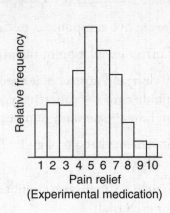

OR:

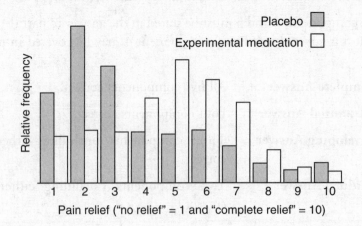

Pain relief ("no relief" = 1 and "complete relief" = 10)

The distribution for pain relief of placebo takers is skewed to the right, while that for experimental medication takers is roughly symmetric and bell-shaped. Both distributions have the same range, with lows of 1 and highs of 10, but the first quartile, median, and third quartile are all lower for the placebo takers. The mean and median pain relief are much higher for the experimental medication takers. Based on the 1.5 IQR definition, both distributions have outliers with values of 10. It appears, in general, that those taking the medication got more relief than those taking a placebo.

(b) The proper hypothesis test is the two-sample *t*-test (unpooled, because there is no reason to suppose the standard deviations are equal). Assumption of independent random samples is satisfied by the given scheme. Since the sample sizes are large, by the Central Limit Theorem the distribution of sample means is approximately normal and a *t*-test may be run.

$H_0: \mu_m - \mu_p = 0$ (Average pain relief of medication takers is the same as that of placebo takers.)

$H_a: \mu_m - \mu_p > 0$ (Average pain relief of medication takers is greater than that of placebo takers.)

On the TI-84, we get $t = 5.66$ and $P = .000$.

Or calculate $\bar{x}_m = 4.92$, $s_m = 2.15$, $\bar{x}_p = 3.76$, and $s_p = 2.43$. Then

$$t = \frac{\bar{x}_m - \bar{x}_p}{\sqrt{\dfrac{s_m^2}{n_m} + \dfrac{s_p^2}{n_p}}} = \frac{4.92 - 3.76}{\sqrt{\dfrac{2.15^2}{236} + \dfrac{2.43^2}{264}}} = 5.66$$

With df = min{236, 264} = 236, we get $P = .000$.

With such a small P-value, we have very strong evidence to reject H_0 and conclude that the average pain relief of the experimental medication takers is greater than that of the placebo takers.

Scoring

There are two components to part (a): an appropriate graph and appropriate comments based on the graph. Give only partial credit to the graphing component if the scales or labels are missing, but credit for labels can be recovered if comments are complete. Comments can be given either full, partial, or no credit depending upon completeness.

There are two components to part (b): one component is to name the appropriate test, give appropriate hypotheses, and name and check assumptions; the second component is to find correct values for t and P, and give a correct conclusion in context. Each component can receive either full, partial, or no credit.

4 Complete Answer   All four components correct OR three correct and fourth partially correct.

3 Substantial Answer   Three components correct OR two correct and other two both partially correct.

2 Developing Answer   Two components correct OR one correct and other three partially correct.

1 Minimal Answer   One component correct OR at least three components partially correct.

---

5. (a) No treatment is being imposed, so this is an observational study.
   (b) This is not a simple random sample of students. All were chosen from a health class, which could well impact a study about eating a good breakfast. The counselor/teacher interviewed her own students, which could have introduced bias in the way they responded. It's not clear that "good breakfast" can be measured by "calories."
   (c) While the residual plot shows no pattern and the histogram of residuals is approximately normal, a further assumption is that the scatterplot looks approximately linear, which is not the case here.

Scoring

Part (a) is partially correct if student states that it is an observational study without giving a reason.

Parts (b) and (c) are essentially correct or partially correct depending upon completeness.

4 Complete Answer   All three parts essentially correct.

3 Substantial Answer   Two parts essentially correct and the third part partially correct.

2 Developing Answer   Two parts essentially correct OR one part essentially correct and the other two parts partially correct.

1 Minimal Answer   One part essentially correct and one part partially correct OR all three parts partially correct.

6. (a) The students did not obtain a simple random sample since all possible samples of size 200 were not equally likely to be picked. For example, there was no chance of picking a sample where all 200 students were from one high school. This answer would not change if we knew that all five high schools had the same number of male students. While this would mean that all male high school students in the city had an equal probability of being selected, each possible sample of size 200 would still not have an equal chance of being picked.

(b) As pointed out in part (a), we don't have a true SRS, but the procedure did involve random selection. Given that the sample is from a large city, it seems reasonable to assume that the sample is less than 10% of the entire population of male high school students. We must also check that the sample is sufficiently large. We have $\hat{p} = \dfrac{142}{200} = .71$, and so $n\,\hat{p} = 200(.71) = 142 \geq 10$ and $n(1 - \hat{p}) = 200(.29) = 58 \geq 10$.

$$\hat{p} \pm 1.645\sqrt{\frac{\hat{p}(1-\hat{p})}{n}} = .71 \pm 1.645\sqrt{\frac{(.71)(.29)}{200}} = .71 \pm .053 \text{ or } (.657, .763)$$

Thus we are 90% confident that the proportion of male high school students in the city for which the right-hand grip strength is greater than that of the left is between .657 and .763.

(c) The distribution of the set of all possible differences of right-hand and left-hand grip strengths is approximately normal with $\mu_{R\text{-}L} = 7.3 - 6.1 = 1.2$ and $\sigma_{R\text{-}L} = \sqrt{1.3^2 + 1.2^2} = 1.769$. So $P(R - L > 0) = P\left(z > \dfrac{0 - 1.2}{1.769} \right) = P(z > -0.678) = .749$.

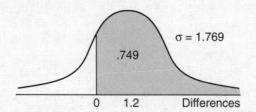

(d) From the answer to part (c), if left-hand and right-hand grip strengths of male students in the city are independent, we would expect approximately 74.9% of male students in the city to have stronger right-hand grip strength than left. From the answer to part (b), we estimate the percent of male students in the city with stronger right-hand grip strength to be between 65.7% and 76.3%. Since 74.9% is in this interval, there is evidence to suggest that left-hand and right-hand grip strengths of male students in the city are independent.

Scoring

Part (a) is essentially correct if both answers are right and partially correct if only one is right.

Part (b) is essentially correct if the student checks that the sample size is large enough, that is, that both np and $n(1 - p)$ are greater than 10; correctly calculates the confidence interval; and interprets the interval in context. Partial correctness should be given for two out of three of these answers.

Part (c) is essentially correct for a complete answer, and partially correct for finding μ_{R-L} and σ_{R-L} but incorrectly finding the probability OR for making a mistake in finding μ_{R-L} and σ_{R-L} but then using a correct procedure to find the probability.

Part (d) is essentially correct for a complete answer and partially correct if the conclusion is right but the links to (b) and (c) are unclear.

Count partially correct answers as one-half an essentially correct answer.

4 Complete Answer   All four parts essentially correct.

3 Substantial Answer   Three parts essentially correct.

2 Developing Answer   Two parts essentially correct.

1 Minimal Answer   One part essentially correct.

Use a holistic approach to decide a score totaling between two numbers.

Answer Sheet

PRACTICE EXAMINATION 2

1. Ⓐ Ⓑ Ⓒ Ⓓ Ⓔ
2. Ⓐ Ⓑ Ⓒ Ⓓ Ⓔ
3. Ⓐ Ⓑ Ⓒ Ⓓ Ⓔ
4. Ⓐ Ⓑ Ⓒ Ⓓ Ⓔ
5. Ⓐ Ⓑ Ⓒ Ⓓ Ⓔ
6. Ⓐ Ⓑ Ⓒ Ⓓ Ⓔ
7. Ⓐ Ⓑ Ⓒ Ⓓ Ⓔ
8. Ⓐ Ⓑ Ⓒ Ⓓ Ⓔ
9. Ⓐ Ⓑ Ⓒ Ⓓ Ⓔ
10. Ⓐ Ⓑ Ⓒ Ⓓ Ⓔ

11. Ⓐ Ⓑ Ⓒ Ⓓ Ⓔ
12. Ⓐ Ⓑ Ⓒ Ⓓ Ⓔ
13. Ⓐ Ⓑ Ⓒ Ⓓ Ⓔ
14. Ⓐ Ⓑ Ⓒ Ⓓ Ⓔ
15. Ⓐ Ⓑ Ⓒ Ⓓ Ⓔ
16. Ⓐ Ⓑ Ⓒ Ⓓ Ⓔ
17. Ⓐ Ⓑ Ⓒ Ⓓ Ⓔ
18. Ⓐ Ⓑ Ⓒ Ⓓ Ⓔ
19. Ⓐ Ⓑ Ⓒ Ⓓ Ⓔ
20. Ⓐ Ⓑ Ⓒ Ⓓ Ⓔ

21. Ⓐ Ⓑ Ⓒ Ⓓ Ⓔ
22. Ⓐ Ⓑ Ⓒ Ⓓ Ⓔ
23. Ⓐ Ⓑ Ⓒ Ⓓ Ⓔ
24. Ⓐ Ⓑ Ⓒ Ⓓ Ⓔ
25. Ⓐ Ⓑ Ⓒ Ⓓ Ⓔ
26. Ⓐ Ⓑ Ⓒ Ⓓ Ⓔ
27. Ⓐ Ⓑ Ⓒ Ⓓ Ⓔ
28. Ⓐ Ⓑ Ⓒ Ⓓ Ⓔ
29. Ⓐ Ⓑ Ⓒ Ⓓ Ⓔ
30. Ⓐ Ⓑ Ⓒ Ⓓ Ⓔ

31. Ⓐ Ⓑ Ⓒ Ⓓ Ⓔ
32. Ⓐ Ⓑ Ⓒ Ⓓ Ⓔ
33. Ⓐ Ⓑ Ⓒ Ⓓ Ⓔ
34. Ⓐ Ⓑ Ⓒ Ⓓ Ⓔ
35. Ⓐ Ⓑ Ⓒ Ⓓ Ⓔ
36. Ⓐ Ⓑ Ⓒ Ⓓ Ⓔ
37. Ⓐ Ⓑ Ⓒ Ⓓ Ⓔ
38. Ⓐ Ⓑ Ⓒ Ⓓ Ⓔ
39. Ⓐ Ⓑ Ⓒ Ⓓ Ⓔ
40. Ⓐ Ⓑ Ⓒ Ⓓ Ⓔ

Practice Examination 2

Section I

━━━

Questions 1–40

Spend 90 minutes on this part of the exam.

> **Directions:** The questions or incomplete statements that follow are each followed by five suggested answers or completions. Choose the response that best answers the question or completes the statement.

1. Suppose that the regression line for a set of data, $y = mx + 3$, passes through the point $(2, 7)$. If $\bar{x}$ and $\bar{y}$ are the sample means of the x- and y-values, respectively, then $\bar{y} =$

   (A) $\bar{x}$
   (B) $\bar{x} - 2$.
   (C) $\bar{x} + 3$.
   (D) $2\bar{x} + 3$.
   (E) $3.5\bar{x} + 3$.

2. A study is made to determine whether more hours of academic studying leads to higher point scoring by basketball players. In surveying 50 basketball players, it is noted that the 25 who claim to study the most hours have a higher point average than the 25 who study less. Based on this study, the coach begins requiring the players to spend more time studying. Which of the following are true statements?

   I. While this study indicates a relation, it does not prove causation.
   II. There could well be a confounding variable responsible for the seeming relationship.
   III. While this is a controlled experiment, the conclusion of the coach is not justified.

   (A) I only
   (B) I and II
   (C) I and III
   (D) II and III
   (E) I, II, and III

GO ON TO THE NEXT PAGE ➤

3. The longevity of people living in a certain locality has a standard deviation of 14 years. What is the mean longevity if 30% of the people live longer than 75 years? Assume a normal distribution for life spans.

(A) 61.00
(B) 67.65
(C) 74.48
(D) 82.35
(E) The mean cannot be computed from the information given.

4. Which of the following are true statements?

I. The correlation r is equal to the slope of the regression line when z-scores for the y-variable are plotted against z-scores for the x-variable.
II. If the slope of the regression line is exactly 1, then the correlation is exactly 1.
III. If the correlation is 0, then the slope of the regression line is 0.

(A) I only
(B) II only
(C) III only
(D) I and III
(E) All are true.

5. Which of the following are affected by outliers?

I. Mean
II. Median
III. Standard deviation
IV. Range
V. Interquartile range

(A) I, III, and V
(B) II and IV
(C) I and V
(D) III and IV
(E) I, III, and IV

6. A company that produces facial tissues continually monitors tissue strength. If the mean strength from sample data drops below a specified level, the production process is halted and the machinery inspected. Which of the following would result from a Type I error?

(A) Halting the production process when sufficient customer complaints are received.
(B) Halting the production process when the tissue strength is below specifications.
(C) Halting the production process when the tissue strength is within specifications.
(D) Allowing the production process to continue when the tissue strength is below specifications.
(E) Allowing the production process to continue when the tissue strength is within specifications.

7. Suppose that for a certain Caribbean island in any 3-year period the probability of a major hurricane is .25, the probability of water damage is .44, and the probability of both a hurricane and water damage is .22. What is the probability of water damage given that there is a hurricane?

(A) .47
(B) .50
(C) .69
(D) .88
(E) .91

8. An engineer wishes to determine the quantity of heat being generated by a particular electronic component. If she knows that the standard deviation is 2.4, how many of these components should she consider to be 99% sure of knowing the mean quantity to within ±0.6?

(A) 27
(B) 87
(C) 107
(D) 212
(E) 425

GO ON TO THE NEXT PAGE ➤

9. Two possible wordings for a questionnaire on a proposed school budget increase are as follows:

    I. This school district has one of the highest per student expenditure rates in the state. This has resulted in low failure rates, high standardized test scores, and most students going on to good colleges and universities. Do you support the proposed school budget increase?

    II. This school district has one of the highest per student expenditure rates in the state. This has resulted in high property taxes, with many people on fixed incomes having to give up their homes because they cannot pay the school tax. Do you support the proposed school budget increase?

    One of these questions showed that 58% of the population favor the proposed school budget increase, while the other question showed that only 13% of the population support the proposed increase. Which produced which result and why?

    (A) The first showed 58% and the second 13% because of the lack of randomization as evidenced by the wording of the questions.
    (B) The first showed 13% and the second 58% because of a placebo effect due to the wording of the questions.
    (C) The first showed 58% and the second 13% because of the lack of a control group.
    (D) The first showed 13% and the second 58% because of response bias due to the wording of the questions.
    (E) The first showed 58% and the second 13% because of response bias due to the wording of the questions.

10. A union spokesperson is trying to encourage a college faculty to join the union. She would like to argue that faculty salaries are not truly based on years of service as most faculty believe. She gathers data and notes the following scatterplot of salary versus years of service.

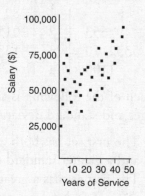

    Which of the following most correctly interprets the overall scatterplot?

    (A) The faculty member with the fewest years of service makes the lowest salary, and the faculty member with the most service makes the highest salary.
    (B) A faculty member with more service than another has the greater salary than the other.
    (C) There is a strong positive correlation with little deviation.
    (D) There is no clear relationship between salary and years of service.
    (E) While there is a strong positive correlation, there is a distinct deviation from the overall pattern for faculty with fewer than ten years of service.

GO ON TO THE NEXT PAGE ➤

11. Two random samples of students are chosen, one from those taking an AP Statistics class and one from those not. The following back-to-back stemplots compare the GPAs.

AP Statistics		No AP Statistics
	1	89
97653	2	015688
98775332110	3	133344777888
1100	4	

Which of the following is true about the ranges and standard deviations?

(A) The first set has both a greater range and a greater standard deviation.
(B) The first set has a greater range, while the second has a greater standard deviation.
(C) The first set has a greater standard deviation, while the second has a greater range.
(D) The second set has both a greater range and a greater standard deviation.
(E) The two sets have equal ranges and equal standard deviations.

12. In a group of 10 scores, the largest score is increased by 40 points. What will happen to the mean?

(A) It will remain the same.
(B) It will increase by 4 points.
(C) It will increase by 10 points.
(D) It will increase by 40 points.
(E) There is not sufficient information to answer this question.

13. Suppose X and Y are random variables with $\mu_x = 32$, $\sigma_x = 5$, $\mu_y = 44$, and $\sigma_y = 12$. Given that X and Y are independent, what are the mean and standard deviation of the random variable $X + Y$?

(A) $\mu_{x+y} = 76$, $\sigma_{x+y} = 8.5$
(B) $\mu_{x+y} = 76$, $\sigma_{x+y} = 13$
(C) $\mu_{x+y} = 76$, $\sigma_{x+y} = 17$
(D) $\mu_{x+y} = 38$, $\sigma_{x+y} = 17$
(E) There is insufficient information to answer this question.

14. Suppose you toss a fair die three times and it comes up an even number each time. Which of the following is a true statement?

(A) By the law of large numbers, the next toss is more likely to be an odd number than another even number.
(B) Based on the properties of conditional probability the next toss is more likely to be an even number given that three in a row have been even.
(C) Dice actually do have memories, and thus the number that comes up on the next toss will be influenced by the previous tosses.
(D) The law of large numbers tells how many tosses will be necessary before the percentages of evens and odds are again in balance.
(E) The probability that the next toss will again be even is .5.

15. A pharmaceutical company is interested in the association between advertising expenditures and sales for various over-the-counter products. A sales associate collects data on nine products, looking at sales (in $1000) versus advertising expenditures (in $1000). The results of the regression analysis are shown below.

```
Dependent variable: Sales

Source          df          Sum of Squares      Mean Square      F-ratio
Regression      1               9576.1             9576.1        1118.45
Residual        7                 59.9                8.6

Variable      Coefficient        SE Coef            t-ratio          P
Constant        123.800           1.798              68.84         0.000
Advertising      12.633           0.378              33.44         0.000
R-Sq = 99.4%     R-Sq (adj) = 99.3%
s = 2.926 with 9-2 = 7 degrees of freedom
```

Which of the following gives a 90% confidence interval for the slope of the regression line?

(A) $12.633 \pm 1.415(0.378)$
(B) $12.633 \pm 1.895(0.378)$
(C) $123.800 \pm 1.414(1.798)$
(D) $123.800 \pm 1.895(1.798)$
(E) $123.800 \pm 1.645(1.798/\sqrt{9})$

16. Suppose you wish to compare the AP Statistics exam results for the male and female students taking AP Statistics at your high school. Which is the most appropriate technique for gathering the needed data?

(A) Census
(B) Sample survey
(C) Experiment
(D) Observational study
(E) None of these is appropriate.

17. Jonathan obtained a score of 80 on a statistics exam, placing him at the 90th percentile. Suppose five points are added to everyone's score. Jonathan's new score will be at the

(A) 80th percentile.
(B) 85th percentile.
(C) 90th percentile.
(D) 95th percentile.
(E) There is not sufficient information to answer this quesiton.

18. To study the effect of music on piecework output at a clothing manufacturer, two experimental treatments are planned: day-long classical music for one group versus day-long light rock music for another. Which one of the following groups would serve best as a control for this study?

(A) A third group for which no music is played
(B) A third group that randomly hears either classical or light rock music each day
(C) A third group that hears day-long R & B music
(D) A third group that hears classical music every morning and light rock every afternoon
(E) A third group in which each worker has earphones and chooses his or her own favorite music

19. Suppose H_0: $p = .6$, and the power of the test for H_a: $p = .7$ is .8. Which of the following is a valid conclusion?

(A) The probability of committing a Type I error is .1.
(B) If H_a is true, the probability of failing to reject H_0 is .2.
(C) The probability of committing a Type II error is .3.
(D) All of the above are valid conclusions.
(E) None of the above are valid conclusions.

GO ON TO THE NEXT PAGE ➤

20. Following is a histogram of the numbers of ties owned by bank executives.

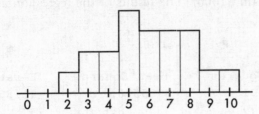

Which of the following statements are true?

I. The median number of ties is five.
II. More than four executives own over eight ties.
III. An executive is equally likely to own fewer than five ties or more than seven ties.

(A) I only
(B) II only
(C) III only
(D) I and III
(E) I, II, and III

21. Which of the following is a binomial random variable?

(A) The number of tosses before a "5" appears when tossing a fair die.
(B) The number of points a hockey team receives in 10 games, where two points are awarded for wins, one point for ties, and no points for losses.
(C) The number of hearts out of five cards randomly drawn from a deck of 52 cards, without replacement.
(D) The number of motorists not wearing seat belts in a random sample of five drivers.
(E) None of the above.

22. Company I manufactures bomb fuses that burn an average of 50 minutes with a standard deviation of 10 minutes, while company II advertises fuses that burn an average of 55 minutes with a standard deviation of 5 minutes. Which company's fuse is more likely to last at least 1 hour? Assume normal distributions of fuse times.

(A) Company I's, because of its greater standard deviation
(B) Company II's, because of its greater mean
(C) For both companies, the probability that a fuse will last at least 1 hour is 15.9%
(D) For both companies, the probability that a fuse will last at least 1 hour is 84.1%
(E) The problem cannot be solved from the information given.

23. Which of the following is *not* important in the design of experiments?

(A) Control of confounding variables
(B) Randomization in assigning subjects to different treatments
(C) Use of a lurking variable to control the placebo effect
(D) Replication of the experiment using sufficient numbers of subjects
(E) All of the above are important in the design of experiments.

GO ON TO THE NEXT PAGE ➤

24. The travel miles claimed in weekly expense reports of the sales personnel at a corporation are summarized in the following boxplot.

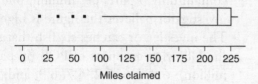

Miles claimed

Which of the following is the most reasonable conclusion?

(A) The mean and median numbers of travel miles are roughly equal.
(B) The mean number of travel miles is greater than the median number.
(C) Most of the claimed numbers of travel miles are in the [0, 200] interval.
(D) Most of the claimed numbers of travel miles are in the [200, 240] interval.
(E) The left and right whiskers contain the same number of values from the set of personnel travel mile claims.

25. Which of the following statements about residuals are true?

I. The mean of the residuals is always zero.
II. Influential scores have large residuals.
III. A definite pattern in the residual plot is an indication that a nonlinear model should be tried.

(A) II only
(B) I and II
(C) I and III
(D) II and III
(E) I, II, and III

26. Four pairs of data are used in determining a regression line $y = 3x + 4$. If the four values of the independent variable are 32, 24, 29, and 27, respectively, what is the mean of the four values of the dependent variable?

(A) 68
(B) 84
(C) 88
(D) 100
(E) The mean cannot be determined from the given information.

27. According to one poll, 12% of the public favor legalizing all drugs. In a simple random sample of six people, what is the probability that at least one person favors legalization?

(A) .380
(B) .464
(C) .536
(D) .620
(E) .844

28. Sampling error occurs

(A) when interviewers make mistakes resulting in bias.
(B) because a sample statistic is used to estimate a population parameter.
(C) when interviewers use judgment instead of random choice in picking the sample.
(D) when samples are too small.
(E) in all of the above cases.

29. A telephone executive instructs an associate to contact 104 customers using their service to obtain their opinions in regard to an idea for a new pricing package. The associate notes the number of customers whose names begin with *A* and uses a random number table to pick four of these names. She then proceeds to use the same procedure for each letter of the alphabet and combines the 4 × 26 = 104 results into a group to be contacted. Which of the following are true statements?

   I. Her procedure makes use of chance.
   II. Her procedure results in a simple random sample.
   III. Each customer has an equal probability of being included in the survey.

  (A) I only
  (B) I and II
  (C) I and III
  (D) II and III
  (E) I, II, and III

30. The graph below shows cumulative proportions plotted against GPAs for high school seniors.

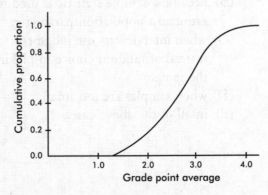

Grade point average

What is the approximate interquartile range?

  (A) 0.85
  (B) 2.25
  (C) 2.7
  (D) 2.75
  (E) 3.1

31. PCB contamination of a river by a manufacturer is being measured by amounts of the pollutant found in fish. A company scientist claims that the fish contain only 5 parts per million, but an investigator believes the figure is higher. The investigator catches six fish that show the following amounts of PCB (in parts per million): 6.8, 5.6, 5.2, 4.7, 6.3, and 5.4. In performing a hypothesis test with H_0: $\mu = 5$ and H_a: $\mu > 5$, what is the test statistic?

  (A) $t = \dfrac{5.67 - 5}{0.763}$

  (B) $t = \dfrac{5.67 - 5}{\sqrt{0.763/5}}$

  (C) $t = \dfrac{5.67 - 5}{\sqrt{0.763/6}}$

  (D) $t = \dfrac{5.67 - 5}{0.763/\sqrt{5}}$

  (E) $t = \dfrac{5.67 - 5}{0.763/\sqrt{6}}$

32. In a certain city 6% of teenagers are married, 25% of married teenagers have children, and 15% of unmarried teenagers have children. If a teenager has a child, what is the probability that the teenager is not married?

  (A) .156
  (B) .200
  (C) .500
  (D) .904
  (E) .940

33. In general, how does tripling the sample size change the confidence interval size?

  (A) It triples the interval size.
  (B) It divides the interval size by 3.
  (C) It multiples the interval size by 1.732.
  (D) It divides the interval size by 1.732.
  (E) This question cannot be answered without knowing the sample size.

GO ON TO THE NEXT PAGE ➤

34. Which of the following are true statements?

   I. While the normal distribution is symmetric, the *t*-distributions are slightly skewed.

   II. The *t*-distributions are lower at the mean and higher at the tails and so are more spread out than the normal distribution.

   III. The greater the df, the closer the *t*-distributions are to the normal distribution.

   (A) III only
   (B) I and II
   (C) I and III
   (D) II and III
   (E) I, II, and III

35. A study on school budget approval among people with different party affiliations resulted in the following segmented bar chart:

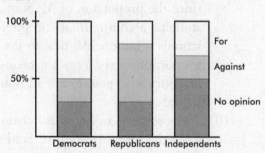

   Which of the following is greatest?

   (A) Number of Democrats who are for the proposed budget
   (B) Number of Republicans who are against the budget
   (C) Number of Independents who have no opinion on the budget
   (D) The above are all equal.
   (E) The answer is impossible to determine without additional information.

36. The sampling distribution of the sample mean is close to the normal distribution

   (A) only if both the original population has a normal distribution and *n* is large.
   (B) if the standard deviation of the original population is known.
   (C) if *n* is large, no matter what the distribution of the original population.
   (D) no matter what the value of *n* or what the distribution of the original population.
   (E) only if the original population is not badly skewed and does not have outliers.

37. What is the probability of a Type II error when a hypothesis test is being conducted at the 10% significance level ($\alpha = .10$)?

   (A) .05
   (B) .10
   (C) .90
   (D) .95
   (E) There is insufficient information to answer this question.

38.

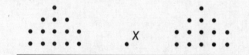

   Above is the dotplot for a set of numbers. One element is labeled *X*. Which of the following are true statements?

   I. *X* has the largest *z*-score, in absolute value, of any element in the set.

   II. A modified boxplot will plot an outlier like *X* as an isolated point.

   III. A stemplot will show *X* isolated from two clusters.

   (A) I only
   (B) II only
   (C) III only
   (D) I and II
   (E) II and III

39. A 1999 survey of 500 households concluded that 82% of the population uses grocery coupons. Which of the following best describes what is meant by the poll having a margin of error of 3%?

(A) Three percent of those surveyed refused to participate in the poll.

(B) It would not be unexpected for 3% of the population to begin using coupons or stop using coupons.

(C) Between 395 and 425 of the 500 households surveyed responded that they used grocery coupons.

(D) If a similar survey of 500 households were taken weekly, a 3% change in each week's results would not be unexpected.

(E) It is likely that between 79% and 85% of the population use grocery coupons.

40.

	College Plans	
	Public	Private
Taking AP Statistics:	18	27
Not taking AP Statistics:	26	40

The above two-way table summarizes the results of a survey of high school seniors conducted to determine if there is a relationship between whether or not a student is taking AP Statistics and whether he or she plans to attend a public or a private college after graduation. Which of the following is the most reasonable conclusion about the relationship between taking AP Statistics and the type of college a student plans to attend?

(A) There appears to be no association since the proportion of AP Statistics students planning to attend public schools is almost identical to the proportion of students not taking AP Statistics who plan to attend public schools.

(B) There appears to be an association since the proportion of AP Statistics students planning to attend public schools is almost identical to the proportion of students not taking AP Statistics who plan to attend public schools.

(C) There appears to be an association since more students plan to attend private than public schools.

(D) There appears to be an association since fewer students are taking AP Statistics than are not taking AP Statistics.

(E) These data do not address the question of association.

STOP

If there is still time remaining, you may review your answers.

SECTION II

PART A

Questions 1–5

Spend about 65 minutes on this part of the exam.
Percentage of Section II grade—75

> You must show all work and indicate the methods you use. You will be graded on the correctness of your methods and on the accuracy of your results and explanations.

1. A brand of refrigerator has a lifetime expectancy that is normally distributed with a mean of 15 years and a standard deviation of 2.5 years.

   (a) The company has a warranty stating that any refrigerator that has problems within the first 10 years will be fixed free of charge. If the contracted cost to the company to repair a refrigerator is $150, what is the expected value of the cost to the company per refrigerator? Show your work.

   (b) Suppose the company wants a warranty that would apply to an additional 5% of the refrigerators. How many years warranty should the company offer to obtain this result? Show your work.

2. A college guidance counselor has decided to study the effect of taking an 8:00 a.m. class on semester grade point average (GPA).

   (a) The counselor wishes to take into account year in college (upperclass versus underclass). Suggest two designs, one an observational study and one a controlled experiment.

   (b) Assume the results indicate that upperclass students who take 8:00 a.m. classes have higher semester GPAs than upperclass students who do not have 8:00 a.m. classes. State a reasonable conclusion for each of the above designs.

   (c) Give an example of a possible lurking variable that demonstrates why the conclusions in part (b) above are different. Explain your answer.

3. A magazine article reports the president's approval rating to be 68%.

   (a) Suppose a pollster suspects the correct rating to be around 68%, but would like to confirm this with a new poll. How many people must be surveyed to obtain a 95% confidence interval estimate with a margin of error ≤1%? Show your work.

   (b) Suppose the pollster samples 500 randomly chosen adults and finds 325 approve of the way the president is doing his job. Is there evidence that the president's approval rating is different from the claimed 68%? Perform an appropriate statistical test.

   (c) In part (b) above, suppose the pollster suspects that a recent presidential blunder has affected his approval rating. Is there evidence that the president's approval rating has gone down from the claimed 68%? Perform an appropriate statistical test.

   (d) Are the answers in parts (b) and (c) contradictory? Explain.

4. I. The average heights in inches of four different species of pre-human cave dwellers are calculated from simple random samples of skeletons from each group. The results are summarized as follows:

   Species

	1	2	3	4
Average height	49	47	50	46

   Archeologists are interested in whether or not there is sufficient evidence that the average heights of the species were different.

   II. Last year, seven local radio stations captured 32%, 25%, 20%, 16%, 4%, 2%, and 1%, respectively, of the listening audience. A reporter interviews a simple random sample of 85 radio listeners to determine whether or not there is evidence that preferences have changed.

   III. A reporter interviews 250 people leaving a local hospital to determine whether or not over-the-counter pain killer preference (aspirin, acetaminophen, or ibuprofen) is independent of education level (no high school degree, high school degree, college degree) among adults over the age of 21.

   (a) Would a chi-square test for goodness-of-fit be appropriate in I? Explain.

   (b) Would a chi-square test for goodness-of-fit be appropriate in II? Explain.

   (c) Would a chi-square test for independence be appropriate in III? Explain.

5. The starting salary of new employees at a firm versus years of education is fitted with a least squares regression line. The graph of the residuals and some computer output for their regression are as follows:

Regression Analysis: Salary (in $1000) Versus Education (in years beyond 8th grade)

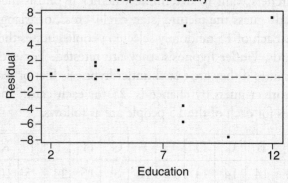

Residuals Versus Education
Response is Salary

```
Predictor        Coef      SE Coef         T         P
Constant        5.840       3.803       1.54     0.163
Education      3.4862      0.5310       6.57     0.000

S = 5.051      R-Sq = 84.3%    R-Sq(adj) = 82.4%

Analysis of Variance

Source          DF         SS         MS        F        P
Regression       1      1099.9     1099.9    43.11    0.000
Residual Error   8       204.1       25.5
Total            9      1304.0
```

(a) Is a line an appropriate model? Explain.
(b) Interpret the slope of the regression line in context.
(c) Interpret the *y*-intercept of the regression line in context.
(d) What is the predicted salary of a new employee with 10 years education beyond 8th grade?
(e) What was the actual salary of the new employee with 10 years education beyond 8th grade?

GO ON TO THE NEXT PAGE ➤

SECTION II

PART B

Question 6

Spend about 25 minutes on this part of the exam.
Percentage of Section II grade—25

6. Researchers want to determine whether hypnosis increases one's ability to correctly guess the picture (star, circle, cross, or triangle) from a set of ESP cards. Each of 15 randomly selected people guesses the picture from each of 40 cards. Under hypnosis, they are retested. For each person, a coin flip determines if they are tested under hypnosis first or second. The probability of a correct guess by chance is .25 for each card. The numbers of correct guesses for each of the 15 people are as follows:

Subject	A	B	C	D	E	F	G	H	I	J	K	L	M	N	O
Without hypnosis	13	14	14	11	12	10	8	15	11	5	10	12	9	12	7
Under hypnosis	11	15	13	9	12	12	8	14	11	7	11	12	10	13	10

(a) Do the data suggest that under hypnosis people are able to correctly guess more ESP cards than expected by chance? Explain.

(b) Do the data suggest that hypnosis improves one's ability to correctly guess ESP cards? Explain.

(c) Does knowing a person's number of correct guesses without hypnosis help predict his/her number correct under hypnosis? Explain. Any necessary assumptions must be stated, but you may assume they are met.

Answer Key

Section I

1. D	9. E	17. C	25. C	33. D
2. B	10. E	18. A	26. C	34. D
3. B	11. D	19. B	27. C	35. E
4. D	12. B	20. C	28. B	36. C
5. E	13. B	21. D	29. A	37. E
6. C	14. E	22. C	30. A	38. C
7. D	15. B	23. C	31. E	39. E
8. C	16. A	24. D	32. D	40. A

Answers Explained

Section I

1. **(D)** Since $(2, 7)$ is on the line $y = mx + 3$, we have $7 = 2m + 3$ and $m = 2$. Thus the regression line is $y = 2x + 3$. The point $(\overline{x}, \overline{y})$ is always on the regression line, and so we have $\overline{y} = 2\overline{x} + 3$.

2. **(B)** It could well be that conscientious students are the same ones who both study and do well on the basketball court. If students could be randomly assigned to study or not study, the results would be more meaningful. Of course, ethical considerations might make it impossible to isolate the confounding variable in this way.

3. **(B)** The critical z-score is 0.525. Thus $75 - \mu = 0.525(14)$ and $\mu = 67.65$.

4. **(D)** The slope of the regression line and the correlation are related by $b_1 = r\dfrac{s_y}{s_x}$. When using z-scores, the standard deviations s_x and s_y are 1.

5. **(E)** The median and interquartile range are specifically used when outliers are suspected of unduly influencing the mean, range, or standard deviation.

6. **(C)** This is a hypothesis test with H_0: tissue strength is within specifications, and H_a: tissue strength is below specifications. A Type I error is committed when a true null hypothesis is mistakenly rejected.

7. **(D)**

$$P\left(\text{water}\,|\,\text{hurricane}\right) = \frac{P\left(\text{water} \cap \text{hurricane}\right)}{P\left(\text{hurricane}\right)} = \frac{.22}{.25} = .88$$

8. **(C)** $2.576\left(\frac{2.4}{\sqrt{n}}\right) \le 0.6$, which gives $\sqrt{n} \ge 10.304$ and $n \ge 106.2$.

9. **(E)** The wording of questions can lead to response bias. The neutral way of asking this question would simply have been, "Do you support the proposed school budget increase?"

10. **(E)** While it is important to look for basic patterns, it is also important to look for deviations from these patterns. In this case, there is an overall positive correlation; however, those faculty with under ten years of service show little relationship between years of service and salary. While (A) is a true statement, it does not give an overall interpretation of the scatterplot.

11. **(D)** The second set has a greater range, $3.8 - 1.8 = 2.0$ as compared to $4.1 - 2.3 = 1.8$, and with its skewness it also has a greater standard deviation.

12. **(B)** With $n = 10$, increasing Σx by 40 increases $\frac{\Sigma x}{n}$ by 4.

13. **(B)** The means and the variances can be added. Thus the new variance is $5^2 + 12^2 = 169$, and the new standard deviation is 13.

14. **(E)** Dice have no memory, so the probability that the next toss will be an even number is .5 and the probability that it will be an odd number is .5. The law of large numbers says that as the number of tosses becomes larger, the proportion of even numbers tends to become closer to .5.

15. **(B)** The critical t-scores for 90% confidence with df = 7 are ±1.895.

16. **(A)** Either directly or anonymously, you should be able to obtain the test results for *every* student.

17. **(C)** Percentile ranking is a measure of relative position. Adding five points to everyone's score will not change the relative positions.

18. **(A)** The control group should have experiences identical to those of the experimental groups except for the treatment under examination. They should not be given a new treatment.

19. **(B)** If H_a is true, the probability of failing to reject H_0 and thus committing a Type II error is 1 minus the power, that is, $1 - .8 = .2$.

20. **(C)** Five does not split the area in half, so 5 is not the median. Histograms such as these show relative frequencies, not actual frequencies. The area from 1.5 to 4.5 is the same as that between 7.5 and 10.5, each being about 25% of the total.

21. **(D)** There must be a fixed number of trials, which rules out (A); only two possible outcomes, which rules out (B); and a constant probability of success on any trial, which rules out (C).

22. **(C)** In both cases 1 hour is one standard deviation from the mean with a right tail probability of .1587.

23. **(C)** Control, randomization, and replication are all important aspects of well-designed experiments. We try to control lurking variables, not to use them to control something else.

24. **(D)** The data are strongly skewed to the left, indicating that the mean is less than the median. The median appears to be roughly 215, indicating that the interval [200, 240] probably has more than 50% of the values. While in a standard boxplot each whisker contains 25% of the values, this is a modified boxplot showing four outliers, and so the left whisker has four fewer values than the right whisker.

25. **(C)** The sum and thus the mean of the residuals are always zero. An influential score may have a small residual but still have a great effect on the regression line. In a good straight-line fit, the residuals show a random pattern.

26. **(C)** $\bar{x} = \frac{32 + 24 + 29 + 27}{4} = 28$. Since $(\bar{x}, \bar{y})$ is a point on the regression line,

    $\bar{y} = 3(28) + 4 = 88$.

27. **(C)** $P(\text{at least } 1) = 1 - P(\text{none}) = 1 - (.88)^6 = .536$

28. **(B)** Different samples give different sample statistics, all of which are estimates for the same population parameter, and so error, called *sampling error*, is naturally present.

29. **(A)** While the associate does use chance, each customer would have the same chance of being selected only if the same number of customers had names starting with each letter of the alphabet. This selection does not result in a simple random sample because each possible set of 104 customers does not have the same chance of being picked as part of the sample. For example, a group of customers whose names all start with A will not be chosen.

30. **(A)** Corresponding to cumulative proportions of 0.25 and 0.75 are $Q_1 = 2.25$ and $Q_3 = 3.1$, respectively, and so the interquartile range is $3.1 - 2.25 = 0.85$.

31. **(E)** The standard deviation of the test statistic is $\sigma_{\bar{x}} = \dfrac{\sigma}{\sqrt{n}} \approx \dfrac{s}{\sqrt{n}}$.

32. **(D)**
$$P(\text{child}) = P(\text{married} \cap \text{child}) + P(\text{not married} \cap \text{child})$$
$$= (.06)(.25) + (.94)(.15)$$
$$= .015 + .141 = .156$$

$$P\left(\text{not married} \middle| \text{child}\right) = \frac{.141}{.156} = .904$$

33. **(D)** Increasing the sample size by a multiple of d divides the interval estimate by $\sqrt{d}$.

34. **(D)** The t-distributions are symmetric; however, they are lower at the mean and higher at the tails and so are more spread out than the normal distribution. The greater the df, the closer the t-distributions are to the normal distribution.

35. **(E)** The given bar chart shows percentages, not actual numbers.

36. **(C)** This follows from the central limit theorem.

37. **(E)** There is a different Type II error for each possible correct value for the population parameter.

38. **(C)** X is close to the mean and so will have a z-score close to 0. Modified boxplots show only outliers that are far from the mean. X and the two clusters are clearly visible in a stemplot of these data.

39. **(E)** Using a measurement from a sample, we are never able to say *exactly* what a population proportion is; rather we always say we have a certain *confidence* that the population proportion lies in a particular *interval*. In this case that interval is 82% ± 3% or between 79% and 85%.

40. **(A)** Whether or not students are taking AP Statistics seems to have no relationship to which type of school they are planning to go to. Chi-square is close to 0.

SECTION II

1. (a) The probability a refrigerator will have a problem within the first 10 years is $P\left(z < \dfrac{10-15}{2.5}\right) = P(z < -2) = .02275$, and so the expected cost to the company per refrigerator is $(.02275)(\$150) = \$3.41$.

   (b) The company is now willing to warranty $.02275 + .05 = .07275$ of the refrigerators. The corresponding z-score from the tables is 1.46 [or use invNorm on the TI-84]. Thus the new warranty should be for $15 - 1.46(2.5) = 11.35$ years.

Scoring

Part (a) is partially correct if the probability, .02275, is correctly calculated, but not the expected cost, OR if the probability is calculated incorrectly, but the correct procedure for calculating the expected cost is demonstrated.

Part (b) is essentially correct if the complete and correct procedure is demonstrated using the probability calculated in part (a). Part (b) is partially correct if the *z*-score is found based on the result from part (a) but then the year is miscalculated, OR if the *z*-score is miscalculated, but the correct procedure is then demonstrated for finding the year.

4 Complete Answer   Both parts essentially correct.

3 Substantial Answer   One part essentially correct and the other part partially correct.

2 Developing Answer   One part essentially correct OR both parts partially correct.

1 Minimal Answer   One part partially correct.

---

2. (a) There are many possible answers. An observational study would be, for example, to interview a simple random sample of students, divide the students into two strata, one upperclass students and the other underclass students, and for each strata note every student's GPA and whether or not that student has an 8:00 a.m. class. An experiment would be, for example, to pick a simple random sample of students, split them into two blocks, one upperclass students and the other underclass students, for each block to randomly pick half the students to assign 8:00 a.m. classes and the rest to restrict from 8:00 a.m. classes, and at the end of the semester to note every student's GPA.

   (b) For the observational study, the counselor can conclude that among upperclassmen, there is an association between taking 8:00 a.m. classes and having higher GPAs. For the experiment, the counselor can conclude that among upperclassmen, there appears to be a cause-and-effect relationship of taking 8:00 a.m. classes leading to higher GPAs.

   (c) There are many possible answers. For example, a lurking variable may be that students who voluntarily sign up for 8:00 a.m. classes are more serious, study harder, and don't like to party late at night. An observational study might then indicate that students who take 8:00 a.m. classes have higher GPAs, but the cause is not the taking of the class, but rather something about the personality of students who voluntarily take these classes. On the other hand, the experiment, in randomly assigning students to 8:00 classes, will control for this lurking variable in that equal proportions of the serious students will end up in each of the classes.

Scoring

Part (a) is essentially correct if designs for both an appropriate observational study and an appropriate experiment are clearly stated. Part (a) is partially correct if only one of the designs is clearly and accurately stated.

Part (b) is essentially correct if there are clear statements relating the observational study to an "association" or similar expression, and relating the experiment to "cause-and-effect" or some similar expression. Part (b) is partially correct for clearly stating one or the other, but not both, of these concepts.

Part (c) is essentially correct for giving an appropriate lurking variable and clearly explaining its effect on the observational study and on the experiment in this problem. Part (c) is partially correct for giving an appropriate lurking variable but giving an incomplete or incorrect explanation of it in context.

4 **Complete Answer**   All three parts essentially correct.

3 **Substantial Answer**   Two parts are essentially correct OR one part is essentially correct and the other two parts are partially correct.

2 **Developing Answer**   One part is essentially correct and one part is partially correct OR all three parts are partially correct.

1 **Minimal Answer**   One part is essentially correct OR at least two parts are partially correct.

3. (a) $1.96\sqrt{\dfrac{(.68)(.32)}{n}} \leq .01$, giving $\sqrt{n} \geq 91.43$ and $n \geq 8359.3$. Thus 8360 people must be surveyed.

(b) This is a two-sided, one proportion z-test with H_0: $p = .68$ and H_a: $p \neq .68$. The assumptions are met that $np = (500)(.68) = 340$ and $n(1-p) = (500)(.32) = 160$ are both greater than 10, and the sample is an SRS.

With $\hat{p} = \dfrac{325}{500}.65$ we get $z = \dfrac{.65 - .68}{\sqrt{\dfrac{(.68)(.32)}{500}}} = -1.438$ and thus

$\dfrac{P}{2} = .075$ and $P = .15$. Since $.15 > .10$ there is *not* sufficient evidence (even at the 10% significance level) to reject H_0, and thus there is *not* sufficient evidence that the president's approval rating has changed.

(c) This is a one-sided, one proportion z-test with H_0: $p = .68$ and H_a: $p < .68$. Again, the assumptions are met that $np = (500)(.68) = 340$ and $n(1-p) = (500)(.32) = 160$ are both greater than 10, and that the sample is an SRS.

With $\hat{p} = \dfrac{325}{500} = .65$ we get $z = \dfrac{.65 - .68}{\sqrt{\dfrac{(.68)(.32)}{500}}} = -1.438$ and thus

$P = .075$.

Since .075 < .10 there *is* some evidence (at least at the 10% significance level) to reject H_0, and thus there *is* some evidence that the president's approval rating has gone down.

(d) There is no contradiction. A two-sided test involves splitting the rejection region, with half on each side. Thus a *z*-score that falls in the rejection region for a one-sided test might not be far enough from the mean to fall in the rejection region for a two-sided test.

Scoring

Part (a) is partially correct for a correct value for *n* but showing no work, OR for working the problem as $1.96\dfrac{5}{n} \le .01$ and getting $n = 9604$.

Parts (b) and (c) involve naming the test, checking the assumptions, stating the hypotheses, calculating *z* and *P*, and giving a correct conclusion in context. Parts (b) and (c) are essentially correct if four out of five of these steps are correct and complete, and are partially correct if two or three out of the five steps are correct and complete.

4 Complete Answer   All four parts essentially correct.

3 Substantial Answer   Three parts essentially correct OR two parts essentially correct and the other two parts partially correct.

2 Developing Answer   Two parts essentially correct OR one part essentially correct and at least two of the other parts partially correct.

1 Minimal Answer   One part essentially correct OR at least two parts partially correct.

---

4. (a) No, chi-square tests deal with "counts," not "measurements."

   (b) No, because no expected counts should be less than 5. In this example, $0.04(85) = 3.4$, $0.02(85) = 1.7$, and $0.01(85) = 0.85$.

   (c) No, because the data are not coming from an SRS of the intended population. Interviewing people leaving a hospital about medications is not the same as interviewing people in the general population.

Scoring

Each part is essentially correct with an answer of "no" together with a clear and correct reason, and partially correct with an answer of "no" without a clear and correct reason.

4 Complete Answer — All three parts essentially correct.

3 Substantial Answer — Two parts essentially correct and one part partially correct.

2 Developing Answer — Two parts essentially correct OR one part essentially correct and the other two parts partially correct.

1 Minimal Answer — One part essentially correct and one part partially correct OR all three parts partially correct.

---

5. (a) The correlation $r = \sqrt{.843} = .918$, indicating a strong, positive, linear relationship. From the computer output, a linear regression t-test with H_0: $\beta = 0$, H_a: $\beta \neq 0$ yields $t = 6.57$ with a P-value of .000, indicating very strong evidence of a linear association. However, since the absolute values of the residuals increase with increasing x, there is some reason to believe that a model other than a line might be a more appropriate model.

(b) The slope of the regression line is 3.486, indicating that, on average, the starting salary at the firm is predicted to increase by $3,486 for each additional year of education past eighth grade.

(c) The y-intercept, 5.840, refers to 0 years of education beyond eighth grade. Thus the predicted starting salary at the firm for someone with only an eighth grade education would be $5,840.

(d) For a new employee with 10 years education beyond 8th grade, $x = 10$, so the predicted salary is 3.486(10) + 5.840 = 40.7, or $40,700.

(e) The residual for $x = 10$ from the residual plot is approximately −8, and since residual = actual−predicted, we estimate the actual starting salary to be 40.7 − 8 = 32.7 ($1,000), or about $32,700.

Scoring

(a) Explanation should refer to the regression plot and to either the correlation r or to the linear regression t-test. Answer may be either "yes" or "no" in accord with the explanation. The answer is either essentially correct or partially correct depending upon completeness.

(b) The answer is essentially correct for noting both that the slope is 3.486 and giving a correct interpretation in context. The answer is partially correct for giving the slope as 3.486 without a correct interpretation in context.

(c) The answer is essentially correct for noting both that the y-intercept is 5.840 and giving a correct interpretation in context. The answer is partially correct for giving the y-intercept as 5.840 without a correct interpretation in context.

(d) The answer is essentially correct for giving both the correct regression equation and for determining the value for $x = 10$. The answer is partially correct for either giving the correct equation without plugging in $x = 10$, or for giving an incorrect equation and plugging $x = 10$ into whatever equation is given.

(e) The answer is either essentially correct or partially correct depending upon completeness.

4	**Complete Answer**	All five parts essentially correct.
3	**Substantial Answer**	Four parts essentially correct OR three parts essentially correct and the other two parts partially correct.
2	**Developing Answer**	Three parts essentially correct OR two parts essentially correct and at least two parts partially correct OR one part essentially correct and the other four parts partially correct.
1	**Minimal Answer**	Two parts essentially correct OR one part essentially correct and two or three parts partially correct OR at least four parts partially correct.

6. (a) To do a one-sample *t*-test on the "under hypnosis" data, we must first check assumptions: (1) random sample (given) and (2) approximately normal population. The student may use a dotplot, boxplot, stemplot, or histogram:

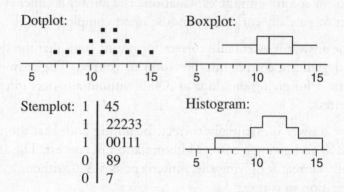

Dotplot: Boxplot:

Stemplot: 1 | 45
          1 | 22233
          1 | 00111
          0 | 89
          0 | 7

Histogram:

The plots and histogram indicate a roughly symmetric, bell-shaped distribution with no outliers and no extreme skewness, so it is reasonable to assume that the sample comes from a population that is approximately normal.

State hypotheses: Let μ be the average number of correct guesses under hypnosis.

H_0: $\mu = 10$ (10 correct out of 40 would be expected by chance.)

H_a: $\mu > 10$ (Under hypnosis, mean correct is over 10.)

t-test: $\bar{x} = 11.2$ $s = 2.178$

$$t = \frac{11.2 - 10}{2.178/\sqrt{15}} = \frac{1.2}{0.5623} = 2.13$$

df = $n - 1 = 15 - 1 = 14$
P-value = .026

With this small a *P*-value, there is sufficient evidence to reject H_0, and so the data do suggest that under hypnosis people are able to correctly guess more ESP cards than expected by chance.

(b) The proper hypothesis test is a matched pairs *t*-test on the set of differences of correctly answered questions, {–2, 1, –1, –2, 0, 2, 0, –1, 0, 2, 1, 0, 1, 1, 3}, with H_0: $\mu_D = 0$, H_a: $\mu_D > 0$. (Or subtract in other order and use H_a: $\mu_D < 0$.)

Assumptions to be checked: (1) random samples (given) and (2) the population of differences is approximately normal (use dotplot, stemplot, boxplot, or histogram of differences).

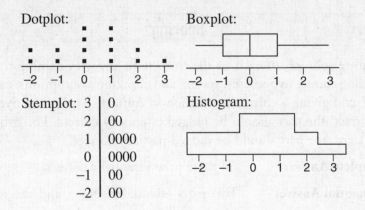

Dotplot: Boxplot:

Stemplot:
```
 3 | 0
 2 | 00
 1 | 0000
 0 | 0000
-1 | 00
-2 | 00
```
Histogram:

The plots and histogram indicate a roughly symmetric, bell-shaped distribution with no outliers and no extreme skewness, so it is reasonable to assume that the sample comes from a population that is approximately normal.

$$t\text{-test:} \quad \bar{x}_D = 0.3333 \qquad s_D = 1.447$$

$$t = \frac{\bar{x}_D - 0}{S_D / \sqrt{n_D}} = \frac{0.3333 - 0}{1.447 / \sqrt{15}} = \frac{0.3333}{0.3736} = 0.892$$

$$\text{df} = n_D - 1 = 15 - 1 = 14$$

The *P*-value = .194, so differences are not significant and the data do not suggest that hypnosis improves one's ability to correctly guess ESP cards.

(c) The student should compare the without hypnosis and under hypnosis scores (not the differences), and the suggested test is the *t*-test for regression. The assumptions include that the scatterplot is roughly linear, the residual plot has no apparent pattern, and the histogram of residuals is approximately normal (all given).

$$H_0: \beta = 0, \qquad H_a: \beta \neq 0$$

The *t*-test as run on the TI-83 gives *t* = 5.97 with *P* = 0.000. With such a small *P*-value, there is very strong evidence to reject H_0 and conclude that knowing a person's number correct without hypnosis helps predict his/her number correct under hypnosis.

Alternatively, full credit should also be given if a student graphs the scatterplot, notes that it appears very linear, calculates the correlation *r* = .856, and notes that this is close to 1.

Scatterplot:

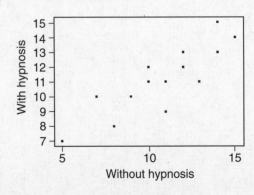

Scoring

A complete solution to each of the three parts involves naming the correct test, stating correct hypotheses, stating and checking assumptions, calculating *t* and *P*, and giving a correct conclusion in context. For four or five of these steps correct, the part should be judged essentially correct. For two or three correct steps, the part should be judged partially correct.

4 Complete Answer   All three parts essentially correct.

3 Substantial Answer   Two parts essentially correct and the remaining part partially correct.

2 Developing Answer   Two parts essentially correct OR one part essentially correct and the two remaining parts partially correct.

1 Minimal Answer   One part essentially correct OR at least two parts partially correct.

Answer Sheet

PRACTICE EXAMINATION 3

1. Ⓐ Ⓑ Ⓒ Ⓓ Ⓔ
2. Ⓐ Ⓑ Ⓒ Ⓓ Ⓔ
3. Ⓐ Ⓑ Ⓒ Ⓓ Ⓔ
4. Ⓐ Ⓑ Ⓒ Ⓓ Ⓔ
5. Ⓐ Ⓑ Ⓒ Ⓓ Ⓔ
6. Ⓐ Ⓑ Ⓒ Ⓓ Ⓔ
7. Ⓐ Ⓑ Ⓒ Ⓓ Ⓔ
8. Ⓐ Ⓑ Ⓒ Ⓓ Ⓔ
9. Ⓐ Ⓑ Ⓒ Ⓓ Ⓔ
10. Ⓐ Ⓑ Ⓒ Ⓓ Ⓔ

11. Ⓐ Ⓑ Ⓒ Ⓓ Ⓔ
12. Ⓐ Ⓑ Ⓒ Ⓓ Ⓔ
13. Ⓐ Ⓑ Ⓒ Ⓓ Ⓔ
14. Ⓐ Ⓑ Ⓒ Ⓓ Ⓔ
15. Ⓐ Ⓑ Ⓒ Ⓓ Ⓔ
16. Ⓐ Ⓑ Ⓒ Ⓓ Ⓔ
17. Ⓐ Ⓑ Ⓒ Ⓓ Ⓔ
18. Ⓐ Ⓑ Ⓒ Ⓓ Ⓔ
19. Ⓐ Ⓑ Ⓒ Ⓓ Ⓔ
20. Ⓐ Ⓑ Ⓒ Ⓓ Ⓔ

21. Ⓐ Ⓑ Ⓒ Ⓓ Ⓔ
22. Ⓐ Ⓑ Ⓒ Ⓓ Ⓔ
23. Ⓐ Ⓑ Ⓒ Ⓓ Ⓔ
24. Ⓐ Ⓑ Ⓒ Ⓓ Ⓔ
25. Ⓐ Ⓑ Ⓒ Ⓓ Ⓔ
26. Ⓐ Ⓑ Ⓒ Ⓓ Ⓔ
27. Ⓐ Ⓑ Ⓒ Ⓓ Ⓔ
28. Ⓐ Ⓑ Ⓒ Ⓓ Ⓔ
29. Ⓐ Ⓑ Ⓒ Ⓓ Ⓔ
30. Ⓐ Ⓑ Ⓒ Ⓓ Ⓔ

31. Ⓐ Ⓑ Ⓒ Ⓓ Ⓔ
32. Ⓐ Ⓑ Ⓒ Ⓓ Ⓔ
33. Ⓐ Ⓑ Ⓒ Ⓓ Ⓔ
34. Ⓐ Ⓑ Ⓒ Ⓓ Ⓔ
35. Ⓐ Ⓑ Ⓒ Ⓓ Ⓔ
36. Ⓐ Ⓑ Ⓒ Ⓓ Ⓔ
37. Ⓐ Ⓑ Ⓒ Ⓓ Ⓔ
38. Ⓐ Ⓑ Ⓒ Ⓓ Ⓔ
39. Ⓐ Ⓑ Ⓒ Ⓓ Ⓔ
40. Ⓐ Ⓑ Ⓒ Ⓓ Ⓔ

Practice Examination 3

Section I

Questions 1–40

Spend 90 minutes on this part of the exam.

> **Directions:** The questions or incomplete statements that follow are each followed by five suggested answers or completions. Choose the response that best answers the question or completes the statement.

1. Which of the following are true statements?

   I. While properly designed experiments can strongly suggest cause-and-effect relationships, a complete census is the only way of establishing such a relationship.
   II. If properly designed, observational studies can establish cause-and-effect relationships just as strongly as properly designed experiments.
   III. Controlled experiments are often undertaken later to establish cause-and-effect relationships first suggested by observational studies.

   (A) I only
   (B) II only
   (C) III only
   (D) I and II
   (E) I, II, and III

2. Two classes take the same exam. Suppose a certain score is at the 40th percentile for the first class and at the 80th percentile for the second class. Which of the following is the most reasonable conclusion?

   (A) Students in the first class generally scored higher than students in the second class.
   (B) Students in the second class generally scored higher than students in the first class.
   (C) A score at the 20th percentile for the first class is at the 40th percentile for the second class.
   (D) A score at the 50th percentile for the first class is at the 90th percentile for the second class.
   (E) One of the classes has twice the number of students as the other.

GO ON TO THE NEXT PAGE ➤

3. In an experiment, the control group should receive

   (A) treatment opposite that given the experimental group.
   (B) the same treatment given the experimental group without knowing they are receiving the treatment.
   (C) a procedure identical to that given the experimental group except for receiving the treatment under examination.
   (D) a procedure identical to that given the experimental group except for a random decision on receiving the treatment under examination.
   (E) none of the procedures given the experimental group.

4. In a random sample of Toyota car owners, 83 out of 112 said they were satisfied with the Toyota front-wheel drive, while in a similar survey of Subaru owners, 76 out of 81 said they were satisfied with the Subaru four-wheel drive. A 90% confidence interval estimate for the difference in proportions between Toyota and Subaru car owners who are satisfied with their drive systems is reported to be −.197 ± .081. Which is a proper conclusion?

   (A) The interval is invalid because probabilities cannot be negative.
   (B) The interval is invalid because it does not contain zero.
   (C) Subaru owners are approximately 19.7% more satisfied with their drive systems than are Toyota owners.
   (D) 90% of Subaru owners are approximately 19.7% more satisfied with their drive systems than are Toyota owners.
   (E) We are 90% confident that the difference in proportions between Toyota and Subaru car owners who are satisfied with their drive systems is between −.278 and −.116.

5. Which of the following statements about the correlation coefficient are true?

   I. The correlation coefficient and the slope of the regression line may have opposite signs.
   II. A correlation of 1 indicates a perfect cause-and-effect relationship between the variables.
   III. Correlations of +.87 and −.87 indicate the same degree of clustering around the regression line.

   (A) I only
   (B) II only
   (C) III only
   (D) I and II
   (E) I, II, and III

6. A computer manufacturer sets up three locations to provide technical support for its customers. Logs are kept noting whether or not calls about problems are solved successfully. Data from a sample of 1000 calls are summarized in the following table:

Location

	1	2	3	Total
Problem solved	325	225	150	700
Problem not solved	125	100	75	300
Total	450	325	225	1000

Assuming there is no association between location and whether or not a problem is resolved successfully, what is the expected number of successful calls (problem solved) from location 1?

   (A) $\dfrac{(325)(450)}{700}$

   (B) $\dfrac{(325)(700)}{450}$

   (C) $\dfrac{(325)(450)}{1000}$

   (D) $\dfrac{(325)(700)}{1000}$

GO ON TO THE NEXT PAGE ➤

(E) $\dfrac{(450)(700)}{1000}$

7. In a study on the effect of music on worker productivity, employees were told that a different genre of background music would be played each day and the corresponding production outputs noted. Every change in music resulted in an increase in production. This is an example of

(A) the effect of a treatment unit.
(B) the placebo effect.
(C) the control group effect.
(D) sampling error.
(E) voluntary response bias.

8. Suppose X and Y are random variables with $E(X) = 780$, var$(X) = 75$, $E(Y) = 430$, and var$(Y) = 25$. Given that X and Y are independent, what are the expected value and variance of the random variable $X - Y$?

(A) $E(X - Y) = 350$, var$(X - Y) = 50$
(B) $E(X - Y) = 350$, var$(X - Y) = 100$
(C) $E(X - Y) = 1210$, var$(X - Y) = 50$
(D) $E(X - Y) = 1210$, var$(X - Y) = 100$
(E) There is insufficient information to answer this question.

9. What is a sample?

(A) A measurable characteristic of a population
(B) A set of individuals having a characteristic in common
(C) A value calculated from raw data
(D) A subset of a population
(E) None of the above

10. A study is carried out noting the numbers of cars passing through an intersection between different hours during the morning rush hour. Following is the

resulting histogram:

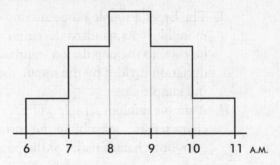

Which of the following statements are true?

I. The number of cars passing through before 8 A.M. is the same as the number passing through after 9 A.M.
II. The numbers of cars passing through in any time interval can be determined by noting the area under the curve above that interval.
III. The ratio of the number of cars passing through between 10 and 11 A.M. to the number of cars passing through between 6 and 7 A.M. is approximately $\dfrac{10.5}{6.5} = 1.62$.

(A) I only
(B) II only
(C) III only
(D) I and II
(E) II and III

11. A soft drink dispenser can be adjusted to deliver any fixed number of ounces. If the machine is operating with a standard deviation in delivery equal to 0.3 ounce, what should be the mean setting so that a 12-ounce cup will overflow less than 1% of the time? Assume a normal distribution for ounces delivered.

(A) 11.23 ounces
(B) 11.30 ounces
(C) 11.70 ounces
(D) 12.70 ounces
(E) 12.77 ounces

GO ON TO THE NEXT PAGE ➤

12. Which of the following are true statements?

I. The larger a simple random sample, the more likely its standard deviation will be close to the population standard deviation divided by the square root of the sample size.

II. A simple random sample will have characteristics resembling the corresponding characteristics of the population.

III. Randomness has less importance when choosing a large sample than when choosing a small sample.

(A) I only
(B) II only
(C) III only
(D) All are true.
(E) None are true.

13. The probability that a person will show a certain gene-transmitted trait is .8 if the father shows the trait and .06 if the father doesn't show the trait. Suppose that the children in a certain community come from families in 25% of which the father shows the trait. Given that a child shows the trait, what is the probability that her father shows the trait?

(A) .245
(B) .250
(C) .750
(D) .816
(E) .860

14. Given an experiment with $H_0: \mu = 10$, H_a: $\mu > 10$, and a possible correct value of 11, which of the following increases as n increases?

I. The probability of a Type I error.
II. The probability of a Type II error.
III. The power of the test.

(A) I only
(B) II only
(C) III only
(D) II and III
(E) None will increase.

15. If all the values of a data set are the same, all of the following must equal zero except for which one?

(A) Mean
(B) Standard deviation
(C) Variance
(D) Range
(E) Interquartile range

16. The number of leasable square feet of office space available in a city on any given day has a normal distribution with mean 640,000 square feet and standard deviation 18,000 square feet. What is the interquartile range for this distribution?

(A) 652,000 − 628,000 = 24,000
(B) 658,000 − 622,000 = 36,000
(C) 667,000 − 613,000 = 54,000
(D) 676,000 − 604,000 = 72,000
(E) 694,000 − 586,000 = 108,000

17. A company has 1000 employees evenly distributed throughout five assembly plants. A sample of 30 employees is to be chosen as follows. Each of the five managers will be asked to place the 200 time cards of their respective employees in a bag, shake them up, and randomly draw out six names. The six names from each plant will be put together to make up the sample. Will this method result in a simple random sample of the 1000 employees?

(A) Yes, because every employee has the same chance of being selected.
(B) Yes, because every plant is equally represented.
(C) Yes, because this is an example of stratified sampling, which is a special case of simple random sampling.
(D) No, because the plants are not chosen randomly.
(E) No, because not every group of 30 employees has the same chance of being selected.

18. Following are parts of the probability distributions for the random variables X and Y.

x	$P(x)$	y	$P(y)$
1	?	1	?
2	?	2	?
3	?		
4	?		

If X and Y are independent and two joint probabilities are $P(X = 3, Y = 1) = .14$ and $P(X = 3, Y = 2) = .26$, what is $P(Y = 2)$?

(A) .35
(B) .40
(C) .50
(D) .65
(E) It cannot be determined from the given information.

19. To determine the average number of minutes it takes to manufacture one unit of a new product, an assembly line manager tracks a random sample of 15 units and records the number of minutes it takes to make each unit. The assembly times are assumed to have a normal distribution. If the mean and standard deviation of the sample are 3.92 and 0.45 minutes respectively, which of the following gives a 90% confidence interval for the mean assembly time, in minutes, for units of the new product?

(A) $3.92 \pm 1.645 \dfrac{0.45}{\sqrt{14}}$

(B) $3.92 \pm 1.753 \dfrac{0.45}{\sqrt{14}}$

(C) $3.92 \pm 1.761 \dfrac{0.45}{\sqrt{14}}$

(D) $3.92 \pm 1.753 \dfrac{0.45}{\sqrt{15}}$

(E) $3.92 \pm 1.761 \dfrac{0.45}{\sqrt{15}}$

20. Consider the following back-to-back stemplot:

	0	348
	1	01256
843	2	29
65210	3	2557
92	4	
7552	5	6
	6	1458
6	7	09
8541	8	
90	9	

Which of the following are true statements?

I. The distributions have the same mean.
II. The distributions have the same range.
III. The distributions have the same standard deviation.

(A) II only
(B) I and II
(C) I and III
(D) II and III
(E) I, II, and III

21. Which of the following are true statements?

I. A study results in a 99% confidence interval estimate of (34.2, 67.3). This means that in about 99% of all samples selected by this method, the sample means will fall between 34.2 and 67.3.

II. A high confidence level may be obtained no matter what the sample size.

III. The central limit theorem is most useful when drawing samples from normally distributed populations.

(A) I only
(B) II only
(C) III only
(D) I and II
(E) I and III

GO ON TO THE NEXT PAGE ➤

22. Which of the following leads to a binomial distribution?

    I. A committee of two is to be randomly selected from among seven juniors and four seniors attending an organizational meeting for a New Year's Eve dance. What are the probabilities that the committee will consist of two juniors, of two seniors, or of exactly one junior and one senior?

    II. As a basketball player continues to improve during the season, her free throw percentage rises. The coach is interested in the probability of the player sinking various numbers of free throws during the season.

    III. An inspection procedure at a chemical production plant calls for a government inspector to note each day whether the amount of pollutants being dumped into a nearby river is less than 10 kilograms, between 10 and 25 kilograms, or more than 25 kilograms.

    (A) I only
    (B) II only
    (C) III only
    (D) All of these.
    (E) None of these.

23. Suppose two events, E and F, have nonzero probabilities p and q, respectively. Which of the following is impossible?

    (A) $p + q > 1$
    (B) $p - q < 0$
    (C) $p/q > 1$
    (D) E and F are neither independent nor mutually exclusive.
    (E) E and F are both independent and mutually exclusive.

24. An inspection procedure at a manufacturing plant involves picking four items at random and accepting the whole lot if at least three of the four items are in perfect condition. If in reality 90% of the whole lot are perfect, what is the probability that the lot will be accepted?

    (A) .2916
    (B) .3439
    (C) .6561
    (D) .7084
    (E) .9477

25. A town has one high school, which buses students from urban, suburban, and rural communities. Which of the following samples is recommended in studying attitudes toward tracking of students in honors, regular, and below-grade classes?

    (A) Convenience sample
    (B) Simple random sample (SRS)
    (C) Stratified sample
    (D) Systematic sample
    (E) Voluntary response sample

26. Suppose there is a correlation of $r = 0.9$ between number of hours per day students study and GPAs. Which of the following is a reasonable conclusion?

    (A) 90% of students who study receive high grades.
    (B) 90% of students who receive high grades study a lot.
    (C) 90% of the variation in GPAs can be explained by variation in number of study hours per day.
    (D) 10% of the variation in GPAs cannot be explained by variation in number of study hours per day.
    (E) 81% of the variation in GPAs can be explained by variation in number of study hours per day.

27. To determine the average number of children living in single-family homes, a researcher picks a simple random sample of 50 such homes. However, even after one follow-up visit the interviewer is unable to make contact with anyone in 8 of these homes. Concerned about nonresponse bias, the researcher picks another simple random sample and instructs the interviewer to keep trying until contact is made with someone in a total of 50 homes. The average number of children is determined to be 1.73. Is this estimate probably too low or too high?

(A) Too low, because of undercoverage bias.

(B) Too low, because convenience samples overestimate average results.

(C) Too high, because of undercoverage bias.

(D) Too high, because convenience samples overestimate average results.

(E) Too high, because voluntary response samples overestimate average results.

28. The graph below shows cumulative proportions plotted against land values (in dollars per acre) for farms on sale in a rural community.

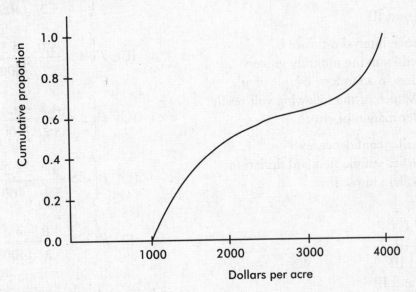

What is the median land value?

(A) $2000
(B) $2250
(C) $2500
(D) $2750
(E) $3000

29. Which of the following are true statements?

  I. Even if the original population is badly skewed, the mean of the set of all sample means from all samples of a given size will equal the mean of the population.

  II. If the original population is very large, it is usually advisable to work with a large sample.

  III. A confidence interval estimate for a population mean is used to eliminate the element of chance from the estimation.

  (A) I only
  (B) II only
  (C) III only
  (D) I and III
  (E) I, II, and III

30. A confidence interval estimate is determined from the monthly grocery expenditures in a random sample of *n* families. Which of the following will result in a smaller margin of error?

  I. A smaller confidence level
  II. A smaller sample standard deviation
  III. A smaller sample size

  (A) II only
  (B) I and II
  (C) I and III
  (D) II and III
  (E) I, II, and III

31. A medical research team claims that high vitamin C intake increases endurance. In particular, 1000 milligrams of vitamin C per day for a month should add an average of 4.3 minutes to the length of maximum physical effort that can be tolerated. Army training officers believe the claim is exaggerated and plan a test on an SRS of 400 soldiers in which they will reject the medical team's claim if the sample mean is less than 4.0 minutes. Suppose the standard deviation of added minutes is 3.2. If the true mean increase is only 4.2 minutes, what is the probability that the officers will fail to reject the false claim of 4.3 minutes?

  (A) $P\left(z < \dfrac{4.0 - 4.3}{3.2/\sqrt{400}}\right)$

  (B) $P\left(z > \dfrac{4.0 - 4.3}{3.2/\sqrt{400}}\right)$

  (C) $P\left(z < \dfrac{4.3 - 4.2}{3.2/\sqrt{400}}\right)$

  (D) $P\left(z > \dfrac{4.3 - 4.2}{3.2/\sqrt{400}}\right)$

  (E) $P\left(z > \dfrac{4.0 - 4.2}{3.2/\sqrt{400}}\right)$

32. Consider the two sets $X = \{10, 30, 45, 50, 55, 70, 90\}$ and $Y = \{10, 30, 35, 50, 65, 70, 90\}$. Which of the following is false?

  (A) The sets have identical medians.
  (B) The sets have identical means.
  (C) The sets have identical ranges.
  (C) The sets have identical boxplots.
  (E) None of the above are false.

GO ON TO THE NEXT PAGE ➤

33. The weight of an aspirin tablet is 300 milligrams according to the bottle label. An FDA investigator weighs a simple random sample of seven tablets, obtains weights of 299, 300, 305, 302, 299, 301, and 303, and runs a hypothesis test of the manufacturer's claim. Which of the following gives the *P*-value of this test?

    (A) $P(t > 1.54)$ with df = 6
    (B) $2P(t > 1.54)$ with df = 6
    (C) $P(t > 1.54)$ with df = 7
    (D) $2P(t > 1.54)$ with df = 7
    (E) $0.5P(t > 1.54)$ with df = 7

34. A teacher believes that giving her students a practice quiz every week will motivate them to study harder, leading to a greater overall understanding of the course material. She tries this technique for a year, and everyone in the class achieves a grade of at least C. Is this an experiment or an observational study?

    (A) An experiment, but with no reasonable conclusion possible about cause and effect
    (B) An experiment, thus making cause and effect a reasonable conclusion
    (C) An observational study, because there was no use of a control group
    (D) An observational study, but a poorly designed one because randomization was not used
    (E) An observational study, and thus a reasonable conclusion of association but not of cause and effect

35. Which of the following is *not* true with regard to contingency tables for chi-square tests for independence?

    (A) The categories are not numerical for either variable.
    (B) Observed frequencies should be whole numbers.
    (C) Expected frequencies should be whole numbers.
    (D) Expected frequencies in each cell should be at least 5, and to achieve this, one sometimes combines categories for one or the other or both of the variables.
    (E) The expected frequency for any cell can be found by multiplying the row total by the column total and dividing by the sample size.

36. Which of the following are true statements?

    I. The probability of a Type II error does not depend on the probability of a Type I error.
    II. In conducting a hypothesis test, it is possible to simultaneously make both a Type I and a Type II error.
    III. A Type II error will result if one incorrectly assumes the data are normally distributed.

    (A) I only
    (B) II only
    (C) III only
    (D) I, II, and III
    (E) None are true.

37.

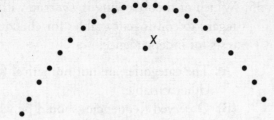

Above is a scatterplot with one point labeled *X*. Suppose you find the least squares regression line. Which of the following are true statements?

   I. *X* has the largest residual, in absolute value, of any point on the scatterplot.
   II. *X* is an influential point.
   III. The residual plot will show a curved pattern.

(A) I only
(B) II only
(C) III only
(D) I and III
(E) II and III

38. A banking corporation advertises that 90% of the loan applications it receives are approved within 24 hours. In a random sample of 50 applications, what is the expected number of loan applications that will be turned down?

(A) $50(.90)$
(B) $50(.10)$
(C) $50(.90)(.10)$
(D) $\sqrt{50(.90)(.10)}$
(E) $\sqrt{\dfrac{(.90)(.10)}{50}}$

GO ON TO THE NEXT PAGE ➤

39. The parallel boxplots below show monthly rainfall summaries for Liberia, West Africa.

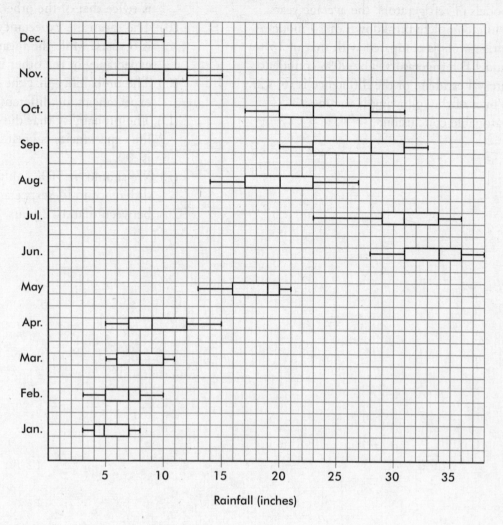

Rainfall (inches)

Which of the following months has the least variability as measured by *interquartile range*?

(A) January
(B) February
(C) March
(D) May
(E) December

40. In comparing the life expectancies of two models of refrigerators, the average years before complete breakdown of 10 model A refrigerators is compared with that of 15 model B refrigerators. The 90% confidence interval estimate of the difference is (6, 12). Which of the following is the most reasonable conclusion?

(A) The mean life expectancy of one model is twice that of the other.

(B) The mean life expectancy of one model is 6 years, while the mean life expectancy of the other is 12 years.

(C) The probability that the life expectancies are different is .90.

(D) The probability that the difference in life expectancies is greater than 6 years is .90.

(E) We should be 90% confident that the difference in life expectancies is between 6 and 12 years.

STOP

If there is still time remaining, you may review your answers.

SECTION II

PART A

Questions 1–5

Spend about 65 minutes on this part of the exam.
Percentage of Section II grade—75

> You must show all work and indicate the methods you use. You will be graded on the correctness of your methods and on the accuracy of your results and explanations.

1. High resting heart rate level can often be reduced by exercise or medication. Researchers would like to test the effectiveness of a new medication to reduce resting heart rate, and to note whether the effectiveness, if any, is enhanced by a weekly regime of three hours exercise. One hundred volunteers with high resting heart rates, all nonexercisers currently not on medication, have been recruited for a study.

   (a) Conclusions will apply to what population?
   (b) Explain how you would design a completely randomized experiment.
   (c) How might you incorporate blocking and for what purpose?
   (d) How is blinding incorporated in this study and why is this important?
   (e) How might you incorporate double blinding and for what purpose?

2. A study comparing commuting distances (in miles) to yearly salary (in $1000) was conducted with 10 randomly selected corporate executives in the New York metropolitan area. A least squares regression line yielded the following computer printout:

Regression Analysis: Salary Versus Distance

```
Dependent variable: Salary
Predictor        Coef      SE Coef         T        P
Constant        285.39      29.50         9.67     0.000
Distance        -8.829       2.637       -3.35     0.010

S = 25.20      R-Sq = 58.4%    R-Sq(adj) = 53.1%

Analysis of Variance

Source          DF       SS         MS        F        P
Regression       1     7115.3     7115.3    11.21    0.010
Residual Error   8     5078.8      634.9
Total            9    12194.1
```

GO ON TO THE NEXT PAGE ➤

Residuals from Salary Versus Distance

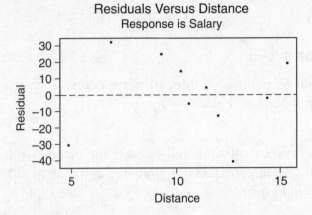

Residuals Versus Distance
Response is Salary

(a) Interpret the slope of the regression line in context.
(b) Interpret the *y*-intercept of the regression line in context.
(c) How reasonable do you think is the answer in part (b)? Explain.
(d) Predict the salary of an executive who commutes 12 miles.
(e) Does this model underestimate or overestimate the actual salary of the executive in the sample who commutes 12 miles? Explain.

3. (a) Give the probability distribution for the number of heads in 5 tosses of a fair coin.

   (b) Draw a cumulative frequency plot of the theoretical probability distribution.

   (c) Following is the cumulative frequency plot for the number of heads in 100 tosses of a coin that may or may not be fair.

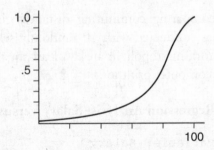

Give a reasonable conclusion about the fairness of this coin. Explain.

   (d) Following is the cumulative frequency plot for the number of heads in 100 tosses of a coin that may or may not be fair.

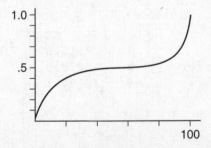

Give a reasonable conclusion about the fairness of this coin. Explain.

GO ON TO THE NEXT PAGE ➤

4. Do home owners have different credit card habits than do apartment renters? One study compared monthly credit card balances of a random sample of home owners with a random sample of apartment renters. Monthly balances averaged $1357 with a standard deviation of $380 for 83 home owners and $1248 with a standard deviation of $330 for 94 apartment renters.

   (a) Is there evidence that the mean monthly credit card balance of homeowners is different from the mean balance of apartment renters? Give statistical justification for your answer.

   (b) Suppose a study using this design resulted in a *P*-value less than .01. Would it be reasonable to conclude that buying a home causes a change in credit card habits? Explain.

   (c) Assuming standard deviations of $380 and $330 as above, how large a sample (same number for each) should be used to be 95% sure of knowing the difference between monthly credit card balances of home versus apartment dwellers to within $100?

5. An AP Statistics teacher believes that high school students who take AP Statistics are more likely to have pets at home than students who are not in AP Statistics. She randomly picks and questions 50 of her AP Stat students and 50 of her other students. The data are summarized below:

	Have pets	Don't have pets
AP Stat students	30	20
Other students	24	26

   (a) Are the intended and sampled populations the same or different, and how does this affect the study?

   (b) Discuss the appropriateness of using a chi-square test.

   (c) Discuss the appropriateness of using a *z*-test for difference of proportions.

   (d) Which would be most appropriate for a quick impression of the data: segmented bar charts, back-to-back stemplots, or parallel boxplots? Draw this visual display.

SECTION II

PART B

Question 6

Spend about 25 minutes on this part of the exam.
Percentage of Section II grade—25

6. A college has four schools: Humanities & Sciences (H&S), Business, Music, and Human Services & Human Performance (HSHP). During the past year the proportions of students applying to H&S, Business, Music, and HSHP were .4, .3, .1, and .2, respectively. A simple random sample of 225 prospective students, out of an estimated 15,000 total applicants, contains 105 H&S, 62 Business, 19 Music, and 39 HSHP applicants.

   (a) Does the number of H&S applicants give evidence that last year's H&S proportion has changed? Give statistical justification for your answer.

   (b) Is there evidence that the overall proportions of students applying to the different schools has changed? Give statistical justification for your answer.

   (c) Compare and comment on the conclusion in parts (a) and (b) above.

   (d) The Dean of HSHP is disturbed by the proportion of HSHP applicants in the sample and asks about the margin of error. What is the margin of error for a 95% confidence interval estimate of the proportion of HSHP applicants among the 225 prospective students? Show your work.

   (e) The Dean of H&S is quite satisfied with the sample results. He would like to report a ±.01 margin of error for the H&S proportion. What would be the underlying level of confidence? Show your work.

STOP

If there is still time remaining, you may review your answers.

Answer Key

Section I

1. **C**	9. **D**	17. **E**	25. **C**	33. **B**
2. **A**	10. **A**	18. **D**	26. **E**	34. **A**
3. **C**	11. **B**	19. **E**	27. **C**	35. **C**
4. **E**	12. **E**	20. **D**	28. **A**	36. **E**
5. **C**	13. **D**	21. **B**	29. **A**	37. **C**
6. **E**	14. **C**	22. **E**	30. **B**	38. **B**
7. **B**	15. **A**	23. **E**	31. **E**	39. **E**
8. **B**	16. **A**	24. **E**	32. **E**	40. **E**

Answers Explained

Section I

1. **(C)** A complete census can give much information about a population, but it doesn't necessarily establish a cause-and-effect relationship among seemingly related population parameters. While the results of well-designed observational studies might suggest relationships, it is difficult to conclude that there is cause and effect without running a well-designed experiment.

2. **(A)** In the first class only 40% of the students scored below the given score, while in the second class 80% scored below the same score.

3. **(C)** The control group should have experiences identical to those of the experimental groups except for the treatment under examination. They should not be given a new treatment.

4. **(E)** The negative sign comes about because we are dealing with the difference of proportions. The confidence interval estimate means that we have a certain *confidence* that the difference in population proportions lies in a particular *interval*.

5. **(C)** The slope and the correlation always have the same sign. Correlation shows association, not causation.

6. **(E)** The proportion of successful calls (problem solved) is $\frac{700}{1000}$, so $\frac{700}{1000}(450)$ is the expected number of calls from location 1 that are successful. Alternatively, the proportion of calls from location 1 is $\frac{450}{1000}$, so $\frac{450}{1000}(700)$ gives the expected number of successful calls from location 1.

7. **(B)** The desire of the workers for the study to be successful led to a placebo effect.

8. **(B)** If two random variables are independent, the mean of the difference of the two random variables is equal to the difference of the two individual means; however, the variance of the difference of the two random variables is equal to the *sum* of the two individual variances.

9. **(D)** A sample is simply a subset of a population.

10. **(A)** Histograms give relative frequencies, not actual frequencies, and these relative frequencies correspond to relative areas.

11. **(B)** With a right tail having probability .01, the critical z-score is 2.326. Thus $\mu + 2.326(.3) = 12$, giving $\mu = 11.3$.

12. **(E)** The larger a random sample, the more likely its standard deviation will be close to the population standard deviation. A sample may not resemble the population at all. Randomness is important when choosing any sample, no matter what its size.

13. **(D)**

$$P\left(\begin{array}{c}\text{child}\\\text{shows}\end{array}\right) = P\left(\begin{array}{c}\text{father}\\\text{shows}\end{array} \cap \begin{array}{c}\text{child}\\\text{shows}\end{array}\right) + P\left(\begin{array}{c}\text{father}\\\text{doesn't}\end{array} \cap \begin{array}{c}\text{child}\\\text{shows}\end{array}\right)$$
$$= (.25)(.8) + (.75)(.06)$$
$$= .200 + .045$$
$$= .245$$

$$P\left(\begin{array}{c}\text{father}\\\text{shows}\end{array} \middle| \begin{array}{c}\text{child}\\\text{shows}\end{array}\right) = \frac{.200}{.245} = .816$$

14. **(C)** As n increases the probabilities of Type I and Type II errors both decrease.

15. **(A)** The mean equals the common value of all the data elements. The other terms all measure variability, which is zero when all the data elements are equal.

16. **(A)** The quartiles Q_1 and Q_3 have z-scores of ±0.67, so $Q_1 = 640,000 - (0.67)18,000 \approx 628,000$, while $Q_3 = 640,000 + (0.67)18,000 \approx 652,000$. The interquartile range is the difference $Q_3 - Q_1$.

17. **(E)** In a simple random sample, every possible group of the given size has to be equally likely to be selected, and this is not true here. For example, with this procedure it is impossible for the employees in the final sample to all be from a single plant. This method is an example of stratified sampling, but stratified sampling does not result in simple random samples.

18. **(D)** By independence we have $P(X = 3)P(Y = 1) = .14$ and $P(X = 3)$ $P(Y = 2) = .26$. Thus $P(X = 3) = .14 + .26 = .4$, $(.4)P(Y = 2) = .26$, and $P(Y = 2) = .65$.

19. **(E)** $\mu_{\bar{x}} \approx \dfrac{s}{\sqrt{n}} = \dfrac{0.45}{\sqrt{15}}$ and with df = 15 − 1 = 14, the critical *t*-scores are ±1.761.

20. **(D)** The two sets have different means, but they have identical shapes and thus the same variability including both range and standard deviation.

21. **(B)** A 99% confidence interval estimate means that in about 99% of all samples selected by this method, the population mean will be included in the confidence interval. The wider the confidence interval, the higher the confidence level. The central limit theorem applies to any population, no matter if it is normally distributed or not.

22. **(E)** None of them leads to a binomial distribution. In the first, there is a lack of independence in choosing the first and second members of the committee; for example, if the first person chosen is a junior, the probability that the second will also be a junior is $\frac{6}{10}$, not $\frac{7}{11}$. In the second, *p* increases throughout the season. The third has three outcomes, not two.

23. **(E)** Independence implies $P(E \cap F) = P(E)P(F)$, while mutually exclusive implies $P(E \cap F) = 0$.

24. **(E)** $4(.9)^3(.1) + (.9)^4 = .9477$.

25. **(C)** In stratified sampling the population is divided into representative groups, and random samples of persons from each group are chosen. In this case it might well be important to be able to consider separately the responses from each of the three groups—urban, suburban, and rural.

26. **(E)** r^2, the coefficient of determination, indicates the percentage of variation in *y* that is explained by variation in *x*.

27. **(C)** It is most likely that the homes at which the interviewer had difficulty finding someone home were homes with fewer children living in them. Replacing these homes with other randomly picked homes will most likely replace homes with fewer children with homes with more children.

28. **(A)** The median corresponds to the 0.5 cumulative proportion.

29. **(A)** The mean of the set of all sample means from all samples of a given size equals the mean of the population regardless of sample size or type of distribution. It is usually advisable to work with a large sample when the original population is more variable. Use of a confidence interval for estimation acknowledges, not eliminates, the element of chance.

30. **(B)** The margin of error varies directly with the critical z-value and directly with the standard deviation of the sample, but inversely with the square root of the sample size.

31. **(E)** $\sigma_{\bar{x}} = \frac{3.2}{\sqrt{400}} = 0.16$. With a true mean increase of 4.2, the z-score for 4.0 is $\frac{4.0-4.2}{0.16} = -1.25$ and the officers fail to reject the claim if the sample mean has z-score greater than this.

32. **(E)** Both have 50 for their means and medians, both have a range of $90 - 10 = 80$, and both have identical boxplots, with first quartile 30 and third quartile 70.

33. **(B)** Since we are not told that the investigator suspects that the average weight is over 300 mg or is under 300 mg, and since a tablet containing too little or too much of a drug clearly should be brought to the manufacturer's attention, this is a two-sided test. Thus the P-value is twice the tail probability obtained (using the t-distribution with df $= n - 1 = 6$.)

34. **(A)** This study was an experiment because a treatment (weekly quizzes) was imposed on the subjects. However, it was a poorly designed experiment with no use of randomization and no control over lurking variables.

35. **(C)** The expected frequencies, as calculated by the rule in (E), may not be whole numbers.

36. **(E)** The probabilities of Type I and Type II error are related; for example, lowering the Type I error increases the probability of a Type II error. A Type I error can be made only if the null hypothesis is true, while a Type II error can be made only if the null hypothesis is false.

37. **(C)** X is probably very close to the least squares regression line and so has a small residual. Removing X will change the regression line very little if at all, and so it is not an influential point.

38. **(B)** The probability of an application being turned down is $1 - .90 = .10$, and the expected value of a binomial with $n = 50$ and $p = .10$ is $np = 50(.10)$.

39. **(E)** A boxplot gives a five-number summary: smallest value, 25th percentile (Q_1), median, 75th percentile (Q_3), and largest value. The interquartile range is given by $Q_3 - Q_1$, or the total length of the two "boxes" minus the "whiskers."

40. **(E)** Using a measurement from a sample, we are never able to say *exactly* what a population mean is; rather we always say we have a certain *confidence* that the population mean lies in a particular *interval*.

SECTION II

1. (a) Conclusions will apply to nonexercisers with high resting heart rates who are not taking medication.

   (b) The design should incorporate random assignment of the volunteers to four treatment groups (medication and exercise, medication and no exercise, placebo and exercise, placebo and no exercise) and a measurement and comparison of resting heart rates.

   A diagram such as the following is sufficient:

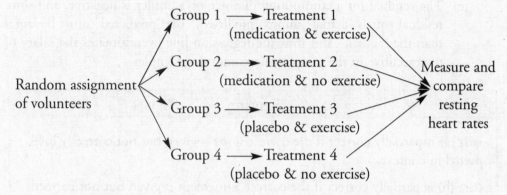

   (c) There are many possible answers, but each should be explained in the context of this problem. For example, you could block on gender or age because men versus women or old versus young might respond differently to medication or exercise. In any case, give the scheme of first splitting volunteers into separate blocks (such as gender or age) and then randomly assigning subjects in each block to the four treatment groups before final measuring and comparison of resting heart rates.

   (d) Blinding is incorporated through the use of a placebo, which, hopefully, looks the same as the medication. The reason for this is that some volunteers might subconsciously influence their heart rates if they knew for sure they were receiving the medication and thus expected results.

   (e) Double blinding would mean that neither the subjects nor those handing out the medication and placebo doses should know who is receiving what. The reason is that the people handing out the doses might inadvertently "give away" who is getting what.

Scoring	
4 **Complete Answer**	All five components correct.
3 **Substantial Answer**	Four components correct.
2 **Developing Answer**	Three components correct.
1 **Minimal Answer**	Two components correct.

2. (a) The slope is −8.829, so on average, every extra mile an executive commutes, his/her salary is expected to decrease by $8,829.

    (b) The *y*-intercept is 285.39, indicating that an executive who commutes 0 miles is expected to have a salary of $285,390.

    (c) Not accurate. Even though the linear regression *t*-test gives a *P*-value of .010, the commuting distances ranged from about 5 to about 15, so a distance of 0 is a real extrapolation. Furthermore, a distance of 0 means the executive lives at the workplace, which itself is unclear.

    (d) −8.829(12) + 285.39 = 179.4, so an executive who commutes 12 miles should be expected to have a salary around $179,400.

    (e) The residual for a commuting distance of 12 miles is negative, and since residual equals "actual" minus "predicted," the "predicted" must be larger than the "actual," and thus the regression line overestimates the salary of the executive in the sample who commutes 12 miles.

Scoring

Part (a) is partially correct if the correct slope is given but not correctly interpreted in context.

Part (b) is partially correct if the correct *y*-intercept is given but not correctly interpreted in context.

Part (c) is partially correct if a clear and correct explanation is not given with the answer.

Part (d) is partially correct if the correct equation of the regression line is indicated, but the correct salary is not calculated, OR if 12 is plugged into an incorrect equation, and the resulting answer is correctly interpreted in terms of dollars.

Part (e) is partially correct if the correct answer of "overestimation" is given, but the explanation is missing or weak.

4 Complete Answer All five parts correct.

3 Substantial Answer Four parts correct OR three parts correct and the other two parts partially correct.

2 Developing Answer Three parts correct OR two parts correct and at least two of the other parts partially correct.

1 Minimal Answer Two parts correct OR one part correct and at least two of the other parts partially correct.

3. (a) $P(0) = \quad (.5)^5 \quad = .03125$
 $P(1) = \ 5(.5)^4(.5) \ = .15625$
 $P(2) = 10(.5)^3(.5)^2 = .31250$
 $P(3) = 10(.5)^2(.5)^3 = .31250$
 $P(4) = \ 5(.5)(.5)^4 \ = .15625$
 $P(5) = \quad (.5)^5 \quad = .03125$

(b)

(c) Note that the median number of heads is about 80. Thus the coin is unfair, with a probability of heads greater than .5.

(d) The graph is steepest (and thus the actual frequencies are greatest) for low numbers of heads and again for high numbers of heads. This is impossible no matter what the probability of getting a head on a coin is. Thus the graph is unrealistic and no conclusion about fairness can be made.

Scoring

Parts (a) and (b) are either essentially correct or not. Do not take off for minor arithmetical errors. Part (b) is correct if it is the correct cumulative plot for distribution (right or wrong) given in part (a).

Parts (c) and (d) are partially correct if the answer is correct but the explanation is weak.

4 Complete Answer — All four parts essentially correct OR three parts essentially correct and one part partially correct.

3 Substantial Answer — Three parts essentially correct OR two parts essentially correct and at least one part partially correct.

2 Developing Answer — Two parts essentially correct OR one part essentially correct and at least one part partially correct.

1 Minimal Answer — One part essentially correct OR two parts partially correct.

4. (a) The proper hypothesis test is the two-sample *t*-test (unpooled, because there is no reason to suppose the standard deviations are equal). Assumption of independent random samples is satisfied by the given scheme. Because of large sample sizes, the Central Limit Theorem tells us that the distribution of sample means will be approximately normal.

$H_0: \mu_h - \mu_a = 0$ (The mean monthly credit card balance for all home owners is the same as the mean monthly credit card balance for all apartment renters.)

$H_a: \mu_h - \mu_a \neq 0$ (The mean monthly credit card balance for all home owners is different from the mean monthly credit card balance for all apartment renters.)

On the TI-83, we get $t = 2.025$ and $P = 0.045$.

Or, calculate $t = \dfrac{\bar{x}_h - \bar{x}_a}{\sqrt{\dfrac{s_h{}^2}{n_h} + \dfrac{s_a{}^2}{n_a}}} = \dfrac{1357 - 1248}{\sqrt{\dfrac{380^2}{83} + \dfrac{330^2}{94}}} = 2.025$.

With df = min{83,94} = 83, we get $P = 2(.023) = .046$.
With this small a *P*-value there is evidence (at the 5% significance level) to reject H_0 and conclude that the mean monthly credit card balance for home owners is different from the mean monthly credit card balance for apartment renters.

(b) Even with a small *P*-value we cannot conclude causation because this is an observational study, not a controlled experiment. For example, it may just be that the type of person who chooses to live in an apartment also is the type who doesn't like to run up credit card balances.

(c) We calculate $1.96\sqrt{\dfrac{380^2}{n} + \dfrac{330^2}{n}} \leq 100$, so $1.96\dfrac{503.3}{\sqrt{n}} \leq 100$ and $n \geq 97.3$. We should choose $n = 98$. (Note that we are using the normal score of 1.96, but given the size of *n*, this is reasonable.)

Scoring

There are two components to part (a): one component is to name the appropriate test, give appropriate hypotheses, and name and check assumptions; the second component is to find correct values for t and P, and give the correct conclusion in context. Each component can receive full, partial, or no credit.

There is one component to part (b), with full credit either through stating that this is an observational study and observational studies do not imply causation OR through mentioning a confounding variable and how the confounding operates in the context of this problem.

There is one component to part (c), with full credit for any answer in the 95 to 100 range together with some calculation or justification.

4 Complete Answer   All four components correct OR three correct and fourth partially correct.

3 Substantial Answer   Three components correct OR two correct and other two both partially correct.

2 Developing Answer   Two components correct OR one correct and two partially correct.

1 Minimal Answer   One component correct OR two components partially correct.

---

5. (a) The intended population is all high school students; however, the sample was gathered from a single teacher's classes. Any conclusion is applicable only to students of this teacher.

 (b) The teacher is interested in a one-sided test, namely whether AP Stat students are *more* likely to have pets than are other students. A chi-square test will only indicate whether there is an association between pet ownership and taking AP Statistics.

 (c) To perform a z-test for differences of proportions, two samples must have been *independently* drawn. Since all the students were from one teacher, there is a lack of independence. For example, the teacher might have encouraged or discouraged students from having pets.

 (d) Segmented bar charts give a good initial impression of this data.

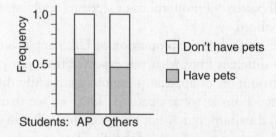

	Scoring	
4 **Complete Answer**	All four questions correct.	
3 **Substantial Answer**	Three questions correct.	
2 **Developing Answer**	Two questions correct.	
1 **Minimal Answer**	One question correct.	

6. (a) For this one-proportion z-test we check the assumptions: $np = (225)(.4) = 90 > 10$ and $n(1 - p) = (225)(.6) = 135 > 10$, and we have a simple random sample that is less than 10% of the population, that is, 10% of 15,000 is 1500 and 225 < 1500.

$H_0: p = .4$ (where p = the proportion of students applying to H&S)

$H_a: p \neq .4$

$$\hat{p} = \frac{105}{225} = .4667 \text{ and } z = \frac{.4667 - .4}{\sqrt{\frac{(.4)(.6)}{225}}} = 2.042, \text{ so } \frac{P}{2} = .0206 \text{ and } P = .0412$$

With this small a P-value, there is evidence at the 5% significance level (because .0412 < .05) to reject H_0 and conclude that last year's H&S proportion has changed.

(b) To do a chi-square test for goodness-of-fit we check the assumptions: the expected cell counts $(.4)(225) = 90$, $(.3)(225) = 67.5$, $(.1)(225) = 22.5$, and $(.2)(225) = 45$ are all >5, and we have a simple random sample that is less than 10% of the population (225 < 1500).

H_0: The observed numbers are consistent with the old proportions.

H_a: The observed numbers are not consistent with the old proportions.

$$\chi^2 = \sum \frac{(\text{obs} - \text{exp})^2}{\text{exp}}$$
$$= \frac{(105 - 90)^2}{90} + \frac{(62 - 67.5)^2}{67.5} + \frac{(19 - 22.5)^2}{22.5} + \frac{(39 - 45)^2}{45} = 4.29$$

With df = 4 − 1 = 3, we get $P = .232$.

With this large a value for P, there is *not* sufficient evidence to reject H_0, and thus we conclude that there is *not* sufficient evidence to say that the overall pattern of proportions of students applying to the different schools has changed.

(c) We see that while the proportion of H&S applicants is statistically significantly different from what was expected from last year's pattern, the overall distribution of applicants is not significantly different from what was expected from last year's pattern. Thus we see that while a change in an overall distribution may not be significant, it is still possible for the change in one of the components of the distribution to be significant.

(d) To find a confidence interval for a population proportion, we check the assumptions: $n\hat{p} = 225\left(\dfrac{39}{225}\right) = 39 > 10$ and $n(1 - \hat{p}) = 186 > 10$, and we have a simple random sample which is less than 10% of the population $(225 < 1500)$.

With $\hat{p} = \dfrac{39}{225} = .173$, the margin of error for 95% confidence is:

$$\pm 1.96\sqrt{\frac{\hat{p}(1 - \hat{p})}{n}} = \pm 1.96\sqrt{\frac{(.173)(.827)}{225}} = \pm .049$$

(e) To find a confidence interval for a population proportion, we check the assumptions:

$$n\hat{p} = 225\left(\frac{105}{225}\right) = 105 > 10,\ n(1 - \hat{p}) = 120 > 10,$$ and we have a simple random sample that is less than 10% of the population $(225 < 1500)$.

With $\hat{p} = \dfrac{105}{225} = .467$, we have $z\sqrt{\dfrac{(.467)(.533)}{225}} = .01$, and so $z = .301$, which results in a confidence of only 24%!

Scoring

Parts (a) and (b) involve five steps: naming a test, checking assumptions, stating hypotheses, calculating t and P, and giving a conclusion in context. Parts (a) and (b) are each essentially correct if at least four of the five steps are correct and complete, and are partially correct if two or three of the five steps are correct and complete.

Part (c) is essentially correct if the answer is reasonable for the answers given in parts (a) and (b).

Parts (d) and (e) are essentially correct for correct final answers together with work showing how these answers were obtained, and partially correct for unclear or incomplete work.

4 **Complete Answer**	All five parts essentially correct.
3 **Substantial Answer**	Four parts essentially correct OR three parts essentially correct and the other two partially correct.
2 **Developing Answer**	Three parts essentially correct OR two parts essentially correct and at least two of the remaining parts partially correct.
1 **Minimal Answer**	Two parts essentially correct OR one part essentially correct and at least two of the remaining parts partially correct OR four parts partially correct.

Answer Sheet

1. Ⓐ Ⓑ Ⓒ Ⓓ Ⓔ
2. Ⓐ Ⓑ Ⓒ Ⓓ Ⓔ
3. Ⓐ Ⓑ Ⓒ Ⓓ Ⓔ
4. Ⓐ Ⓑ Ⓒ Ⓓ Ⓔ
5. Ⓐ Ⓑ Ⓒ Ⓓ Ⓔ
6. Ⓐ Ⓑ Ⓒ Ⓓ Ⓔ
7. Ⓐ Ⓑ Ⓒ Ⓓ Ⓔ
8. Ⓐ Ⓑ Ⓒ Ⓓ Ⓔ
9. Ⓐ Ⓑ Ⓒ Ⓓ Ⓔ
10. Ⓐ Ⓑ Ⓒ Ⓓ Ⓔ

11. Ⓐ Ⓑ Ⓒ Ⓓ Ⓔ
12. Ⓐ Ⓑ Ⓒ Ⓓ Ⓔ
13. Ⓐ Ⓑ Ⓒ Ⓓ Ⓔ
14. Ⓐ Ⓑ Ⓒ Ⓓ Ⓔ
15. Ⓐ Ⓑ Ⓒ Ⓓ Ⓔ
16. Ⓐ Ⓑ Ⓒ Ⓓ Ⓔ
17. Ⓐ Ⓑ Ⓒ Ⓓ Ⓔ
18. Ⓐ Ⓑ Ⓒ Ⓓ Ⓔ
19. Ⓐ Ⓑ Ⓒ Ⓓ Ⓔ
20. Ⓐ Ⓑ Ⓒ Ⓓ Ⓔ

21. Ⓐ Ⓑ Ⓒ Ⓓ Ⓔ
22. Ⓐ Ⓑ Ⓒ Ⓓ Ⓔ
23. Ⓐ Ⓑ Ⓒ Ⓓ Ⓔ
24. Ⓐ Ⓑ Ⓒ Ⓓ Ⓔ
25. Ⓐ Ⓑ Ⓒ Ⓓ Ⓔ
26. Ⓐ Ⓑ Ⓒ Ⓓ Ⓔ
27. Ⓐ Ⓑ Ⓒ Ⓓ Ⓔ
28. Ⓐ Ⓑ Ⓒ Ⓓ Ⓔ
29. Ⓐ Ⓑ Ⓒ Ⓓ Ⓔ
30. Ⓐ Ⓑ Ⓒ Ⓓ Ⓔ

31. Ⓐ Ⓑ Ⓒ Ⓓ Ⓔ
32. Ⓐ Ⓑ Ⓒ Ⓓ Ⓔ
33. Ⓐ Ⓑ Ⓒ Ⓓ Ⓔ
34. Ⓐ Ⓑ Ⓒ Ⓓ Ⓔ
35. Ⓐ Ⓑ Ⓒ Ⓓ Ⓔ
36. Ⓐ Ⓑ Ⓒ Ⓓ Ⓔ
37. Ⓐ Ⓑ Ⓒ Ⓓ Ⓔ
38. Ⓐ Ⓑ Ⓒ Ⓓ Ⓔ
39. Ⓐ Ⓑ Ⓒ Ⓓ Ⓔ
40. Ⓐ Ⓑ Ⓒ Ⓓ Ⓔ

Practice Examination 4

Section I

Questions 1–40

Spend 90 minutes on this part of the exam.

> **Directions:** The questions or incomplete statements that follow are each followed by five suggested answers or completions. Choose the response that best answers the question or completes the statement.

1. A company wishes to determine the relationship between the number of days spent training employees and their performances on a job aptitude test. Collected data result in a least squares regression line, $\hat{y} = 12.1 + 6.2x$, where x is the number of training days and $\hat{y}$ is the predicted score on the aptitude test. Which of the following statements best interprets the slope and y-intercept of the regression line?

   (A) The base score on the test is 12.1, and for every day of training one would expect, on average, an increase of 6.2 on the aptitude test.

   (B) The base score on the test is 6.2, and for every day of training one would expect, on average, an increase of 12.1 on the aptitude test.

   (C) The mean number of training days is 12.1, and for every additional 6.2 days of training one would expect, on average, an increase of one unit on the aptitude test.

   (D) The mean number of training days is 6.2, and for every additional 12.1 days of training one would expect, on average, an increase of one unit on the aptitude test.

   (E) The mean number of training days is 12.1, and for every day of training one would expect, on average, an increase of 6.2 on the aptitude test.

GO ON TO THE NEXT PAGE ➤

2. To survey the opinions of the students at your high school, a researcher plans to select every twenty-fifth student entering the school in the morning. Assuming there are no absences, will this result in a simple random sample of students attending your school?

(A) Yes, because every student has the same chance of being selected.

(B) Yes, but only if there is a single entrance to the school.

(C) Yes, because the 24 out of every 25 students who are not selected will form a control group.

(D) Yes, because this is an example of systematic sampling, which is a special case of simple random sampling.

(F) No, because not every sample of the intended size has an equal chance of being selected.

3. Consider a hypothesis test with $H_0 : \mu = 70$ and $H_a : \mu < 70$. Which of the following choices of significance level and sample size results in the greatest power of the test when $\mu = 65$?

(A) $\alpha = 0.05$, $n = 15$

(B) $\alpha = 0.01$, $n = 15$

(C) $\alpha = 0.05$, $n = 30$

(D) $\alpha = 0.01$, $n = 30$

(E) There is no way of answering without knowing the strength of the given power.

4. Suppose that 60% of a particular electronic part last over 3 years, while 70% last less than 6 years. Assuming a normal distribution, what are the mean and standard deviation with regard to length of life of these parts?

(A) $\mu = 3.677$, $\sigma = 3.561$

(B) $\mu = 3.977$, $\sigma = 3.861$

(C) $\mu = 4.177$, $\sigma = 3.561$

(D) $\mu = 4.377$, $\sigma = 3.261$

(E) The mean and standard deviation cannot be computed from the information given.

5. The graph below shows cumulative proportions plotted against numbers of employees working in mid-sized retail establishments.

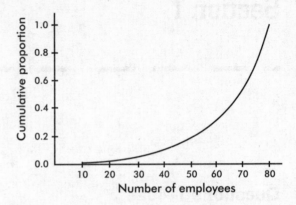

What is the approximate interquartile range?

(A) 18

(B) 35

(C) 57

(D) 68

(E) 75

6. The label on a package of cords claims that the breaking strength of a cord is 3.5 pounds, but a hardware store owner believes the real value is less. She plans to test 36 such cords; if their mean breaking strength is less than 3.25 pounds, she will reject the claim on the label. If the standard deviation for the breaking strengths of all such cords is 0.9 pounds, what is the probability of mistakenly rejecting a true claim?

(A) .05

(B) .10

(C) .15

(D) .45

(E) .94

7. What is a placebo?

(A) A method of selection

(B) An experimental treatment

(C) A control treatment

(D) A parameter

(E) A statistic

GO ON TO THE NEXT PAGE ➤

8. To study the effect of alcohol on reaction time, subjects were randomly selected and given three beers to consume. Their reaction time to a simple stimulus was measured before and after drinking the alcohol. Which of the following statements are true?

   I. This study was an observational study.
  II. This study was an experiment with no controls.
 III. This study was an experiment in which the subjects were used as their own controls.
 IV. The researcher should be concerned about a placebo effect.

   (A) I only
   (B) II only
   (C) III only
   (D) II and IV
   (E) III and IV

9. Suppose that the regression line for a set of data, $y = 7x + b$, passes through the point $(-2, 4)$. If $\bar{x}$ and $\bar{y}$ are the sample means of the x- and y-values, respectively, then $\bar{y} =$

   (A) $\bar{x}$
   (B) $\bar{x} + 2$
   (C) $\bar{x} - 4$
   (D) $7\bar{x}$
   (E) $7\bar{x} + 18$

10. Which of the following is a false statement about simple random samples?

   (A) A sample must be reasonably large to be properly considered a simple random sample.
   (B) Inspection of a sample will give no indication of whether or not it is a simple random sample.
   (C) Attributes of a simple random sample may be very different from attributes of the population.
   (D) Every element of the population has an equal chance of being picked.
   (E) Every sample of the desired size has an equal chance of being picked.

11. A local school has seven math teachers and seven English teachers. When comparing their mean salaries, which of the following is most appropriate?

   (A) A two-sample z-test of population means.
   (B) A two-sample t-test of population means.
   (C) A one-sample z-test on a set of differences.
   (D) A one-sample t-test on a set of differences.
   (E) None of the above are appropriate.

12. Following is a histogram of ages of people applying for a particular high school teaching position.

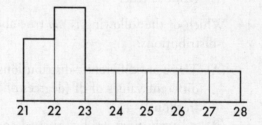

Which of the following statements are true?

   I. The median age is between 24 and 25.
  II. The mean age is between 22 and 23.
 III. The mean age is greater than the median age.

   (A) I only
   (B) II only
   (C) III only
   (D) All are true.
   (E) None is true.

GO ON TO THE NEXT PAGE ➤

13. To conduct a survey of which long distance carriers are used in a particular locality, a researcher opens a telephone book to a random page, closes his eyes, puts his finger down on the page, and then calls the next 75 names. Which of the following are true statements?

   I. The survey design incorporates chance.
   II. The procedure results in a simple random sample.
   III. The procedure could easily result in selection bias.

   (A) I and II
   (B) I and III
   (C) II and III
   (D) All are true.
   (E) None is true.

14. Which of the following is *not* true about *t*-distributions?

   (A) There are different *t*-distributions for different values of df (degrees of freedom).
   (B) *t*-distributions are bell-shaped and symmetric.
   (C) *t*-distributions always have mean 0 and standard deviation 1.
   (D) *t*-distributions are more spread out than the normal distribution.
   (E) The larger the df value, the closer the distribution is to the normal distribution.

15. Suppose the probability that a person picked at random has lung cancer is .035 and the probability that the person both has lung cancer and is a heavy smoker is .014. Given that someone picked at random has lung cancer, what is the probability that the person is a heavy smoker?

   (A) .021
   (B) .049
   (C) .244
   (D) .250
   (F) .400

16. Taxicabs in a metropolitan area are driven an average of 75,000 miles per year with a standard deviation of 12,000 miles. What is the probability that a randomly selected cab has been driven less than 100,000 miles if it is known that it has been driven over 80,000 miles? Assume a normal distribution of miles per year among cabs.

   (A) .06
   (B) .34
   (C) .66
   (D) .68
   (E) .94

17. A plant manager wishes to determine the difference in number of accidents per day between two departments. How many days' records should be examined to be 90% certain of the difference in daily averages to within 0.25 accidents per day? Assume standard deviations of 0.8 and 0.5 accidents per day in the two departments, respectively.

   (A) 39
   (B) 55
   (C) 78
   (D) 109
   (E) 155

18. Suppose the correlation between two variables is $r = .19$. What is the new correlation if .23 is added to all values of the *x*-variable, every value of the *y*-variable is doubled, and the two variables are interchanged?

   (A) .19
   (B) .42
   (C) .84
   (D) −.19
   (E) −.84

19. Two commercial flights per day are made from a small county airport. The airport manager tabulates the number of on-time departures for a sample of 200 days.

Number of on-time departures	0	1	2
Observed number of days	10	80	110

GO ON TO THE NEXT PAGE ➤

What is the χ^2 statistic for a goodness-of-fit test that the distribution is binomial with probability equal to .8 that a flight leaves on time?

(A) $\dfrac{(10-8)^2}{8} + \dfrac{(80-64)^2}{64} + \dfrac{(110-128)^2}{128}$

(B) $\dfrac{(10-8)^2}{10} + \dfrac{(80-64)^2}{80} + \dfrac{(110-128)^2}{110}$

(C) $\dfrac{(10-10)^2}{10} + \dfrac{(80-30)^2}{30} + \dfrac{(110-160)^2}{160}$

(D) $\dfrac{(10-10)^2}{10} + \dfrac{(80-30)^2}{80} + \dfrac{(110-160)^2}{110}$

(E) $\dfrac{(10-66)^2}{10} + \dfrac{(80-67)^2}{80} + \dfrac{(110-67)^2}{110}$

20. A company has a choice of three investment schemes. Option I gives a sure $25,000 return on investment. Option II gives a 50% chance of returning $50,000 and a 50% chance of returning $10,000. Option III gives a 5% chance of returning $100,000 and a 95% chance of returning nothing. Which option should the company choose?

(A) Option II if it wants to maximize expected return

(B) Option I if it needs at least $20,000 to pay off an overdue loan

(C) Option III if it needs at least $80,000 to pay off an overdue loan

(D) All of the above answers are correct.

(E) Because of chance, it really doesn't matter which option it chooses.

21. Suppose $P(X) = .35$ and $P(Y) = .40$. If $P(X|Y) = .28$, what is $P(Y|X)$?

(A) .14
(B) .32
(C) .50
(D) .70
(E) .80

22. To test whether extensive exercise lowers the resting heart rate, a study is performed by randomly selecting half of a group of volunteers to exercise 1 hour each morning, while the rest are instructed to perform no exercise. Is this study an experiment or an observational study?

(A) An experiment with a control group and blinding

(B) An experiment with blocking

(C) An observational study with comparison and randomization

(D) An observational study with little if any bias

(E) None of the above

23. Consider the following parallel boxplots of the effective durations of three over-the-counter pain relievers:

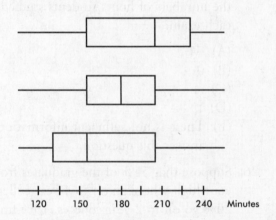

Which of the following are true statements?

I. All three have the same range.

II. All three have the same interquartile range.

III. The difference in the medians between the first and third distributions is equal to the interquartile range of the second distribution.

(A) I and II
(B) I and III
(C) II and III
(D) I, II, and III
(E) None of the above gives the complete set of true responses.

GO ON TO THE NEXT PAGE ➤

24. The waiting times for a new roller coaster ride are normally distributed with a mean of 35 minutes and a standard deviation of 10 minutes. If there are 150,000 riders the first summer, which of the following is the shortest time interval associated with 100,000 riders?

   (A) 0 to 31.7 minutes
   (B) 31.7 to 39.3 minutes
   (C) 25.3 to 44.7 minutes
   (D) 25.3 to 35 minutes
   (E) 39.3 to 95 minutes

25. For their first exam, students in an AP Statistics class studied an average of 4 hours with a standard deviation of 1 hour. Almost everyone did poorly on the exam, and so for the second exam every student studied 10 hours. What is the correlation between the numbers of hours students studied for each exam?

   (A) −1
   (B) 0
   (C) .4
   (D) 1
   (E) There is not sufficient information to answer this question.

26. Suppose that 54% of the graduates from your high school go on to 4-year colleges, 20% go on to 2-year colleges, 19% find employment, and the remaining 7% search for a job. If a randomly selected student is not going on to a 2-year college, what is the probability she will be going on to a 4-year college?

   (A) .460
   (B) .540
   (C) .630
   (D) .675
   (E) .730

27. We are interested in the proportion p of people who are unemployed in a large city. Eight percent of a simple random sample of 500 people are unemployed. What is the midpoint for a 95% confidence interval estimate of p?

   (A) .012
   (B) .025
   (C) .475
   (D) p
   (E) None of the above.

28. In the following table, what value for n results in a table showing perfect independence?

n	52
35	10

   (A) 10
   (B) 17
   (C) 27
   (D) 77
   (E) 182

29. A survey was conducted to determine the percentage of parents who would support raising the legal driving age to 18. The results were stated as 67% with a margin of error of ±3%. What is meant by ±3%?

   (A) Three percent of the population were not surveyed.
   (B) In the sample, the percentage of parents who would support raising the driving age is between 64% and 70%.
   (C) The percentage of the entire population of parents who would support raising the driving age is between 64% and 70%.
   (D) It is unlikely that the given sample proportion result could be obtained unless the true percentage was between 64% and 70%.
   (E) Between 64% and 70% of the population were surveyed.

30. Random samples of size n are drawn from a population. The mean of each sample is calculated, and the standard deviation of this set of sample means is found. Then the procedure is repeated, this time with samples of size $4n$. How does the standard deviation of the second group compare with the standard deviation of the first group?

    (A) It will be the same.
    (B) It will be twice as large.
    (C) It will be four times as large.
    (D) It will be half as large.
    (E) It will be one-quarter as large.

31. It is estimated that 30% of all cars parked in a metered lot outside City Hall receive tickets for meter violations. In a random sample of 5 cars parked in this lot, what is the probability that at least one receives a parking ticket?

    (A) $1 - (.3)^5$
    (B) $1 - (.7)^5$
    (C) $5(.3)(.7)^4$
    (D) $5(.3)^4(.7)$
    (E) $5(.3)^4(.7) + 10(.3)^3(.7)^2 + 10(.3)^2(.7)^3 + 5(.3)(.7)^4 + (.7)^5$

32. Given that the sample has a standard deviation of zero, which of the following statements are true?

    I. The standard deviation of the population is also zero.
    II. The sample mean and sample median are equal.
    III. The sample may have outliers; however, the distribution must be symmetric.

    (A) I only
    (B) II only
    (C) III only
    (D) All are true.
    (E) None is true.

33. In leaving for school on an overcast April morning you make a judgment on the null hypothesis: The weather will remain dry. What would the results be of Type I and Type II errors?

    (A) Type I error: get drenched
    Type II error: needlessly carry around an umbrella
    (B) Type I error: needlessly carry around an umbrella
    Type II error: get drenched
    (C) Type I error: carry an umbrella, and it rains
    Type II error: carry no umbrella, but weather remains dry
    (D) Type I error: get drenched
    Type II error: carry no umbrella, but weather remains dry
    (E) Type I error: get drenched
    Type II error: carry an umbrella, and it rains

34. A poll shows that 65% of the registered voters in a county are Democrats and the rest are Republicans. Suppose that 60% of the Democrats and 50% of the Republicans support a bond issue. What is the probability that a registered voter supports the bond issue? What is the probability that a voter chosen at random who turns out to support the bond issue is a Democrat?

    (A) .535, .650
    (B) .550, .610
    (C) .550, .650
    (D) .550, .690
    (E) .565, .690

GO ON TO THE NEXT PAGE ➤

35. In one study half of a class were instructed to watch exactly 1 hour of television per day, the other half were told to watch 5 hours per day, and then their class grades were compared. In a second study students in a class responded to a questionnaire asking about their television usage and their class grades.

    (A) The first study was an experiment without a control group, while the second was an observational study.
    (B) The first study was an observational study, while the second was a controlled experiment.
    (C) Both studies were controlled experiments.
    (D) Both studies were observational studies.
    (E) Each study was part controlled experiment and part observational study.

36. Data are collected on income levels x versus number of bank accounts y. Summary calculations give $\bar{x} = 32{,}000$; $s_x = 11{,}500$; $\bar{y} = 1.7$; $s_y = 0.4$, and $r = .42$. What is the slope of the least squares regression line of number of bank accounts on income level?

    (A) 0.000015
    (B) 0.000022
    (C) 0.000035
    (D) 0.000053
    (E) 0.000083

37. The mean thrust of a certain model jet engine is 9500 pounds. Concerned that a production process change might have lowered the thrust, an inspector tests a sample of units, calculating a mean of 9350 pounds with a
    z-score of −2.46 and a P-value of .0069. Which of the following is the most reasonable conclusion?

    (A) 99.31% of the engines produced under the new process will have a thrust under 9350 pounds.
    (B) 99.31% of the engines produced under the new process will have a thrust under 9500 pounds.
    (C) 0.69% of the time an engine produced under the new process will have a thrust over 9500 pounds.
    (D) There is evidence to conclude that the new process is producing engines with a mean thrust under 9350 pounds.
    (E) There is evidence to conclude that the new process is producing engines with a mean thrust under 9500 pounds.

38. A reading specialist in a large public school system believes that the more time students spend reading, the better they will do in school. She plans a middle school experiment in which an SRS of 30 eighth graders will be assigned four extra hours of reading per week, an SRS of 30 seventh graders will be assigned two extra hours of reading per week, and an SRS of 30 sixth graders with no extra assigned reading will be a control group. After one school year, the mean GPAs from each group will be compared. Is this a good experimental design?

    (A) Yes
    (B) No, because while this design may point out an association between reading and GPA, it cannot establish a cause-and-effect relationship.
    (C) No, because without blinding, there is a strong chance of a placebo effect.
    (D) No, because any conclusion would be flawed because of blocking bias.
    (E) No, because grade level is a lurking variable which may well be confounded with the variables under consideration.

GO ON TO THE NEXT PAGE ➤

39. A study at 35 large city high schools gives the following back-to-back stemplot of the percentages of students who say they have tried alcohol.

School Year 1995–96		School Year 1998–99
	0	
2	1	
9	2	
6 6 3	3	1 8
4 3 1 1	4	0 2 9
9 9 8 6 5 3 2 2 0	5	3 3 4 6
9 8 7 6 6 5 1 1	6	1 2 2 2 7
5 4 3 2 2	7	0 1 3 3 5 5 6 7 8 9
5 4 0	8	2 3 4 5 8 8 9
0	9	0 1 1 2

Which of the following does *not* follow from the above data?

(A) In general, the percentage of students trying alcohol seems to have increased from 1995–96 to 1998–99.

(B) The median alcohol percentage among the 35 schools increased from 1995–96 to 1998–99.

(C) The spread between the lowest and highest alcohol percentages decreased from 1995–96 to 1998–99.

(D) For both school years in most of the 35 schools, most of the students said they had tried alcohol.

(E) The percentage of students trying alcohol increased in each of the schools between 1995–96 and 1998–99.

40. All of the following statements are true for all discrete random variables except for which one?

(A) The possible outcomes must all be numerical.

(B) The possible outcomes must be mutually exclusive.

(C) The mean (expected value) always equals the sum of the products obtained by multiplying each value by its corresponding probability.

(D) The standard deviation of a random variable can never be negative.

(E) Approximately 95% of the outcomes will be within two standard deviations of the mean.

STOP

If there is still time remaining, you may review your answers.

SECTION II

PART A

Questions 1–5

Spend about 65 minutes on this part of the exam.
Percentage of Section II grade—75

> You must show all work and indicate the methods you use. You will be graded on the correctness of your methods and on the accuracy of your results and explanations.

1. An efficiency expert conducts a study to determine the effect of sleep on work production in an electronics assembly plant. She interviews a simple random sample of 30 employees who claim to sleep under 6 hours per night and a simple random sample of 30 employees who claim to sleep over 6 hours per night. In each group she calculates the mean number of electronic devices assembled per day.

   (a) Explain why this is an observational study and not an experiment.
   (b) Give an example of a possible confounding variable with an explanation in the context of this study.
   (c) If the mean number of electronic devices assembled is statistically greater for those who sleep over 6 hours per night than for those who sleep under 6 hours, is it reasonable to encourage all employees to sleep over 6 hours? Explain.
   (d) How could the efficiency expert design a related experiment to study the effect of sleep on work production?

2. The following scatterplot shows the results of a random sample of 25 major cities from around the world, where standard of living (on a 1 to 10 scale, with 10 the highest) is plotted against percentage of population with HIV/AIDS.

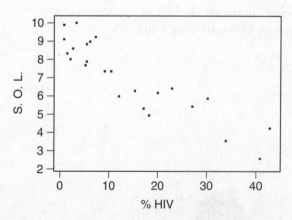

Computer output is shown below, along with a plot of the residuals:

GO ON TO THE NEXT PAGE ➤

Regression Analysis: S. O. L. Versus % HIV

```
Dependent variable: S.O.L.

Predictor          Coef      SE Coef          T          P
Constant         9.0303       0.2590      34.86      0.000
% HIV           -0.14320      0.01397     -10.25      0.000

S = 0.8870      R-Sq = 82.0%

Analysis of Variance

Source            DF          SS          MS          F          P
Regression         1       82.632      82.632     105.02      0.000
Residual Error    23       18.097       0.787
Total             24      100.703
```

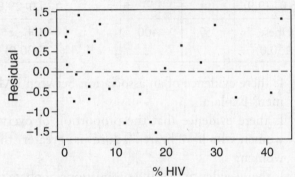

Residuals Versus % HIV
Response is S. O. L.

(a) What is the equation of the regression line that predicts standard of living as a function of HIV/AIDS percentage?

(b) Interpret the slope of the regression line in context.

(c) Interpret the *y*-intercept of the regression line in context.

(d) Interpret R-Sq in context.

(e) What is the value of the correlation coefficient?

(f) Is there evidence of an association between standard of living and percentage of population with HIV/AIDS? Explain. You may assume that a histogram of the residuals is approximately normal.

GO ON TO THE NEXT PAGE ➤

3. Extreme obesity has long been recognized as a risk factor for heart failure. A study reported in the August 1, 2002 *New England Journal of Medicine* aimed to see if lesser degrees of obesity also are a risk factor for heart failure. Some of the results are summarized in the following tables, where BMI, or body mass index (weight in kilograms divided by the square of height in meters), is cross-classified against whether or not heart failure occurs.

Men

BMI		Heart failure	Healthy heart
	Normal (≤24.9)	57	812
	Overweight (25.0–29.9)	123	1255
	Obese (≥30.0)	57	400

Women

	Heart failure	Healthy heart
Normal (≤24.9)	107	1622
Overweight (25.0–29.9)	86	869
Obese (≥30.0)	65	428

(a) Is there evidence of an association between BMI and heart failure for men? Explain.

(b) Is there evidence that the proportion of overweight, but not obese, women who have heart failure is greater than that for normal women?

(c) Is there evidence that the proportions of obese men and obese women who have heart failure are different?

4. A tetrahedron (a solid with four congruent triangular sides) has three 5's and one 8 on its four faces, while a die has two 3's and four 6's on its six faces. Assume both are "fair." One player rolls the tetrahedron, while a second player simultaneously rolls the die. The winner is the player with the higher number on the bottom.

(a) If you want to win, would you rather roll the tetrahedron or the die? Explain.

(b) Suppose after each player rolls, one of the visible "up" sides of the tetrahedron shows a 5 and one of the visible "up" sides of the die shows a 6. Which player has the greater probability of winning that roll?

(c) Suppose in the original game, the winner wins points (no one loses points) equal to the number showing on the bottom of the loser's toss. What is the expected value for one roll of this new game to each player?

(d) With this last set of rules, suppose the overall winner will be the player who ends up with the most points. Would the die roller rather play a game of 10 or 20 rolls? Explain.

5. An engineer wishes to test which of two drills can more quickly bore holes in various materials. He assembles a random sampling of 10 materials of various hardnesses and thicknesses.

   (a) Given that a drill's efficiency is influenced by how long and how hard it has been operating, the engineer decided to randomly choose the order in which the materials will be tested. Design and implement a scheme to place the materials in random order using the following random number table:

   $$84177 \quad 06757 \quad 17613 \quad 15582 \quad 51506$$
   $$81435 \quad 41050 \quad 92031 \quad 06449 \quad 05059$$

   (b) Suppose the drilling times (in seconds) are summarized as follows:

Material	A	B	C	D	E	F	G	H	I	J
Drill 1	4.2	5.1	8.8	1.5	0.8	7.4	3.4	4.7	6.2	4.8
Drill 2	4.0	5.3	8.7	1.8	1.0	8.0	3.7	4.6	6.5	4.7

   What is the mean drilling time for each drill? Is the difference significant? Justify your answer.

SECTION II

PART B

Question 6

Spend about 25 minutes on this part of the exam.
Percentage of Section II grade—25

6. To compare annual family medical expenditures, simple random samples of people in each of three cities were surveyed. The results are summarized in the following table:

	Capitaltown	Healthville	Statsburg
Sample size	100	100	100
Mean expenditure	$1800	$1810	$1820
Median expenditure	$1800	$1800	$1800
Minimum expenditure	$1600	$1600	$1300
Maximum expenditure	$2000	$2400	$2400
1st quartile expenditure	$1700	$1700	$1700
3rd quartile expenditure	$1900	$1900	$1955
Standard deviation	$120	$150	$200

(a) Draw parallel boxplots of the distributions of family medical expenditures in these three cities, and write a few sentences comparing the variability of the distributions in context.

(b) Is there evidence that any or all of these samples are drawn from populations whose distributions are normal? Explain your answer.

(c) Construct a 95% confidence interval estimate of the mean family medical expenditures in Statsburg.

(d) Construct a 95% confidence interval estimate of the mean difference in family medical expenditures between Capitaltown and Healthville.

(e) Assume that all three samples are drawn from populations whose distributions are approximately normal. If a family budgets $1850 for medical expenditures, in which city would the greatest percentage of annual family medical expenditures be within their budget? Explain.

STOP

If there is still time remaining, you may review your answers.

Answer Key

Section I

1. **A**	9. **E**	17. **A**	25. **B**	33. **B**
2. **E**	10. **A**	18. **A**	26. **D**	34. **E**
3. **C**	11. **E**	19. **A**	27. **E**	35. **A**
4. **B**	12. **C**	20. **D**	28. **E**	36. **A**
5. **A**	13. **B**	21. **B**	29. **D**	37. **E**
6. **A**	14. **C**	22. **E**	30. **D**	38. **E**
7. **C**	15. **E**	23. **B**	31. **B**	39. **E**
8. **E**	16. **E**	24. **C**	32. **B**	40. **E**

Answers Explained

Section I

1. **(A)** The slope, 6.2, gives the predicted increase in the *y*-variable for each unit increase in the *x*-variable.

2. **(E)** For a simple random sample, every possible group of the given size has to be equally likely to be selected, and this is not true here. For example, with this procedure it will be impossible for all the early arrivals to be together in the final sample. This procedure is an example of systematic sampling, but systematic sampling does not result in simple random samples.

3. **(C)** Power $= 1 - \beta$, and β is smallest when α is more and n is more.

4. **(B)** The critical z-scores for 60% to the right and 70% to the left are -0.253 and 0.524, respectively. Then $\{\mu - 0.253\sigma = 3, \mu + 0.524\sigma = 6\}$ gives $\mu = 3.977$ and $\sigma = 3.861$.

5. **(A)** The cumulative proportions of 0.25 and 0.75 correspond to $Q_1 = 57$ and $Q_3 = 75$, respectively, and so the interquartile range is $75 - 57 = 18$.

6. **(A)** We have H_0: $\mu = 3.5$ and H_a: $\mu < 3.5$. Then $\sigma_{\bar{x}} = \frac{0.9}{\sqrt{36}} = 0.15$, the z-score of 3.25 is $\frac{3.25-3.5}{0.15} = -1.67$, and Table A gives .0475. (A t-test on the TI-83 gives .0522.)

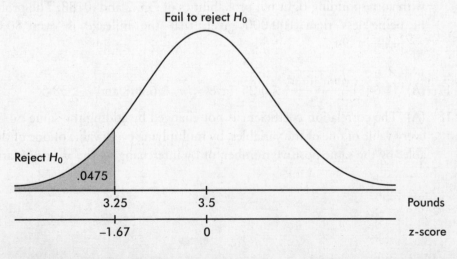

7. **(C)** A placebo is a control treatment in which members of the control group do not realize whether or not they are receiving the experimental treatment.

8. **(E)** In experiments on people, subjects can be used as their own controls, with responses noted before and after the treatment. However, with such designs there is always the danger of a placebo effect. In this case, subjects might well have slower reaction times after drinking the alcohol because they think they should. Thus the design of choice would involve a separate control group to use for comparison.

9. **(E)** Since $(-2, 4)$ is on the line $y = 7x + b$, we have $4 = -14 + b$ and $b = 18$. Thus the regression line is $y = 7x + 18$. The point $(\bar{x}, \bar{y})$ is always on the regression line, and so we have $\bar{y} = 7\bar{x} + 18$.

10. **(A)** A simple random sample can be any size.

11. **(E)** With such small populations, censuses instead of samples are used, and there is no resulting probability statement about the difference.

12. **(C)** Half the area is on either side of 23, so 23 is the median. The distribution is skewed to the right, and so the mean is greater than the median.

13. **(B)** While the procedure does use some element of chance, all possible groups of size 75 do not have the same chance of being picked, so the result is not a simple random sample. There is a real chance of selection bias. For example, a number of relatives with the same name and all using the same long-distance carrier might be selected.

14. **(C)** While t-distributions do have mean 0, their standard deviations are greater than 1.

15. **(E)**
$$P(\text{smoker}|\text{cancer}) = \frac{P(\text{smoker} \cap \text{cancer})}{P(\text{cancer})} = \frac{.014}{.035} = .4$$

16. **(E)** The critical z-scores are $\frac{80,000 - 75,000}{12,000} = 0.42$ and $\frac{100,000 - 75,000}{12,000} = 2.08$, with corresponding right tail probabilities of .3372 and .0188. The probability of being less than 100,000 given that the mileage is over 80,000 is $\frac{.3372 - .0188}{.3372} = .94$.

17. **(A)** $(1.645)\frac{\sqrt{(0.8)^2 + (0.5)^2}}{\sqrt{n}} \le 0.25$ gives $\sqrt{n} \ge 6.208$ and $n \ge 38.5$.

18. **(A)** The correlation coefficient is not changed by adding the same number to every value of one of the variables, by multiplying every value of one of the variables by the same positive number, or by interchanging the x- and y-variables.

19. **(A)** The binomial distribution with $n = 2$ and $p = .8$ is $P(0) = (.2)^2 = .04$, $P(1)$ $= 2(.2)(.8) = .32$, and $P(2) = (.8)^2 = .64$, resulting in expected numbers of $.04(200) = 8$, $.32(200) = 64$, and $.64(200) = 128$. Thus,

$$\chi^2 = \sum \frac{(obs - exp)^2}{exp} = \frac{(10 - 8)^2}{8} + \frac{(80 - 64)^2}{64} + \frac{(110 - 128)^2}{128}$$

20. **(D)** Option II gives the highest expected return: $(50{,}000)(.5) + (10{,}000)(.5) = 30{,}000$, which is greater than $25{,}000$ and is also greater than $(100{,}000)(.05) = 5000$. Option I guarantees that the \$20,000 loan will be paid off. Option III provides the only chance of paying off the \$80,000 loan. The moral is that the highest expected value is not automatically the "best" answer.

21. **(B)** $P(X \cap Y) = P(X|Y)\,P(Y) = (.28)(.40) = .112$. Then $P(Y|X) =$

$\frac{P(X \cap Y)}{P(X)} = \frac{.112}{.35} = .32$.

22. **(E)** This study is an experiment because a treatment (extensive exercise) is imposed. There is no blinding because subjects clearly know whether or not they are exercising. There is no blocking because subjects are not divided into blocks before random assignment to treatments. For example, blocking would have been used if subjects had been separated by gender or age before random assignment to exercise or not.

23. **(B)** The range is the distance between the tips of the two whiskers, so all three ranges are equal. The interquartile range is the length of the box, so they are not all equal. The median of the first distribution is equal to Q_3 of the second, while the median of the third distribution is equal to Q_1 of the second, so the difference of the medians equals $Q_3 - Q_1$, which is the interquartile range of the second.

24. **(C)** From the shape of the normal curve, the answer is in the middle. The middle two-thirds is between z-scores of ± 0.97, and $35 \pm 0.97(10)$ gives $(25.3, 44.7)$.

25. **(B)** A scatterplot would be horizontal; the correlation is zero.

26. **(D)** $\frac{.54}{.54 + .19 + .07} = .675$

27. **(E)** The midpoint of the confidence interval is .08.

28. **(E)** Relative frequencies must be equal. Either looking at rows gives $\frac{n}{n+52} = \frac{35}{45}$ or looking at columns gives $\frac{n}{n+35} = \frac{52}{62}$. We could also set up a proportion $\frac{n}{52} = \frac{35}{10}$ or $\frac{n}{35} = \frac{52}{10}$. Solving any of these equations gives $n = 182$.

29. **(D)** While the sample proportion is between 64% and 70% (more specifically, it is 67%), this is not the meaning of ±3%. While the percentage of the entire population is likely to be between 64% and 70%, this is not known for certain.

30. **(D)** Increasing the sample size by a multiple of d^2 divides the standard deviation of the set of sample means by d.

31. **(B)** In this binomial situation, the probability that a car does not receive a ticket is $1 - 3 = .7$, the probability that none of the five cars receives a ticket is $(.7)^5$, and thus the probability that at least one receives a ticket is $1 - (.7)^5$.

32. **(B)** If the standard deviation of a set is zero, all the values in the set are equal. The mean and median would both equal this common value and so would equal each other. If all the values are equal, there are no outliers. Just because the sample happens to have one common value, there is no reason for this to be true for the whole population.

33. **(B)** A Type I error means that the null hypothesis is correct (the weather will remain dry), but you reject it (thus you needlessly carry around an umbrella). A Type II error means that the null hypothesis is wrong (it will rain), but you fail to reject it (thus you get drenched).

34. **(E)** $P(\text{supports}) = P(\text{Dem} \cap \text{supports}) + P(\text{Rep} \cap \text{supports})$

$$= (.65)(.60) + (.35)(.50)$$
$$= .390 + .175$$
$$= .565$$
$$P(\text{Dem|supports}) = \frac{.390}{.565}$$
$$= .690$$

35. **(A)** The first study is an experiment with two treatment groups and no control group. The second study is observational; the researcher did not randomly divide the subjects into groups and have each group watch a designated number of hours of television per night.

36. **(A)** $b_1 = r\frac{s_y}{s_x} = .42\frac{1.7}{11,500} = 0.000015$

37. **(E)** If the sample statistic is far enough away from the claimed population parameter, we say that there is sufficient evidence to reject the null hypothesis. In this case the null hypothesis is that $\mu = 9500$. The P-value is the probability of obtaining a sample statistic as extreme as the one obtained if the null hypothesis is assumed to be true. The smaller the P-value, the more significant the difference between the null hypothesis and the sample results. With $P = .0069$, there is strong evidence to reject H_0.

38. **(E)** Good experimental design aims to give each group the same experiences except for the treatment under consideration, Thus, all three SRSs should be picked from the same grade level.

39. **(E)** The stemplot does not indicate what happened for any individual school.

40. **(E)** This refers only to very particular random variables, for example, random variables whose values are the numbers of successes in a binomial probability distribution with large n.

SECTION II

1. (a) No treatment is being imposed on anyone.
   (b) There are many possible answers. For example, it is possible that people with a certain personality trait both enjoy sleep and enjoy detailed assembly work. It is possible that impatient people feel that they are wasting time if they sleep too much and also don't have the poise or tenacity to concentrate on detailed assembly work. If a student names a reasonable confounding variable, but does not give an explanation in context, the answer is partially correct.
   (c) No, because cause-and-effect conclusions cannot be drawn from observational studies.
   (d) Randomly select a group of employees who are told they must sleep over 6 hours a night and randomly select another group who are told they must set an early alarm to make sure they sleep under 6 hours a night. After some specified number of days, observe their work and calculate and then compare the mean number of pieces assembled per day for each group. If a student's scheme splits employees into the two called-for groups but does not involve randomness in the selection process the answer is partially correct.

Scoring

Partially correct answers for both parts (b) and (d) count as one part correct.

4 Complete Answer   All four parts correct.

3 Substantial Answer   Three parts correct.

2 Developing Answer   Two parts correct.

1 Minimal Answer   One part correct.

If exactly one of parts (b) and (d) is partially correct, the set of all answers must be looked at holistically to determine which score is to be given.

---

2. (a) From the computer output, S.O.L. = −0.1432 (%HIV) + 9.0303.
   (b) The slope of the regression line is −0.1432, meaning that, on the average, the standard of living drops 0.1432 for every one point rise in HIV/AIDS percentage.

(c) The *y*-intercept is 9.0303, meaning that a city with 0% HIV/AIDS would be expected to have a standard of living of 9.0303.

(d) R-Sq = 82.0% means that 82.0% of the variation in standard of living is related to variation in HIV/AIDS percentage.

(e) In absolute value the correlation is $\sqrt{.82} = .906$, and since the slope is negative, so is the correlation. Thus $r = -.906$.

(f) A test of significance of correlation corresponds to a test of significance of the slope of the regression line: H_0: $\beta = 0$, H_a: $\beta \neq 0$.

Assumptions: The scatterplot approximately linear, there is no apparent pattern in the residuals plot, and the histogram of residuals is given to be approximately normal.

$t = -10.25$ and $P = .000$ immediately from the computer output.

With this small a *P*-value there is very strong evidence to reject H_0 and conclude that the slope $\neq 0$, and the relationship between standard of living and HIV/AIDS percentage is significant.

Scoring

Parts (b) and (c) should be based on the answer to part (a), whether that answer is right or wrong.

Part (e) is partially correct if missing the negative sign.

Part (f) has five steps including identification of test, statement of hypotheses, check of assumptions, noting of *t* and *P* from the computer output, and a correct conclusion in context. Part (f) is essentially correct if four out of the five steps are answered correctly. Part (f) is partially correct if two or three of the five steps are answered correctly.

4 Complete Answer	All six parts essentially correct.
3 Substantial Answer	Five parts are essentially correct OR four parts including part (f) are essentially correct OR part (f) is partially correct and at least four of the remaining parts are essentially correct.
2 Developing Answer	Four parts are essentially correct OR three parts including part (f) are essentially correct OR part (f) is partially correct and at least three of the remaining parts are essentially correct.
1 Minimal Answer	Three parts are essentially correct OR two parts including part (f) are essentially correct OR part (f) is partially correct and at least two of the remaining parts are correct.

3. (a) Use the chi-square test for independence. The expected cell counts are as follows:

76.2	792.8
120.8	1257.2
40.0	417.0

The condition that all cell counts are greater than 5 is met.

H_0: Body mass index and heart failure are independent.

H_a: There is a relationship between BMI and heart failure.

Using the TI-83, a chi-square test gives $\chi^2 = 13.19$ and $P = .0014$.

Alternatively,

$$\chi^2 = \sum \frac{(\text{obs} - \text{exp})^2}{\text{exp}}$$
$$= \frac{(57 - 76.2)^2}{76.2} + \frac{(812 - 792.8)^2}{792.8} + \frac{(123 - 120.8)^2}{120.8}$$
$$+ \frac{(1255 - 1257.2)^2}{1257.2} + \frac{(57 - 40.0)^2}{40.0} + \frac{(400 - 417.0)^2}{417.0} = 13.19$$

With df $= (r - 1)(c - 1) = 2$, we get $P = .0014$.

Since $.0014 < .01$, the data provide strong evidence (at the 1% significance level) to reject H_0 and conclude that there is evidence of a relationship between body mass index and heart failure.

(b) A two-proportion z-test is indicated. We must check that n is large enough: $n_1 p_1 = 86 > 10$, $n_1(1 - p_1) = 869 > 10$, $n_2 p_2 = 107 > 10$, $n_2(1 - p_2) = 1622 > 10$. We must assume simple random samples from the target population, since this is not given.

$H_0: p_1 - p_2 = 0$ (where p_1 is the proportion of women with heart failure among overweight, but not obese women, and p_2 is the proportion of women with heart failure among normal BMI women.)

$H_a: p_1 - p_2 = 0$ (the proportion of women with heart failure among overweight, but not obese, women is higher than the proportion of women with heart failure among normal BMI women.)

The TI-83 gives $z = 2.70$ and $P = .0034$.

Alternatively,

$$\hat{p}_1 = \frac{86}{955} = .0901, \quad \hat{p}_2 = \frac{107}{1729} = .0619, \quad \text{and}$$

$$\sigma_d = \sqrt{\hat{p}(1-\hat{p})\left(\frac{1}{955} + \frac{1}{1729}\right)} = .0104, \quad \text{where}$$

$$\hat{p} = \frac{86 + 107}{955 + 1729} = .0719$$

So $z = \frac{.0901 - .0619}{.0104} = 2.71$, and again we get $P = .0034$.

With such a small P-value there is very strong evidence to reject H_0 and conclude that the proportion of women with heart failure among overweight, but not obese, women is higher than the proportion of women with heart failure among normal BMI women.

(c) A two-proportion z-test is indicated. We must check that n is large enough: $n_1 p_1 = 57 > 10$, $n_1(1 - p_1) = 400 > 10$, $n_2 p_2 = 65 > 10$, $n_2(1 - p_2) = 428 > 10$. We must assume simple random samples from the target population, since this is not given.

$H_0: p_1 - p_2 = 0$ (where p_1 is the proportion of obese men with heart failure and p_2 is the proportion of obese women with heart failure.)

$H_a: p_1 - p_2 \neq 0$ (the proportion of obese men with heart failure is different from the proportion of obese women with heart failure.)

The TI-84 gives $z = -0.328$ and $P = .743$.

Alternatively,

$$\hat{p}_1 = \frac{57}{457} = .125, \quad \hat{p}_2 = \frac{65}{493} = .132, \quad \text{and}$$

$$\sigma_d = \sqrt{\hat{p}(1-\hat{p})\left(\frac{1}{457} + \frac{1}{493}\right)} = .0217, \quad \text{where}$$

$$\hat{p} = \frac{57 + 65}{457 + 493} = .1284$$

So $z = \frac{.125 - .132}{.0217} = -.323$, and again we get $P = .747$.

With such a large P-value there is no evidence to reject H_0 and so we conclude that there is no evidence that the proportion of obese men with heart failure is different from the proportion of obese women with heart failure.

Scoring

Each part involves five steps: naming a test, checking assumptions, stating hypotheses, calculating t (or z) and P, and giving a conclusion in context. Each part is essentially correct if at least four of the five steps are correct and complete, and is partially correct if two or three of the five steps are correct and complete.

Give full credit in parts (b) and (c) for using $\sigma_d = \sqrt{\dfrac{\hat{p}_1(1-\hat{p}_1)}{n_1} + \dfrac{\hat{p}_2(1-\hat{p}_2)}{n_2}}$.

4 Complete Answer   All three parts essentially correct.

3 Substantial Answer   Two parts essentially correct and the other part partially correct.

2 Developing Answer   Two parts essentially correct OR one part essentially correct and the other two parts partially correct.

1 Minimal Answer   One part essentially correct and one part partially correct OR all three parts partially correct.

4. (a) The possible outcomes (showing on the bottoms) with their probabilities are as follows:

Tetrahedron	Die	Winner	Probability
5	3	Tetrahedron	$\left(\frac{3}{4}\right)\left(\frac{1}{3}\right) = \frac{1}{4}$
5	6	Die	$\left(\frac{3}{4}\right)\left(\frac{2}{3}\right) = \frac{1}{2}$
8	3	Tetrahedron	$\left(\frac{1}{4}\right)\left(\frac{1}{3}\right) = \frac{1}{12}$
8	6	Tetrahedron	$\left(\frac{1}{4}\right)\left(\frac{2}{3}\right) = \frac{1}{6}$

The probability that the tetrahedron roller wins is $\frac{1}{4} + \frac{1}{12} + \frac{1}{6} = \frac{1}{2}$, and the probability that the die roller wins is also $\frac{1}{2}$. So both players have an equal chance of winning.

(b) With the 5 and 6 showing the probabilities of winning are shown below:

$P(\text{tetrahedron wins}) = \left(\frac{2}{3}\right)\left(\frac{2}{5}\right) + \left(\frac{1}{3}\right)\left(\frac{2}{5}\right) + \left(\frac{1}{3}\right)\left(\frac{3}{5}\right) = \frac{3}{5}$

$P(\text{die wins}) = \left(\frac{2}{3}\right)\left(\frac{3}{5}\right) = \frac{2}{5}$

So with the 5 and 6 showing, the tetrahedron roller has the greater chance of winning.

(c) Find the expected value of one roll of this game for each player, using the probabilities from the table above:

$E(\text{tetrahedron roller}) = \Sigma x\, P(x) = 3(\frac{1}{4}) + 3(\frac{1}{12}) + 6(\frac{1}{6}) = 2$

$E(\text{die roller}) = \Sigma x\, P(x) = 5(\frac{1}{2}) = 2.5$

(d) Since he has a higher expected value per roll, the die roller would want a game with more rolls, so, in this case, 20.

Scoring

Do not take off for minor arithmetical errors.

(a) Must give correct answer including calculations based on probability to receive credit.

(b) Must give correct answer including calculations based on probability to receive credit.

(c) Give full credit for correct procedure but using incorrect probabilities from part (a).

(d) Give full credit for correct reasoning but using incorrect calculations from part (c).

4 Complete Answer   All four parts correct.

3 Substantial Answer   Three parts correct.

2 Developing Answer   Two parts correct.

1 Minimal Answer   One part correct.

---

5. (a) Different schemes are possible. For example, assign each material a single-digit number, such as A-0, B-1, C-2, D-3, E-4, F-5, G-6, H-7, I-8, J-9. Then read off the digits from the random number list, one at a time, throwing away any repeats, until each of the materials have been picked (or nine have been picked, as the one left over will go last). The order of picking then gives the order of being tested.

Using this scheme we would get the following order:

```
IEBH  AG F D    CJ

84177 06757 17613 15582 51506 81435 41050 92031 06449
```

(b) The mean drilling times in the 10 materials are 4.69 seconds for Drill 1 and 4.83 seconds for Drill 2.

The proper hypothesis test is a matched pairs t-test on the set of differences, {0.2, −0.2, 0.1, −0.3, −0.2, −0.6, −0.3, 0.1, −0.3, 0.1}, with $H_0: \mu_D = 0$, $H_a: \mu_D \neq 0$.

Assumptions to be checked: (1) random samples (given) and (2) the population of differences is approximately normal (use dotplot, stemplot, boxplot, or histogram of differences).

Dotplot:

Boxplot:

Stemplot:

0.	1112
-0.	22333
-0.	6

Histogram:

The plots and histogram indicate a very roughly symmetric, bell-shaped distribution with no outliers and no extreme skewness, so the set of differences appears approximately normal.

t-test: $\bar{x}_D = -0.14$ $s_D = 0.2547$

$$t = \frac{\bar{x}_D - 0}{s_D/\sqrt{n_D}} = \frac{-0.14 - 0}{0.2547/\sqrt{10}} = \frac{-0.14}{0.08054} = -1.738$$

$\text{df} = n_D - 1 = 10 - 1 = 9$

The *P*-value = .116, and since $P > .10$, there is not sufficient evidence to reject H_0, the differences are not significant, and the data do *not* suggest that there is a significant difference in the mean drilling times of the two drills through different materials.

Scoring

For part (a) a complete solution involves an assignment scheme, noting that duplicate digits are ignored, telling when to stop, and noting the resulting ordering. For all four items correct, part (a) should be judged essentially correct, and for three items correct, part (a) should be judged partially correct.

For part (b) a complete solution involves calculating the two mean drilling times, naming the correct test, stating correct hypotheses, stating and checking assumptions, calculating t and P, and giving a correct conclusion in context. For five out of these six items correct, part (b) should be judged essentially correct. For three or four correct items, part (b) should be judged partially correct.

4 Complete Answer   Both parts essentially correct.

3 Substantial Answer   One part essentially correct and the other part partially correct.

2 Developing Answer   One part essentially correct OR both parts partially correct.

1 Minimal Answer   One part partially correct.

6. (a)

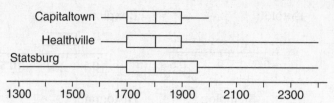

In all three cities the median family medical expenditures are identical, and there does not appear to be much difference between the cities with regard to the middle 50% of family expenditures. However, while Capitaltown has small ranges in expenditures among both its lower and upper quarters in the distribution, Statsburg has large ranges in expenditures among both its lower and upper quarters, and Healthville has a small range in its lower quarter and a large range in its upper quarter.

(b) For Capitaltown, the minimum, 1600, and maximum, 2000, are both only $\frac{200}{120} = 1.67$ standard deviations from the mean, a very unlikely occurrence for a normal distribution.

For Healthville, the distribution does not appear symmetric, instead looking skewed to the right.

For Statsburg, the minimum, 1st quartile, 3rd quartile, and maximum have z-scores of $-\frac{520}{200} = -2.6$, $-\frac{120}{200} = -0.6$, $\frac{135}{200} = 0.675$, and $\frac{580}{200} = 2.9$, respectively, while under a normal curve, almost all values are between -3 and $+3$, and the quartiles have z-scores of ± 0.67.

Thus, there is evidence that the Statsburg distribution is roughly normal, but not for the other two cities.

(c) To do a confidence interval of a population mean, we check the assumptions: a simple random sample (given) and either n is large (it's 100 here) or the population is roughly normal (true for Statsburg as discussed above).

$$\bar{x} \pm t\sigma_{\bar{x}} = \bar{x} \pm t\frac{s}{\sqrt{n}} = 1820 \pm 1.984\frac{200}{\sqrt{100}} = 1820 \pm 39.68$$

So we are 95% confident that the mean family annual medical expenditure in Statsburg is between $1780.32 and $1859.68.

(d) To do a confidence interval for the difference of two population means, we check the assumptions: two independent simple random samples (given) and both sample sizes are large (both are 100).

$$\left(\bar{x}_1 - \bar{x}_2\right) \pm t\sqrt{\frac{s_1^2}{n_1} + \frac{s_2^2}{n_2}} = \left(1800 - 1810\right) \pm 1.97\sqrt{\frac{120^2}{100} + \frac{150^2}{100}}$$
$$= -10 \pm 37.84$$

So we are 95% confident that the difference in medical expenditures between Capitaltown and Healthville is between $-$47.84 and $27.84.

(e) For Capitaltown, $z = \frac{1850-1800}{120} = 0.4167$, which gives .662 or 66.2%.

For Healthville, $z = \frac{1850-1810}{150} = 0.2667$, which gives .605 or 60.5%.

For Statsburg, $z = \frac{1850-1820}{200} = 0.15$, which gives .560 or 56.0%.

So for a family with a budget of $1850 for medical expenditures, the greatest percentage of annual family expenditures within their budget would be in Capitaltown.

Scoring

Part (a) is partially correct if the boxplots are accurate but the written description is incomplete or unclear.

Part (b) is partially correct if an argument is made that both Capitaltown and Statsburg have normal distributions.

Parts (c), (d), and (e) are essentially correct for correct final answers in context together with work showing how these answers were obtained, and partially correct for unclear or incomplete work.

4 Complete Answer	All five parts essentially correct.
3 Substantial Answer	Four parts essentially correct OR three parts essentially correct and the other two partially correct.
2 Developing Answer	Three parts essentially correct OR two parts essentially correct and at least two of the remaining parts partially correct.
1 Minimal Answer	Two parts essentially correct OR one part essentially correct and at least two of the remaining parts partially correct OR four parts partially correct.

Answer Sheet

PRACTICE EXAMINATION 5

1. Ⓐ Ⓑ Ⓒ Ⓓ Ⓔ
2. Ⓐ Ⓑ Ⓒ Ⓓ Ⓔ
3. Ⓐ Ⓑ Ⓒ Ⓓ Ⓔ
4. Ⓐ Ⓑ Ⓒ Ⓓ Ⓔ
5. Ⓐ Ⓑ Ⓒ Ⓓ Ⓔ
6. Ⓐ Ⓑ Ⓒ Ⓓ Ⓔ
7. Ⓐ Ⓑ Ⓒ Ⓓ Ⓔ
8. Ⓐ Ⓑ Ⓒ Ⓓ Ⓔ
9. Ⓐ Ⓑ Ⓒ Ⓓ Ⓔ
10. Ⓐ Ⓑ Ⓒ Ⓓ Ⓔ

11. Ⓐ Ⓑ Ⓒ Ⓓ Ⓔ
12. Ⓐ Ⓑ Ⓒ Ⓓ Ⓔ
13. Ⓐ Ⓑ Ⓒ Ⓓ Ⓔ
14. Ⓐ Ⓑ Ⓒ Ⓓ Ⓔ
15. Ⓐ Ⓑ Ⓒ Ⓓ Ⓔ
16. Ⓐ Ⓑ Ⓒ Ⓓ Ⓔ
17. Ⓐ Ⓑ Ⓒ Ⓓ Ⓔ
18. Ⓐ Ⓑ Ⓒ Ⓓ Ⓔ
19. Ⓐ Ⓑ Ⓒ Ⓓ Ⓔ
20. Ⓐ Ⓑ Ⓒ Ⓓ Ⓔ

21. Ⓐ Ⓑ Ⓒ Ⓓ Ⓔ
22. Ⓐ Ⓑ Ⓒ Ⓓ Ⓔ
23. Ⓐ Ⓑ Ⓒ Ⓓ Ⓔ
24. Ⓐ Ⓑ Ⓒ Ⓓ Ⓔ
25. Ⓐ Ⓑ Ⓒ Ⓓ Ⓔ
26. Ⓐ Ⓑ Ⓒ Ⓓ Ⓔ
27. Ⓐ Ⓑ Ⓒ Ⓓ Ⓔ
28. Ⓐ Ⓑ Ⓒ Ⓓ Ⓔ
29. Ⓐ Ⓑ Ⓒ Ⓓ Ⓔ
30. Ⓐ Ⓑ Ⓒ Ⓓ Ⓔ

31. Ⓐ Ⓑ Ⓒ Ⓓ Ⓔ
32. Ⓐ Ⓑ Ⓒ Ⓓ Ⓔ
33. Ⓐ Ⓑ Ⓒ Ⓓ Ⓔ
34. Ⓐ Ⓑ Ⓒ Ⓓ Ⓔ
35. Ⓐ Ⓑ Ⓒ Ⓓ Ⓔ
36. Ⓐ Ⓑ Ⓒ Ⓓ Ⓔ
37. Ⓐ Ⓑ Ⓒ Ⓓ Ⓔ
38. Ⓐ Ⓑ Ⓒ Ⓓ Ⓔ
39. Ⓐ Ⓑ Ⓒ Ⓓ Ⓔ
40. Ⓐ Ⓑ Ⓒ Ⓓ Ⓔ

Practice Examination 5

Section I

Questions 1–40

Spend 90 minutes on this part of the exam.

> **Directions:** The questions or incomplete statements that follow are each followed by five suggested answers or completions. Choose the response that best answers the question or completes the statement.

1. The mean and standard deviation of the population {1, 5, 8, 11, 15} are $\mu = 8$ and $\sigma = 4.8$, respectively. Let S be the set of the 125 *ordered* triples (repeats allowed) of elements of the original population. Which of the following is a correct statement about the mean $\mu_{\bar{x}}$ and standard deviation $\sigma_{\bar{x}}$ of the means of the triples in S?

   (A) $\mu_{\bar{x}} = 8$, $\sigma_{\bar{x}} = 4.8$
   (B) $\mu_{\bar{x}} = 8$, $\sigma_{\bar{x}} < 4.8$
   (C) $\mu_{\bar{x}} = 8$, $\sigma_{\bar{x}} > 4.8$
   (D) $\mu_{\bar{x}} < 8$, $\sigma_{\bar{x}} = 4.8$
   (E) $\mu_{\bar{x}} > 8$, $\sigma_{\bar{x}} > 4.8$

2. A survey to measure job satisfaction of high school mathematics teachers was taken in 1993 and repeated 5 years later in 1998. Each year a random sample of 50 teachers rated their job satisfaction on a 1-to-100 scale with higher numbers indicating greater satisfaction. The results are given in the following back-to-back stemplot.

1993		1998
	0	
98775	1	
98530	2	1
65210	3	3589
96430	4	01122233455667889999
87421	5	035667899
99877555322100	6	1344789
976442	7	22689
7511	8	138
0	9	7

What is the trend from 1993 to 1998 with regard to the standard deviation and range of the two samples?

(A) Both the standard deviation and range increased.
(B) The standard deviation increased, while the range decreased.
(C) The range increased, while the standard deviation decreased.
(D) Both the standard deviation and range decreased.
(E) Both the standard deviation and range remained unchanged.

GO ON TO THE NEXT PAGE ➤

523

3. Consider the following studies being run by three different AP Statistics instructors.

   I. One rewards students every day with lollipops for relaxation, encouragement, and motivation to learn the material.

   II. One promises that all students will receive A's as long as they give their best efforts to learn the material.

   III. One is available every day after school and on weekends so that students with questions can come in and learn the material.

   (A) None of these studies use randomization.
   (B) None of these studies use control groups.
   (C) None of these studies use blinding.
   (D) Important information can be found from all these studies, but none can establish causal relationships.
   (E) All of the above.

4. The number of days it takes to build a new house has a variance of 386. A sample of 40 new homes shows an average building time of 83 days. With what confidence can we assert that the average building time for a new house is between 80 and 90 days?

   (A) 15.4%
   (B) 17.8%
   (C) 20.0%
   (D) 38.8%
   (E) 82.2%

5. A shipment of resistors have an average resistance of 200 ohms with a standard deviation of 5 ohms, and the resistances are normally distributed. Suppose a randomly chosen resistor has a resistance under 194 ohms. What is the probability that its resistance is greater than 188 ohms?

   (A) .07
   (B) .12
   (C) .50
   (D) .93
   (E) .97

6. Suppose 4% of the population have a certain disease. A laboratory blood test gives a positive reading for 95% of people who have the disease and for 5% of people who do not have the disease. What is the probability of testing positive? If a person tests positive, what is the probability the person has the disease?

   (A) .086, .442
   (B) .086, .500
   (C) .086, .914
   (D) .500, .950
   (E) .914, .950

7. For which of the following is it appropriate to use a census?

   (A) A 95% confidence interval of mean height of teachers in a small town.
   (B) A 95% confidence interval of the proportion of students in a small town who are taking some AP class.
   (C) A two-tailed hypothesis test where the null hypothesis was that the mean expenditure on entertainment by male students at a high school is the same as that of female students.
   (D) Calculation of the standard deviation in the number of broken eggs per carton in a truckload of eggs.
   (E) All of the above.

8. On the same test, Mary and Pam scored at the 64th and 56th percentiles, respectively. Which of the following is a true statement?

   (A) Mary scored eight more points than Pam.
   (B) Mary's score is 8% higher than Pam's.
   (C) Eight percent of those who took the test scored between Pam and Mary.
   (D) Thirty-six people scored higher than both Mary and Pam.
   (E) None of the above.

GO ON TO THE NEXT PAGE ➤

9. Which of the following are true statements?

   I. While observational studies gather information on an already existing condition, they still often involve intentionally forcing some treatment to note the response.
   II. In an experiment, researchers decide on the treatment but typically allow the subjects to self-select into the control group.
   III. If properly designed, either observational studies or controlled experiments can easily be used to establish cause and effect.

   (A) I only
   (B) II only
   (C) III only
   (D) All are true.
   (E) None is true.

10. In a random sample of 500 spectators who attend either high school football or basketball games but not both, food was purchased in the following amounts.

	Hot dogs	Popcorn	No purchase
Football	240	80	30
Basketball	50	90	10

   What percentage of spectators at football games bought popcorn? What percentage of hot dog purchasers were at basketball games?

   (A) 16%, 10%
   (B) 16%, 17%
   (C) 23%, 10%
   (D) 23%, 17%
   (E) 47%, 33%

11. To determine the mean cost of groceries in a certain city, an identical grocery basket of food is purchased at each store in a random sample of ten stores. If the average cost is $47.52 with a standard deviation of $1.59, find a 98% confidence interval estimate for the cost of these groceries in the city.

   (A) $47.52 \pm 2.33\sqrt{1.59}$

   (B) $47.52 \pm 2.33\left(\dfrac{1.59}{\sqrt{10}}\right)$

   (C) $47.52 \pm 2.33\left(\sqrt{\dfrac{1.59}{10}}\right)$

   (D) $47.52 \pm 2.821\left(\dfrac{1.59}{\sqrt{10}}\right)$

   (E) $47.52 \pm 2.821\sqrt{\dfrac{1.59}{10}}$

12. A set consists of four numbers. The largest value is 200, and the range is 50. Which of the following statements is true?

   (A) The mean is less than 185.
   (B) The mean is greater than 165.
   (C) The median is less than 195.
   (D) The median is greater than 155.
   (E) The median is the mean of the second and third numbers if the set is arranged in ascending order.

13. A telephone survey of 400 registered voters showed that 256 had not yet made up their minds 1 month before the election. How sure can we be that between 60% and 68% of the electorate were still undecided at that time?

   (A) 2.4%
   (B) 8.0%
   (C) 64.0%
   (D) 90.5%
   (E) 95.3%

GO ON TO THE NEXT PAGE ➤

14. Suppose we have a random variable X where the probability associated with the value k is

$$\binom{15}{k}(.29)^k(.71)^{15-k} \text{ for } k = 0, \ldots, 15.$$

What is the mean of X?

(A) 0.29
(B) 0.71
(C) 4.35
(D) 10.65
(E) None of the above

15. The financial aid office at a state university conducts a study to determine the total student costs per semester. All students are charged $4500 for tuition. The mean cost for books is $350 with a standard deviation of $65. The mean outlay for room and board is $2800 with a standard deviation of $380. The mean personal expenditure is $675 with a standard deviation of $125. Assuming independence among categories, what is the standard deviation of the total student costs?

(A) $24
(B) $91
(C) $190
(D) $405
(E) $570

16. Suppose X and Y are random variables with $E(X) = 312$, $\text{var}(X) = 6$, $E(X) = 307$, and $\text{var}(Y) = 8$. What are the expected value and variance of the random variable $X + Y$?

(A) $E(X + Y) = 619$, $\text{var}(X + Y) = 7$
(B) $E(X + Y) = 619$, $\text{var}(X + Y) = 10$
(C) $E(X + Y) = 619$, $\text{var}(X + Y) = 14$
(D) $E(X + Y) = 309.5$, $\text{var}(X + Y) = 14$
(E) There is insufficient information to answer this question.

17. Consider the following three events:

I. In a survey about former drug use, an embarrassed politician deliberately gives the wrong answers.
II. A surveyor mistakenly records answers to one question in the wrong place.
III. Although 6% of the adult population is unemployed, in a random sample of ten adults, all are employed.

Which of the following correctly characterizes the above?

(A) I, response bias; II, human mistake; III, sampling error
(B) I, nonresponse bias; II, hidden error; III, sampling error
(C) I, voluntary sample bias; II, sampling error; III, hidden bias
(D) I, voluntary error; II, unintentional error; III, undercoverage error
(E) I, deliberate error; II, mistaken error; III, small sample error

18. The following histogram gives the shoe sizes of people in an elementary school building one morning.

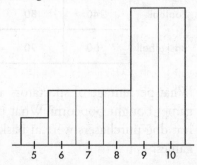

Which of the following statements are true?

I. The same number of people had size 9 shoes as had size 10 shoes.
II. The median shoe size was $7\frac{1}{2}$.
III. The mean shoe size was less than the median shoe size.

(A) II only
(B) I and II
(C) I and III
(D) II and III
(E) I, II, and III

GO ON TO THE NEXT PAGE ➤

19. When comparing the standard normal (z) distribution to the t-distribution with df = 30, which of (A)–(D), if any, are false?

    (A) Both are symmetric.
    (B) Both are bell-shaped.
    (C) Both have center 0.
    (D) Both have standard deviation 1.
    (E) All the above are true statements.

20. Given a probability of .65 that interest rates will jump this year, and a probability of .72 that if interest rates jump the stock market will decline, what is the probability that interest rates will jump and the stock market will decline?

    (A) .070
    (B) .097
    (C) .468
    (D) .532
    (E) .903

21. Sampling error is

    (A) the mean of a sample statistic.
    (B) the standard deviation of a sample statistic.
    (C) the standard error of a sample statistic.
    (D) the result of bias.
    (E) the difference between a population parameter and an estimate of that parameter.

22. Suppose that the weights of trucks traveling on the interstate highway system are normally distributed. If 70% of the trucks weigh more than 12,000 pounds and 80% weigh more than 10,000 pounds, what are the mean and standard deviation for the weights of trucks traveling on the interstate system?

    (A) $\mu = 14,900$; $\sigma = 6100$
    (B) $\mu = 15,100$; $\sigma = 6200$
    (C) $\mu = 15,300$; $\sigma = 6300$
    (D) $\mu = 15,500$; $\sigma = 6400$
    (E) The mean and standard deviation cannot be computed from the information given.

23. If the correlation coefficient $r = .78$, what percentage of variation in y is explained by variation in x?

    (A) 22%
    (B) 39%
    (C) 44%
    (D) 61%
    (E) 78%

24. Consider the following scatterplot:

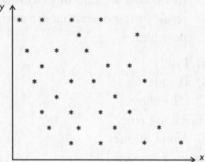

    Which of the following is the best estimate of the correlation between x and y?

    (A) −.95
    (B) −.15
    (C) 0
    (D) .15
    (E) .95

25. For one NBA playoff game the actual percentage of the television viewing public who watched the game was 24%. If you had taken a survey of 50 television viewers that night and constructed a confidence interval estimate of the percentage watching the game, which of the following would have been true?

    I. The center of the interval would have been 24%.
    II. The interval would have contained 24%.
    III. A 99% confidence interval estimate would have contained 24%.

    (A) I and II
    (B) I and III
    (C) II and III
    (D) All are true.
    (E) None is true.

GO ON TO THE NEXT PAGE ➤

26. Which of the following are true statements?

    I. In a well-designed, well-conducted sample survey, sampling error is effectively eliminated.
    II. In a well-designed observational study, responses are influenced through an orderly, carefully planned procedure during the collection of data.
    III. In a well-designed experiment, the treatments are carefully planned to result in responses that are as similar as possible.

    (A) I only
    (B) II only
    (C) III only
    (D) All are true.
    (E) None is true.

27. Consider the following scatterplot showing the relationship between caffeine intake and job performance.

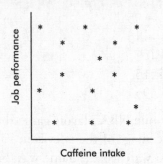

    Which of the following is a reasonable conclusion?

    (A) Low caffeine intake is associated with low job performance.
    (B) Low caffeine intake is associated with high job performance.
    (C) High caffeine intake is associated with low job performance.
    (D) High caffeine intake is associated with high job performance.
    (E) Job performance cannot be predicted from caffeine intake.

28. An author of a new book claims that anyone following his suggested diet program will lose an average of 2.8 pounds per week. A researcher believes that the true figure will be lower and plans a test involving a random sample of 36 overweight people. She will reject the author's claim if the mean weight loss in the volunteer group is less than 2.5 pounds per week. Assume that the standard deviation among individuals is 1.2 pounds per week. If the true mean value is 2.4 pounds per week, what is the probability that the researcher will mistakenly fail to reject the author's false claim of 2.8 pounds?

    (A) $P\left(z > \dfrac{2.5 - 2.4}{1.2/\sqrt{36}}\right)$

    (B) $P\left(z < \dfrac{2.5 - 2.4}{1.2/\sqrt{36}}\right)$

    (C) $P\left(z < \dfrac{2.8 - 2.5}{1.2/\sqrt{36}}\right)$

    (D) $P\left(z > \dfrac{2.8 - 2.4}{1.2/\sqrt{36}}\right)$

    (E) $P\left(z < \dfrac{2.8 - 2.4}{1.2/\sqrt{36}}\right)$

GO ON TO THE NEXT PAGE ➤

29. Which of the following is the central limit theorem?

    (A) No matter how the population is distributed, as the sample size increases, the mean of the sample means becomes closer to the mean of the population.

    (B) No matter how the population is distributed, as the sample size increases, the standard deviation of the sample means becomes closer to the standard deviation of the population divided by the square root of the sample size.

    (C) If the population is normally distributed, then as the sample size increases, the sampling distribution of the sample mean becomes closer to a normal distribution.

    (D) All of the above together make up the central limit theorem.

    (E) The central limit theorem refers to something else.

30. What is a sampling distribution?

    (A) A distribution of all the statistics that can be found in a given sample

    (B) A histogram, or other such visual representation, showing the distribution of a sample

    (C) A normal distribution of some statistic

    (D) A distribution of all the values taken by a statistic from all possible samples of a given size

    (E) All of the above

31. A judge chosen at random reaches a just decision roughly 80% of the time. What is the probability that in randomly chosen cases at least two out of three judges reach a just decision?

    (A) .384
    (B) .488
    (C) .512
    (D) .616
    (E) .896

32. Miles per gallon versus speed (miles per hour) for a new model automobile is fitted with a least squares regression line. Following is computer output of the statistical analysis of the data.

    ```
    Dependent variable: Miles per gallon

    Source          df      Sum of Squares  Mean Square  F-ratio
    Regression      1          199.34         199.34       3.79
    Residual        6          315.54           5.59

    Variable   Coefficient    SE Coef      t-ratio    P
    Constant     38.929        5.651         6.89     0.000
    Speed        -0.2179       0.112        -1.95     0.099

    R-Sq = 38.7%      R-Sq(adj) = 28.5%
    s = 7.252 with 8 - 2 = 6 degrees of freedom
    ```

    Which of the following gives a 99% confidence interval for the slope of the regression line?

    (A) $-0.2179 \pm 3.707(0.112)$

    (B) $-0.2179 \pm 3.143(0.112/\sqrt{8})$

    (C) $-0.2179 \pm 3.707(0.112/\sqrt{8})$

    (D) $38.929 \pm 3.143(3.651/\sqrt{8})$

    (E) $38.929 \pm 3.707(5.651\sqrt{8})$

GO ON TO THE NEXT PAGE ➤

33. What fault do all these sampling designs have in common?

   I. The Parent-Teacher Association (PTA), concerned about rising teenage pregnancy rates at a high school, randomly picks a sample of high school students and interviews them concerning unprotected sex they have engaged in during the past year.

   II. A radio talk show host asks people to phone in their views on whether the United States should keep troops in Bosnia indefinitely to enforce the cease-fire.

   III. The *Ladies Home Journal* plans to predict the winner of a national election based on a survey of its readers.

   (A) All the designs make improper use of stratification.
   (B) All the designs have errors that can lead to strong bias.
   (C) All the designs confuse association with cause and effect.
   (D) All the designs suffer from sampling error.
   (E) None of the designs makes use of chance in selecting a sample.

34. Hospital administrators wish to determine the average length of stay for all surgical patients. A statistician determines that for a 95% confidence level estimate of the average length of stay to within ±0.50 days, 100 surgical patients' records would have to be examined. How many records should be looked at for a 95% confidence level estimate to within ±0.25 days?

   (A) 25
   (B) 50
   (C) 200
   (D) 400
   (E) There is not enough information given to determine the necessary sample size.

35. Following are parts of the probability distributions for the random variables X and Y.

x	$P(x)$	y	$P(y)$
1	?	1	?
2	?	2	?
3	?	3	.1

If X and Y are independent and the joint probabilities $P(X=1, Y=2) = .186$, $P(X=2, Y=2) = .198$, and $P(X=3, Y=2) = .216$, what is $P(X=1, Y=1)$?

   (A) .031
   (B) .036
   (C) .093
   (D) .099
   (E) .108

36. Which of the following statements are true?

   I. If the population mean is known, there is no reason to run a hypothesis test on the population mean.

   II. The P-value is usually chosen before an experiment is conducted.

   III. The P-value is based on a specific test statistic and thus should not be used in a two-sided test.

   (A) I only
   (B) II only
   (C) III only
   (D) I, II, and III
   (E) None are true.

37. An assembly line machine is supposed to turn out ball bearings with a diameter of 1.25 centimeters. Each morning the first 30 bearings produced are pulled and measured. If their mean diameter is under 1.23 centimeters or over 1.27 centimeters, the machinery is stopped and an engineer is called to make adjustments before production is resumed. The quality control procedure may be viewed as a hypothesis test with the null hypothesis $H_0: \mu = 1.25$ and the alternative hypothesis $H_a: \mu \neq 1.25$. The engineer is asked to make adjustments when the null hypothesis is rejected. In test terminology, what would a Type II error result in?

   (A) A warranted halt in production to adjust the machinery
   (B) An unnecessary stoppage of the production process
   (C) Continued production of wrong size ball bearings
   (D) Continued production of proper size ball bearings
   (E) Continued production of ball bearings that randomly are the right or wrong size

38. Both over-the-counter niacin and the prescription drug Lipitor are known to lower blood cholesterol levels. In one double-blind study Lipitor outperformed niacin. The 95% confidence interval estimate of the difference in mean cholesterol level lowering was (18, 41). Which of the following is a reasonable conclusion?

   (A) Niacin lowers cholesterol an average of 18 points, while Lipitor lowers cholesterol an average of 41 points.
   (B) There is a .95 probability that Lipitor will outperform niacin in lowering the cholesterol level of any given individual.
   (C) There is a .95 probability that Lipitor will outperform niacin by at least 23 points in lowering the cholesterol level of any given individual.
   (D) We should be 95% confident that Lipitor will outperform niacin as a cholesterol-lowering drug.
   (E) None of the above.

39. The following parallel boxplots show the average daily hours of bright sunshine in Liberia, West Africa:

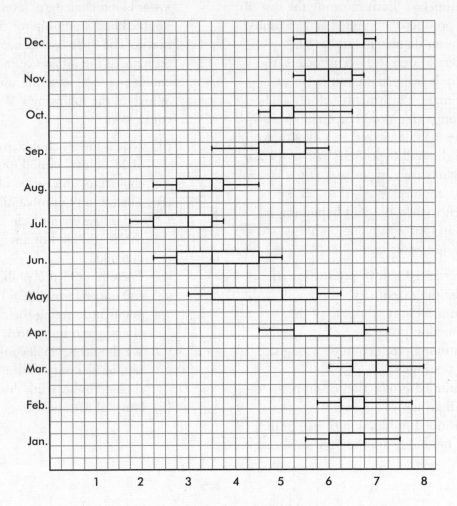

Average daily hours of bright sunshine

For how many months is the median below 4 hours?

(A) One
(B) Two
(C) Three
(D) Four
(E) Five

GO ON TO THE NEXT PAGE ➤

40. Following is a cumulative probability graph for the number of births per day in a city hospital.

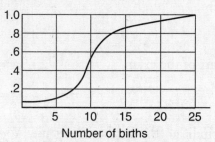

Number of births

Assuming that a birthing room can be used by only one woman per day, how many rooms must the hospital have available to be able to meet the demand at least 90 percent of the days?

(A) 5
(B) 10
(C) 15
(D) 20
(E) 25

STOP

If there is still time remaining, you may review your answers.

SECTION II

PART A

Questions 1–5

Spend about 65 minutes on this part of the exam.
Percentage of Section II grade—75

> You must show all work and indicate the methods you use. You will be graded on the correctness of your methods and on the accuracy of your results and explanations.

1. At a particular college, honor roll status used to be awarded when a student scored above a 3.0 average and high honor roll status when a student scored above 3.5 in any given semester. Over the years, more and more students were making the honor rolls. After consulting a statistics professor, the Dean's office decides to make a change and bestow honor roll status to any student with a semester average more than 1.5 standard deviations above the mean student average that semester, and high honor roll status if the student's average is an outlier, that is, more than 1.5 interquartile ranges above the upper quartile. Students may be listed on both lists if they satisfy both criteria. Assume the set of all students' averages is approximately normally distributed.

   (a) What proportion of students will be on the honor roll?
   (b) What proportion of students will be on the high honor roll?
   (c) What proportion of students will be on the honor roll, but not on the high honor roll?
   (d) What is the probability that at least 1 out of 3 randomly chosen students will be on the honor roll?

2. In past years, 4% of the applicants to a large state university have been National Merit finalists. An admissions counselor believes the true figure is now higher.

   (a) Explain why a simple random sample of 75 applicants would not be appropriate to run a 1-proportion z-test.
   (b) What is the minimum sample size needed to run this hypothesis test? Show your work.
   (c) Suppose the admissions counselor uses your result from part (b) and finds 6% National Merit applicants in the sample. Is this sufficient evidence at the 5% significance level to say that the percentage of National Merit finalists among applicants to this university is now more than 4%? Give statistical justification for your answer.

GO ON TO THE NEXT PAGE ➤

(d) Suppose the admissions counselor wants to establish a 95% confidence interval estimate of the current percentage of National Merit finalists among applicants. What is the minimum sample size necessary for a margin of error of only ±1% if it is believed that the true answer will be around 4%? Show your work.

3. To compare two levels of treatment with a new fertilizer, cherry tomatoes are to be grown in each of eight test plots. Tall windows line the north side of the hothouse and a breezy doorway is located on the east side.

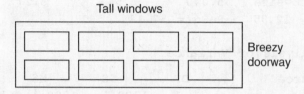

(a) Suppose you decide to block using the scheme below (one block is white, one gray). How would you use randomization, and what is the purpose of the randomization?

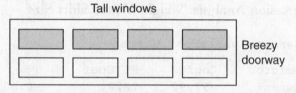

(b) Comment on the strength and weakness of the above scheme as compared to the following blocking scheme (one block is white, one gray).

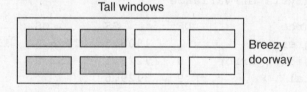

4. A professional golfer drives 25 balls off a tee, and an electric timer measuring ball speed off the tee shows a mean speed of 173.5 mph with a standard deviation of 2.4 mph. Assume the golfer never tires and there is a consistent normal distribution of ball speeds off the tee.

(a) Would you question the timer if it registered 170 mph on the next drive? Explain.

(b) Would you question the timer if the next 25 drives averaged 170 mph? Explain.

GO ON TO THE NEXT PAGE ➤

5. In a study of weight, belt size, and shirt size, measurements on a simple random sample of ten men yield the following computer output:

Regression Analysis: Weight Versus Belt Size

```
Dependent variable: Weight

Predictor    Coef       SE Coef      T           P
Constant     3.27       59.41        0.06        0.957
Belt size    5.040      1.543        3.27        0.011
S = 12.95    R-Sq = 57.1%
```

Analysis of Variance

```
Source           DF      SS        MS        F       P
Regression       1       1788.1    1788.1    10.66   0.011
Residual Error   8       1341.5    167.7
Total            9       3129.6
```

Regression Analysis: Weight Versus Shirt Size

```
Dependent variable: Weight

Predictor    Coef       SE Coef      T           P
Constant     -57.92     72.24        -0.80       0.446
Shirt size   15.438     4.372        3.53        0.008
S = 12.36    R-Sq = 60.9%
```

Analysis of Variance

```
Source           DF      SS        MS        F       P
Regression       1       1906.5    1906.5    12.47   0.008
Residual Error   8       1223.1    152.9
Total            9       3129.6
```

Residuals

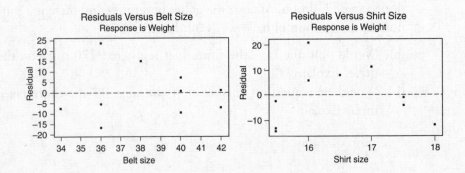

GO ON TO THE NEXT PAGE ➤

(a) Compare the correlation of weight versus belt size to the correlation of weight versus shirt size.

(b) Which of the two regression lines is a better linear fit? Explain.

(c) Use the two regression lines to obtain two estimates of the weight of a man with a shirt size of 17 and a belt size of 38. Show your work.

(d) One person in the sample had a shirt size of 17 and a belt size of 38. Which of the estimates in part (c) were overestimates and which were underestimates? Explain.

GO ON TO THE NEXT PAGE ➤

SECTION II

PART B

Question 6

Spend about 25 minutes on this part of the exam.
Percentage of Section II grade—25

6. A provost at a large state university issues a statement criticizing the faculty
    for grade inflation. A psychology professor concerned as to whether her
    department has succumbed to inflating grades checks a random sample of
    grades given in her department 10 years ago and during the past year. The
    sample results are as follows:

Frequency of students receiving grades

Grade	10 years ago	Last year
A (4.0)	10	12
B (3.0)	14	10
C (2.0)	34	16
D (1.0)	9	6
F (0.0)	8	6

(a) Is there evidence that the proportion of A's given last year in the
     Psychology Department is greater than the proportion of A's given 10
     years ago? Give a statistical justification.

(b) Counting A as 4.0 quality points, B as 3.0, etc., is there evidence that
     the average grade given in her department last year is higher than the
     average given 10 years ago? Give a statistical justification for your
     answer.

(c) If we assume that the distribution in the 10-year old sample is actu-
     ally the true aggregate distribution from 10 years ago, is there evi-
     dence that the overall distribution of grades last year is different from
     that of 10 years ago? Give a statistical justification for your answer.

(d) How would any seeming contradiction in parts (a), (b), and (c) above
     be explained to a layperson?

(e) Write a brief report for the psychology professor to the provost,
     including any relevant graphs, summarizing your findings. Assume
     that the provost knows nothing technical about statistics.

STOP

If there is still time remaining, you may review your answers.

Answer Key

Section I

1. **B**	9. **E**	17. **A**	25. **E**	33. **B**
2. **C**	10. **D**	18. **C**	26. **E**	34. **D**
3. **E**	11. **D**	19. **D**	27. **E**	35. **C**
4. **E**	12. **E**	20. **C**	28. **A**	36. **A**
5. **D**	13. **D**	21. **E**	29. **E**	37. **C**
6. **A**	14. **C**	22. **C**	30. **D**	38. **E**
7. **D**	15. **D**	23. **D**	31. **E**	39. **C**
8. **C**	16. **E**	24. **B**	32. **A**	40. **D**

Answers Explained

Section I

1. **(B)** $\mu_{\bar{x}} = \mu = 8$, and $\sigma_{\bar{x}} = \frac{\sigma}{\sqrt{n}} = \frac{4.8}{\sqrt{3}} < 4.8$

2. **(C)** The range increased from $90 - 15 = 75$ to $97 - 21 = 76$, while the standard deviation decreased (note how the values are bunched together more closely in 1998).

3. **(E)** None of the studies have any controls such as randomization, a control group, or blinding, and so while they may give valuable information, they cannot establish cause and effect.

4. **(E)** $\sigma_{\bar{x}} = \frac{\sqrt{386}}{\sqrt{40}} = 3.106$. The critical z-scores are $\frac{80-83}{3.106} = -0.97$ and $\frac{90-83}{3.106} = 2.25$, resulting in a probability of $.3340 + .4878 = .8218$.

5. **(D)** The critical z-scores are $\frac{188-200}{5} = -2.4$ and $\frac{194-200}{5} = -1.2$, with corresponding left tail probabilities of $.0082$ and $.1151$, respectively. The probability of being greater than 188 given that it is less than 194 is $\frac{.1151-.0082}{.1151} = .93$.

6. **(A)**

$$P(\text{pos test}) = P(\text{disease} \cap \text{pos}) + P(\text{healthy} \cap \text{pos})$$
$$= (.04)(.95) + (.96)(.05)$$
$$= .038 + .048 = .086$$

$$P(\text{disease} \mid \text{pos test}) = \frac{.038}{.086} = .442$$

7. **(D)** Given a census, the population parameter is known, and there is no need to use the techniques of inference.

8. **(C)** Sixty-four percent of the students scored below Mary and 56% scored below Pam, and so 8% must have scored between them.

9. **(E)** Intentionally forcing some treatment to note the response is associated with controlled experiments, not with observational studies. In experiments, the researchers decide how people are placed in different groups; self-selection is associated with observational studies. Results of observational studies may suggest cause-and-effect relationships; however, controlled studies are used to establish such relationships.

10. **(D)** $P\left(\text{popcorn} \mid \text{football}\right) = \frac{80}{350} = 23\%$, and $P\left(\text{basketball} \mid \text{hot dogs}\right) = \frac{50}{290}$ $= 17\%$.

11. **(D)** df $= 9$, and $47.52 \pm 2.821\left(\frac{1.59}{\sqrt{10}}\right)$.

12. **(E)** The set could be {150, 150, 150, 200} with mean 162.5 and median 150. It might also be {150, 200, 200, 200} with mean 187.5 and median 200. The median of a set of four elements is the mean of the two middle elements.

13. **(D)** $\hat{p} = \frac{256}{400} = .64$, and $\sigma_{\hat{p}} = \sqrt{\frac{(.64)(.36)}{400}} = .024$. The z-scores of .60 and .68 are $\pm\frac{.04}{.024} = \pm1.67$ for a probability of $.9525 - .0475 = .9050$.

14. **(C)** This is a binomial with $n = 15$ and $p = .29$, and so the mean is $np = 15(.29)$ $= 4.35$.

15. **(D)** With independence, variances add, so $\sqrt{65^2 + 380^2 + 125^2} = 405$.

16. **(E)** Without independence we cannot determine var$(X + Y)$ from the information given.

17. **(A)** Embarrassing questions and resulting untruthful answers are an example of response bias. Inaccuracies and mistakes due to human error are a real concern to researchers. The natural variation in samples is called *sampling error*.

18. **(C)** Relative frequency is given by relative area. The distribution is skewed to the left, and so the mean is less than the median.

19. **(D)** While the standard deviation of the z-distribution is 1, the standard deviation of the t-distribution is greater than 1.

20. **(C)**

$$P\left(\begin{array}{c}\text{market}\\\text{decline}\end{array} \cap \begin{array}{c}\text{interest}\\\text{jumps}\end{array}\right) = P\left(\begin{array}{c}\text{market}\\\text{decline}\end{array} \middle| \begin{array}{c}\text{interest}\\\text{jumps}\end{array}\right) P\left(\begin{array}{c}\text{interest}\\\text{jumps}\end{array}\right)$$
$$= (.72)(.65)$$
$$= .468$$

21. **(E)** Different samples give different sample statistics, all of which are estimates for the same population parameter, and so error, called *sampling error*, is naturally present.

22. **(C)** With four-digit accuracy as found on the TI-83, the critical z-scores for 70% to the right and 80% to the right are -0.5244 and -0.8416, respectively. Then $\{\mu - 0.5244\sigma = 12{,}000, \mu - 0.8416\sigma = 10{,}000\}$ gives $\mu = 15{,}306$ and $\sigma = 6305$. Using two-digit accuracy as found in Table A, that is, -0.52 and -0.84, results in $\mu = 15{,}250$ and $\sigma = 6250$.

23. **(D)** The percentage of the variation in y explained by the variation in x is given by the coefficient of determination r^2. In this example, $(.78)^2 = .61$.

24. **(B)** There is a weak negative correlation, and so $-.15$ is the only reasonable possibility among the choices given.

25. **(E)** There is no guarantee that 24 is anywhere near the interval, and so none of the statements are true.

26. **(E)** Sampling error relates to natural variation between samples, and it can never be eliminated. In good observational studies, responses are not influenced during the collection of data. In good experiments, treatments are compared as to the differences in responses.

27. **(E)** The scatterplot suggests a zero correlation.

28. **(A)** $\sigma_{\bar{x}} = \frac{1.2}{\sqrt{36}} = 0.2$. The z-score of 2.5 is $\frac{2.5 - 2.4}{0.2} = 0.5$, and to the right of 2.5 is the probability of failing to reject the false claim.

29. **(E)** The central limit theorem says that no matter how the original population is distributed, as the sample size increases, the sampling distribution of the sample mean becomes closer to a normal distribution.

30. **(D)** A sampling distribution is the distribution of all the values taken by a statistic, such as sample mean or sample proportion, from all possible samples of a given size.

31. **(E)** $3(.8)^2(.2) + (.8)^3 = .896$

32. **(A)** The critical t-scores for 99% confidence with df $= 6$ are ± 3.707.

33. **(B)** The PTA survey has strong response bias in that students may not give truthful responses to a parent or teacher about their engaging in unprotected sex. The talk show survey results in a voluntary response sample, which typically gives too much emphasis to persons with strong opinions. The *Ladies Home Journal* survey has strong selection bias; that is, people who read the *Journal* are not representative of the general population.

34. **(D)** To divide the interval estimate by d, the sample size must be increased by a multiple of d^2.

35. **(C)** $P(Y = 2) = .186 + .198 + .216 = .6$
$P(X = 1)(.6) = .186$, and so $P(X = 1) = .31$
$P(Y = 1) = 1 - (.6 + .1) = .3$
$P(X = 1, Y = 1) = (.31)(.3) = .093$

36. **(A)** There is nothing to hypothesize about if the population parameter is already known. The *P*-value depends on the sample chosen. The *P*-value in a two-sided test is calculated by doubling the indicated tail probability.

37. **(C)** A Type II error is a mistaken failure to reject a false null hypothesis or, in this case, a failure to realize that the machinery is turning out wrong size ball bearings.

38. **(E)** Using a measurement from a sample, we are never able to say *exactly* what a population mean is; rather we always say we have a certain *confidence* that the population mean lies in a particular *interval*. In this case we are 95% confident that Lipitor will outperform niacin by 18 to 41 points.

39. **(C)** The median number of hours is less than four for June, July, and August.

40. **(D)** A horizontal line drawn at the .9 probability level corresponds to roughly 19 rooms.

SECTION II

1. (a) From Table A or a calculator, we get $P(z > 1.5) = .0668$, so 6.68% of the students will be on the honor roll.
   (b) From Table A the quartiles are at ± 0.674 standard deviations from the mean. Thus IQR $= 2(0.674) = 1.348$ standard deviations. The cutoff for the high honor roll is at $z = 0.674 + 1.5(1.348) = 2.696$. Because $P(z > 2.696) = .0035$, 0.35% of the students will make the high honor roll (use of Table A gives 0.37%).
   (c) $2.696 > 1.5$, and so we can simply subtract: $.0668 - .0035 = .0633$, so 6.33% of the students will make the honor roll but not the high honor roll.
   (d) P(at least 1 of 3 on honor roll) $= 1 - P$(none on honor roll) $= 1 - (.9332)^3 = .1873$

Scoring

Do not take off for minor arithmetic errors.

Give one point each for parts (a), (c), and (d) if essentially correct. Count part (d) correct if it makes proper use of whatever answer was obtained in part (a), and similarly, part (c) correct if it makes proper use of whatever answers were obtained in parts (a) and (b).

Grade part (b) holistically, with two points for essentially correct and one point for partially correct.

4 Complete Answer   All five points.

3 Substantial Answer   Four points.

2 Developing Answer   Three points.

1 Minimal Answer   One or two points.

2. (a) The sample size n must be large enough for both np and $n(1-p)$ to be larger than 10. In this case $np = (75)(.04) = 3$.

   (b) For $np > 10$, n must be greater than $\frac{10}{p} = \frac{10}{.04} = 250$, and so the minimum n is 251.

   (c) The student should do a z-test for proportion. The assumptions concerning the sample size being large enough are met: $np = (251)(.04) = 10.04 > 10$ and $n(1-p) = (251)(.96) = 240.96 > 10$. We must assume that the sample is less than 10% of the entire population, which seems reasonable given that this is a large state university.

   H_0: $p = .04$ (where p is the proportion of the applicants who are

   H_a: $p > .04$ National Merit finalists)

   $z = \dfrac{.06 - .04}{\sqrt{\dfrac{(.04)(.96)}{251}}} = 1.617$, and thus $P = .053$.

   Since $.053 > .05$, there is *not* sufficient evidence at the 5% significance level to say that the percentage of National Merit finalists among applicants to this university is now more than 4%.

   (d) $1.96\sqrt{\dfrac{(.04)(.96)}{n}} < .01$, so $\sqrt{n} > 38.4$ and $n > 1474.56$. Thus, choose $n = 1475$.

Scoring

Do not take off for minor arithmetical mistakes.

Part (a) is only partially correct if the student states that $np > 10$ isn't satisfied, but doesn't show that it is only 3.

Part (b) is only partially correct if the correct final answer is given but work is not shown. Full credit should be given for any answer reasonably close to 251 if the work is shown.

A complete answer to part (c) involves a statement of the test to be used, a check of assumptions, a statement of hypotheses, a calculation of z and P, and a correct conclusion. Part (c) is essentially correct if four out of these five items are complete and correct, while the fifth is not incorrect but is incomplete or unclear. With fewer than four items correct, part (c) should be graded holistically and scored either incorrect or partially correct. Part (c) should be worked with the answer from part (b) as the sample size.

Part (d) is only partially correct if the correct final answer is given but work is not shown. Full credit should be given for any answer reasonably close to 1475 if the work is shown.

4	**Complete Answer**	All four parts correct.
3	**Substantial Answer**	Three parts including part (c) correct OR parts (a), (b), and (d) correct and (c) partially correct OR two parts correct and the other two parts partially correct.
2	**Developing Answer**	Parts (a), (b), and (d) correct but part (c) incorrect OR two parts including (c) correct OR part (c) correct and two of the other parts partially correct.
1	**Minimal Answer**	One part correct OR two parts partially correct.

---

3. (a) In each block, two of the plots will be randomly assigned to receive one level of the fertilizer treatment, while the remaining two plots in the block will receive the other level of the fertilizer treatment. Randomization of levels of fertilizer to the plots within each block should reduce bias due to confounding variables associated with the plots such as soil and moisture that might be related to productivity of the tomato plants. In particular, the randomization in blocks in scheme (a) should even out the effect of the distance plants are from the breezy doorway.

   (b) Scheme (a) creates homogeneous blocks with respect to window exposure, while scheme (b) creates homogeneous blocks with respect to doorway exposure. Randomization of fertilizer levels to plots within blocks in scheme (a) should even out effects of doorway exposure, while randomization of fertilizer levels to plots within blocks in scheme (b) should even out effects of window exposure.

Students must explain why randomization is important within the context of this problem. Use of terms like bias and confounding must be within the context of this problem. Students must explain the importance of homogeneous experimental units (plots, not fertilizer levels) within blocks.

Scoring	
4 Complete Answer	Correct explanation of use of blocking and randomization in context of this problem.
3 Substantial Answer	Correct explanation of use of either blocking or randomization in context and partial explanation of other.
2 Developing Answer	Correct explanation of use of either blocking or randomization in context OR partial explanations of both.
1 Minimal Answer	Partial explanation of either blocking or randomization in context.

4. (a) Since 170 is only $\frac{173.5-170}{2.4} = 1.5$ standard deviations away from the mean, it is not an unreasonable occurrence, so the answer is "no," the timer should not be questioned.

(b) Since the standard deviation of the sampling distribution is

$s_{\bar{x}} = \frac{s}{\sqrt{n}} = \frac{2.4}{\sqrt{25}} = 0.48$, the sample mean of 170 is $\frac{173.5-170}{0.48} = 7.3$ standard deviations away from the mean, a very suspicious result, and so the answer is "yes," the timer should be questioned.

Scoring	
4 Complete Answer	Answers to both parts correct with correct explanations.
3 Substantial Answer	Answers to both parts correct with incomplete explanations.
2 Developing Answer	Answers to both parts correct with incorrect or missing explanations.
1 Minimal Answer	One answer correct with correct explanation.

5. (a) For weight versus belt size, R-Sq = 57.1%, so the correlation is $r = \sqrt{.571} = .756$, and for weight versus shirt size, R-Sq = 60.9%, so the correlation is $r = \sqrt{.609} = .780$, and we see that weight versus shirt size has a slightly higher correlation than weight versus belt size.

(b) Weight versus belt size is a better linear fit than weight versus shirt size, as can be seen in the two residual plots, one of which (weight versus shirt size) shows a distinct curved pattern going from negative to positive and then back to negative. Such a distinct pattern indicates that a fit other than linear is probably a better model. The residual plot for weight versus belt size shows no such pattern.

(c) An estimate for the weight of a man with a shirt size of 17 is 15.4(17) − 57.9 ≈ 204, while an estimate for the weight of a man with a belt size of 38 is 5.04(38) + 3.3 ≈ 195.

(d) Residual = actual − predicted. Since the residuals for a shirt size of 17 and for a belt size of 38 are both positive, the actual must be greater than the predicted in both cases. So both estimates using the regression lines give underestimates.

Scoring

Part (b) is essentially correct only if a clearly explained reference to the residual plots is given, and is partially correct if a correct answer with a weak explanation is given.

Part (c) is essentially correct if 17 and 38 are plugged into the correct linear equations even if an arithmetical mistake results in the wrong answer. Part (c) is partially correct if 17 and 38 are plugged into incorrect, but consistent with each other, linear equations.

Part (d) is partially correct if a correct answer is given without a proper explanation.

4 Complete Answer   All four parts essentially correct.

3 Substantial Answer   Three parts essentially correct OR two parts essentially correct and the other two parts partially correct.

2 Developing Answer   Two parts essentially correct OR one part essentially correct and at least two of the other parts partially correct.

1 Minimal Answer   One part essentially correct OR at least two parts partially correct.

6. (a) The proper test is a two-proportion z-test.

   Assumptions: Simple random samples (given) and sample sizes <10% of population (reasonable since sample is from entire department's grades). We must confirm that "successes" and "failures" are all ≥ 10: $n_1 p_1 = 12$, $n_1(1 - p_1) = 38$, $n_2 p_2 = 10$, $n_2(1 - p_2) = 65$, and all are ≥ 10.

   $H_0: p_1 - p_2 = 0$ (where p_1 is the proportion of A's last year and p_2 is the proportion of A's 10 years ago.)

   $H_a: p_1 - p_2 > 0$ (the proportion of A's given last year in the Psychology Department is greater than the proportion of A's given 10 years ago.)

   $$\hat{p}_1 = \frac{12}{50} = .24, \quad \hat{p}_2 = \frac{10}{75} = .133, \quad \text{and} \quad \hat{p} = \frac{12+10}{50+75} = .176$$

   $$z = \frac{.24 - .133}{\sqrt{(.176)(.824)\left(\frac{1}{50} + \frac{1}{75}\right)}} = 1.539 \text{ and } P = .062.$$

   With this small a P-value, there is some evidence (at the 10% significance level because $P < .10$) to reject H_0 and conclude that there is evidence that the proportion of A's given last year in the Psychology Department is greater than the proportion of A's given 10 years ago.

   (b) The proper hypothesis test is a two-sample t-test (unpooled, because there is no reason to suppose that the standard deviations are equal).

   An assumption of independent random samples is satisfied by the given scheme. An assumption of normality is satisfied by looking at the histograms, which are roughly bell-shaped with no outliers or extreme skewness:

   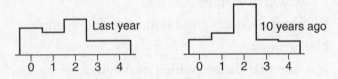

   $H_0: \mu_1 - \mu_2 = 0$ (average grade given in her department last year equals the average given 10 years ago.)

   $H_a: \mu_1 - \mu_2 > 0$ (average grade given in her department last year is higher than the average given 10 years ago.)

   On the TI-83 we get $t = 0.888$ and $P = .188$.

   Alternatively, we can calculate $\bar{x}_1 = 2.32$, $s_1 = 1.30$, $\bar{x}_2 = 2.12$, and $s_2 = 1.13$. Then

   $$t = \frac{\bar{x}_1 - \bar{x}_2}{\sqrt{\dfrac{s_1^2}{n_1} + \dfrac{s_2^2}{n_2}}} = \frac{2.32 - 2.12}{\sqrt{\dfrac{1.30^2}{50} + \dfrac{1.13^2}{75}}} = 0.887$$

   With df $= \min\{50, 75\} = 50$, we get $P = 0.190$.

   With such a large P-value, there is no evidence to reject H_0, and so we conclude that there is no evidence that the average grade given in her department last year is higher than the average given 10 years ago.

(c) The proportions from 10 years ago are shown in the table below:

Grade	A	B	C	D	F
Proportion	$\frac{10}{75}$ = .133	$\frac{14}{75}$ = .187	$\frac{35}{75}$ = .467	$\frac{9}{75}$ = .120	$\frac{7}{75}$ = .093

To do a chi-square test for goodness-of-fit, we check the assumptions: the expected cell counts $(.133)(50) = 6.65$, $(.187)(50) = 9.35$, $(.453)(50) = 22.65$, $(.120)(50) = 6$, and $(.107)(50) = 5.35$ are all >5, and we have a simple random sample (given) that is less than 10% of the population (reasonable since she sampled from the entire department's grades).

H_0: The observed numbers are consistent with the old proportions.

H_a: The observed numbers are not consistent with the old proportions.

$$\chi^2 = \sum \frac{(obs - exp)^2}{exp} = \frac{(12 - 6.65)^2}{6.65} + \frac{(10 - 9.35)^2}{9.35}$$
$$+ \frac{(16 - 22.65)^2}{22.65} + \frac{(6 - 6)^2}{6} + \frac{(6 - 5.35)^2}{5.35} = 6.38$$

With df $= 5 - 1 = 4$, we get $P = .173$.

With this large a value for P, there is *not* sufficient evidence to reject H_0, and thus we conclude that there is *not* sufficient evidence to say that the overall distribution of grades last year is different from that of 10 years ago.

(d) There is no contradiction. It is possible for a significant change in one component of a distribution without a significant change in the distribution as a whole.

(e) There are many possible answers. For example:

Dear Provost,

It is true that the proportion of A's given by the Psychology Department was significantly greater last year than 10 years ago; however, both the average grade given and the overall distribution of grades given have not changed significantly from 10 years ago to last year.

Following are pie charts and segmented bar charts to illustrate the above.

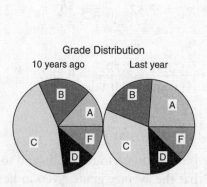

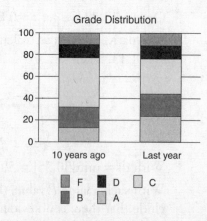

Scoring

Parts (a), (b), and (c) involve five steps: naming a test, checking assumptions, stating hypotheses, calculating t (or z) and P, and giving a conclusion in context. Parts (a), (b), and (c) are each essentially correct if at least four of the five steps are correct and complete, and are partially correct if two or three of the five steps are correct and complete.

Part (e) is essentially correct for a reasonable summary illustrated by a reasonable graph, and partially correct for only one of the summary and graph correct and clear.

4 Complete Answer   All five parts essentially correct.

3 Substantial Answer   Four parts essentially correct OR three parts essentially correct and the other two partially correct.

2 Developing Answer   Three parts essentially correct OR two parts essentially correct and at least two of the remaining parts partially correct.

1 Minimal Answer   Two parts essentially correct OR one part essentially correct and at least two of the remaining parts partially correct OR four parts partially correct.

Answer Sheet
PRACTICE EXAMINATION 6

1. Ⓐ Ⓑ Ⓒ Ⓓ Ⓔ
2. Ⓐ Ⓑ Ⓒ Ⓓ Ⓔ
3. Ⓐ Ⓑ Ⓒ Ⓓ Ⓔ
4. Ⓐ Ⓑ Ⓒ Ⓓ Ⓔ
5. Ⓐ Ⓑ Ⓒ Ⓓ Ⓔ
6. Ⓐ Ⓑ Ⓒ Ⓓ Ⓔ
7. Ⓐ Ⓑ Ⓒ Ⓓ Ⓔ
8. Ⓐ Ⓑ Ⓒ Ⓓ Ⓔ
9. Ⓐ Ⓑ Ⓒ Ⓓ Ⓔ
10. Ⓐ Ⓑ Ⓒ Ⓓ Ⓔ

11. Ⓐ Ⓑ Ⓒ Ⓓ Ⓔ
12. Ⓐ Ⓑ Ⓒ Ⓓ Ⓔ
13. Ⓐ Ⓑ Ⓒ Ⓓ Ⓔ
14. Ⓐ Ⓑ Ⓒ Ⓓ Ⓔ
15. Ⓐ Ⓑ Ⓒ Ⓓ Ⓔ
16. Ⓐ Ⓑ Ⓒ Ⓓ Ⓔ
17. Ⓐ Ⓑ Ⓒ Ⓓ Ⓔ
18. Ⓐ Ⓑ Ⓒ Ⓓ Ⓔ
19. Ⓐ Ⓑ Ⓒ Ⓓ Ⓔ
20. Ⓐ Ⓑ Ⓒ Ⓓ Ⓔ

21. Ⓐ Ⓑ Ⓒ Ⓓ Ⓔ
22. Ⓐ Ⓑ Ⓒ Ⓓ Ⓔ
23. Ⓐ Ⓑ Ⓒ Ⓓ Ⓔ
24. Ⓐ Ⓑ Ⓒ Ⓓ Ⓔ
25. Ⓐ Ⓑ Ⓒ Ⓓ Ⓔ
26. Ⓐ Ⓑ Ⓒ Ⓓ Ⓔ
27. Ⓐ Ⓑ Ⓒ Ⓓ Ⓔ
28. Ⓐ Ⓑ Ⓒ Ⓓ Ⓔ
29. Ⓐ Ⓑ Ⓒ Ⓓ Ⓔ
30. Ⓐ Ⓑ Ⓒ Ⓓ Ⓔ

31. Ⓐ Ⓑ Ⓒ Ⓓ Ⓔ
32. Ⓐ Ⓑ Ⓒ Ⓓ Ⓔ
33. Ⓐ Ⓑ Ⓒ Ⓓ Ⓔ
34. Ⓐ Ⓑ Ⓒ Ⓓ Ⓔ
35. Ⓐ Ⓑ Ⓒ Ⓓ Ⓔ
36. Ⓐ Ⓑ Ⓒ Ⓓ Ⓔ
37. Ⓐ Ⓑ Ⓒ Ⓓ Ⓔ
38. Ⓐ Ⓑ Ⓒ Ⓓ Ⓔ
39. Ⓐ Ⓑ Ⓒ Ⓓ Ⓔ
40. Ⓐ Ⓑ Ⓒ Ⓓ Ⓔ

Practice Examination 6

Section I

Questions 1–40

Spend 90 minutes on this part of the exam.

> **Directions:** The questions or incomplete statements that follow are each followed by five suggested answers or completions. Choose the response that best answers the question or completes the statement.

1. The boxplots below summarize the distributions of SAT verbal and math scores among students at an upstate New York high school.

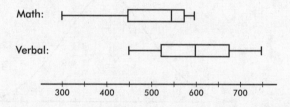

Which of the following statements are true?

I. The range of the math scores equals the range of the verbal scores.

II. The highest math score equals the median verbal score.

III. The verbal scores appear to be roughly symmetric, while the math scores appear to be skewed to the right.

(A) I only
(B) III only
(C) I and II
(D) II and III
(E) I, II, and III

2. Which of the following are true statements?

I. The significance level of a test is the probability of a Type II error.

II. Given a particular alternative, the power of a test against that alternative is 1 minus the probability of the Type II error associated with that alternative.

III. If the significance level remains fixed, increasing the sample size will reduce the probability of a Type II error.

(A) II only
(B) III only
(C) I and II
(D) I and III
(E) II and III

GO ON TO THE NEXT PAGE ➤

3. One national test has a normal distribution with a mean of 500 and a standard deviation of 100, while a second test has a normal distribution with a mean of 18 and a standard deviation of 6. A student scored 677 on one test and 29 on the other. Relative to the respective distributions, which score is better?

(A) The score of 677 is better.
(B) The score of 29 is better.
(C) The scores are equally strong.
(D) A comparison cannot be made without knowing the number of students who took each exam.
(E) It is improper to compare scores for the same student on different exams.

4. A survey is to be taken to estimate the proportion of people who support the NATO decision to be actively involved in the Balkans. Among the following proposed sample sizes, which is the smallest that will still guarantee a margin of error of at most 0.03 for a 95% confidence interval?

(A) 35
(B) 70
(C) 800
(D) 1100
(E) 4300

5. Which among the following is most useful in establishing cause-and-effect relationships?

(A) A complete census
(B) A least squares regression line showing high correlation
(C) A simple random sample (SRS)
(D) A well-designed, well-conducted survey incorporating chance to ensure a representative sample
(E) A controlled experiment

6. A cumulative frequency plot of the numbers of ounces in a sample of 200 half-gallon orange juice containers is shown below.

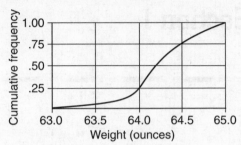

Which of the following is a valid conclusion?

(A) Only 25% of the containers had at least 64.0 ounces.
(B) The median number of ounces is 64.0 ounces.
(C) The interquartile range is ≥ 1.0 ounce.
(D) The upper 100 weights show more variability than the lower 100 weights.
(E) None of the above are valid conclusions.

7. A histogram of the cholesterol levels of all employees at a large law firm is as follows:

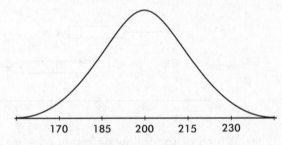

Which of the following is the best estimate of the standard deviation of this distribution?

(A) $\frac{(230-170)}{6} = 10$
(B) 15
(C) $\frac{200}{6} = 33.3$
(D) $230 - 170 = 60$
(E) $245 - 155 = 90$

GO ON TO THE NEXT PAGE ➤

8. A cross-country coach who is also the school's AP Statistics teacher calculates that one runner, Bill, had a standardized score (*z*-score) of 1.6 for his time in a 5 K event as compared to the entire group of runners in the race. Which of the following is the best interpretation of Bill's *z*-score?

   (A) Bill's time was 16 minutes.
   (B) Only 1.6% of the runners did better than Bill.
   (C) Bill's time is 1.6 times the average time of all the runners.
   (D) Bill's time is 1.6 minutes above the average time of all the runners.
   (E) Bill's time is 1.6 standard deviations above the average time of all the runners.

9. A basketball player makes one out of his first two free throws. From that point on, the probability he makes the next shot is equal to the proportion of shots made up to that point. If he takes two more shots, what is the probability he ends up making a total of one free throw? Two free throws? Three free throws?

   (A) $P(1 \text{ free throw}) = \frac{1}{3}$, $P(2 \text{ free throws}) = \frac{1}{3}$, $P(3 \text{ free throws}) = \frac{1}{3}$
   (B) $P(1 \text{ free throw}) = \frac{1}{4}$, $P(2 \text{ free throws}) = \frac{1}{2}$, $P(3 \text{ free throws}) = \frac{1}{4}$
   (C) $P(1 \text{ free throw}) = \frac{1}{6}$, $P(2 \text{ free throws}) = \frac{2}{3}$, $P(3 \text{ free throws}) = \frac{1}{6}$
   (D) $P(1 \text{ free throw}) = \frac{1}{8}$, $P(2 \text{ free throws}) = \frac{3}{4}$, $P(3 \text{ free throws}) = \frac{1}{8}$
   (E) $P(1 \text{ free throw}) = \frac{1}{8}$, $P(2 \text{ free throws}) = \frac{1}{4}$, $P(3 \text{ free throws}) = \frac{5}{8}$

10. Hurricanes are categorized by their severity on a 1 to 5 scale, with Category 1 referring to hurricanes with sustained winds between 74 and 95 mph and Category 5 referring to hurricanes with sustained winds over 155 mph. The following chart shows the frequency distribution for the number of major hurricanes (Categories 3, 4, and 5) to strike an island group each year during the past 80 years.

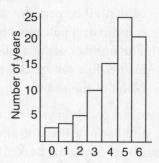

If we want to minimize the influence of the most extreme values, which of the following are the most appropriate measures of center and spread for the number of major hurricanes per year?

   (A) Mean and standard deviation
   (B) Mean and variance
   (C) Mean and range
   (D) Median and range
   (E) Median and interquartile range

11. What is an outlier?

   (A) An observation, sufficiently different from the others, that should be ignored.
   (B) An observation more than three standard deviations from the mean.
   (C) An observation that affects the mean but not the median.
   (D) An observation that is either greater or less than all the others.
   (E) An observation significantly different from the others.

12. The binomial distribution is an appropriate model for which of the following?

    (A) The number of minutes in an hour for which the Dow-Jones average is above its beginning average for the day.
    (B) The number of cities among the 10 largest in New York State for which the weather is cloudy for most of a given day.
    (C) The number of drivers wearing seat belts if 10 consecutive drivers are stopped at a police roadblock.
    (D) The number of A's a student receives in his/her five college classes.
    (E) None of the above.

13. There is a rough linear relationship between the number of assists and the number of steals by professional basketball players. A least squares fit results in the model $\hat{y} = 0.52x - 0.43$, $x \geq 1$, where x is the number of assists and $\hat{y}$ is the estimated number of steals. What is the estimated increase in steals that corresponds to an increase of three assists?

    (A) 0.27
    (B) 0.52
    (C) 1.13
    (D) 1.56
    (E) 3.09

14. A survey is conducted to determine the percentage of students at state universities that change their major at least once. In an SRS of 100 students, 78% indicated that they graduated with a major different from the one with which they entered college. Which of the following is the proper meaning of a 95% confidence level estimate for the percentage of students who change their major?

    (A) We are 95% certain that 78% of students change their major.
    (B) Between 69.9% and 86.1% of students change their majors 95% of the time.
    (C) 95% of the students change their majors between 69.9% and 86.1% of the time.
    (D) In repeated samplings of 100 students, there is a 95% chance that the proportion of students who change their major in any given sample is between 69.9% and 86.1%.
    (E) In repeated samplings of 100 students, 95% of the resulting intervals will contain the true proportion of students who change their major.

GO ON TO THE NEXT PAGE ➤

15. A survey is to be taken to ascertain student opinions about the quality of teaching at a high school. Consider the following survey methods of picking 100 students out of the 2000 students registered at the school.

    I. Randomly pick one day of the week. As the students arrive at school that day, they are asked to write their names on slips of paper and place these slips in a box. The principal then reaches in and pulls out 100 names.
    II. Using an official school roster of the 2000 students, pick every 20th name.
    III. Using a random number table, 25 names are chosen from each of separate lists of freshmen, sophomores, juniors, and seniors.
    IV. Randomly pick 5 freshmen homerooms, 5 sophomore homerooms, 5 junior homerooms, and 5 senior homerooms. Then randomly pick 5 students from each of these 20 homerooms.

    How many of the above survey methods will result in a simple random sample of 100 students?

    (A) None
    (B) One
    (C) Two
    (D) Three
    (E) Four

16. Which of the following are true statements about experimental design?

    I. Randomization refers to the random selection of subjects to be used in an experiment.
    II. Replication refers to repeating an experiment a sufficient number of times.
    III. The primary method of control of lurking variables is use of a control group.

    (A) I only
    (B) II only
    (C) III only
    (D) All are true.
    (E) None are true.

17. Which of the following are true?

    I. The power of a test concerns its ability to detect an alternative hypothesis.
    II. The significance level of a test is the probability of rejecting a true null hypothesis.
    III. The probability of a Type I error plus the probability of a Type II error always equals 1.

    (A) None are true.
    (B) I and II
    (C) I and III
    (D) II and III
    (E) I, II, and III

18. In early 1999 when the NBA lockout finally ended, one national poll reported that 73% of the public were less interested in watching a professional basketball game than they were before the lockout. Which of the following best explains what is meant by the statement that the margin of error was ±3%.

    (A) The true proportion of people who were less interested in watching a pro game can reasonably be expected to be within .03 or so from the sample finding of .73.
    (B) Three percent of those polled had no opinion.
    (C) In all polls using the same sample size and the same sampling technique, a wrong answer will result about 3% of the time.
    (D) The true proportion of people less interested than before in watching a pro game is either .70 or .76.
    (E) Three percent of the population were not polled.

19. In one experiment to test the effect of alcohol on large motor skills, volunteers were randomly placed in two groups, A and B. Everyone in group A drank 2 ounces of alcohol, and 20 minutes later everyone in both groups was timed on a manual dexterity test. The average completion times for the group A and B volunteers were 38 and 31 seconds, respectively. The 90% confidence interval estimate for the mean difference is (3, 11). If μ_A and μ_B are the true mean completion times, respectively, for people who have and have not drunk 2 ounces of alcohol, how many of the following statements are reasonable conclusions?

I. $\mu_A > \mu_B$ with probability .90.
II. There is a .90 probability that $3 < \mu_A - \mu_B < 11$.
III. The interval (3, 11) was calculated by a method that gives correct results (for where $\mu_A - \mu_B$ lies) in 90% of all possible samples.
IV. We are 90% confident that $\mu_A - \mu_B$ lies between 3 and 11 seconds.

(A) None
(B) One
(C) Two
(D) Three
(E) Four

Questions 20–21 refer to the following: A diner advertises that for the month of January the price of soup in cents will be the temperature at 10 A.M. that morning. If the 10 A.M. temperature is 20 degrees Fahrenheit, soup will sell for 20 cents a cup! The store owner, whose hobby is statistics, notes that the temperatures seem normally distributed and calculates that the expected value of a cup of soup any day in January is 28 cents with a standard deviation of 7 cents. Suppose a customer buys a cup of soup on ten random days in January. (Assume the temperature on any day is independent of the temperature on any other day, which is probably an unreasonable assumption!)

20. What is the total amount the customer should expect to pay?

(A) $2.10
(B) $2.45
(C) $2.80
(D) $3.15
(E) $3.50

21. What is the approximate probability that the total amount the customer pays will exceed $3.00?

(A) .06
(B) .18
(C) .28
(D) .39
(E) .90

22. There are 16,253 men and 21,784 women at a large state university. If 28% of the men and 44% of the women are social science majors, what is the expected number of social science majors in a random sample of 100 students?

(A) 28
(B) 36
(C) 37
(D) 44
(E) 72

GO ON TO THE NEXT PAGE ➤

23. A test is run to compare two calculus textbooks, one with a traditional approach and one with a reform approach. A state university agrees to use the traditional text in all of its 30 classes, each with approximately 50 students. A private liberal arts college agrees to use the reform text in its 4 calculus classes, each with 20 students. An SRS of 30 calculus students at each school will take a standardized final exam, and the mean scores of the two groups will be compared. Is this a good experimental design?

(A) Yes

(B) No, because proportions of students passing would be the more appropriate variable.

(C) No, because the standardized exam should have been given to all the calculus students.

(D) No, because so many more students take calculus at the university than at the college.

(E) No, because there is a confounding variable: large public university versus small private college.

24. When making an inference about a population mean, which of the following suggests the use of *t*-scores rather than *z*-scores?

(A) Outliers are present.

(B) The population is not normal.

(C) The sample size is over 30.

(D) The population is strongly skewed.

(E) The population standard deviation is unknown.

25. In which of the following is the median greater than the mean?

(A)

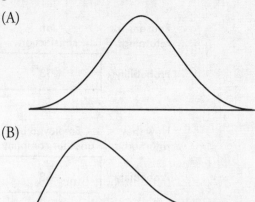

(B)

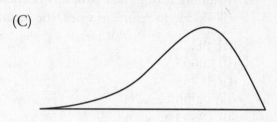

(C)

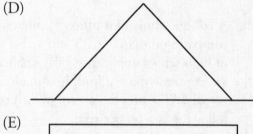

(D)

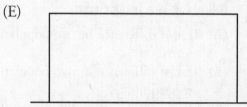

(E)

26. The distribution of blood cholesterol levels among females between 30 and 40 years old is roughly normal. If 10% have levels above 238 and 20% have levels below 187, what is the mean of this distribution?

(A) 207.2

(B) 208.9

(C) 212.5

(D) 217.8

(E) The mean cannot be calculated from the given information.

GO ON TO THE NEXT PAGE ➤

27. One-third of retired senior executives return to work within 1.5 years of retirement. According to Russel Reynolds Association, the reasons for and path of return are as follows:

Reasons for returning:	Job satisfaction	Avoid boredom	Contribute to society
Probability:	.53	.29	.18

How they returned:	Employed by another company	Self-employed	Consultant	Started a new company
Probability:	.38	.32	.23	.07

Assuming reasons and path are independent, what is the probability that a retired senior executive will decide to return to work, the reason being to avoid boredom, on a path as a consultant?

(A) .0222
(B) .0667
(C) .1733
(D) .5200
(E) .8533

28. A college admissions officer is interested in comparing the SAT math and verbal scores of high school applicants. Fifty applicants are randomly picked and their math and verbal SAT scores are noted. Which of the following is a proper test?

(A) Test of difference in two population means.
(B) Test of difference in two population proportions.
(C) One sample test on differences of paired data.
(D) Chi-square goodness-of-fit test.
(E) Chi-square test for homogeneity.

29. A reporter from a consumer magazine notes that the mean price of 30 selected grocery items at a particular supermarket is $0.75 with a standard deviation of $0.20. Suppose the following week the store raises all prices by 5 cents, and then the next week they lower all prices by 5%. What are the new mean and standard deviation for the 30 selected items?

(A) $0.76, $0.19
(B) $0.76, $0.20
(C) $0.76, $0.24
(D) $0.80, $0.20
(E) $0.80, $0.24

Questions 30–31 refer to the following: In a general market survey of buying habits, 15,000 out of 50,000 town residents are identified as purchasers of vitamin supplements.

30. In an SRS of five town residents, what is the expected value for the number who purchase vitamin supplements?

    (A) 0.3
    (B) 1.0
    (C) 1.5
    (D) 2.0
    (E) 5.0

31. Researchers for a national health food chain wish to estimate the total potential market for a new store. Plan 1 is to pick an SRS of 50 town residents, calculate a confidence interval estimate for the average amount spent on vitamin supplements by these 50 people, and multiply both ends of this interval by 50,000, the number of town residents. Plan 2 is to pick an SRS of 50 of the 15,000 people identified as vitamin supplement purchasers, calculate a confidence interval estimate for the average amount spent on vitamin supplements by these 50 people, and multiply both ends of this interval by 15,000. Which plan should be chosen to best estimate the total spent on vitamin supplements in the town?

    (A) Plan 1
    (B) Plan 2
    (C) Both plans should produce good results.
    (D) Neither plan will give a meaningful estimate.
    (E) Without additional information, there is no way to decide between the two plans.

32. Treatment of breast cancer is a controversial issue with treatment varying from radical mastectomy to radiation and chemotherapy combined with simple lumpectomy. One long-term study involved noting survival times of all breast cancer patients from over 100 hospitals and comparing the percentage of women still alive 5 years after radical mastectomies with the percentage of women still alive 5 years after radiation and chemotherapy combined with simple lumpectomy. The results seemed to show a better outcome for women who received lumpectomy treatment. Which of the following statements are appropriate?

    I. This appears to be a well-conducted controlled experiment.
    II. This observational study should be faulted for a lack of blinding.
    III. A lurking variable involving who received what treatment makes any conclusion suspect.

    (A) I only
    (B) II only
    (C) III only
    (D) I and II
    (E) II and III

33. An SRS of 20 patients undergoing laboratory work at a city hospital showed an average blood potassium level of 3.1 with a standard deviation of 0.3. Which of the following is a 99% confidence interval estimate for the average blood potassium level for patients undergoing laboratory work at this hospital?

    (A) $3.1 \pm 2.576\sqrt{\frac{0.3}{20}}$

    (B) $3.1 \pm 2.576\frac{\sqrt{0.3}}{20}$

    (C) $3.1 \pm 2.576\frac{0.3}{\sqrt{20}}$

    (D) $3.1 \pm 2.861\sqrt{0.3}$

    (E) $3.1 \pm 2.861\frac{0.3}{\sqrt{20}}$

GO ON TO THE NEXT PAGE ➤

34. In testing the null hypothesis H_0: $\mu = 50$ against the alternative hypothesis H_a: $\mu <$ 50, a sample from a normal population has a mean of 48.1 with a corresponding z-score of -2.23 and a P-value of .0129. Which of the following are reasonable conclusions?

   I. If the null hypothesis is assumed to be true, the probability of obtaining a sample mean as extreme as 48.1 is only .0129.
   II. In all samples using the same sample size and the same sampling technique, the null hypothesis will be wrong 1.29% of the time.
   III. There is sufficient evidence to reject the null hypothesis.

   (A) III only
   (B) I and II
   (C) I and III
   (D) II and III
   (E) I, II, and III

35.

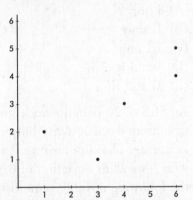

The least squares regression line for the above scatterplot is $\hat{y} = 0.61x + 0.56$. What is the residual for the point (3, 1)?

   (A) -1.00
   (B) -1.39
   (C) -1.83
   (D) -2.39
   (E) -3.00

36. A piece of medical equipment is not functioning properly; however, in running operational checks, a lab technician does not find evidence of the malfunction. The lab technician has committed

   (A) a Type I error.
   (B) a Type II error.
   (C) both a Type I and a Type II error.
   (D) neither a Type I nor a Type II error.
   (E) a random sampling error.

37. The back-to-back stemplot below gives the average bowling scores of male and female participants in the finals of a national tournament.

Males		Females
997632	28	6
88421	27	25
80	26	378
3	25	0267
1	24	028
	23	89

Which of the following is *not* true?

   (A) The same number of male and female bowlers participated.
   (B) The male scores are skewed while the female scores are roughly symmetric.
   (C) The median and mean female scores are roughly equal.
   (D) The mean male score is greater than the median male score.
   (E) The range of the male scores equals the range of the female scores.

38. The Wechsler Adult Intelligence Scale results in a normal distribution with a mean of 110 and a standard deviation of 25. If someone tests at the 80th percentile, what score did that individual have?

   (A) 89
   (B) 113
   (C) 118
   (D) 130
   (E) 131

Questions 39–40 refer to the following calculator output:

```
LinReg
   y=ax+b
   a=-23.82160513
   b=267.1903472
   r=-.8732584155
```

where *x* is the yearly liters per person of alcohol from wine consumption and *y* is the yearly deaths per 100,000 people from heart disease for a sample of countries.

39. For every additional liter per person of alcohol from wine consumption, what is the anticipated change in heart disease deaths per 100,000 people?

(A) 24 fewer
(B) 27 more
(C) 87 fewer
(D) 267 more
(E) 267 fewer

40. What fraction of the difference in heart disease deaths per 100,000 people among various countries can be explained by differences in wine consumption among those countries?

(A) 8.7%
(B) 23.8%
(C) 63.5%
(D) 76.2%
(E) 87.3%

STOP

If there is still time remaining, you may review your answers.

SECTION II

PART A

Questions 1–5

Spend about 65 minutes on this part of the exam.
Percentage of Section II grade—75

> You must show all work and indicate the methods you use. You will be graded on the correctness of your methods and on the accuracy of your results and explanations.

1. A mall manager in Buffalo, New York, must choose between two snow plowing services, Frosty's and Rudolph's. Frosty's charges a yearly $150 fee plus $500 per month for every month with measurable snowfall and covers an unlimited number of plowings. Rudolph's charges $600 per plowing plus a yearly $100 fee. The probability of measurable snowfall in any month is summarized in the following table (assume that the probability of measurable snowfall in any month is independent of what happens in any other month):

Month	Jan = 1	2	3	4	5	6	7	8	9	10	11	12
Probability of measurable snowfall	.9	.9	.6	.2	.05	0	0	0	0	.05	.2	.6

The yearly distribution of number of snowfalls that actually lead to plowings is as follows:

Annual number of plowings	0	1	2	3	4	5
Probability	.05	.10	.15	.35	.25	.10

Which service should the manager choose? Justify your answer.

GO ON TO THE NEXT PAGE ➤

2. To explore the average number of hours per week students at her high school use for study, a high school principal randomly selects 25 students among all those found in the school library after school on one randomly selected day. She interviews the students and gives the data to the AP Statistics teacher, who computes the following:

Sample size:	25	Minimum:	6
Mean:	17.8	First quartile:	9.5
Standard deviation:	8.8	Median:	15
		Third quartile:	24.5
		Maximum:	35

The principal asks the teacher to calculate a confidence interval estimate of the average number of hours students at this high school study per week. Comment on this study. Include a discussion of necessary assumptions to make this calculation and how well they are met. (You are *not* asked to calculate the confidence interval.)

3. An insecticide is tested at two different concentrations and at 30 times of the day. Sixty identical containers of insects are used and the percentages of insects killed are noted. The results are summarized in the following scatterplots:

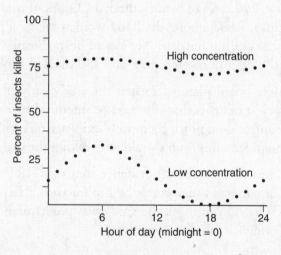

(a) Describe the effect of using the insecticide at low concentration and how it is related to time of day.

(b) Describe the effect of using the insecticide at high concentration and how it is related to time of day.

(c) Which strength and at what time of day is the choice if the object is to kill the highest percentage of insects?

For parts (d) and (e), assume that the high concentration is twice as expensive to produce as the low concentration.

(d) Assuming we wish to minimize cost, which strength and at what time of day is the choice if the object is to kill at least 25% of insects? Explain.

(e) What strength and what time of day give the best percent killed per dollar cost ratio? Explain.

GO ON TO THE NEXT PAGE ➤

4. To test their new bowling ball, a manufacturer randomly selects 10 bowling alleys in a large city, and then randomly selects one bowler at each alley. The bowlers are instructed to bowl one game with their old balls and one with the new balls. The results are given in the following table:

Bowler	1	2	3	4	5	6	7	8	9	10
Old ball	180	225	178	225	224	148	194	192	209	250
New ball	209	215	185	223	223	146	187	201	224	263

(a) Comment on the design of this experiment.

(b) Is the manufacturer justified in claiming that the new ball yields higher bowling scores? Support your answer with appropriate statistical evidence.

5. The Women's Health Initiative (WHI), begun in 1991, recruited a large number of post-menopausal women for a variety of tests to find ways of preventing heart disease, cancer, and osteoporosis (bone disease). Final results were due out in 2005, but one part of the study was halted in 2002. This part involved testing whether a particular hormone-replacement therapy (HRT) resulted in measurable benefits. By 2002, among the 8506 women taking this HRT, there had been 106 cases of hip fracture, 286 cases of heart disease, 290 cases of breast cancer, 112 cases of colon cancer, and 212 cases of stroke, while among the 8102 women taking a placebo, there had been 161 cases of hip fracture, 22 cases of heart disease, 181 cases of breast cancer, 178 cases of colon cancer, and 150 cases of stroke.

A safety board planned to halt the study early if there were evidence of such clear benefits that it would be unethical to withhold this HRT from the control group or if there were evidence of such clear risks that women in the treatment group should stop taking the drugs.

(a) Was there statistical evidence that this HRT is effective in reducing osteoporosis (detectable by hip fractures)? Explain.

(b) Why do you think the study was prematurely halted? Give statistical evidence for your conclusion.

SECTION II

PART B

Question 6

Spend about 25 minutes on this part of the exam.
Percentage of Section II grade—25

6. An accounting firm asks job applicants for salary expectation and requires a math aptitude test, but by law cannot ask about gender, race, age, or marital status. The Human Resources department (HR) is interested in information about gender bias and gathers the following data from 15 randomly chosen past applicants for whom gender is known. The math aptitude score is on a 0 to 4.1 scale.

Male applicants	A	B	C	D	E	F	G	H	I	J
Math aptitude	2.9	4.1	2.4	3.2	4.0	2.5	3.6	3.0	3.2	2.5
Salary expectation (in $1000)	35	45	30	40	45	30	40	35	35	30

Female applicants	K	L	M	N	O
Math aptitude	3.9	3.0	3.8	3.4	4.1
Salary expectation (in $1000)	40	30	40	30	45

The scatterplots (for male and female applicants) of salary expectation versus math aptitude are both roughly linear. The graph of residuals, histogram of residuals, and some computer output for the regression for male and female applicants separately are as follows:

Male applicants:

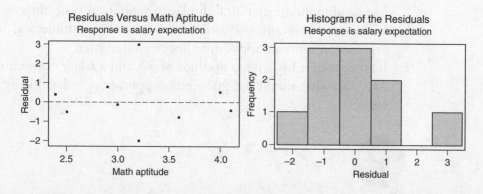

Regression Analysis: Salary Expectation Versus Math Aptitude

Predictor	Coef	SE Coef	T	P
Constant	7.310	2.418	3.02	0.016
Math Apt.	9.2960	0.7574	12.27	0.000

S = 1.381   R–Sq = 95.0%   R–Sq(adj) = 94.3%

Analysis of Variance

Source	DF	SS	MS	F	P
Regression	1	287.25	287.25	150.66	0.000
Residual Error	8	15.25	1.91		
Total	9	302.50			

Female applicants:

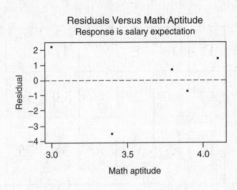

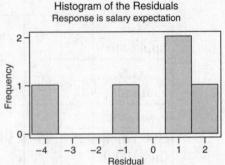

Regression Analysis: Salary Expectation Versus Math Aptitude

Predictor	Coef	SE Coef	T	P
Constant	−15.34	10.87	−1.41	0.253
Math Apt.	14.378	2.968	4.84	0.017

S = 2.608   R–Sq = 88.7%   R–Sq(adj) = 84.9%

Analysis of Variance

Source	DF	SS	MS	F	P
Regression	1	159.60	159.60	23.47	0.017
Residual Error	3	20.40	6.80		
Total	4	180.00			

(a) Compare and comment on the math aptitudes for the two groups. Use an appropriate graphical display in support of your observations.

(b) Is there a significant association between math aptitude and salary expectation for male applicants? Justify your answer.

(c) If an applicant has a math aptitude of 3.5 and a salary expectation of $36,000, what would you guess is the applicant's gender? Justify your answer.

If there is still time remaining, you may review your answers.

Answer Key

Section I

1. **C**	9. **A**	17. **B**	25. **C**	33. **E**
2. **E**	10. **E**	18. **A**	26. **A**	34. **C**
3. **B**	11. **E**	19. **C**	27. **A**	35. **B**
4. **D**	12. **E**	20. **C**	28. **C**	36. **B**
5. **E**	13. **D**	21. **B**	29. **A**	37. **D**
6. **E**	14. **E**	22. **C**	30. **C**	38. **E**
7. **B**	15. **A**	23. **E**	31. **B**	39. **A**
8. **E**	16. **E**	24. **E**	32. **C**	40. **D**

Answers Explained

Section I

1. **(C)** The ranges, 600–300 and 750–450, are both equal to 300. The highest math score and the median verbal score are both 600. While the verbal scores do appear to be roughly symmetric, the math scores appear to be skewed to the left, not the right.

2. **(E)** The significance level is the probability of a Type I error, not of a Type II error.

3. **(B)** To compare z-scores we calculate $\frac{(677-500)}{100} = 1.77$ and $\frac{(29-18)}{6} = 1.83$

4. **(D)** $\frac{1.96(.5)}{\sqrt{n}} \leq 0.03$ gives $n = 1067$.

5. **(E)** Regression lines show association, not causation. Surveys suggest relationships, which controlled experiments can help show to be cause and effect.

6. **(E)** 75% had at least 64.0 ounces, the median is approximately 64.2 ounces, the interquartile range is approximately $64.5 - 64.0 = 0.5$ ounces, and the upper 100 weights range from 64.2 to 65.0 ounces while the lower 100 weights range from 63.0 to 64.2 ounces.

7. **(B)** The markings, spaced 15 apart, clearly look like the standard deviation spacings associated with a normal curve.

8. **(E)** The z-score is defined to be the number of standard deviations from the mean.

9. **(A)**

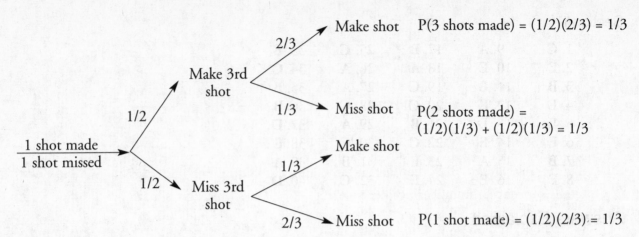

$P(\text{3 shots made}) = (1/2)(2/3) = 1/3$

$P(\text{2 shots made}) = (1/2)(1/3) + (1/2)(1/3) = 1/3$

$P(\text{1 shot made}) = (1/2)(2/3) = 1/3$

10. **(E)** The mean, standard deviation, variance, and range are all influenced by extreme values.

11. **(E)** Outliers are significantly different from the other observations. Sometimes, a rough rule of thumb is used that scores more than 1.5 interquartile ranges below Q_1 or above Q_3 are considered to be outliers.

12. **(E)** In none of the above are the trials independent. For example, as each consecutive person is stopped at a roadblock, the probability the next person has a seat belt on will quickly increase; if a student has one A, the probability is increased that he/she has another A.

13. **(D)** $3(0.52) = 1.56$

14. **(E)** The meaning of a 95% confidence level is that if the same procedure is followed many times, 95% of the resulting intervals will contain the true population parameter.

15. **(A)** Students with high absentee rates have less chance to be picked using method I, thus resulting in undercoverage bias. Systematic sampling (method II), stratified sampling (method III), and multistage sampling (method IV) are useful time- and cost-saving procedures, but they do not result in simple random sample. A simple random sample is one in which every possible sample of the desired size has an equal chance of being selected.

16. **(E)** Randomization refers to random allocation of subjects to treatment groups. Replication refers to use of a sufficient number of subjects. Randomization and blocking help control for lurking variables; control groups are used to compare the responses of subjects who receive a treatment to subjects who don't receive the treatment.

17. **(B)** In general there is one probability of a Type I error that is at the significance level, while there is a different probability of a Type II error associated with each possible correct answer.

18. **(A)** It is reasonable to estimate that the true proportion is within a couple of standard deviations of the sample proportion.

19. **(C)** Only III and IV are reasonable conclusions. We cannot say there is a 0.90 probability that $3 < \mu_A - \mu_B < 11$, because for given sample means, the difference $\mu_A - \mu_B$ either is or isn't within the specified interval, and so the probability is either 1 or 0.

20. **(C)** 10($0.28) = $2.80

21. **(B)** The probability that the average amount paid during the ten days exceeds 30 cents is found using the z-score $\frac{30-28}{\frac{7}{\sqrt{10}}} = 0.90$, which gives a tail probability of .18.

22. **(C)** The probability a student is male is $\frac{16,253}{38,037} = .427$. We thus calculate $100(.427)(.28) + 100(.573)(.44) \approx 37$.

23. **(E)** If one group does better on the standardized test, there is no way of knowing if this is due to the different textbook or to some confounding variable such as different caliber of student, different class sizes, or differently motivated teachers found at the two institutions.

24. **(E)** In the presence of strong skewness or outliers, the t-procedures are contraindicated. The t-procedures assume normality of the original population. When the population standard deviation σ is unknown and the standard error $\frac{s}{\sqrt{n}}$ is substituted for $\frac{\sigma}{\sqrt{n}}$, then t-scores rather than z-scores are the proper choice.

25. **(C)** For skewed distributions the mean is found farther out in the tail than the median.

26. **(A)** Using Table A we see that the relevant z-scores are 1.28 and -0.84. Solving the system of equations $\{\mu + 1.28\sigma = 238, \ \mu - 0.84\sigma = 187\}$ results in $\mu = 207.2$.

27. **(A)** $(\frac{1}{3})(.29)(.23) = .0222$

28. **(C)** Two-sample tests require that the two samples being compared be independent of each other. However, in this case the data occur in related pairs, namely the SAT math and SAT verbal scores of each individual student. The proper procedure is to run a one-sample test on the single variable consisting of the difference between math and verbal SAT scores for each student.

29. **(A)** Raising all prices by 5 cents will raise the mean by 5 cents but will leave the standard deviation unchanged. Lowering all prices by 5% will lower both the mean and the standard deviation by 5%.

30. **(C)** $5(\frac{15,000}{50,000}) = 1.5$

31. **(B)** With plan 1 the expected number of vitamin supplement purchasers is only 15. Plan 2 allows an estimate to be made using a full 50 purchasers.

32. **(C)** This was an observational study, not a controlled experiment. Blinding makes no sense here. It could well be that women with more advanced cancers were more likely to undergo complete mastectomies, and this lurking variable makes the conclusion very suspect.

33. **(E)** With df = 20 − 1 = 19 and tail probabilities of .005, the critical *t*-values are ±2.861.

34. **(C)** Conclusion I comes directly from the definition of a *P*-value. With *P* this small, there is sufficient evidence to reject the null hypothesis.

35. **(B)** The residual is the difference between the actual and the predicted values. In this case the actual value is 1, while the predicted value is 0.61(3) + 0.56 = 2.39, and so the residual is −1.39.

36. **(B)** The null hypothesis is "proper functioning of the equipment." A Type II error is a mistaken failure to reject a false null hypothesis.

37. **(D)** The male scores are skewed toward the low end, and in such distributions the mean is found farther out in the tail than the median.

38. **(E)** Using 0.84 from Table A or 0.8416 from invNorm(.8) on a TI-83, we calculate 110 + 0.84(25) = 131.

39. **(A)** For each unit change in the *x*-direction, the predicted change in the *y*-direction is given by the slope. In this output the slope is $a = -23.8$.

40. **(D)** The coefficient of determination, r^2, gives the proportion of variation in *y* that is explained by the variation in *x*. In this case $r^2 = (-.873)^2 = .762$.

SECTION II

1. For Frosty's:

   The expected cost for January is ($500)(.9) + ($0)(.1), . . . , and for December is ($500)(.6) + ($0)(.4). Thus,

   Expected annual cost = $150 + $500(.9 + .9 + .6 + .2 + .05 + .05 + .2 + .6) = $1900

   For Rudolph's:

   Expected number of plowings per year = 0(.05) + 1(.1) + 2(.15) + 3(.35) + 4(.25) + 5(.1) = 2.95

   Expected annual cost = $100 + 2.95($600) = $1870

Choice: Choose Rudolph's because it has a lower expected or average yearly cost. Frosty's will average $1900 − $1870 = $30 more per year than Rudolph's.

Scoring

There are four components to the answer:

1. Correct calculation of expected annual cost for Frosty's service.

2. Correct calculation of expected number of plowings per year.

3. Correct calculation of expected annual cost for Rudolph's service.

4. Choice of Rudolph's together with an explanation based on expected annual cost calculations, noting that calculations and comparison are for expected or average annual costs.

Note that components 2 and 3 can be combined into a single calculation:
$100 + $600[0(.05) + 1(.1) + 2(.15) + 3(.35) + 4(.25) + 5(.1)] = $1870

4 Complete Answer   All four components correct.

3 Substantial Answer   Three components correct.

2 Developing Answer   Two components correct.

1 Minimal Answer   One component correct.

---

2. There is a real problem with possible response bias. Students may not answer the school principal honestly about how much they study.

Necessary assumption: random sample from population of all students at that high school. This is not reasonable, as the sample was taken from those using the library after school. These students probably study more than the typical student.

Necessary assumption: distribution of number of hours studied per week for all students is approximately normal. This is not reasonable, as a boxplot of the sample seems to indicate a skewed distribution (although it is reasonable to argue that it is not too skewed):

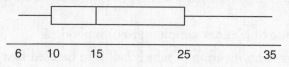

Also, the minimum value is only $\frac{17.8-6}{8.8} = 1.34$ standard deviations below

the mean and the maximum is only $\frac{35-17.8}{8.8} = 1.95$ standard deviations above

the mean, both of which are less than would be expected with a normal distribution.

	Scoring	

There are four components to a complete answer: stating the possible source of bias, stating both assumptions, explaining why the random sample assumption is not met, and giving a reasonable argument for whether or not the normality assumption is met.

4 Complete Answer   All four components correct.

3 Substantial Answer   Three components correct.

2 Developing Answer   Two components correct.

1 Minimal Answer   One component correct.

3. (a) At low concentration the effectiveness varies cyclically from a high between 30% and 35% around 6 A.M. to a low of almost 0% about 6 P.M.
   (b) At high concentration the effectiveness also varies cyclically but with very little variance, reaching a high of almost 80% around 6 A.M., and a low of approximately 70% at around 6 P.M.
   (c) Apply high concentration at 6 A.M.
   (d) Apply low concentration at 6 A.M. (No need to use the twice as expensive high concentration to achieve a 25% kill ratio.)
   (e) High concentration at 6 A.M. (Twice the cost as the low concentration but more than twice the effectiveness.)

	Scoring	

4 Complete Answer   All five components correct.

3 Substantial Answer   Four components correct.

2 Developing Answer   Three components correct.

1 Minimal Answer   Two components correct.

4. (a) The use of random samples (given) is good.

There is no mention of which ball is to be used first by each bowler. This could make a difference. For example, after bowling a first game, some bowlers might fatigue, and so their second game scores would naturally be lower. For some bowlers, it might take one game to warm up, so their second game scores would naturally be higher. Some random method (such as a coin toss) should be employed for each bowler to decide which ball is to be used for the first game.

There is no mention of blinding. Bias could result if bowlers expect to do better with their old comfortable balls, or if they expect to do better with something newly designed. If possible, the balls should be doctored to look alike.

(b) There are several components to a complete answer, including correct statement of hypotheses, identifying the correct test to be used, checking of assumptions, correct mechanics in determining the *P*-value or rejection region, and stating a correct conclusion in context of the problem.

The proper hypothesis test is a matched pairs *t*-test on the set of differences, {−29, 10, −7, 2, 1, 2, 7, −9, −15, −13}, with H_0: $\mu_D = 0$, H_a: $\mu_D < 0$. (Or use "new" minus "old," changing the signs of the elements in the set and switching to H_a: $\mu_D > 0$.)

Assumptions to be checked: (1) random samples (given) and (2) the population of differences is approximately normal (use stemplot, boxplot, or histogram of differences).

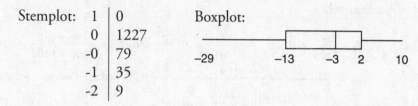

Stemplot:

```
 1 | 0
 0 | 1227
-0 | 79
-1 | 35
-2 | 9
```

Boxplot:

−29   −13   −3   2   10

The plots indicate a very roughly symmetric, bell-shaped distribution with no outliers (by IQR test, no outliers) and no extreme skewness, so the set of differences appears approximately normal.

$$t\text{-test:} \quad \overline{x}_D = -5.1 \quad s_D = 11.85$$
$$t = \frac{\overline{x}_D - 0}{s_D/\sqrt{n_D}} = \frac{-5.1 - 0}{11.85/\sqrt{10}} = \frac{-5.1}{3.75} = -1.36$$
$$df = n_D - 1 = 10 - 1 = 9$$

The *P*-value = .103 so the differences are not significant at the 10% significance level. Thus, at the 10% significance level (or because *P* is relatively large), the manufacturer is not justified in claiming that the new ball yields higher bowling scores.

Scoring	
4 **Complete Answer**	All components correct.
3 **Substantial Answer**	Incomplete part (a), but complete and correct part (b), OR complete part (a) and substantially, but not fully, complete part (b).
2 **Developing Answer**	Incomplete part (a) and one component of part (b) only partially correct, OR complete part (a) and one component of part (b) missing or incorrect.
1 **Minimal Answer**	Incomplete part (a) and one component of part (b) missing or incorrect, OR complete part (a) and two components of part (b) missing or incorrect.

5. (a) A two-proportion z-test is indicated. We must check that n is large enough: $n_1p_1 = 106 > 10$, $n_1(1-p_1) = 8400 > 10$, $n_2p_2 = 161 > 10$, $n_2(1-p_2) = 7941 > 10$. We must assume simple random samples from the target population, since this is not given.

$H_0: p_1 - p_2 = 0$ (where p_1 is the proportion of all women taking HRT who have hip fractures and p_2 is the proportion of all women not taking HRT who have hip fractures.)

$H_a: p_1 - p_2 < 0$ (the proportion of women on HRT who have hip fractures is lower than the proportion of women not on HRT who have hip fractures.)

The TI-83 gives $z = -3.795$ and $P = .000$.

Alternatively,

$$\hat{p}_1 = \frac{106}{8506} = .0125, \quad \hat{p}_2 = \frac{161}{8102} = .0199, \quad \text{and} \quad \hat{p} = \frac{106+161}{8506+8102} = .0161$$

$$z = \frac{.0125 - .0199}{\sqrt{(.0161)(.9839)\left(\frac{1}{8506} + \frac{1}{8102}\right)}} = -3.79, \text{ and again we get } P = .000.$$

With such a small P-value there is very strong evidence to reject H_0 and conclude that the proportion of women on HRT who have hip fractures is lower than the proportion of women not on HRT who have hip fractures.

(b) While women on HRT did have lower rates of osteoporosis and colon cancer, unfortunately, they also had statistically significantly higher rates of heart disease and breast cancer, and because of these risks, the study was halted early. The rate of stroke was higher, but not statistically significantly higher, with women taking HRT.

Heart disease: $n_1p_1 = 286$, $n_1(1-p_1) = 8220$, $n_2p_2 = 222$, and $n_2(1-p_2) = 7880$ are all >10. $H_0: p_1 - p_2 = 0$, $H_a: p_1 - p_2 > 0$, $t = 2.328$, $P = .010$

Breast cancer: $n_1p_1 = 290$, $n_1(1-p_1) = 8216$, $n_2p_2 = 181$, and $n_2(1-p_2) = 7921$ are all >10. $H_0: p_1 - p_2 = 0$, $H_a: p_1 - p_2 > 0$, $t = 4.561$, $P = .000$

Stroke: $n_1p_1 = 212$, $n_1(1-p_1) = 8294$, $n_2p_2 = 150$, and $n_2(1-p_2) = 7952$ are all >10. $H_0: p_1 - p_2 = 0$, $H_a: p_1 - p_2 > 0$, $t = 1.095$, $P = .137$

Scoring

(a) One component is naming the right test and checking the size assumptions $(np > 10$, etc.). One component is calculating P and stating a correct conclusion in context.

(b) One component is noting that heart disease OR breast cancer rates are statistically significantly higher for women on HRT. One component is showing the work for the hypothesis test with reference to either heart disease or breast cancer.

The student receives one point for each completely correct component and one-half point for each partially correct component. If the total score is between two values, grade holistically.

4 Complete Answer   4 points

3 Substantial Answer   3 points

2 Developing Answer   2 points

1 Minimal Answer   1 point

6. (a) The comparison can be made using dotplots, back-to-back stemplots, or parallel boxplots:

Dotplot:

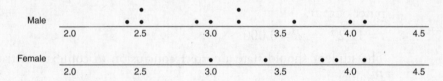

Back-to-back stemplot:

Male		Female
9554	2.	
6220	3.	0489
10	4.	1

Parallel boxplots:

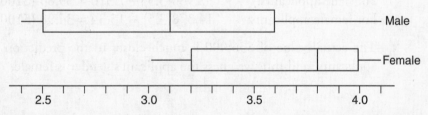

Whichever graphical display is presented, the student must still comment to the effect that although the maximum math aptitude is the same for both genders, the math aptitude for females tend to be higher than the math aptitude for males (or the minimum, lower quartile, median, and upper quartile are all higher for females than those for males.)

(b) Note that the correlation $r = \sqrt{.950} = .975$ is very close to +1, indicating a very strong positive, linear relationship.

To check the significance:

The student should state proper hypotheses:

H_0: There is no relationship between math aptitude and salary expectancy for male applicants.

H_a: There is a relationship between math aptitude and salary expectancy for male applicants.

OR

H_0: $\beta = 0$

H_a: $\beta \neq 0$

The student should identify the proper test, "linear regression t-test" or "t-test for slope," and check the necessary assumptions for inference:

There is a random sample (given).

The scatterplot of math aptitude and salary expectation for male applicants is approximately linear (given).

The graph of residuals has consistent spread and shows no apparent pattern.

The histogram of residuals is roughly normal.

The student should note the t-statistic and P-value as provided by the computer output:

$t = 12.27$ $P = 0.000$

The student should state a correct conclusion in context, referring to the P-value:

Since the P-value is 0.000, there is sufficient evidence to reject the null hypothesis, and conclude that for male applicants there is a significant relationship between math aptitude and salary expectancy.

(c) The student should combine both math aptitude and salary expectation information and compare to both male and female applicants.

Using the two regression lines, the predicted salary expectation of an applicant with a math aptitude of 3.5 follows:

For male applicants: $9.2960(3.5) + 7.310 = 39.85$ ($1000)
For female applicants: $14.378(3.5) - 15.34 = 34.98$ ($1000)

The actual value of $36,000 is much closer to the prediction for female applicants, and thus we guess the applicant's gender is female.

Scoring

Part (a) is correct if there is both a correct graphical display and correct commentary based on the graph. A correct graph with no commentary or correct comments with no graph give a partially correct answer.

A complete response to part (b) must include a calculation and comment on the correlation, proper statement of hypotheses, identification of the proper test, checking of correct assumptions, observation of the correct *t*-statistic and *P*-value from the computer output, and giving the correct conclusion in context with reference to *P*.

A complete response to part (c) must look at both math aptitude and salary expectations and be checked against both male and female applicants. Checking against only one group (male or female) gives a partially correct answer.

4 Complete Answer   Everything correct except at most one partially correct answer in parts (a), (c), or a section of (b).

3 Substantial Answer   Everything correct except one incorrect answer or two partially correct answers from among parts (a), (c), and sections of (b).

2 Developing Answer   Everything correct except one incorrect answer and one or two partially correct answers or three or four partially correct answers from among parts (a), (c), and sections of (b).

1 Minimal Answer   Everything correct except two incorrect answers or one incorrect and three or four partially correct answers from among parts (a), (c), and sections of (b).

Appendix

ASSUMPTIONS

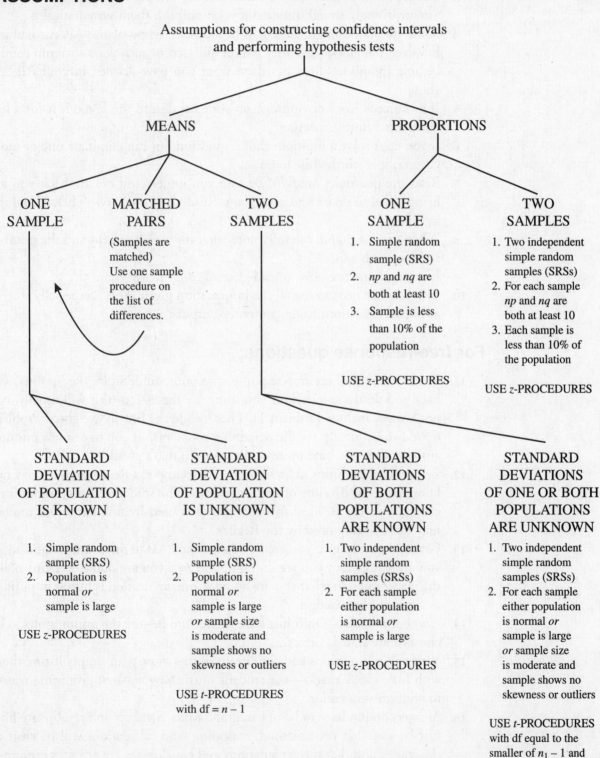

Assumptions for constructing confidence intervals
and performing hypothesis tests

MEANS

PROPORTIONS

ONE SAMPLE

MATCHED PAIRS

(Samples are matched)
Use one sample procedure on the list of differences.

TWO SAMPLES

ONE SAMPLE

1. Simple random sample (SRS)
2. np and nq are both at least 10
3. Sample is less than 10% of the population

USE z-PROCEDURES

TWO SAMPLES

1. Two independent simple random samples (SRSs)
2. For each sample np and nq are both at least 10
3. Each sample is less than 10% of the population

USE z-PROCEDURES

STANDARD DEVIATION OF POPULATION IS KNOWN

1. Simple random sample (SRS)
2. Population is normal *or* sample is large

USE z-PROCEDURES

STANDARD DEVIATION OF POPULATION IS UNKNOWN

1. Simple random sample (SRS)
2. Population is normal *or* sample is large *or* sample size is moderate and sample shows no skewness or outliers

USE t-PROCEDURES
with df = $n - 1$

STANDARD DEVIATIONS OF BOTH POPULATIONS ARE KNOWN

1. Two independent simple random samples (SRSs)
2. For each sample either population is normal *or* sample is large

USE z-PROCEDURES

STANDARD DEVIATIONS OF ONE OR BOTH POPULATIONS ARE UNKNOWN

1. Two independent simple random samples (SRSs)
2. For each sample either population is normal *or* sample is large *or* sample size is moderate and sample shows no skewness or outliers

USE t-PROCEDURES
with df equal to the smaller of $n_1 - 1$ and $n_2 - 1$ or df calculated by calculator

AP EXAM HINTS

Basics:

1. Insert new calculator batteries or carry replacement batteries with you.
2. Underline key words and phrases while reading questions.
3. Some problems look scary on first reading but are not overly difficult and are surprisingly straightforward if you approach them systematically.
4. Some questions might take you beyond the scope of the AP curriculum; however, remember that they will be phrased in such ways that you should be able to answer them based on what you have learned in your AP Stat class.
5. If a choice is "too" obvious, sit on your hands and think about it for a few moments before answering.
6. If you can't solve a multiple-choice question but can eliminate one or more choices, it is worthwhile to guess.
7. Read the questions *carefully!* Be sure you understand exactly what you are being asked to do or find or explain. Read all of the answers before making a choice.
8. When given a graph, carefully note what the axes represent and the number scales of each axis.
9. Learn and practice how to read generic computer output.
10. You may not need to use all the information given. This is especially true in answering questions using computer outputs.

For free-response questions:

11. Read through all six free-response questions, underlining key points. Go back and do those questions you think are the easiest (this will usually, but not always, include Problem 1). Then tackle the heavily weighted Problem 6 for awhile. Finally, try the remaining problems. If you have a few minutes at the end to try some more on Problem 6, that's great!
12. Be careful about using abbreviations. For example, a Reader may or may not know what LOBF (line of best fit) means. The student must communicate clearly with the Reader, and abbreviations used by one particular teacher may not be recognized by the Reader.
13. Graders want to give you credit—help them! Make them understand *what* you are doing, *why* you are doing it, and *how* you are doing it. Don't make the reader guess at what you are doing. Communication is just as important as statistical knowledge!
14. *Check assumptions.* Don't just state them! And be sure the assumptions to be checked are stated correctly.
15. Verifying assumptions and conditions means more than simply listing them with little check marks—the student must show work or give some reason to confirm verification.
16. Answers do not have to be in paragraph form. Symbols and algebra are fine. Just be sure that your method, reasoning, and calculations will be clear to the reader, and that the explanations and conclusions are given *in context of the problem.*

17. Clearly identify your variables. For example, in confidence interval and hypothesis test problems involving the difference between two means or two proportions, be sure it is clear which of the two groups goes with which of the variables.

18. If you show calculations carefully, a wrong answer due to a computational error might still result in full credit.

19. When using a formula, write down the formula and then substitute the values.

20. Conclusions should be stated in proper English. For example, don't use multiple negatives.

21. Know that the language of experiments is different from the language of observational studies; for example, blocking versus stratification.

22. If you are asked to design an experiment, understand the importance of, and how to incorporate, control, randomization, and replication.

23. Read carefully and recognize that sometimes very different tests are required in different parts of the same problem.

24. Realize that there may be several reasonable approaches to a given problem. In such cases pick either the one you feel most comfortable with or the one you feel will require the least time.

25. Realize that there may not be one clear correct answer. Some questions are designed to give you an opportunity to creatively synthesize a relationship among the problem's statistical components.

26. When using calculators avoid "calculator-speak." For example, do not simply write "2-SampTTest . . ." or "binomcdf . . ." Again, you must explain *what* you are doing, *why* you are doing it, and *how* you are doing it.

27. Punching a long list of data into the calculator doesn't show statistical knowledge. Be sure it's necessary!

28. Don't give more than one solution to the same problem. You will receive credit only for the weaker one.

29. If you can't solve part of a problem, but that solution is necessary to proceed, make up a reasonable answer to the first part and use this to proceed to the remaining parts of the problem.

For linear regression and correlation problems:

30. Scatterplots and residual plots of data are not the same. Understand the difference!

31. When describing residual graphs, "randomly scattered" does *not* mean the same as "half below and half above." You should comment on whether or not there are nonlinear patterns, increasing or decreasing spread, and uneven variation about the line, and on whether the residuals are small or large compared to the associated *y*-values.

32. When one variable is "year," be careful if you are using, for example, 1998 or 98 or 8 or are dating from some other point.

33. Note any transformation of the original data (e.g., by logarithms.)

34. Define explanatory and response variables.

35. If the slope is close to ±1, it does *not* follow that the correlation is strong.

36. When making predictions and interpreting slopes and intercepts, don't become confused between your original data and your regression model.

37. Simply using a calculator to find a regression line is not enough; you must understand it (for example, be able to interpret the slope and intercepts in context).

For free-response hypothesis test problems:

38. The hypotheses should be stated in terms of the problem, *with all variables clearly defined.* For example:

    Because the researcher wishes to determine whether student scores improve over the previous average of 78, the test should be of the null hypothesis, H_0: $\mu = 78$, against the alternative hypothesis, H_a: $\mu > 78$, where μ is the true mean student score.

39. Specify the test used, why it was chosen, and how the test assumptions are met. For example:

    The standard deviation of the population is unknown and the sample shows no skewness or outliers, and so the one-sample *t*-test is appropriate. We check for and confirm that the data are said to come from a *simple random sample.*

40. Calculate the test statistic and associated *P*-value. For example:

    The sample mean is $\bar{x} = 83$ with a *t*-score of $\frac{83-78}{\frac{21.3}{\sqrt{41}}} = 1.50$. With

    df = 41 − 1 = 40, we note that the *P*-value is between .05 and .10. [Or using a TI-84, the one-sample *t*-test with μ: 78, $\bar{x}$: 83, S_x: 21.3, *n*: 41, and μ: $>\mu_0$ gives *t* = 1.50 and *P* = .070.]

41. With regard to *P*-values, you should indicate how you interpret the *P*-value, that is, you need *linkage.* So, "Given that *P* = .007, I reject ..." isn't enough. You need something like "Because *P* = .007 is so small, there is sufficient evidence to reject ..." Of course, in addition to linkage and a *decision* (to reject or fail to reject), one needs a conclusion *in the context* of the problem.

42. Write a conclusion in the context of the problem, linking the test statistic and *P*-value to the conclusion. For example:

    With .05 < *P* < .10, there is *some* evidence that the students' mean test score is now greater than 78. The sample mean test score of 83 *is* significant at the 10% significance level but *not* at the 5% level.

43. You need to know, and show that you know, the meaning of the *P*-value. You might consider concluding with a statement such as "Under the assumption of (H_0 *in words*), the chance of observing our result or a more extreme result is (*P-value*). Therefore, ..."

44. *Show* calculator graphs where appropriate! For example, draw graphs showing and labeling rejection regions. Be sure to label axes and show number scales on all graphs.

45. If you refer to a graph, whether it is a histogram, boxplot, stemplot, scatterplot, residuals plot, normal probability plot, or some other kind of graph, you should roughly draw it. It is not enough to simply say, "I did a normal probability plot of the residuals on my calculator and it looked linear."

Further suggestions:

46. A simple random sample (SRS) and a random assignment of treatments to subjects both have to do with randomness, but they are not the same. Understand the difference!

47. When checking conditions for a chi-square test, check "expected" cells, not "observed" cells.

48. Replication on many subjects to reduce chance variation on results is different from replication of the experiment itself to achieve validation. Understand the difference!

49. Blinding and placebos in experiments are important but are not always feasible. You can still have "experiments" without these.

50. The histogram for a particular sample is not the same as the sampling distribution of a statistic. Understand the difference!

51. Remember that uniform and symmetric are not the same, and that not all symmetric unimodal distributions are normal.

52. When interpreting histograms, while center, spread, and shape (in context!) are of course important, be sure that you spend your time answering (in context!) the question that is being asked.

53. Simply saying to "randomly assign" subjects to treatment groups is usually an incomplete response. You should carefully explain how to make the assignments, for example, using a random number table or through generating random numbers on a calculator.

54. Use proper terminology! Stratification is a technique found in sampling design; blocking is a technique in experimental design. Replication in experiments does not mean repeating an experiment; rather, it refers to use of a sufficient number of subjects.

55. Whether $\bar{x} > \mu$ or $\bar{x} < \mu$ has nothing to do with the size of the sample or with the skewness of the population.

56. Just because a sample is large does not imply that the distribution of the sample will be close to a normal distribution.

57. For any inference problem, be sure to ask yourself whether the variable is categorical (usually leading to proportions) or quantitative (usually leading to means), whether there is a single population of interest or two populations are being compared, and in the case of comparison whether there are independent samples or a paired comparison.

58. Understand the relationship between confidence intervals and two-sided significance tests. For example, whether a particular value is located in a 90% confidence interval is related to a two-sided hypothesis test around that value at the .10 significance level.

59. An observed value will be statistically significant at some level and not at another level, and thus it is usually more informative to consider *P*-values. And don't forget the usefulness of confidence intervals used in conjunction with hypothesis tests.

AP SCORING GUIDE

The Multiple-Choice and Free-Response sections are weighted equally.

As a penalty for guessing in the Multiple-Choice section, the number of correct responses is reduced by one-quarter the number of wrong responses.

The Investigative Task counts for 25% of the Free-Response section. Each of the Free-Response questions has a possible four points.

To find your score use the following guide:

<u>Multiple-Choice section</u> (40 questions)

(Number correct minus $\frac{1}{4}$ number incorrect) $\times 1.25$ = _____

<u>Free-Response section</u> (5 Open-Ended questions plus an Investigative Task)

Question 1 _____ $\times 1.875$ = _____
_{out of 4}

Question 2 _____ $\times 1.875$ = _____
_{out of 4}

Question 3 _____ $\times 1.875$ = _____
_{out of 4}

Question 4 _____ $\times 1.875$ = _____
_{out of 4}

Question 5 _____ $\times 1.875$ = _____
_{out of 4}

Question 6 _____ $\times 3.125$ = _____
_{out of 4}

Total points from Multiple-Choice and Free-Response sections = _____

Conversion chart based on a recent AP exam (subject to future modification)

Total points	AP Score
68–100	5
53–67	4
40–52	3
29–39	2
0–28	1

In the past, roughly 10% of students scored 5, 20% scored 4, 25% scored 3, 20% scored 2, and 25% scored 1 on the AP Statistics exam. Colleges generally require a score of at least 3 for a student to receive college credit.

BASIC USES OF THE TI-83/TI-84

There are many more useful features than introduced below – see the guidebook that comes with the calculator. For reference, following is a listing of the basic uses with which all students should be familiar. However, always remember that the calculator is only a tool, that it will find minimal use in the multiple-choice section, and that "calculator talk" (calculator syntax) should NOT be used in the free-response section.

Plotting statistical data:

> **STAT PLOT** allows one to show scatterplots, histograms, modified boxplots, and regular boxplots of data stored in lists. Note the use of **TRACE** with the various plots.

Numerical statistical data:

> **1-Var Stats** gives the mean, standard deviation, and 5-number summary of a list of data.

Binomial probabilities:

> **binompdf** (n, p, x) gives the probability of exactly x successes in n trials where p is the probability of success on a single trial.
> **binomcdf** (n, p, x) gives the cumulative probability of x or fewer successes in n trials where p is the probability of success on a single trial.

Geometric probabilities:

> **geometpdf** (p, x) gives the probability that the first success occurs on the x-th trial, where p is the probability of success on a single trial.
> **geometcdf** (p, x) gives the cumulative probability that the first success occurs on or before the x-th trial, where p the probability of success on a single trial.

The normal distribution:

> **normalcdf** (lowerbound, upperbound, μ, σ) gives the probability that a score is between the two bounds for the designated mean μ and standard deviation σ. The defaults are $\mu = 0$ and $\sigma = 1$.
> **InvNorm** (area, μ, σ) gives the score associated with an area (probability) to the left of the score for the designated mean μ and standard deviation σ. The defaults are $\mu = 0$ and $\sigma = 1$.

The t-distribution:

> **tcdf** (lowerbound, upperbound, df) gives the probability a score is between the two bounds for the specified df (degrees of freedom).
> **invT**(area, df) gives the t-score associated with an area (probability) to the left of the score under the student t-probability function for the specified df (degrees of freedom). [Note: this is available on the new operating system for the TI-84+.]

The chi-square distribution:

> χ^2**cdf** (lowerbound, upperbound, df) gives the probability a score is between the two bounds for the specified df (degrees of freedom).

χ^2**GOF-Test** is a chi-square goodness-of-fit test to confirm whether sample data conforms to a specified distribution. [Note: this is available on the new operating system for the TI-84+.]

Linear regression and correlation:

LinReg (ax + b) fits the equation $y = ax + b$ to the data in lists L1 and L2 using a least-squares fit. When **DiagnosticOn** is set, the values for r^2 and r are also displayed.

Confidence intervals:

For proportions—

1-PropZInt gives a confidence interval for a proportion of successes. **2-PropZInt** gives a confidence interval for the difference between the proportion of successes in two populations.

For means—

TInterval gives a confidence interval for a population mean (use the t-distribution because population variances are never really known). **2-SampTInt** gives a confidence interval for the difference between two population means.

Hypothesis tests:

For proportions—

1-PropZTest
2-PropZTest compares the proportion of successes from two populations (making use of the pooled sample proportion).

For means—

T-Test
2-SampTTest

For chi-square test for association—

χ^2**-Test** gives the χ^2-value and P-value for the null hypothesis H_0: no association between row and column variables, and the alternative hypothesis H_a: the variables are related. The observed counts must first be entered into a matrix.

For linear regression—

LinRegTTest calculates a linear regression and performs a t-test on the null hypothesis H_0: $\beta = 0$ (H_0: $\rho = 0$). The regression equation is stored in **RegEQ** (under **VARS Statistics EQ**) and the list of residuals is stored in **RESID** (under **LIST NAMES**).

FORMULAS

The following are in the form and notation as recommended by the College Board.

Descriptive Statistics

$$\overline{x} = \frac{\sum x_i}{n}$$

$$s = \sqrt{\frac{1}{n-1}\sum\left(x_i - \overline{x}\right)^2}$$

$$s_p = \sqrt{\frac{\left(n_1 - 1\right)s_1^2 + \left(n_2 - 1\right)s_2^2}{\left(n_1 - 1\right) + \left(n_2 - 1\right)}}$$

$$\hat{y} = b_0 + b_1 x$$

$$b_1 = \frac{\sum\left(x_i - \overline{x}\right)\left(y_i - \overline{y}\right)}{\sum\left(x_i - \overline{x}\right)^2}$$

$$b_0 = \overline{y} - b_1\overline{x}$$

$$r = \frac{1}{n-1}\sum\left(\frac{x_i - \overline{x}}{s_x}\right)\left(\frac{y_i - \overline{y}}{s_y}\right)$$

$$b_1 = r\frac{s_y}{s_x}$$

$$s_{b1} = \frac{\sqrt{\dfrac{\sum\left(y_i - \hat{y}_i\right)^2}{n-2}}}{\sqrt{\sum\left(x_i - \overline{x}\right)^2}}$$

Probability

$$P(A \cup B) = P(A) + P(B) - P(A \cap B)$$

$$P\left(A \mid B\right) = \frac{P\left(A \cap B\right)}{P\left(B\right)}$$

$$E\left(X\right) = \mu_x = \sum x_i p_i$$

$$\mathrm{var}\left(X\right) = \sigma_x^2 = \sum \left(x_i - \mu_x\right)^2 p_i$$

If X has a binomial distribution with parameters n and p, then:

$$P\left(X = k\right) = \binom{n}{k} p^k \left(1 - p\right)^{(n-k)}$$

$$\mu_x = np$$

$$\sigma_x = \sqrt{np\left(1 - p\right)}$$

$$\mu_{\hat{p}} = p$$

$$\sigma_{\hat{p}} = \sqrt{\frac{p\left(1 - p\right)}{n}}$$

If X has a normal distribution with mean μ and standard deviation σ, then:

$$\mu_{\bar{x}} = \mu$$

$$\sigma_{\bar{x}} = \frac{\sigma}{\sqrt{n}}$$

Inferential Statistics

Standardized statistic: $\dfrac{\text{Estimate} - \text{parameter}}{\text{Standard deviation of the estimate}}$

Confidence interval:
   Estimate ± (critical value) · (standard deviation of the estimate)

SINGLE SAMPLE

Statistic	Standard Deviation
Mean	$\dfrac{\sigma}{\sqrt{n}}$
Proportion	$\sqrt{\dfrac{p(1-p)}{n}}$

TWO SAMPLE

Statistic	Standard Deviation
Difference of means (unequal variances)	$\sqrt{\dfrac{\sigma_1^2}{n_1} + \dfrac{\sigma_2^2}{n_2}}$
Difference of means (equal variances)	$\sigma\sqrt{\dfrac{1}{n_1} + \dfrac{1}{n_2}}$
Difference of proportions (unequal variances)	$\sqrt{\dfrac{p_1(1-p_1)}{n_1} + \dfrac{p_2(1-p_2)}{n_2}}$
Difference of proportions (equal variances)	$\sqrt{p(1-p)}\sqrt{\dfrac{1}{n_1} + \dfrac{1}{n_2}}$

Chi-square statistic $= \sum \dfrac{\left(\text{observed} - \text{expected}\right)^2}{\text{expected}}$

GRAPHICAL DISPLAYS

	Right-skewed	Symmetric	Left-skewed

Dotplots

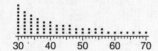
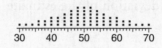
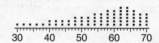

Stemplots

3	00000000022222222444444666668888
4	000022244466688
5	002244668
6	02468
7	0

3	0246688
4	000222444466666888888
5	0000002222222444446666888
6	000224468
7	0

3	02468
4	00224466888
5	0002224446666688888
6	0000022222224444444666668888
7	0000

Histograms

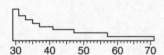
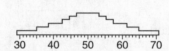
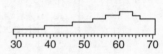

Cumulative
Frequency
Plots

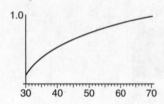
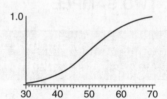
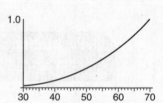

Boxplots

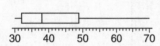
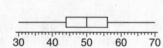
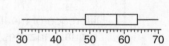

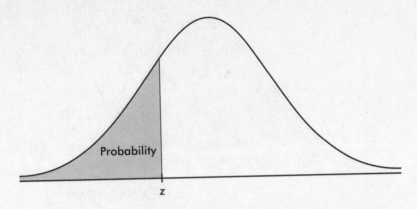

TABLE A: STANDARD NORMAL PROBABILITIES

z	.00	.01	.02	.03	.04	.05	.06	.07	.08	.09
−3.4	.0003	.0003	.0003	.0003	.0003	.0003	.0003	.0003	.0003	.0002
−3.3	.0005	.0005	.0005	.0004	.0004	.0004	.0004	.0004	.0004	.0003
−3.2	.0007	.0007	.0006	.0006	.0006	.0006	.0006	.0005	.0005	.0005
−3.1	.0010	.0009	.0009	.0009	.0008	.0008	.0008	.0008	.0007	.0007
−3.0	.0013	.0013	.0013	.0012	.0012	.0011	.0011	.0011	.0010	.0010
−2.9	.0019	.0018	.0018	.0017	.0016	.0016	.0015	.0015	.0014	.0014
−2.8	.0026	.0025	.0024	.0023	.0023	.0022	.0021	.0021	.0020	.0019
−2.7	.0035	.0034	.0033	.0032	.0031	.0030	.0029	.0028	.0027	.0026
−2.6	.0047	.0045	.0044	.0043	.0041	.0040	.0039	.0038	.0037	.0036
−2.5	.0062	.0060	.0059	.0057	.0055	.0054	.0052	.0051	.0049	.0048
−2.4	.0082	.0080	.0078	.0075	.0073	.0071	.0069	.0068	.0066	.0064
−2.3	.0107	.0104	.0102	.0099	.0096	.0094	.0091	.0089	.0087	.0084
−2.2	.0139	.0136	.0132	.0129	.0125	.0122	.0119	.0116	.0113	.0110
−2.1	.0179	.0174	.0170	.0166	.0162	.0158	.0154	.0150	.0146	.0143
−2.0	.0228	.0222	.0217	.0212	.0207	.0202	.0197	.0192	.0188	.0183
−1.9	.0287	.0281	.0274	.0268	.0262	.0256	.0250	.0244	.0239	.0233
−1.8	.0359	.0351	.0344	.0336	.0329	.0322	.0314	.0307	.0301	.0294
−1.7	.0446	.0436	.0427	.0418	.0409	.0401	.0392	.0384	.0375	.0367
−1.6	.0548	.0537	.0526	.0516	.0505	.0495	.0485	.0475	.0465	.0455
−1.5	.0668	.0655	.0643	.0630	.0618	.0606	.0594	.0582	.0571	.0559
−1.4	.0808	.0793	.0778	.0764	.0749	.0735	.0721	.0708	.0694	.0681
−1.3	.0968	.0951	.0934	.0918	.0901	.0885	.0869	.0853	.0838	.0823
−1.2	.1151	.1131	.1112	.1093	.1075	.1056	.1038	.1020	.1003	.0985
−1.1	.1357	.1335	.1314	.1292	.1271	.1251	.1230	.1210	.1190	.1170
−1.0	.1587	.1562	.1539	.1515	.1492	.1469	.1446	.1423	.1401	.1379
−0.9	.1841	.1814	.1788	.1762	.1736	.1711	.1685	.1660	.1635	.1611
−0.8	.2119	.2090	.2061	.2033	.2005	.1977	.1949	.1922	.1894	.1867
−0.7	.2420	.2389	.2358	.2327	.2296	.2266	.2236	.2206	.2177	.2148
−0.6	.2743	.2709	.2676	.2643	.2611	.2578	.2546	.2514	.2483	.2451
−0.5	.3085	.3050	.3015	.2981	.2946	.2912	.2877	.2843	.2810	.2776
−0.4	.3446	.3409	.3372	.3336	.3300	.3264	.3228	.3192	.3156	.3121
−0.3	.3821	.3783	.3745	.3707	.3669	.3632	.3594	.3557	.3520	.3483
−0.2	.4207	.4168	.4129	.4090	.4052	.4013	.3974	.3936	.3897	.3859
−0.1	.4602	.4562	.4522	.4483	.4443	.4404	.4364	.4325	.4286	.4247
−0.0	.5000	.4960	.4920	.4880	.4840	.4801	.4761	.4721	.4681	.4641

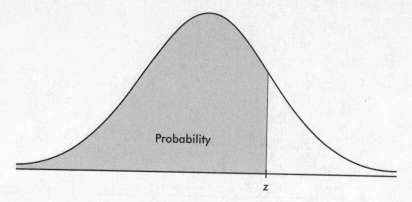

Probability

TABLE A: STANDARD NORMAL PROBABILITIES (CONTINUED)

z	.00	.01	.02	.03	.04	.05	.06	.07	.08	.09
0.0	.5000	.5040	.5080	.5120	.5160	.5199	.5239	.5279	.5319	.5359
0.1	.5398	.5438	.5478	.5517	.5557	.5596	.5636	.5675	.5714	.5753
0.2	.5793	.5832	.5871	.5910	.5948	.5987	.6026	.6064	.6103	.6141
0.3	.6179	.6217	.6255	.6293	.6331	.6368	.6406	.6643	.6480	.6517
0.4	.6554	.6591	.6628	.6664	.6700	.6736	.6772	.6808	.6844	.6879
0.5	.6915	.6950	.6985	.7019	.7054	.7088	.7123	.7157	.7190	.7224
0.6	.7257	.7291	.7324	.7357	.7389	.7422	.7454	.7486	.7517	.7549
0.7	.7580	.7611	.7642	.7673	.7704	.7734	.7764	.7794	.7823	.7852
0.8	.7881	.7910	.7939	.7967	.7995	.8023	.8051	.8078	.8106	.8133
0.9	.8159	.8186	.8212	.8238	.8264	.8289	.8315	.8340	.8365	.8389
1.0	.8413	.8438	.8461	.8485	.8508	.8531	.8554	.8577	.8599	.8621
1.1	.8643	.8665	.8686	.8708	.8729	.8749	.8770	.8790	.8810	.8830
1.2	.8849	.8869	.8888	.8907	.8925	.8944	.8962	.8980	.8997	.9015
1.3	.9032	.9049	.9066	.9082	.9099	.9115	.9131	.9147	.9162	.9177
1.4	.9192	.9207	.9222	.9236	.9251	.9265	.9279	.9292	.9306	.9319
1.5	.9332	.9345	.9357	.9370	.9382	.9394	.9406	.9418	.9429	.9441
1.6	.9452	.9463	.9474	.9484	.9495	.9505	.9515	.9525	.9535	.9545
1.7	.9554	.9564	.9573	.9582	.9591	.9599	.9608	.9616	.9625	.9633
1.8	.9641	.9649	.9656	.9664	.9671	.9678	.9686	.9693	.9699	.9706
1.9	.9713	.9719	.9726	.9732	.9738	.9744	.9750	.9756	.9761	.9767
2.0	.9772	.9778	.9783	.9788	.9793	.9798	.9803	.9808	.9812	.9817
2.1	.9821	.9826	.9830	.9834	.9838	.9842	.9846	.9850	.9854	.9857
2.2	.9861	.9864	.9868	.9871	.9875	.9878	.9881	.9884	.9887	.9890
2.3	.9893	.9896	.9898	.9901	.9904	.9906	.9909	.9911	.9913	.9916
2.4	.9918	.9920	.9922	.9925	.9927	.9929	.9931	.9932	.9934	.9936
2.5	.9938	.9940	.9941	.9943	.9945	.9946	.9948	.9949	.9951	.9952
2.6	.9953	.9955	.9956	.9957	.9959	.9960	.9961	.9962	.9963	.9964
2.7	.9965	.9966	.9967	.9968	.9969	.9970	.9971	.9972	.9973	.9974
2.8	.9974	.9975	.9976	.9977	.9977	.9978	.9979	.9979	.9980	.9981
2.9	.9981	.9982	.9982	.9983	.9984	.9984	.9985	.9985	.9986	.9986
3.0	.9987	.9987	.9987	.9988	.9988	.9989	.9989	.9989	.9990	.9990
3.1	.9990	.9991	.9991	.9991	.9992	.9992	.9992	.9992	.9993	.9993
3.2	.9993	.9993	.9994	.9994	.9994	.9994	.9994	.9995	.9995	.9995
3.3	.9995	.9995	.9995	.9996	.9996	.9996	.9996	.9996	.9996	.9997
3.4	.9997	.9997	.9997	.9997	.9997	.9997	.9997	.9997	.9997	.9998

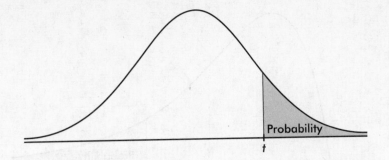

TABLE B: *t*-DISTRIBUTION CRITICAL VALUES

						Tail probability *p*						
df	.25	.20	.15	.10	.05	.025	.02	.01	.005	.0025	.001	.0005
1	1.000	1.376	1.963	3.078	6.314	12.71	15.89	31.82	63.66	127.3	318.3	636.6
2	.816	1.061	1.386	1.886	2.920	4.303	4.849	6.965	9.925	14.09	22.33	31.60
3	.765	.978	1.250	1.638	2.353	3.182	3.482	4.541	5.841	7.453	10.21	12.92
4	.741	.941	1.190	1.533	2.132	2.776	2.999	3.747	4.604	5.598	7.173	8.610
5	.727	.920	1.156	1.476	2.015	2.571	2.757	3.365	4.032	4.773	5.893	6.869
6	.718	.906	1.134	1.440	1.943	2.447	2.612	3.143	3.707	4.317	5.208	5.959
7	.711	.896	1.119	1.415	1.895	2.365	2.517	2.998	3.499	4.029	4.785	5.408
8	.706	.889	1.108	1.397	1.860	2.306	2.449	2.896	3.355	3.833	4.501	5.041
9	.703	.883	1.100	1.383	1.833	2.262	2.398	2.821	3.250	3.690	4.297	4.781
10	.700	.879	1.093	1.372	1.812	2.228	2.359	2.764	3.169	3.581	4.144	4.587
11	.697	.876	1.088	1.363	1.796	2.201	2.328	2.718	3.106	3.497	4.025	4.437
12	.695	.873	1.083	1.356	1.782	2.179	2.303	2.681	3.055	3.428	3.930	4.318
13	.694	.870	1.079	1.350	1.771	2.160	2.282	2.650	3.012	3.372	3.852	4.221
14	.692	.868	1.076	1.345	1.761	2.145	2.264	2.624	2.977	3.326	3.787	4.140
15	.691	.866	1.074	1.341	1.753	2.131	2.249	2.602	2.947	3.286	3.733	4.073
16	.690	.865	1.071	1.337	1.746	2.120	2.235	2.583	2.921	3.252	3.686	4.015
17	.689	.863	1.069	1.333	1.740	2.110	2.224	2.567	2.898	3.222	3.646	3.965
18	.688	.862	1.067	1.330	1.734	2.101	2.214	2.552	2.878	3.197	3.611	3.922
19	.688	.861	1.066	1.328	1.729	2.093	2.205	2.539	2.861	3.174	3.579	3.883
20	.687	.860	1.064	1.325	1.725	2.086	2.197	2.528	2.845	3.153	3.552	3.850
21	.686	.859	1.063	1.323	1.721	2.080	2.189	2.518	2.831	3.135	3.527	3.819
22	.686	.858	1.061	1.321	1.717	2.074	2.183	2.508	2.819	3.119	3.505	3.792
23	.685	.858	1.060	1.319	1.714	2.069	2.177	2.500	2.807	3.104	3.485	3.768
24	.685	.857	1.059	1.318	1.711	2.064	2.172	2.492	2.797	3.091	3.467	3.745
25	.684	.856	1.058	1.316	1.708	2.060	2.167	2.485	2.787	3.078	3.450	3.725
26	.684	.856	1.058	1.315	1.706	2.056	2.162	2.479	2.779	3.067	3.435	3.707
27	.684	.855	1.057	1.314	1.703	2.052	2.158	2.473	2.771	3.057	3.421	3.690
28	.683	.855	1.056	1.313	1.701	2.048	2.154	2.467	2.763	3.047	3.408	3.674
29	.683	.854	1.055	1.311	1.699	2.045	2.150	2.462	2.756	3.038	3.396	3.659
30	.683	.854	1.055	1.310	1.697	2.042	2.147	2.457	2.750	3.030	3.385	3.646
40	.681	.851	1.050	1.303	1.684	2.021	2.123	2.423	2.704	2.971	3.307	3.551
50	.679	.849	1.047	1.299	1.676	2.009	2.109	2.403	2.678	2.937	3.261	3.496
60	.679	.848	1.045	1.296	1.671	2.000	2.099	2.390	2.660	2.915	3.232	3.460
80	.678	.846	1.043	1.292	1.664	1.990	2.088	2.374	2.639	2.887	3.195	3.416
100	.677	.845	1.042	1.290	1.660	1.984	2.081	2.364	2.626	2.871	3.174	3.390
1000	.675	.842	1.037	1.282	1.646	1.962	2.056	2.330	2.581	2.813	3.098	3.300
∞	.674	.841	1.036	1.282	1.645	1.960	2.054	2.326	2.576	2.807	3.091	3.291
	50%	60%	70%	80%	90%	95%	96%	98%	99%	99.5%	99.8%	99.9%
						Confidence level *C*						

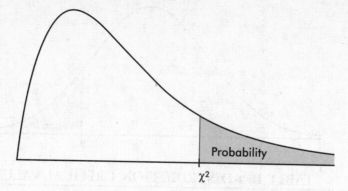

TABLE C: χ^2 CRITICAL VALUES

df	Tail probability p										
	.25	.20	.15	.10	.05	.025	.02	.01	.005	.0025	.001
1	1.32	1.64	2.07	2.71	3.84	5.02	5.41	6.63	7.88	9.14	10.83
2	2.77	3.22	3.79	4.61	5.99	7.38	7.82	9.21	10.60	11.98	13.82
3	4.11	4.64	5.32	6.25	7.81	9.35	9.84	11.34	12.84	14.32	16.27
4	5.39	5.99	6.74	7.78	9.49	11.14	11.67	13.28	14.86	16.42	18.47
5	6.63	7.29	8.12	9.24	11.07	12.83	13.39	15.09	16.75	18.39	20.51
6	7.84	8.56	9.45	10.64	12.59	14.45	15.03	16.81	18.55	20.25	22.46
7	9.04	9.80	10.75	12.02	14.07	16.01	16.62	18.48	20.28	22.04	24.32
8	10.22	11.03	12.03	13.36	15.51	17.53	18.17	20.09	21.95	23.77	26.12
9	11.39	12.24	13.29	14.68	16.92	19.02	19.68	21.67	23.59	25.46	27.88
10	12.55	13.44	14.53	15.99	18.31	20.48	21.16	23.21	25.19	27.11	29.59
11	13.70	14.63	15.77	17.28	19.68	21.92	22.62	24.72	26.76	28.73	31.26
12	14.85	15.81	16.99	18.55	21.03	23.34	24.05	26.22	28.30	30.32	32.91
13	15.98	16.98	18.20	19.81	22.36	24.74	25.47	27.69	29.82	31.88	34.53
14	17.12	18.15	19.41	21.06	23.68	26.12	26.87	29.14	31.32	33.43	36.12
15	18.25	19.31	20.60	22.31	25.00	27.49	28.26	30.58	32.80	34.95	37.70
16	19.37	20.47	21.79	23.54	26.30	28.85	29.63	32.00	34.27	36.46	39.25
17	20.49	21.61	22.98	24.77	27.59	30.19	31.00	33.41	35.72	37.95	40.79
18	21.60	22.76	24.16	25.99	28.87	31.53	32.35	34.81	37.16	39.42	42.31
19	22.72	23.90	25.33	27.20	30.14	32.85	33.69	36.19	38.58	40.88	43.82
20	23.83	25.04	26.50	28.41	31.41	34.17	35.02	37.57	40.00	42.34	45.31
21	24.93	26.17	27.66	29.62	32.67	35.48	36.34	38.93	41.40	43.78	46.80
22	26.04	27.30	28.82	30.81	33.92	36.78	37.66	40.29	42.80	45.20	48.27
23	27.14	28.43	29.98	32.01	35.17	38.08	38.97	41.64	44.18	46.62	49.73
24	28.24	29.55	31.13	33.20	36.42	39.36	40.27	42.98	45.56	48.03	51.18
25	29.34	30.68	32.28	34.38	37.65	40.65	41.57	44.31	46.93	49.44	52.62
26	30.43	31.79	33.43	35.56	38.89	41.92	42.86	45.64	48.29	50.83	54.05
27	31.53	32.91	34.57	36.74	40.11	43.19	44.14	46.96	49.64	52.22	55.48
28	32.62	34.03	35.71	37.92	41.34	44.46	45.42	48.28	50.99	53.59	56.89
29	33.71	35.14	36.85	39.09	42.56	45.72	46.69	49.59	52.34	54.97	58.30
30	34.80	36.25	37.99	40.26	43.77	46.98	47.96	50.89	53.67	56.33	59.70
40	45.62	47.27	49.24	51.81	55.76	59.34	60.44	63.69	66.77	69.70	73.40
50	56.33	58.16	60.35	63.17	67.50	71.42	72.61	76.15	79.49	82.66	86.66
60	66.98	68.97	71.34	74.40	79.08	83.30	84.58	88.38	91.95	95.34	99.61
80	88.13	90.41	93.11	96.58	101.9	106.6	108.1	112.3	116.3	120.1	124.8
100	109.1	111.7	114.7	118.5	124.3	129.6	131.1	135.8	140.2	144.3	149.4

TABLE D: RANDOM NUMBER TABLE

84177	06757	17613	15582	51506	81435	41050	92031	06449	05059
59884	31180	53115	84469	94868	57967	05811	84514	75011	13006
63395	55041	15866	06589	13119	71020	85940	91932	06488	74987
54355	52704	90359	02649	47496	71567	94268	08844	26294	64759
08989	57024	97284	00637	89283	03514	59195	07635	03309	72605
29357	23737	67881	03668	33876	35841	52869	23114	15864	38942

INDEX

Parentheses indicate problem numbers

A

Alternative hypothesis, 324
AP exam hints, 582–585
AP Scoring Guide, 586
Assumptions for confidence intervals
    and hypothesis tests, 581

B

Bar charts (bar graph), 2
    double bar charts, 52, 56(1)
    segmented bar charts, 118–119,
        123–124(6-10)
Bell-shaped data, 9–10,
Bias, 129, 132(2), 141, 143(1, 2),
    144(3, 4), 146(8), 147(11),
    148(15), 150(1), 151(3, 5),
    168(9)
Binomial probability, 183–187,
    194–196(3–6, 11),
    204–205(42, 43, 46, 47),
    211(4), 215(1, 2), 255(1)
Binomial random variable, 191–193,
    194–195(1, 7), 255(1)
Blinding, 157–159, 168(5, 10)
    double blinding, 158, 167(2)
Blocking, 160, 163(4, 5)
Boxplots (box and whisker display),
    32–34, 36–39(3–8), 106(9)
    parallel boxplots, 53–54,
        58–59(5–7), 62(3), 69(1)

C

Census, 131–132, 136(6–8)
Central limit theorem, 264–266
Changing units, 34–35
Chi-square, 272–273
    goodness-of-fit test, 367–371,
        381–384(1, 3, 7, 9),
        388–389(1, 3, 4), 392(1),
        397(6), 411(17), 446(4),
        478(6), 496(19)
    homogeneity test, 375–376
    independence test, 371–375,
        381–384(2, 4–6, 8), 389(2),
        393(2), 395(4), 416(1), 504(3)

Clusters, 7–8, 40(15), 74
Coefficient of determination, 75, 78,
    92(15)
Computer printout, 82–83, 106(8),
    379, 393–397(3, 5), 418(5),
    439(15), 447(5), 475(2),
    502(2), 529(32), 536(5),
    567(6)
Confidence interval estimate:
    of difference of two means,
        298–299, 309–310(21, 22),
        312(34), 317(3)
    of difference of two proportions,
        292–293, 309(18, 19)
    of population mean, 294–298,
        307–308(12–15),
        310–313(24–26, 29, 31–33,
        35, 36), 317(4–6), 359(3)
    of population proportion,
        287–292, 305–306(2, 4, 5,
        7–9), 316–317(1, 2)
    of slope of least squares line,
        301–303, 318(7, 8)
Confounding, 157, 162(1)
Contingency tables, 115
Control group, 157–158, 163(6)
Correlation, 74–75, 89(4, 5),
    91(8–11), 94–98(19–21,
    23–27, 30), 102–103(2–5),
    105(7), 106(8), 107(10)
Cumulative frequency, 5–6, 32,
    44–45(29–31), 55, 63(4),
    410(13), 442(30), 469(28),
    494(5), 533(40), 554(6)

D

Dotplots, 1–2, 51, 443(38)

E

Empirical rule, 28–29
Ethical considerations, 167(2),
    168(7), 566(5)
Experiment, 132, 134–137(1–10),
    155–161, 162–165(1–13),
    167–169(1–11), 172–173(1, 2)

Experimental units, 158

F

Factors, 156
Five-number summary, 32–34
Frequency:
    marginal frequencies, 115–116,
        126–127(1, 2)
    relative frequencies, 117–120,
        126–127(1, 2), 175

G

Gaps, 7–8, 40(15)
Generalizability, 160
Geometric distribution, 186–188,
    215(1), 216(5)
Grading the exam, 586

H

Histogram, 2–4, 11(3), 12(4), 13(9),
    14(10, 11), 15(13), 17(2),
    29–32, 36–39(3–8), 49(3),
    57(2, 3), 62(1), 106(9)
Hypothesis test:
    for slope of least squares line,
        377–380, 389(5), 393–397(3,
        5), 448(6), 567(6)
    of difference of two means,
        332–333, 339(8), 342–343(17,
        20, 22), 352(5), 358–359(1,
        2), 411(15), 417(4), 477(4),
        538(6)
    of difference of two proportions,
        326–329, 340(11), 342(18),
        351–352(1, 6), 477(5), 538(6),
        566(5)
    of population mean, 329–331,
        338–342(6, 7, 12–15, 19),
        344(23), 351–352(2, 4, 7, 9),
        358–360(1, 3,), 442(31),
        448(6), 471(33), 500(37)
    of population proportion,
        324–326, 339(10), 341(16),
        344(24), 351–352(3, 8),

The following documentation applies if you purchased *AP Statistics*, book with CD-ROM. Please disregard this information if your version does not contain the CD-ROM.

System Requirements
Windows: Intel Pentium II 450 MHz or faster, 128MB RAM
1024×768 display resolution
Windows 2000, XP, Vista
CD ROM Player

Macintosh: PowerPC G3 500MHz or faster, 128MB RAM
1024×768 display resolution
Mac OS v.9.0 through 10.4
CD ROM Player

Installation Instructions
The product is not installed on the end-user's computer; it is run directly from the CD-ROM.

Windows:
   Insert the CD-ROM. The program will launch automatically. If it doesn't launch automatically:

   1. Click "Run" from the start menu.

   2. Browse to the CD-ROM drive. Select the file *AP_Statistics.exe* and click "Open."

Macintosh®:

   1. Insert the CD-ROM and open the CD-ROM icon.

   2. Double-click the file *AP_Statistics*.